MINERAL IMPURITIES
IN COAL COMBUSTION

MINERAL IMPURITIES IN COAL COMBUSTION
Behavior, Problems, and Remedial Measures

Erich Raask

Central Electricity Generating Board
Technical Planning and Research Division
Leatherhead, England, U.K.

⊙ HEMISPHERE PUBLISHING CORPORATION

Washington   New York   London

DISTRIBUTION OUTSIDE NORTH AMERICA

SPRINGER–VERLAG

Berlin   Heidelberg   New York   Tokyo

MINERAL IMPURITIES IN COAL COMBUSTION
Behavior, Problems, and Remedial Measures

1 2 3 4 5 6 7 8 9 0   B C B C   8 9 8 7 6 5

This book was set in Press Roman by Hemisphere Publishing Corporation. The editors were Elizabeth Dugger, Brenda Brienza, and Roberta Robey; the production supervisor was Miriam Gonzalez; and the typesetter was Sandra Watts. BookCrafters, Inc. was printer and binder.

Library of Congress Cataloging in Publication Data

Raask, Erich.
  Mineral impurities in coal combustion.

  Bibliography: p.
  Includes index.
  1. Furnaces–Combustion. 2. Coal–Combustion.
3. Coal-fired power plants. 4. Coal–Mineral inclusions.
I. Title.
TJ324.7.R33   1985   621.402'3   83-26400
ISBN 0-89116-362-X   Hemisphere Publishing Corporation

DISTRIBUTION OUTSIDE NORTH AMERICA:
ISBN 3-540-13817-X   Springer-Verlag   Berlin

to Audrey and Karen-Mary

CONTENTS

Foreword   xiii
Preface   xv
Nomenclature   xvii
Glossary of Institutions   xxi

1   INTRODUCTION   1

2   INFLUENCE OF COAL MINERAL MATTER
    ON BOILER DESIGN   5

3   MINERAL IMPURITIES IN COAL   9

    3.1   Chemical (Element) and Mineralogical (Species) Analyses   9
    3.2   Coal Mineral Matter in Relation to the Geological Environment   11
    3.3   Clay Minerals (Aluminosilicates) and Silica (Quartz) in Coal   15
    3.4   Introduction and the Mode of Occurrence of Sulfur in Coal   18
    3.5   Carbonate Inclusions in the First and Second Stages
          of Coalification   19
    3.6   The Origin and Mode of Occurrence of Chlorine in Coal   20

4   QUALITY OF COAL UTILIZED IN POWER STATIONS   23

    4.1   Heat Content and Combustibility of Coals of Different Rank   23
    4.2   Inherent and Adventitious Mineral Matter in Coal   29
    4.3   Distribution of Particulate Mineral Impurities in Pulverized Coal 30
    4.4   Relationship between the Mineral Matter Content of Coal
          and Its Ash Residue   35
    4.5   Laboratory-Prepared Ash Content and the Amount of Ash, Clinker,
          and Slag in Coal-Fired Boilers   36

5   COAL GRINDING, ABRASIVE FUEL MINERALS,
    AND PLANT WEAR   41

    5.1   Coal Grindability   41
    5.2   Abrasive Minerals in Coal   43
    5.3   Characteristics of Sliding Abrasion Wear   49
    5.4   Particle Impaction Erosion   51
    5.5   Measures to Combat Abrasion and Erosion Wear in the Coal Plant 56

6   PARTICULATE SILICATE MINERALS IN BOILER FLAME   61

    6.1   Shape Transformation of Flame-Borne Silicate-Ash Particles   61
    6.2   Slag/Coal Interface Phenomena: Wetting, Surface Tension,
          and Combustion of Ash-Rich Coal Particles   67
    6.3   Formation of Ash Cenospheres   71
    6.4   Release of Potassium from Aluminosilicate Clay Minerals   77
    6.5   Volatilization of Silica in Coal Gasification and Combustion Systems 80

7   REACTIONS OF NONSILICATE IMPURITIES
    IN COAL FLAMES   85

    7.1   Dissociation and Oxidation of Pyrites   85
    7.2   Volatilization and Sulfation of Chlorides   89
    7.3   Dissociation and Sulfation of Carbonates   94
    7.4   Distribution of Alkalis in Immiscible Sulfate and Silicate Phases 97

8   CREATION, CAPTURE, AND COALESCENCE
    OF PARTICULATE ASH IN BOILER FLAME   103

    8.1   Fragmentation by Thermal Shock and by Gas Evolution   103
    8.2   Formation of Silica Fume from Silicon Monoxide Vapor   106
    8.3   Formation of Sulfate Fume in Boiler-Flue Gas   108
    8.4   Captive Formation of Sulfate on the Surface of Flue-Gas-Borne Ash 111
    8.5   Coalescence of the Inherent Ash in Burning Coal Particles   113
    8.6   Size Characteristics of Gas-Borne Ash in Pulverized-Coal-
          and Cyclone-Fired Boilers   115

9   SLAG VISCOSITY   121

    9.1   Methods of Viscosity Measurements   121
    9.2   The Newtonian and Non-Newtonian Flow Characteristics
          of Viscous Melts   124
    9.3   The Relationship between Slag Viscosity and Coal-Ash
          Composition   127
    9.4   Viscosity as a Rate-Controlling Parameter in Formation
          of Sintered Deposit and Molten Slag   131

10 SINTERING, FUSION, AND SLAGGING PROPENSITIES
    OF COAL ASHES   137

    10.1 Sintering Model   137
    10.2 Sintering Rate Measurements   140

10.3 Sintered Ash Strength Measurements 152
10.4 A Rapid Test for Initial Sintering 154
10.5 Conventional Ash Fusion Tests 156
10.6 Empirical Fouling and Slagging Indices Based on Ash Composition 160

11 ADHESION OF ASH DEPOSIT ON BOILER TUBES AND REFRACTORY MATERIALS 169

11.1 Van der Waals, Electrostatic, and Surface-Tension Forces 169
11.2 The Temperature Environment for Adhesion of Ash Deposits in Coal Combustion Systems 173
11.3 Thermal and Chemical Compatibility of Ash Deposits with Boiler-Tube Surfaces 175
11.4 Mechanical and Chemical Bonding at Boiler-Tube/Ash-Deposit Interface 178
11.5 Bond-Strength Measurements 180

12 DEPOSITION MECHANISMS, RATE MEASUREMENTS, AND THE MODE OF FORMATION OF BOILER DEPOSITS 189

12.1 Deposition by Vapor and Small-Particle Diffusion 189
12.2 Thermophoresic and Electrophoresic Deposition 191
12.3 Deposition by Inertial Impaction and Summary of Different Mechanisms 194
12.4 Retention of Impacted Particles on Captive Surfaces 198
12.5 The Rate Measurements and Mode of Deposition of Sulfates on Cooled Targets in Boiler Plant 200
12.6 Rate Measurements of Silicate-Ash Deposition and Formation of Monolithic and Layer-Structured Deposits 205
12.7 The Role of Minor Coal Impurity Constituents in Formation of Boiler Deposits 214

13 THERMAL RADIATION AND HEAT TRANSFER PROPERTIES OF BOILER DEPOSITS 217

13.1 Emittance, Absorptance, and Reflectance Measurements 217
13.2 Emittance Characteristics and Particulate Ash and Boiler Deposits 220
13.3 Thermal Conductivity of Particulate Ash and Boiler Deposits 224
13.4 Thermal Barrier Effect of Ash Deposits on Heat Transfer 230
13.5 Heat-Flux Probes and Measurements in Coal-Fired Boilers 232

14 MEASURES TO COMBAT BOILER FOULING AND SLAGGING 239

14.1 Choice of Boiler Systems: Pulverized Fuel or Cyclone Firing 239
14.2 Boiler Temperature Regime Consistent with Ash Slagging Characteristics 241
14.3 Prevention of Deposit Buildup in Burner Quarls 246
14.4 Possible Application of Austenitic-Steel Tubes and Surface Treatment to Reduce Deposit Adhesion 248

14.5 The Scope for Reducing Boiler Slagging by Coal Cleaning
and Blending   249
14.6 Combustion Control for Combating Boiler Slagging   251
14.7 Steam and Air Sootblowers   254
14.8 Water-Jet Sootblowers   257
14.9 Novel Suggestions for Combating Boiler Slagging   259

15 SOME SPECIFIC ASH–RELATED PROBLEMS WITH U.S. COALS   261

15.1 Composition Characteristics of the Mineral Matter in Low-
and High-Rank Coals   261
15.2 The Mode of Occurrence of Sodium and Calcium
in U.S. Low-Rank Fuels   264
15.3 The Mechanism of Formation of Sintered Ash Deposits with Lignite
and Sub-bituminous Coals   266
15.4 The Mode of Occurrence of Iron in U.S. Coals   272
15.5 The Mechanism of Formation of Iron-Rich Slag Deposits   275
15.6 Ash Slagging Propensity Based on the Pyritic Iron Content   277

16 USE OF ADDITIVES IN COAL–FIRED BOILERS   283

16.1 The Aims and Premise   283
16.2 Electrical Resistivity of Particulate Coal Minerals
and Precipitator Ash   284
16.3 Additives for Improving the Performance of Electrical
Precipitators   289
16.4 Devitrification-Enhancing Additives to Combat Boiler Slagging   300
16.5 Use of Alkaline-Earth Additives in Cyclone-Fired and Pulverized-
Coal-Fired Boilers   307
16.6 Intermittent Application of Additives   310

17 HIGH–TEMPERATURE CORROSION IN COAL–FIRED PLANTS   313

17.1 The Fuel Impurity Environment and Temperature Regime   313
17.2 Formation of Corrosive Deposits   316
17.3 Diagnostic Analysis of Corrosive Deposits   320
17.4 The Mechanism of Tube-Metal Corrosion in Molten Sulfate   324
17.5 Chloride-Enhanced Corrosion   330
17.6 Enrichment of Coal Trace Elements in Corrosive Deposits   334
17.7 Assessment of Corrosion Propensity of Coals   342
17.8 The Scope for Reducing High-Temperature Corrosion by Coal
Cleaning and Blending   346
17.9 High-Chromium Steels, Coextruded Tubes, Shields, and Coatings
for Combating High-Temperature Corrosion   348
17.10 Boiler Design and Combustion Control Measures to Alleviate
Furnace-Wall and Superheater Corrosion   356
17.11 Quest for Corrosion-Preventing Additives in Coal-Fired
Boilers   357

18   ASH IMPACTION EROSION WEAR   361

18.1   Changes in the Abrasive Characteristics of Particulate Coal
       Minerals in the Boiler Flame   361
18.2   Abrasive Characteristics of Pulverized-Coal Ash   365
18.3   Temperature Regime and Velocity Parameter Governing Boiler-Tube
       Erosion by Ash Impaction   369
18.4   Erosion Wear Damage in Coal-Fired Boilers   377
18.5   Maximum Flue-Gas Velocity Compatible with Ash Erosion
       Propensity   380
18.6   Measures to Combat Ash Impaction Erosion Wear
       of Boiler Tubes   382
18.7   Combating Wear Damage of Ash Transport Lines   385
18.8   Erosion Wear in Oil-to-Coal Converted Boilers   386
18.9   Erosion Wear in Coal-Gas Turbines Coupled to Fluidized-Bed
       Combustors   390

19   LOW-TEMPERATURE FOULING AND CORROSION   397

19.1   The Temperature Regime and Design of Air Heaters   397
19.2   Acid Deposition Measurements in Coal-Fired Boilers   400
19.3   Acidity of the Flue Gas and Coal Quality   402
19.4   The Mechanism of Air-Heater Fouling and Corrosion   404
19.5   Operation Experience of Fouling and Corrosion of Coal-Fired
       Boiler Air Heaters   408
19.6   Measures to Combat Air-Heater Fouling and Corrosion   411
19.7   Monitoring Acid Penetration in Concrete Structures   416

**20   COMPARISON OF ASH-RELATED PROBLEMS
        IN PULVERIZED-FUEL-FIRED AND OTHER
        COAL-FIRED SYSTEMS**   421

20.1   Boiler Operation Difficulties Related to Specific Coal
       Mineral Species   421
20.2   Benefits and Some Adverse Effects of Coal Cleaning
       on Boiler Operation   423
20.3   Ash-Related Problems in Oil-to-Coal Converted Boilers   426
20.4   Characteristics of Fluidized-Bed Ash   430
20.5   Sintering Propensity of Fluidized-Bed Ash   432
20.6   Slag Agglomeration in a Fluidized-Bed Gasifier   434

References   439
Index   469

FOREWORD

Mineral matter has always been the nemesis of coal-burning industries. Since the dawn of the industrial revolution, the "impurities" in coal have had a major effect on the design of boiler furnaces and how they are operated. An estimated 100 million tons of mineral matter, converted to "coal ash," will pass through the pulverized-coal-fired boiler furnaces in the United States in 1984. The worldwide figure is less certain. This huge mass of troublesome material will lead to some of the most serious operational problems facing the utility industry, and it will cost the industry tremendously in capital investments and in availability because of ash-limited generation. In industrial applications, too, clinker formation in fixed-fuel beds, as in stokers, will continue to limit burning rates, just as it has done for more than a hundred years. In the huge utility-operated, pulverized-coal-fired units, slagging will still be a major determinant in fixing furnace size, and hence relative cost, for a given output of steam, just as it has done since 1920. Fouling in these same steam generators will dictate the spacing and the location of convective tube banks for superheating and reheating steam. In short, the mineral matter in coal continues, as it always has, to be a dominating factor in deciding on the dimensions of boiler furnaces for a given steam output; in proportioning the heat-receiving surfaces for design steam production and properties; in setting limits for flue-gas velocity and flow patterns to minimize metal loss by erosion; and in influencing the physical and chemical properties of the fly ash leaving the stack to ensure its capture by emission-control systems.

It is the mineral matter in coal that plays the dominant role in fuel selection, in setting the design and size of the furnace, and in establishing how that boiler furnace will be operated.

The literature on steam generation in coal-burning boiler furnaces is replete with accounts of operational problems with coal ash. Perhaps there are as many as 5000 published technical papers in the literature, by or aimed mainly at mechanical engineers, covering some aspect of coal-ash behavior in large boiler furnaces. These range from highly technical accounts by specialists in coal-ash behavior to generalized summaries by power-plant operating personnel. Chemists and physicists have contributed, too, but engineering approaches are more common. Yet with all this effort for at least half a century, few comprehensive reviews have been published covering the field adequately that lack bias, are technically sound, and yet are expressed in terms meaningful to a fuels-technology specialist as well as to the mechanical engineer charged with operating and maintaining a power plant. This book, written with a background of a quarter century in solving problems brought about by coal ash in large boiler furnaces, and yet covering the chemistry and physics of ash behavior, will aid tremendously in closing the gap between "practice" and "research."

The state of the art in this field has been based largely on empiricism, even though science and theory have been applied in a widely varying degree. For example, determining the fusibility of coal ash, a basic property in assessing coal behavior, as done most commonly has little relation to the rheology of silicate systems. And yet this basic property of ash at high temperatures almost certainly must influence the thickness of slag on heat-receiving surfaces and so affect heat transfer. Hence, applying empirical measurements to predict the behavior of an unfamiliar coal in a given boiler furnace can be unreliable—and often is. Yet the more technically sound methods of measuring fundamental flow characteristics of coal-ash slags has not contributed much as yet to predicting ash behavior, except under limited conditions. It has been realized for many years that ash characteristics must be related to boiler configuration and to operating conditions in assessing the likelihood of ash-related problems, yet such three-pronged studies are few, and generalized relationships are not yet available. Perhaps the growing realization that coal will be our main source of energy for generating electricity for the coming century will lend emphasis to our need for more basic information on coal-ash behavior under real-life conditions.

This book will go far in satisfying that need. Erich Raask brings to it a quarter century of experience in day-to-day laboratory research on fuel-ash problems and operating experience in large generating plants, coupled with the expertise of a science graduate. He is well known in engineering circles for his extraordinary ability to couple technology and practice. He has the respect of his peers. The detail he provides here on the mineral matter in coal, on the behavior of these minerals during combustion, on the bonding of slag to boiler surfaces, and on the great number of operational problems resulting from coal ash in boiler furnaces will be as remarkably helpful to researchers in this field as to the engineers who plan and operate large power plants.

Texts of this scope and this wide coverage on this important subject are few indeed. This one will serve well for many years.

William T. Reid
Battelle Columbus Laboratories, Columbus, Ohio

PREFACE

The principal aim in writing this book was to present information on many and varied boiler operation problems associated with coal mineral impurities in a form that should assist plant design and operation engineers in their endeavor to produce electrical energy from coal with a maximum efficiency. Also, it is hoped that young scientists and engineers who are embarking on careers in the field of energy sciences and application may find the book useful in bridging the gap between research findings and practical requirements.

The book is based largely on the author's own work in the field of coal utilization research, extended over a period of 25 years. The continuity and scope of the author's experience, which includes both laboratory research and plant investigations, together with a selection of the more significant published data from literature, should serve as a review of various difficulties encountered in operation of coal-fired boilers.

The author would like to record his appreciation and gratitude to scientists, fuel technologists, and engineers whose work has been referred to in this book. Further, it has been the author's privilege to meet and talk to many research workers and boiler-plant design and operation engineers whose names are not recorded in the book. The author is greatly indebted to them for much valuable information and for insight into a variety of the problems encountered at coal-fired power stations.

The author received from his colleagues valuable comments, corrections, and suggestions for changes that have been incorporated in the text. However, the choice and arrangement of the subject matter and the opinions expressed in the book rest solely with the author.

Erich Raask

NOMENCLATURE

SYMBOLS OF SI UNITS
AND CONVERSION TO OTHER UNITS

Basic SI Units

1 meter (m) = 3.28 ft
1 kilogram (kg) = 2.205 lb
1 tonne (t) = 1 Mg = 2205 lb = 0.984 long ton = 1.102 short ton
1 second (s)
1 kelvin (K) = $1°C = 1.8°F$
1 ampere (A)
1 mole (mol) = 6.02×10^{26} molecules
1 radian (rad) = 57.3 angular degrees (°)

Derived SI Units

Area: $1 \ m^2 = 10.76 \ ft^2$
Volume: $1 \ m^3 = 10^3$ liter (l) $= 35.3 \ ft^3 = 264$ U.S. gal $= 220$ U.K. gal
Density: $1 \ kg \ m^{-3} = 0.0624 \ lb/ft$
Force (Newton): $1 \ N = 1 \ kg \ m \ s^{-2} = 10^5$ dyne $= 0.2248$ lbf
Pressure (Pascal): $1 \ Pa = 1 \ N \ m^{-2} = 10^{-5}$ bar $= 1.45 \times 10^{-4}$ psi
Energy (Joule): $1 \ J = 1 \ kg \ m^2 \ s^{-2} = 10^7$ erg $= 0.239$ cal $= 0.738$ ft lb $=$
     9.48×10^{-4} Btu
Power (Watt): $1 \ W = 1 \ J \ s^{-1} = 10^7$ erg $s^{-1} = 0.86$ kcal/h $= 1.34 \times 10^{-3}$ hp $=$
     3.41 Btu/h
Combustion intensity (heat relase): $1 \ W \ m^{-3} = 0.0967$ Btu/ft^3 h

Heat flux: $1 \text{ W m}^{-2} = 0.317 \text{ Btu/ft}^2 \text{ h}$

Heat transfer coefficient: $1 \text{ W m}^{-2} \text{ K}^{-1} = 0.176 \text{ Btu/ft}^2 \text{ h } °\text{F}$

Thermal conductivity: $1 \text{ W m}^{-1} \text{ K}^{-1} = 0.578 \text{ Btu/ft h } °\text{F}$

Calorific value: $1 \text{ KJ kg}^{-1} = 0.239 \text{ kcal/kg} = 0.430 \text{ Btu/lb}$

Specific heat capacity: $1 \text{ J kg}^{-1} \text{ K}^{-1} = 2.39 \times 10^{-4} \text{ cal/g } °\text{C} = 2.39 \times 10^{-4} \text{ Btu/lb } °\text{F}$

Entropy: $1 \text{ J K}^{-1} = 0.239 \text{ cal/}°\text{C}^1 = 5.27 \times 10^{-4} \text{ Btu/}°\text{F}$

Dynamic viscosity: $1 \text{ N s m}^{-2} = 1 \text{ Pa s} = 2420 \text{ lb/ft h}$

Kinematic viscosity: $1 \text{ m}^2 \text{ s}^{-1} = 3.88 \times 10^4 \text{ ft}^2/\text{h}$

Surface tension: $1 \text{ N m}^{-1} = 10^3 \text{ dyne/cm}$

Electrical potential: $1 \text{ V} = 1 \text{ W A}^{-1}$

Electrical resistance: $1 \text{ } \Omega = 1 \text{ W A}^{-2}$

Electrical resistivity $= 1 \text{ } \Omega \text{ m} = 3.28 \text{ } \Omega \text{ A}$

Electric field strength $= 1 \text{ V m}^{-1} = 0.305 \text{ V ft}^{-1}$

Electric conductance: 1 siemen (S) $= 1 \text{ } \Omega^{-1}$

Electric conductivity: $1 \text{ S m}^{-1} = 0.305 \text{ } \Omega^{-1} \text{ ft}^{-1}$

Periodic frequency $= 1$ hertz (Hz) $= 1 \text{ s}^{-1} = \text{vibration/s}$

SI Factors printed before basic units: exa (10^{18}), E; peta (10^{15}), P; tera (10^{12}), T; giga (10^9), G; mega (10^6), m: kilo (10^3), k; milli (10^{-3}), m; micro (10^{-6}), μ; nano (10^{-9}), n; pico (10^{-12}), p; femto (10^{-15}), f; atto (10^{-18}), a

Conversion of Other Units to SI Units

Length: 1 foot (ft) $= 12 \text{ in} = 0.305 \text{ m}$

Mass: 1 pound (lb) $= 0.4536 \text{ kg}$

    1 short ton $= 2000 \text{ lb} = 0.907 \text{ tonne (t)} = 907 \text{ kg}$

    1 long ton $= 2240 \text{ lb} = 1.016 \text{ tonne (t)} = 1016 \text{ kg}$

Area: $1 \text{ ft}^2 = 0.093 \text{ m}^2$

Volume: $1 \text{ ft}^3 = 0.0283 \text{ m}^3 = 28.3 \text{ l (liter)}$

    1 U.K. gal $= 1.201$ U.S. gal $= 4.546 \text{ l} = 0.004546 \text{ m}^3$

    1 bbl $= 159 \text{ l} = 0.159 \text{ m}^3$

Density: $1 \text{ lb/ft}^3 = 0.134 \text{ lb/U.S. gal} = 0.1605 \text{ lb/U.K. gal} = 16.02 \text{ kg m}^{-3}$

Force: 1 lbf $= 4.44 \text{ N}$

Pressure: 1 atm $= 101.3 \text{ kPa}$

    1 psi $= 6.895 \text{ kPa}$

    1 in wg $= 249 \text{ Pa}$

    1 torr (mmHg) $= 133 \text{ Pa}$

Temperature: $1°\text{F} = 0.5556°\text{C} = 0.5556 \text{ K}$

Energy: 1 Btu $= 10^{-5} \text{ therm} = 1.055 \text{ kJ}$

    1 ft lbf $= 1.355 \text{ J}$

    1 tce $= 1$ tonne coal equivalent $= 28.8 \text{ GJ}$

Power: 1 Btu/h $= 0.293 \text{ W}$

    1 hp $= 0.745 \text{ kW}$

Combustion intensity: $1 \text{ Btu/ft}^3 \text{ h} = 10.35 \text{ W m}^{-3}$

Heat flux: $1 \text{ Btu/ft}^2 \text{ h} = 3.155 \text{ W m}^{-2}$

Heat transfer coefficient: $1 \text{ Btu/ft}^2 \text{ h } ° \text{ F} = 5.675 \text{ W m}^{-2} \text{ K}^{-1}$

Thermal conductivity: $1 \text{ Btu/ft h } °\text{F} = 1.731 \text{ W m}^{-1} \text{ K}$

Calorific value: 1 Btu/lb = 2.326 kJ kg^{-1}

1 Btu/ft^3 = 37.26 kJ m^{-3}

Specific heat capacity: 1 Btu/lb °F = 4.187 kJ kg^{-1} K^{-1}

Dynamic viscosity: 1 lb/ft h = 0.413 mPa s

1 cP = 1 mPa s

Kinematic viscosity: 1 ft^2/h = 25.8 mm^2 s^{-1}

1 cSt = 1 mm^2 s^{-1}

Surface tension: 1 dyne/cm = 10^{-3} N m^{-1}

Electrical resistivity: 1 Ω ft = 0.305 Ω m

Electric field strength: 1 V/ft = 3.28 V m^{-1}

Electric conductivity: 1/Ω ft = 3.28 S m^{-1}

Temperature Conversion

$$°F \longleftarrow K \longrightarrow °C$$
$$(1.8x - 459.67)°F = xK = (x - 273.15)°C$$

$$°F \longrightarrow K \longleftarrow °C$$
$$y°F = (0.5556y + 255.15)K; (z + 273.15)K = z°C$$

GLOSSARY OF INSTITUTIONS

ASTM American Society for Testing Materials, Philadelphia, Pennsylvania

BCURA British Coal Utilization Research Association, Leatherhead, U.K.; now part of the National Coal Board, London, England

CEGB Central Electricity Generating Board, London, England

CSIRO Commonwealth Scientific and Industrial Research Organization, Sydney, Australia

DIN Deutsche Industrie Normen (German Industry Standards), West Berlin, Germany

EPA Environmental Protection Agency, Washington, D.C.

EPRI Electric Power Research Institute, Palo Alto, California

IGT Institute of Gas Technology, Chicago, Illinois

JANAF Joint Army, Navy, Air Force Thermochemical Data Bank, Washington, D.C.

NCB National Coal Board, London, England

VGB Vereinigung der Grosskraftwerksbetreiber (Association of Power Generation Undertakings), Essen, West Germany

MINERAL IMPURITIES
IN COAL COMBUSTION

INTRODUCTION

In the 1960s coal was losing ground as the principal primary fuel for electricity generation. Oil was plentiful and cheap, and nuclear power appeared to be poised for rapid growth. By the mid-1970s the fuel situation had undergone a drastic change. The price of liquid fuel had increased sharply and, with the exception of the oil-exporting countries, hardly any new oil-fired power stations have been built anywhere else since that date. It was inevitable, therefore, that there should have been renewed interest in utilizing coal for power generation. A number of countries are currently planning to exploit new coal fields for their own use and for export. Some countries where there are no significant deposits of indigenous coal, notably Japan and the Scandinavian countries, are in the process of building up a substantial electricity-generating capacity based on imported coals. Future coal prospects on a world wide basis have been assessed in the *Report of World Coal Study*, edited by Greene and Gallagher (1980), and by Ion (1980).

More extensive use of solid fuel for power generation will accentuate the boiler operation problems associated with mineral impurities in coal, and there is a need for a comprehensive source of information on various difficulties. First, an account is given of changes in the design of coal-fired utility boilers from the small, 1- to 2-MW capacity, stoker-fired boilers to the present-day large pulverized-fuel-fired units of 500- to 1300-MW output. The need for more efficient utilization of fossil fuel has always been the driving force for the large and more economical power plant. However, difficulties caused by coal mineral impurities, in particular those of ash deposits on heat-exchange surfaces, have influenced the changes in boiler design.

In spite of extensive research and the wealth of practical experience there remain some enigmatic aspects in the subject of boiler slagging. For example, the massive build-up of ash deposits can take design and boiler operation engineers by surprise.

The nature of different mineral species of solid fuels, bituminous and non-bituminous coals, and lignite is evaluated in this book. The difficulties of coal grinding and the wear on fuel handling and milling plants are discussed in relation to the abrasive minerals in coal. This is followed by a synopsis of the physical changes and chemical reactions of different mineral species in coal flame; changes and reactions of great importance, not only to formation of boiler deposits but also in relation to other salient properties of ash. These properties are the corrosion propensity of ash, the abrasiveness of ash related to erosion wear of boiler tubes, the electrical resistivity of ash (which is relevant to the efficient working of the electrical precipitators), and the pozzolanic (cementitious) properties of ash when it is used in concretes and in grouts as a cement extender.

The chapters on the flame reactions are followed by discussions on the viscosity as a rate-controlling parameter in ash sintering and fusion, an assessment of the slagging propensity of ashes, the formation and adhesion of deposits in boiler plant, and the effects of boiler deposits on heat transfer. Under the heading of counter measures to combat boiler slagging, the topics discussed are design considerations, combustion control, and conventional and unconventional methods of boiler cleaning. A separate chapter is devoted for discussion on the specific ash-related problems with U.S. low- and high-rank coals. A wide variety of solid fuels is utilized for power generation in the United States, and the deposit-forming characteristics of the lignite and sub-bituminous coal ashes can be markedly different from those of bituminous coal ashes.

The literature on low-temperature additives used to improve the ash capture performance of the electrical precipitators is extensive. In contrast, less information is available on the use of high-temperature additives to combat boiler slagging.

The manifestation of boiler slagging can be immediate, and large quantities of ash deposits can form in the combustion chamber within a few hours, whereas boiler tube corrosion is usually a long-term event determined largely by the "quality" rather than by the quantity of the fuel impurity deposit. The mode of formation of potentially corrosive deposits of molten alkali-metal sulfates is discussed, followed by an assessment of the corrosion propensities of different coals.

High-temperature slagging and corrosion by ash deposits cease to be serious problems in the middle section of boiler plant, but the economizers and the hot sections of air heaters can be plugged by the high-temperature deposit debris and ash compacts. There is another problem that occurs in this boiler section, namely, the tube erosion wear by ash impaction when the impact velocity exceeds that compatible with the abrasive property of ash. In this book the abrasive property of ash in different coals will be discussed with the intention of helping design engineers to arrive at an optimum velocity for the ash-laden flue gas in boiler ducts without causing significant erosion wear.

Condensation of sulfuric acid in air-heaters and in chimneys of pulverized-coal-fired boilers constitutes a lesser problem than it does in residual oil-fired boilers.

However, acid-bonded deposits and some corrosion of air-heater elements are known to occur in coal fired boilers. As a precautionary measure, concrete chimney flues are protected, usually with an acid-resistant brick lining, but other materials have also been used. A monitoring system is described for assessing the efficiency of an acid-resistant lining.

The final chapter summarizes all the mineral-impurity-related problems coal-fired boilers face in the form of a table with references to previous chapters. This is followed by a comparison assessment of the severity of boiler problems with cleaned and noncleaned coals. The comparative approach is extended to consider the similarities and differences in the behavior of coal mineral species in pulverized-coal-fired boilers, in the systems burning coal/oil mixtures, and in fluidized-bed combustors.

INFLUENCE OF COAL MINERAL MATTER
ON BOILER DESIGN

The design of combustion systems has undergone marked changes since the time coal was first used for electricity generation. The earliest utility plants were based on the principle of burning coal in fixed and moving grates, and Mayers (1945) commented when discussing combustion in fuel beds that it was not an exaggeration to say that the great (U.S.) coal industry existed primarily for the purpose of supplying the nation's fuel beds, in electricity utility and industrial boilers, and in coke ovens. However, since that time there has been an extensive changeover to pulverized fuel firing for large electricity utility boilers.

The mode of combustion where there is a hot bed of burning coal was undesirable when considered from the point of view of the behavior of coal mineral species at high temperatures. Extensive volatization of the mineral matter can occur in the reducing atmosphere either in the elemental state, or in the form of sulfides and chlorides. Subsequent deposition, oxidation, and sulfation of the condensable material on boiler tubes can result in a formation of strongly bonded deposits. In particular, coals of high chlorine content and phosphorous-rich fuels are difficult to burn successfully in stoker-fired boilers.

From the point of view of reducing the rate of build-up of boiler deposits, the inorganic material in coal should remain at high temperatures for the shortest possible time, and the hot atmosphere should be consistently oxidizing.

The development of a pulverized-coal-fired system where coal is burned in a cloud rather than in a thick bed was a major step towards meeting these two

5

requirements. The residence time of ash in this mode of combustion is cut to a few seconds, and there is less chance for the mineral species to undergo reduction reactions in the flame when oxygen is present in excess of that required to burn the flame-borne coal particles. Introduction of the system of pulverized coal combustion paved the way for a rapid increase in the capacity of boiler units, and for raising the temperature and pressure of superheated steam. Thus, there has been a significant increase in the generating efficiency of electricity utility plants.

With the advent of pulverized-coal firing, the design and combustion engineers thought that they had a panacea for all fouling and slagging problems. They were delighted to find that high-ash coals could be burned successfully, and it was thought that the nature of mineral species in coal did not matter to any significant degree when the fuel was burned in the form of a flame-borne cloud. However, as the combustion intensity was increased, resulting in higher rates of heat release and temperatures, large amounts of running slag formed on furnace walls with some coals.

The large amounts of running slag led to the next logical step in the development of boiler design, that is, construction of slag-tap and cyclone-fired boilers. Systems were designed where 50 to 85 percent of the total ash was removed as a continuous stream of molten slag. The design was taken up particularly in the United States and in Germany, where large numbers of slag discharge boilers were built.

In Britain, however, only a few small slag discharge boilers were built in the 1950s, and the idea was soon abandoned. It was claimed that a successful operation of a slag tap or cyclone boiler could not have been guaranteed with British coals. The fuel for a power station is usually supplied from a number of pits, and therefore the variable slag flow characteristics are not consistent with a trouble-free slag discharge.

British boiler design engineers therefore retained the original concept of pulverized coal, "dry-bottom" or "dry-ash" boilers, and utility undertakings in other countries are returning to these dry-ash boilers for their 500-, 600-, 900-, and 1200-MW units. The experience with slag discharge boilers has shown that these units are less tolerant than the dry-ash discharge boilers of the inevitable changes in the quantity and quality of mineral matter in coal. For example, a 900-MW unit requires between 300 and 400 t coal per hour. Assuming that the unit will generate electricity for 25 years and that it will be in service on average 6000 hours a year, the unit will consume about 50 million t coal in its lifetime. There are few coal mines in the world that could provide such amounts of coal of a consistent quality.

For the next few decades pulverized-coal-fired boilers are likely to remain the principal generating units for converting coal to electricity. The chief advantage of the system is its versatility, and boilers can be designed to burn all types of coals, as discussed by Bennett and Bannister (1981). The quality of solid fuels covers a wide spectrum from low-volatile content anthracite, through the range of bituminous and sub-bituminous coals to lignite and peat. The mineral matter in each type of solid fuel brings to bear its own particular boiler plant operation problems.

The surveys on boiler plant failures carried out in the United States and

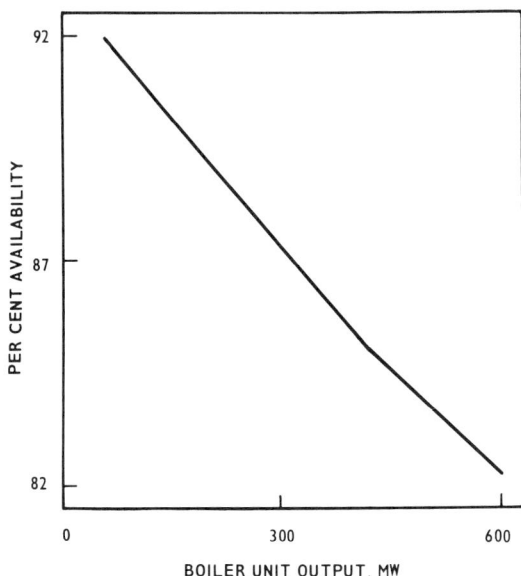

Figure 2.1 Boiler availability as a function of size of fossil-fuel-fired units.

reported by Koppe and Olson (1979) and by Armor et al. (1981) have shown that the development from small to large units has significantly reduced the availability of the generating plant, as depicted in Fig. 2.1. Some decrease in the availability of a large boiler plant is not entirely unexpected, since it takes just a single failure of a high-pressure water or steam tube to take a 50-MW or 500-MW unit out of service, and chances of tube failures in large units are correspondingly higher. It is therefore imperative that the design and operation engineers should have the support they need in their task of maintaining the electricity generating units in an efficient operation.

It appears that the system of fluidized-bed combustion is the strongest candidate for the next stage in the development of converting coal to electricity. From the point of view of problems associated with coal mineral matter, that would be a logical step ahead. The main advantage would be a much-reduced combustion temperature, 1100 to 1200 K, compared with the peak temperature of 1750 to 2000 K of the pulverized coal flame. The low temperature would ensure that a great deal of potentially reactive mineral species remains dormant in the ash. As a result, boiler corrosion and fouling should be markedly reduced, but some ash sintering may occur when the deposit temperature exceed 1100 K. Also, there may be some erosion and abrasion wear damage of the heat exchange tubes caused by the coarse mineral particles in the fluidized bed. It is therefore inevitable that there will be always some problems with ash, whatever the system of coal combustion.

THREE

MINERAL IMPURITIES IN COAL

3.1 CHEMICAL (ELEMENT) AND MINERALOGICAL (SPECIES) ANALYSES

Composition of the noncombustible fraction and the mode of distribution of the mineral species in coals are based on the results of chemical and mineralogical analyses. A prerequisite to the analyses is the removal of the combustible matter, and it is usually carried out by burning coal samples in a laboratory muffle furnace. According to the ASTM Standard (1969) or the British Standard (1970) on coal analysis, the ashing procedure is carried out at comparatively high temperatures, above 900 K. This method of ashing is widely used for routine analysis of nonvolatile elements of coal minerals.

In order to minimize the loss of volatile elements and thermal changes of the nonvolatile species, Hicks and Nagelschmidt (1943), Nelson (1953), and Dixon et al. (1964) reduced the temperature of combustion in a stream of oxygen to 650 K. However, the process of combustion was slow and it required 100 to 250 h to complete. Subsequently, an ashing device for combusting organic substances below 475 K was introduced by Gleit and Holland (1962). In this method an electrodeless discharge takes place in an atmosphere of oxygen, and the gas is said to contain atomic and ionic species of oxygen in an electronically and vibrationally excited state, which can initiate a low-temperature oxidation of organic matter (Gleit, 1963). Gluskoter (1965) used the technique for obtaining low-temperature ashes from bituminous coals, and since that date the method has been used extensively in the mineralogical analysis for identification of inorganic species in coal.

Classical methods of "wet" chemical analysis were the first to be used for coal-ash analysis. These have been largely superseded by more rapid colorimetric methods as described by Archer et al. (1958), and by the spectrochemical methods as used by Dixon (1958) and Zinc et al. (1967). Introduction of flame photometric, atomic absorption, x-ray fluorescence, and neutron activation techniques have speeded up the analytical procedures. Different analytical techniques available for the major constituent and trace-element analysis are described in the book edited by Babu (1975).

For mineralogical analysis, the optical microscope was the earliest tool used to identify different mineral species found in coal. Thiessen (1919) was the first to use petrographic thin-section techniques to identify different mineral species in coal. Optical microscope techniques have been extensively used by Mackowsky and Kirsch, and their methods and results can be found in the books published by Gumz et al. (1958) and Stach et al. (1975).

Development of the scanning electron microscope by Oatley et al. (1965) has greatly increased the scope of identification of sub-micron-size mineral impurities in coal. Since the early 1970s, the electron microscope fitted with the energy dispersive analysis of x-rays (EDAX) has been used by Gluskoter and Lindahl (1973) and Ruch et al. (1973) to identify different mineral species in U.S. coals. Earlier, Dutcher et al. (1964) had used the electron probe microscope with some success to investigate the distribution of mineral impurities in different coal strata. Finkelman and Stanton (1978) have used the scanning electron microscope to study the dispersion of mineral grains in polished blocks and pellets of coal.

Of the nonmicroscopic methods, x-ray diffraction analysis has been most widely used, and Gumz et al. (1958) have given an excellent introduction to x-ray analysis of coal minerals. Notable results of x-ray diffraction analysis have been published by Dixon et al. (1964), Gluskoter (1967), Rao and Gluskoter (1973), O'Gorman and Walker (1972), and Russell and Rimmer (1979).

Differential thermal analysis (DTA) has been used extensively by Warne (1965, 1970, 1975, 1979) to identify different mineral species in Australian coals. O'Gorman and Walker (1972) have used the DTA method as well as thermogravimetric analysis (TGA) and derivative thermogravimetric analysis (DTGA) in their work on mineral matter in U.S. coals. Further notable publications on identification of the silicate, sulfide, and carbonate mineral species in coal are those of Francis (1961), Ode (1963), and Watt (1968). Rekus and Haberkorn (1966) have used the x-ray diffraction technique to identify different coal mineral species in individual particles of coal, 150 to 250 μm in size. They used the results to discuss the mode of dispersion of the identified mineral species in coal. A computer-controlled method using the scanning electron microscope with the x-ray fluorescence emission and energy dispersive analyzer attachment has been developed and used for the analysis of particulate minerals in coal by Lebiedzik et al. (1973) and by Moza et al. (1979). A comprehensive treatise on analysis of coal, coal minerals, trace elements, coke, and combustion products has been compiled by a team of internationally known experts, edited by Karr (1979).

Usual methods of analysis of coal mineral matter rely on taking representative samples of the supply delivery, but more recently on-line techniques have been developed. The most favored technique is that of spectrometry of gamma-ray emission resulting from neutron capture, as described by Stewart and Hall (1978). An encapsulated source of fast neutrons, such as californium-252, is placed in a coal bed, and the neutron stream is slowed down by multiple collisions in a solid fuel. This produces a flux of thermal neutrons subsequently captured by the atoms of coal mineral elements in proportion to their neutron cross sections. The neutron capture results in an emission of gamma rays with energies characteristic of each element. An on-line coal-ash monitor has been developed by the U.K. National Coal Board (NCB, 1979) and has been extensively tested with satisfactory results. Brown et al. (1978) have made a survey of the techniques which have been used to determine the moisture content of coal. Three electromagnetic techniques were identified as promising for eventual on-line application; these are the capacitance, microwave attenuation, and nuclear magnetic resonance (NMR) techniques.

3.2 COAL MINERAL MATTER IN RELATION TO THE GEOLOGICAL ENVIRONMENT

The mode of occurrence and relative concentration of mineral impurities in different coal strata reflect the geological environment in which the fuel deposits were laid down. An informative account of the geological basis for an assessment of the distribution of mineral matter in different coal deposits is given by Teichmuller and Teichmuller in the book by Stach et al. (1975). Other notable publications on this topic are those of Ball (1935), Lowry (1945), Nelson (1953), Gumz et al. (1958), Juranek et al. (1958), Alekseev (1959), Francis (1961), Sen and Roy (1962), Kemezys and Taylor (1964), Murchison and Westoll (1968), Watt (1968), and Rao and Gluskoter (1973).

The surface skin of the earth's crust consists chiefly of soils, sediments, and sedimentary rocks. Below the surface skin, the rocks are composed chiefly of quartz, feldspar, mica, and hornblende. Silicon (Si) combined with oxygen in the form of SiO_2 is the most abundant element in the crust and rock strata known as the earth's sialic layer. Sedimentary deposits and rocks, which are the main source of mineral impurities in coal, are made up chiefly from degraded sialic materials and carbonate precipitates. Table 3.1 gives a chemical analysis of some typical sedimentary deposits and igneous rocks (Wright, 1972) arranged according to their silica (SiO_2) content. Sandstone is a sedimentary rock composed chiefly of quartz grains. Granite is another mineral species of high silica content. Shale, a mixture of sedimentary deposit and weathered rock, consists chiefly of clay minerals. Basalt can have chemical composition similar to that of clays. Peridotite is a basic rock of high magnesium content. Limestone consists largely of calcium carbonate ($CaCO_3$) formed from the shells of animals or by chemical precipitation from sea water.

Mixtures of clay species and basaltic-type rock together with some sedimentary sandstone and carbonate sediment constitute a mineral matter that has the chemical composition similar to that of typical bituminous coal ashes. This is illustrated by

Table 3.1 Chemical analysis of earth's sediment and igneous rock (percent by weight)

Mineral	Sandstone	Granite	Shale	Basalt	Peridotite	Limestone
SiO_2	95.4	70.8	55.1	49.0	45.5	5.2
Al_2O_3	1.1	14.6	16.3	18.2	4.0	0.8
Fe_2O_3	0.4	1.6	4.2	3.2	2.5	0.5
FeO	0.2	1.8	1.9	6.0	9.8	
CaO	1.6	2.0	4.9	11.2	3.5	42.6
MgO	0.1	0.9	2.5	7.6	34.0	7.9
Na_2O	0.1	3.5	0.7	2.6	0.6	0.1
K_2O	0.2	4.2	3.0	0.9	0.2	0.3
TiO_2	0.1	0.4	0.9	1.0	0.8	0.1
H_2O	0.3	0.3	5.2	0.2	0.8	0.7
CO_2	1.1	Trace	4.0	Trace	Trace	41.6

the results given in the first half of Table 3.2, showing representative ash compositions taken from over 100 sample analyses that covered a wide range of British bituminous coals (BCURA, 1963). The ash analysis results for a selection of U.S. bituminous coals are given in the second half of Table 3.2 (O'Gorman and Walker, 1972). The first two samples represent low-SiO_2 ashes in bituminous coals.

Table 3.2 Composition[a] of a selection of bituminous coal ashes

	Low silica		Medium silica			High silica	
Coals	Sample 1	Sample 2	Sample 3	Sample 4	Sample 5	Sample 6	Sample 7
British							
SiO_2	29.6	35.1	40.8	45.3	50.4	55.5	59.2
Al_2O_3	19.8	20.3	19.4	25.5	24.7	30.0	26.7
Fe_2O_3 + FeO	12.7	24.8	8.0	8.0	5.3	2.6	4.3
CaO	24.5	7.2	19.9	11.6	8.1	3.5	1.6
MgO	5.4	7.4	5.1	3.1	3.3	0.4	0.8
Na_2O	0.9	0.3	0.9	0.9	1.5	1.3	1.6
K_2O	1.1	1.1	2.1	3.9	3.0	2.7	2.8
TiO_2	0.9	0.6	0.9	0.8	1.0	1.2	1.1
SO_3	4.4	2.6	2.2	1.1	2.3	0.9	0.8
United States							
SiO_2	25.6	37.0	41.2	46.6	49.0	55.1	59.5
Al_2O_3	21.0	17.8	32.0	27.8	32.6	23.9	26.7
Fe_2O_3 + FeO	20.5	23.3	14.4	17.8	3.7	2.2	4.2
CaO	15.8	5.6	1.8	1.7	3.4	7.3	2.3
MgO	0.7	1.2	1.1	0.6	0.9	0.4	0.7
Na_2O	2.0	1.8	0.7	0.4	0.4	0.3	0.8
K_2O	0.2	0.2	0.7	1.1	2.1	0.4	0.1
TiO_2	0.6	0.5	1.5	1.1	0.9	1.7	3.4
SO_3	10.9	10.9	3.6	0.6	3.9	6.8	1.8

[a]Given as weight percent of ash.

Samples 3, 4, and 5 cover the range of medium-SiO_2 ashes in large number of bituminous coals mined in Britain, the United States, and many other countries, and samples 6 and 7 are examples of high-SiO_2 ash.

The sub-bituminous coals and lignites can have a mineral matter either of the clay type as in Table 3.2, or of entirely different as shown by examples in Table 3.3. Some specific characteristics of the ash composition and mineral species in U.S. lignites, sub-bituminous and bituminous coals, and anthracites are discussed in Chap. 15 (Tables 15.1 to 15.4). Occasionally bituminous coals have mineral matter in an unusual composition where the clay minerals do not constitute the bulk of ash (samples 2, 3, and 4 in Table 3.3). Sample 1 is SiO_2-rich ash of lignite from the Conceicas district of Brazil (Peel, 1978). Sample 2 is SiO_2-rich ash of bituminous coal from the Tarong district of Australia (McLaughlin, 1980). Sample 3 is SiO_2-rich ash of anthracite from Deane, Kentucky (O'Gorman and Walker, 1972). The fourth sample is an iron oxide-rich ash of bituminous coal from Carrier Mills, Illinois (O'Gorman and Walker, 1972). For sample 5 we have calcium-rich ash of sub-bituminous coal from Germany (Lissner, 1956). Sample 6 is calcium- and sodium-rich ash of lignite from Savage, North Dakota district (Tufte and Beckering, 1975). Sample 7 is calcium-, magnesium-, and sodium-rich ash of sub-bituminous coal from the Morwell (Victoria) district of Australia (Durie, 1961).

The above cursory review of the variability of mineral matter in different coals shows that the most abundant elements of the earth's crust—Si, Al, Fe, Ca, Mg, Na, K, and Ti, in combination with oxygen—also constitute the noncombustible fraction of solid fuels. Relative concentrations of the impurity elements depend on local geological environment and on the age of the solid fuel deposits. Lignites and sub-bituminous (brown) coals are the first products of coalification processes of the original vegetable matter, and these deposits lie usually close to the earth's surface. Consequently, the mineral matter in sub-bituminous solid fuels can vary over a wide range, and the impurity species reflect sharply the local mineralogical

Table 3.3 Exceptionally high- and low-silica ashes (composition in weight percent)

Mineral	High-silica			Low-silica			
	Sample 1	Sample 2	Sample 3	Sample 4	Sample 5	Sample 6	Sample 7
SiO_2	80.0	73.8	68.5	17.5	7.2	19.0	1.8
Al_2O_3	6.0	21.6	20.8	9.2	9.4	12.0	5.9
Fe_2O_3 + FeO	5.6	2.0	2.6	64.1	7.9	9.0	5.4
CaO	3.1	0.7	1.4	4.1	28.1	20.0	21.7
MgO	1.8	0.2	0.4	0.4	1.4	8.0	25.9
Na_2O	1.3	0.1	0.6	0.3	2.9	8.0	13.5
K_2O	0.4	0.2	0.1	1.0	ND[a]	0.3	0.3
TiO_2	ND	1.2	3.6	0.4	0.5	0.5	0.2
SO_3	ND	Nil	1.7	1.8	40.3	23.1	23.9

[a]ND denotes not determined.

environment that existed during the period of initial formation of peat and lignite deposits. By contrast, the initial mineralogical regimes are less discernible in the case of bituminous coals. Usually the increase of coal rank from lignite via sub-bituminous to bituminous coal and anthracite has resulted in a gradual process of enrichment of clay mineral dispersions in fuel deposits.

Tables 3.4 and 3.5 list mineral species that have been identified in different coal deposits. The data were compiled from publications by Parks (1952), Gumz et

Table 3.4 Silicate and oxide mineral species in coal

Species	Chemical formula	Specific[a] gravity (kg m⁻³)	Melting point (K)
	Silica and silicates—common occurrence		
Quartz	SiO_2	2650	1983
Kaolinite	$Al_2O_3 \cdot 2SiO_2 \cdot 2H_2O$		2083
Muscovite	$K_2O \cdot 3Al_2O_3 \cdot 6SiO_2 \cdot 2H_2O$	2900	(Mullite)
Illite	As muscovite with Fe, Ca, and Fe		
Montmorillonite	$(1-x)Al_2O_3 \cdot x(MgO,Na_2O) \cdot 4SiO_2 \cdot nH_2O$		
Chlorite	$Al_2O_3 \cdot 5(FeO,MgO) \cdot 3.5SiO_2 \cdot 7.5H_2O$		
Orthoclase	$K_2O \cdot Al_2O_3 \cdot 6SiO_2$	2500	
Plagioclase	$Na_2O \cdot Al_2O_3 \cdot 6SiO_2$ — Albite		
	$CaO \cdot Al_2O_3 \cdot 2SiO_2$ — Anorthite		
	Silicates—less common occurrence		
Augite	$Al_2O_3 \cdot Ca(Mg,Fe,Al,Ti) \cdot 0.2SiO_2$		
Amphibole	Augite + Na,F,P	3100	
Biotite	$Al_2O_3 \cdot 6(MgO \cdot FeO) \cdot 6SiO_2 \cdot 4H_2O$	3100	
Granite	$Al_2O_3 \cdot 3(CaO,MgO,FeO,MnO) \cdot 3SiO_2$		
Epidote	$4CaO \cdot 3(Al,Fe)O_3 \cdot 6SiO_2 \cdot H_2O$	3350	
Kyanite	$Al_2O_3 \cdot SiO_2$	3550	2083 (Mullite)
Sanidine	$K_2O_3 \cdot Al_2O_3 \cdot 6SiO_2$	2570	
Straurolite	$Al_2O_3 \cdot FeO \cdot 2SiO_2 \cdot H_2O$		
Tourmaline	$Na(Fe,Mn)_3 \cdot 3Al_2O_3 \cdot 6SiO_2 \cdot 3BO \cdot 2H_2O$	3100	
Zircon	$ZrO_2 \cdot SiO_2$	4500	2825
	Oxides and hydrated oxides		
Rutile	TiO_2[b]	4200	2100
Mangetite	Fe_3O_4	5140	1865
Hematite	Fe_2O_3	5200	1840
Limonite	$Fe_2O_3 \cdot H_2O$	4300	675[c]
Diaspore	$Al_2O_3 \cdot H_2O$	3400	425[c]

[a]The specific gravity of silicate minerals is in the range of 2500 to 3500 kg m⁻³; it increases with Al_2O_3/SiO_2 ratio and decreases with H_2O content. The silicates containing Na, K, Ca, Mg, and Fe do not have a definite melting point temperature (Chap. 9).

[b]With the exception of rutile, the oxide minerals rarely occur in coal.

[c]Denotes loss of water.

al. (1958), Marshall and Tompkins (1964), Dixon et al. (1964, 1970a, b), O'Gorman and Walker (1972), Rao and Gluskoter (1973), and Stach et al. (1975). The specific gravity and melting point data were taken from publications of Rigby (1953) and Weast (1977). Decomposition temperatures of coal carbonate and sulfide species were taken from the data quoted in Chap. 7.

3.3 CLAY MINERALS (ALUMINOSILICATES) AND SILICA (QUARTZ) IN COAL

By far the most abundant noncombustible species in coal are those of aluminosilicate clay minerals. Together with quartz, they usually account for between 60 and 90 percent of the total mineral matter in coal. Most common species of clay minerals are muscovite–illites (potassium aluminosilicates), kaolinites (aluminosilicates), and mixed layer illite-montmorillonites of variable composition. Rao and Gluskoter (1973), in an investigation of 65 coals from the Illinois Basin, reported a mean value of 52 percent for clay in the mineral matter. O'Gorman and Walker (1972) also found that the clay minerals constituted the greater part of mineral matter in the majority of U.S. coals. Barwood et al. (1982) have described a selective cation saturation method for identifying different clay minerals in coal. The method, based on selective cation saturation with potassium and magnesium, has been shown to allow identification of several clay species in the low-temperature ash of coal.

Much of the clay material dispersed in coal seams is made up of clastic material brought in by water. It therefore follows that the coal deposits laid down in areas that were subject to periodic flooding usually contain a great deal of clay dispersion intimately mixed with the coal. Dixon et al. (1970a) have studied the distribution of mineral species in various coal strata of East Midlands (U.K.) coal fields, and Table 3.6 gives some of their findings.

Dixon et al. (1970a) have listed potassium aluminosilicates as micas, whereas usually these are said to belong to muscovite–illite group of minerals. Other minerals identified included pyrite, siderite, iron oxide, and rutile. The authors, after examining more than 50 seam sections from various localities in the coalfield, have arrived at the following conclusions.

1. Quartz, kaolin, and mica are the principal silica-bearing minerals in all sections of the seam profiles, potassium aluminosilicates frequently being the more abundant species in all strata with the exception of the coal seam.
2. Quartz is more abundant in the roof and floor strata and least in the intraseam dirt and coal layers. The immediate roof layer is frequently rich in carbonate, and the floor material has high kaolin and mica content.
3. The highest kaolin concentrations (relative to the total mineral matter) occur in the coal and in some intraseam dirt bands; the kaolin-to-mica ratio is on average much higher for coal (between 5:1 and 10:1) than for the other strata. For cannel and run-of-mine coal, the average kaolin to mica ratio is respectively 1:2.5 and 1:1.5.

Table 3.5 Carbonate, sulfide, sulfate, phosphate, and chloride minerals in coal

Species	Chemical formula	Specific gravity $(kg\ m^{-3})$	Melting/decomposition temperature (K)
Carbonates			
Calcite	$CaCO_3$	2710	1200^b
Aragonite	$CaCO_3$	2710	1150^b
Dolomite	$CaCO_3 \cdot MgCO_3$	2850	1050^b
Ankerite	$CaCO_3 \cdot FeCO_3$		1000^b
Siderite	$FeCO_3$	3830	800^b
Sulfides			
Pyrite	FeS_2	5000	1075^b
Marcasite	FeS_2	4870	1075^b
Pyrrhotite	FeS_x	4600	1300
Chalcopyrite	$CuFeS$	4100	1300
Melnikovite	$FeS_2 + (As, FeS, H_2O)$	~5000	1075^b
Galena	PbS	7500	1370
Mispickel	$FeS_2 \cdot FeAs_2$	~5000	1075^b
Sphalerite	ZnS		
Sulfates			
Barytes	$BaSO_4$	4500	1855
Gypsum	$CaSO_4 \cdot 2H_2O$	2320	1725
Kieserite	$MgSO_4 \cdot H_2O$	2450	1395^b
Thenardite	Na_2SO_4	2680	1157
Mirahilite	$Na_2SO_4 \cdot 10H_2O$	1460	1157
Melanterite	$FeSO_4 \cdot 7H_2O$	1900	755^b
Keramolite	$Al_2(SO_4)_3 \cdot 16H_2O$	1690	945^b
Jarosite	$K_2SO_4 \cdot xFe_2(SO_4)_3$	2500	900^b
Phosphates			
Apatite	$Ca_5F(PO_4)_3$	3100	>1500
Evansite	$3Al_2O_3 \cdot P_2O_5 \cdot 18H_2O$	2560	>1775
Chlorides			
Halite	$NaCl$	2170	1074
Sylvite	KCl	1980	1043
Bischofite	$MgCl_2 \cdot 6H_2O$	1570	987

[a]Calcite, dolomite, ankerite, siderite, pyrite, barytes, and apatite minerals occur frequently. Gypsum and other sulfates are found mainly in low-rank and weathered coals; other mineral species are rarely found.

[b]Denotes the decomposition temperature.

Table 3.6 Mineral species in East Midlands (U.K.) coal strata (weight percent of the total mineral matter)

Coal seam strata	Mica	Kaolin	Quartz	Chlorite	Feldspar
Floor	38	25	33	2	2
Immediate floor	52	21	4	1	<1
Roof	41	18	27	3	<1
Immediate roof	47	28	18	<1	<1
Intraseam dirt	54	31	9	1	1
Cannel	32	13	10	<1	<1
Coal adjacent to cannel	25	30	8	0	0
Coal	4	35	2	0	0
Run-of-mine coal	38	26	13	<1	<1

4. The nonsilicate minerals constitute from 20 to 90 percent of the mineral matter in the coal seam but are usually absent in the floor material. Siderite may constitute up to 10 percent of mineral matter in the coal seam roof, where it frequently occurs in discrete bands or nodules.

Nelson (1953) classified quartz as a mineral of minor importance, but his survey work was based on early studies of U.S. coal seams by Gauger et al. (1934) and by Ball (1935). More recent analytical data by O'Gorman and Walker (1972) show that the majority of U.S. coals have a quartz content between 1 and 20 percent of the total mineral matter, and there are some coal ashes that have over 30 percent of quartz.

Quartz content of ashed coal residues can be determined by x-ray diffraction analysis, differential thermal analysis (DTA), and infrared absorption (Hicks and Nagelschmidt, 1943; Gumz et al., 1958; Francis, 1961; Dixon et al., 1964; Warne, 1979; O'Gorman and Walker, 1972). When there are no facilities available for a mineralogical analysis for quartz its approximate value can be deduced from the chemical analysis of ash.

Chemically analyzable or total silica, $(SiO_2)_t$, includes quartz, which is said to be "free" silica $(SiO_2)_q$, and "combined" silica $(SiO_2)_c$, when present with Al_2O_3 and other oxides in clay minerals. That is:

$$(SiO_2)_t = (SiO_2)_q + (SiO_2)_c \qquad (3.1)$$

Dixon et al. (1964) have shown that the aluminosilicate clay fraction of coal, chiefly kaolinites and muscovite–illites, has SiO_2 and Al_2O_3 constituents in a weight ratio between 1.4 and 1.6 with an average value of 1.5; that is,

$$(SiO_2)_c = 1.5(Al_2O_3) \qquad (3.2)$$

The quartz content is thus given by:

$$(SiO_2)_q = (SiO_2)_t - 1.5(Al_2O_3) \qquad (3.3)$$

where the concentrations of quartz, total SiO_2, and Al_2O_3 are given in weight percent of either ash or coal.

Equation (3.3) was derived from the results of the chemical and mineralogical analysis of British (bituminous) coal ashes, but the comparative data given in Table 3.7 show that the method gives useful estimates of the quartz content of coals mined in the United States.

The x-ray analysis for coal samples 2 and 3 in Table 3.7 gave unrealistically high quartz values that are not compatible with moderately high alumina content. An exceptionally high quartz content of ash, above 30 percent, is usually accompanied by a low, below 15 percent, alumina content as in sample 139.

Equation (3.3) cannot be used to calculate the quartz content of sub-bituminous coals or lignites when those have a mineral matter with the SiO_2 content below 30 percent. A part of the aluminum in these fuels can be present as precipitates of humic acids (Francis, 1961). Exceptionally high amounts of alumina, above 35 percent by weight, in some coal ashes may indicate that a part of the

Table 3.7 Comparison quartz contents of some U.S. coals, calculated using Eq. (3.3) and x-ray analysis (weight percent of total mineral matter)

Mineral	Sample[a]					
	2	3	4	6	99	139
SiO_2	59.2	68.5	37.0	56.1	39.8	57.3
Al_2O_3	26.7	20.8	17.8	29.1	19.4	14.3
$SiO_2 \equiv Al_2O_3$	40.1	31.2	26.7	43.7	29.1	21.5
Quartz, calculated	19.1	37.3	10.3	12.4	10.7	35.8
Quartz, x-ray analysis	30–40	40–50	1–10	10–20	10–20	30–40

[a]The sample numbers refer to those given in the work of O'Gorman and Walker (1972) on the chemical and mineralogical analysis of U.S. fuels. Samples 2 to 6 were anthracites from Kentucky, sample 99 was sub-bituminous from Wyoming and sample 139 was lignite from Texas.

alumina is present as $Al_2O_3 \cdot xH_2O$ (bauxite). In these cases the quartz contents calculated according to Eq. (3.3) would be low.

3.4 INTRODUCTION AND THE MODE OF OCCURRENCE OF SULFUR IN COAL

Sulfur is a vital element for sustenance of plant life, and it is therefore not surprising to find a relative abundance of sulfur compounds in coal deposits and in associated mineral-matter strata. The coal deposits laid down in brackish marine areas are usually rich in ash, sulfur, and nitrogen (Williams and Keith, 1963). Periodic flooding in these areas reduced the originally high acidity of plant deposits by dilution, resulting in an enhanced environment for bacterial sulfur fixation.

Neavel (1966) considered that iron sulfide (pyrites, FeS_2) found in peat and coal deposits had been largely formed as the result of activity of anaerobic bacteria. Sulfur originated from plant and animal protein (organic sulfur), or it was brought in by sea water in the form of sulfates. Iron originated from weathered silicate minerals; consequently, syngenetic pyrites deposits appear frequently in clay-bearing sediment.

Coals deposited in calcium-rich swamps also show a high degree of sulfidation. Calcareous basements or influx of calcium-rich waters effectively increased the pH of the aqueous environment, thus creating favorable conditions for a high rate of sulfur fixation. An extreme example of this type of coal is that found in Istria district of Greece, which contains 11 percent by weight of sulfur (Petrascheck, 1950).

By contrast, the peat deposits and subsequently formed coal deposits laid down in raised bogs of inland areas are usually low both in ash (clay minerals) and in sulfur. In the absence of flooding, the acidity of peat water had remained high with a pH value between 3 and 5, and such an acidic medium did not sustain a high degree of sulfidation bacterial activity; the bulk of sulfur therefore was returned to the atmosphere in the form of H_2S and SO_2. Coals laid down in those areas have a low sulfur content, around 0.5 percent by weight. Currently, such coals are in great demand for electricity generation because of the low level of SO_2 emission and the low-sulfur fuels. However, on a worldwide basis, occurrences of low-sulfur coals are comparatively rare, and the vast reserves of bituminous coals have a sulfur content between 1.0 and 4.5 percent. For example, the average sulfur content of British coals is 1.6 percent (CEGB, 1977), which may be taken as a typical figure for coals of medium sulfur content.

Wandless (1959) concluded after his extensive study of the mode of distribution of sulfur in British coals that on average the coals contained 0.8 percent by weight organically compounded sulfur, and the remainder was present as pyrites (FeS_2). The amounts of sulfatic sulfur in British coals, as in the majority of all bituminous coals, were negligible. Neavel (1966) arrived at a similar conclusion, that there was a relationship between the relative abundance of pyritic sulfur and that of organic sulfur. Gluskoter and Simon (1968) reported a mean value of 1.56 for the ratio of pyritic to organic sulfur in 473 face-channel samples of Illinois coal. Arguably the most thoroughly investigated high sulfur coals are those mined in Illinois, where the sulfide species constitute about 25 percent by weight of the total mineral matter, as discussed by Rao and Gluskoter (1973), Kuhn et al. (1973), and Shimp et al. (1975).

Pyrite and marcasite, both of an approximate chemical formula of FeS_2, are the principal pyritic minerals found in coal. Pyrrhotite (FeS_x), chalcopyrite ($CuFeS_2$), mispickel ($FeS_2 \cdot FeFS_2$), galena (PbS), and sphalerite (ZnS) as listed in Table 3.5 are other sulfur-containing mineral species that have been identified in coal (Gumz et al., 1958; Watt, 1968). Microstructural dispersion of the pyritic mineral species in coal seams and clay sediment-strata has been studied in the scanning electron microscope, as discussed by Greer (1977), Finkleman and Stanton (1978), and Moza et al. (1979).

Freshly mined coals usually do not contain any significant quantity of sulfates, but a variety of iron sulfates will form when the sulfides are oxidized in air. Rao and Gluskoter (1973) and Gluskoter (1977) have identified a number of hydrated ferrous and ferric sulfates in weathered coals, but the amount of sulfates in coals is usually insufficient to make it worthwhile considering their behavior in coal-cleaning processes or in the boiler flame.

3.5 CARBONATE INCLUSIONS IN THE FIRST AND SECOND STAGES OF COALIFICATION

Carbonate minerals were incorporated in the first as well as in the second stage of coalification processes, as described by Mackowsky in the book published by Stach et al. (1975). The species of the syngenetic (first stage) process consists chiefly of siderite ($FeCO_3$) and dolomite ($MgCO_3 \cdot CaCO_3$). The carbonate products of the second stage of the coalification process are calcite ($CaCO_3$) and ankerite

($MgCO_3 \cdot CaCO_3 \cdot xFeCO_3$). Dixon et al. (1970b), after studying a number of coal samples from the East Midlands coal fields, concluded that calcite and ankerite were the principal carbonate minerals present. The latter has a relatively constant composition, namely 54 percent by weight $CaCO_3$, 24 percent $MgCO_3$, 20 percent $FeCO_3$, and 2 percent $MnCO_3$. This composition is close to that accorded usually to ankerite mineral. Their results were in line with those obtained earlier by Pringle and Bradburn (1958).

Siderite, ankerite, and calcite are commonly occurring carbonate minerals in Australian coals, as determined both by microscopic observations (Kemezys and Taylor, 1964) and by chemical analyses (Brown et al., 1960). Rao and Gluskoter (1973) concluded, however, that the carbonate minerals in Illinois coals are distinctly different from those in British and Australian coals. Calcite was the dominant carbonate mineral in Illinois coals, siderite and dolomite were observed in a few samples in trace amounts, and ankerite was not detected. Unusually high amounts of carbonate minerals can occur in some coals, as for example in Oklahoma coals (O'Gorman and Walker, 1972). In these coals the carbonate minerals, siderite and dolomite, were the main impurity constituents, with small amounts of clay minerals, kaolinite, illite, and chlorite.

3.6 THE ORIGIN AND MODE OF OCCURRENCE OF CHLORINE IN COAL

The origin and the distribution of chlorine in coal have been investigated by many researchers, particularly in Britain, since British coals show a large variation in their chlorine content. Skipsey (1974), after reviewing the statistical data on chlorine in coal accumulated by the National Coal Board (NCB) and the previously published data, concluded that the chlorine content of British low-rank coals varied from 0.05 to 1.0 percent depending on their geographic locality. However, with high-rank coals of carbon contents in excess of 85 percent (dry ash-free basis) the chlorine content rarely exceeds 0.2 percent.

Undoubtedly much of the chlorine found in British coals came from sea water, but its introduction into the fuel deposit cannot be correlated unequivocally with any specific episode in the coalification process. Crumley and McCamley (1958) have suggested that the chlorine was derived from salt absorbed from the aqueous environment in which the coal vegetation grew. According to Wandless (1957), the entry of chlorine may have been by infiltration from the overlying Permo–Triassic rocks. Daybell (1967) has stated that extensive borings in Britain have shown that there is usually a gradual increase in the chloride content of coal seams with increase in depth. Any association of the chlorine distribution pattern with the overlying Permo–Triassic beds would be expected to result in the highest chlorine values occurring in seams adjacent to the salt-bearing beds.

This apparent anomaly was explained by Gluskoter and Rees (1964), who postulated, after studying the chlorine distribution in coals of the Illinois basin, that the chlorine content of coals was related to the salinity of the ground waters.

Skipsey (1975) found that the distribution of chlorine in British coals was closely related to the occurrence of brines in the deep coal measures. Hypersaline brines with concentrations of dissolved solids up to 200 kg m^{-3} occur in several of the Northern and Midlands coalfields. The mode of formation of hypersaline brines has been discussed by Dunham (1970), who concluded that they have been evolved from connate waters of marine origin by osmotic filtration through clays and shales that acted as membranes. The salinity of the brine ground waters increases with depth, and when they are in contact with fuel-bearing strata, correspondingly more chloride is taken up by the fuel. However, according to Skipsey (1975), the high-rank coals, because of their low porosity, are unable to take up large amounts of the chloride and associated cations when in contact with hypersaline waters; thus the chlorine content rarely exceeds 0.2 percent.

The mode of occurrence of chlorine in coal has been subject to many investigations. Watt (1968) summarized the viewpoint of different researches up to that date, and the subject has been discussed further by Gluskoter and Ruch (1971), Skipsey (1974), and Saunders (1980). The main argument has centered around the question of whether NaCl, the main chlorine constitutent, can exist as a discrete mineral in coal, or whether the chloride and sodium ions are held separately in the coal substance. Caswell (1981) has suggested that the water-soluble chlorine component, chiefly NaCl, is present as a solution held within coal pores. This is likely to be the case with coals delivered to power stations, as these contain usually between 8 and 15 percent of evaporable water. Table 3.8 shows the relative concentration of chloride and different water-leachable cations in a high chlorine coal (Raask, 1958, unpublished results). For a comparison, analyses of the mine saline water (Skipsey, 1975) and "standard" sea water (Weast, 1977) are also given.

The results in Table 3.8 show that the relative amounts of different leachable chlorides in the high-chlorine coal, as indicated by the chemical equivalent ratios, were close to those of the highly saline mine water. In contrast, the relative concentrations of sodium and magnesium in the sea water were higher and that of calcium lower. It appears that sodium and magnesium in the sea water were partly replaced by calcium as a result of exchange reactions when the saline water perculated through calcium-rich deposits.

Hodges et al. (1983) have carried out a review of published data on the origin and mode of occurrence of chlorine in coal. They stated that there exists overwhelming evidence to suggest that the major source of chlorine in coals is brine

Table 3.8 Chloride and associated cations in coal, saline mine water, and sea water

Substance	Constituent (percent by weight)					Chemical equivalent ratios, Cl:Na:Ca:Mg:K
	Cl	Na	Ca	Mg	K	
Coal	1.00	0.43	0.14	0.03	0.006	1:0.65:0.21:0.09:0.01
Saline mine water	9.18	4.15	1.12	0.28	–	1:0.70:0.22:0.09:–
Sea water	1.90	1.06	0.04	0.13	0.04	1:0.86:0.04:0.20:0.22

with which the seams have been in contact. They concluded that although a number of theories have been expressed about the bonding of chlorine to coal—for example that it is trapped within the porous structure or associated with basic groups—the nature of the chlorine-coal bond is still uncertain.

NOMENCLATURE

Al_2O_3 alumina content of coal mineral matter
$(SiO_2)_c$ combined silica content of coal mineral matter
$(SiO_2)_q$ quartz content of coal mineral matter
$(SiO_2)_t$ total silica content of coal mineral matter

FOUR

QUALITY OF COAL UTILIZED
IN POWER STATIONS

4.1 HEAT CONTENT AND COMBUSTIBILITY
OF COALS OF DIFFERENT RANK

Electricity utilities purchase their coals on the basis of the energy stored in the solid fuel, but the heat content and combustibility characteristics have also a bearing on the ash-related problems in the large boiler plant. The amount of ash mineral matter passing through the coal milling and boiler plant has an inverse relationship with the heat content of coal, and the increased amounts of ash can markedly enhance coal-plant wear, furnace slagging, and boiler-tube erosion, as discussed in Chaps. 5, 14, and 18. The rate of combustion of coal in the form of pulverized fuel cloud can markedly influence the slagging characteristics of ash and build-up furnace-wall deposits, and high-temperature corrosion, as discussed in Chaps. 9, 14, and 17.

The calorific value and combustion characteristics depend on the degree of coalification (rank) and on the amount of noncombustible ballast—i.e., ash and water—in the fuel. Table 4.1 summarizes the characteristic data of different rank coals in respect to their calorific value and chemical analysis of the combustible matter. The data were compiled from publications by Seyler (1900), Gruner and Bousquet (1911), Parr (1928), Patteisky and Teichmuller (1958), Francis (1961, 1965), Fenton et al. (1962), and O'Gorman and Walker (1972).

The principal changes that occur with increase of coal rank can be summarized as follows:

1. There is a progressive increase in carbon and a decrease in hydrogen when the fuel rank changes from lignite to anthracite.
2. There is a gradual decrease in the volatile matter with increase of coal rank.
3. There is a gradual decrease in the inherent or "hydroscopic" moisture content until the carbonaceous rank is reached.
4. There is an increase in calorific value until the hydrogen content has decreased to about 4.5 percent.

Heat energy stored in coal is based on the principal combustion reactions:

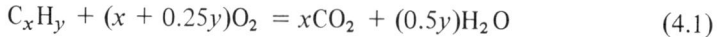

$$C_x H_y + (x + 0.25y)O_2 = xCO_2 + (0.5y)H_2O \qquad (4.1)$$

where $\Delta H_{CO_2} = 393$ MJ kg mol^{-1} and $\Delta H_{H_2O} = 285$ MJ kg mol^{-1} (JANAF, 1971).

The presence of oxygen in coal substance can be taken as an indication of the degree of partial oxidation of the fuel, thus reducing the amount of heat released on combustion. The mode of occurrence of oxygen in the organic substance of coal is discussed by Volborth (1979). Mott and Spooner (1940) introduced two formulas for calculating the calorific value from the results of chemical analyses for carbon, hydrogen, oxygen, and sulfur. When coals contain above 15 percent by weight oxygen, the equation is:

$$Q_g = 0.336\ C + 1.42\ H_2 - (0.153 - 0.00072\ O_2)O_2 + 0.0943 \qquad (4.2)$$

and the corresponding equation for coals containing less than 15 percent oxygen is:

$$Q_g = 0.336\ C + 1.42\ H_2 - 0.1450\ O_2 + 0.0941\ S \qquad (4.3)$$

where Q_g is the gross calorific value of coal in MJ kg^{-1} and C, H_2, O_2, and S denote the weight percent of carbon, hydrogen, oxygen, and sulfur, respectively.

Table 4.1 Gross calorific value, volatile matter, carbon, hydrogen, oxygen, and moisture in coals of different rank (ash-free bases)

Coal rank	Calorific value (MJ kg^{-1})	Volatile matter	Percent by weight				
			Total carbon	Fixed carbon	Hydrogen	Oxygen	Bed moisture
Lignitous, unconsolidated	24–25	50–55	59–68	45–50		20–25	35–50
Lignitous, consolidated	25–30	45–50	68–75	50–55		20–25	20–35
Sub-bituminous	30–32	40–45	75–80	55–60	4.5–5.5	15–20	8–10
Bituminous 1	33–36	32–40	80–84	60–68	5–5.8	5–10	2.5–5
Bituminous 2	36–37	26–32	84–87	68–74	5–5.8	2.5–5	2.5–5
Bituminous 3	37–39	22–26	87–89	74–78	4.5–5.5	2.5–5	1.0–2.5
Bituminous 4	39–40	18–22	89–91	78–82	4.0–4.5	2.5–5	0.8–1.0
Carbonaceous	38–40	10–18	91–93	82–90	2.5–4	2.5	1.0–1.5
Anthracitous	37–39	8–10	93–95	90–92	2.5	2.5	1.5–2.5

Table 4.2 Routine analysis data of British power-station coals (weighted mean values)

| Data | Power station | | | National average |
	Drax	Eggborough	Ratcliffe	
Calorific value (MJ kg^{-1})	23.9	24.1	23.9	24.2
Combustible volatile matter (wt %)	26.4	27.7	27.8	27.2
Ash (wt %)	18.9	18.5	15.6	16.4
Moisture (wt %)	10.6	10.4	13.4	12.0
Sulfur (wt %)	1.7	1.7	1.5	1.6
Chlorine (wt %)	0.25	0.23	0.29	0.21

Alternative correlations between the calorific value and coal composition have been published by Boie (1953):

$$Q_g = 0.36 \, C + 1.16 \, H_2 - 0.111 \, O_2 + 0.105 \, S + 0.063 \, N_2 \qquad (4.4)$$

and by Schuyer and van Krevelen (1954):

$$Q_g = 0.29 \, C + 1.48 \, H_2 - 0.102 \, O_2 + 0.135 \, S + 0.048 \, N_2 - 0.035 \, V_m$$

$$(4.5)$$

where the additional symbols N_2 and V_m denote the percent by weight of nitrogen and volatile matter, respectively.

Equations (4.2) to (4.5) give the gross calorific value of coals on ash- and moisture-free basis, whereas the values quoted in Table 4.1 are those of moist coals, i.e., coals with the typical amounts of inherent (fuel-bed) water depending on their rank. Power-station coals frequently go through a wash-cleaning process or are exposed to rain, so the fuel usually contains a significant amount of adventitious water, between 5 and 15 percent by weight.

Coals delivered to British power stations are usually supplied with the following analytical data carried out on a routine basis: the gross calorific value, the volatile matter, ash, moisture, and sulfur and chlorine contents. The sulfur and chlorine compounds do not make a significant contribution to the total heat release in boiler flame, but they have a marked influence in the formation of slagging and corrosive deposits, and they constitute the source of acidic gases, chiefly SO_2, SO_3, and HCl, formed in the flame as discussed in Chap. 7. Table 4.2 shows the quality of British coals delivered to the CEGB power stations in a typical year (CEGB, 1977).

The gross calorific value of British coals on an "as received" basis varies between 20 and 32 MJ kg^{-1}, as shown on Fig. 4.1. The plot of calorific value against coal ballast—i.e., ash plus water—when extrapolated to zero ballast gives the mean calorific value within the range of 34.3–36.2 MJ kg^{-1} when expressed on ash and moisture free basis. The straight line plot in Fig. 4.1 shows that for a given type of fuel, in this case low-rank bituminous coals, the ash (A) and water (W)

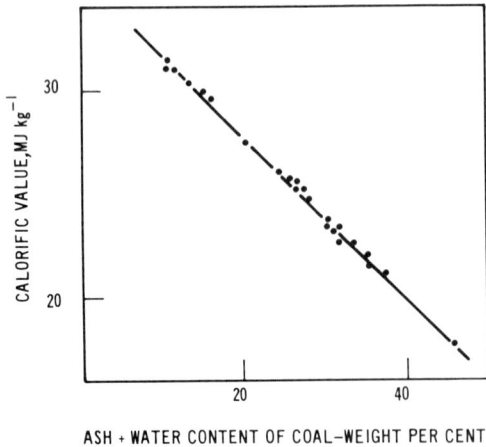

Figure 4.1 The relationship between gross calorific value and the ballast content of power-station coals (Barnsley area, U.K.).

contents can be used to obtain an approximate calorific value of coals. The relationship is in the form

$$Q_{ga} = Q_{go} - m(A + W) \tag{4.6}$$

where Q_{ga} and Q_{go} denote the calorific value of "as received" and of dry, ash-free coals, respectively. The value of m depends on the quality (rank) of the given fuel supply. For coals in the Barnsley area of the United Kingdom, the value of m obtained from the slope of the plot in Fig. 4.1 was 0.38; thus Eq. (4.6) becomes

$$Q_{ga} = 35.5 - 0.38(A + W) \tag{4.6a}$$

Once the Q_{go} and m parameters in Eq. (4.6) have been established for a power-station coal supply, the ash and water data would be useful in monitoring coal energy input to boilers. The ash and moisture contents of coal feedstock can be measured by neutron activation and infrared absorption techniques, as described by Stewart and Hall (1978), Rhodes (1978), Brown et al. (1978a), and the National Coal Board (NCB, 1979); thus the energy input to boilers can be computed from the weight of the coal. Brown (1970) has discussed the relationship between calorific value and ash content of British bituminous coals.

With sub-bituminous and lignitic coals of low and variable calorific values it is sometimes necessary to blend the coals to give a fuel feedstock of more consistent quality, as described by Loeffler and Vaka (1979). They describe the coal blending in the Navaja Mine located in New Mexico, which produces for the local electricity utility sub-bituminous fuel of a calorific value between 16 and 24 MJ kg^{-1}. A variation in heat content of this magnitude would normally cause serious operational problems at the power plant. A solution was found by automated coal blending to give a boiler fuel supply with a calorific value of around 21 MJ kg^{-1}.

In practice, the heat content of coal (gross calorific value) is determined by measuring the heat liberated when a sample of fuel burns in a closed-vessel (bomb) calorimeter (British Standard, 1970; ASTM, 1972). Heat transfer from the combustion chamber of the calorimeter to cooling water takes place near ambient temperature. It therefore follows that the gross calorific value of coal thus measured includes the heat of condensation of water. In a boiler plant, however, useful heat transfer from the flue gas ceases to be practicable at about 375 K, and at this temperature the water originally present in coal and that formed during combustion of coal hydrogen remain in the vapor phase. Thus the net calorific value of coal Q_n of interest to combustion engineers could be defined as

$$Q_n = Q_g - (\Delta H_e)W_t \tag{4.7}$$

where Q_g = the gross calorific value of coal

ΔH_e = the latent heat of evaporation of water (43.7 MJ kg mol^{-1} or 2.43 MJ kg^{-1}

W_t = the total amount of water vapor produced in coal combustion

W_t includes the water originally present in coal W and that produced from burning hydrogen:

$$W_t = W + 9 \ H_2 \tag{4.8}$$

where H_2 denotes the amount of hydrogen in coal substance. Thus, the net calorific value of coal is given by:

$$Q_n = Q_g - 0.0243(W + 9 \ H_2) \tag{4.9}$$

where the net and gross calorific values are given in MJ kg^{-1} units, and the water and hydrogen contents are given in weight percent of coal.

The net calorific value of power-station coals of bituminous rank is usually between 2 and 5 percent lower than the gross calorific value, depending on the amount of water in coal and the hydrogen content of combustible substance, as shown in Table 4.1. This difference between the gross and net calorific values is much higher for water- and hydrogen-rich sub-bituminous coals and lignites. For example, with a low-rank coal that contains 40 percent water and has a hydrogen content of around 10 percent, as much as 15 percent of the coal energy measured as the ambient temperature is required to evaporate that amount of water.

Low-rank lignitious and sub-bituminous coals, including types 1 and 2 of bituminous coals in Table 4.1, have one desirable feature when utilized in pulverized-fuel-fired boilers. That is, their combustible volatile matter content is high, over 25 percent by weight, and consequently reactive coke particles are formed on evolution of the volatile matter in the boiler flame. The coke particles thus formed (Fig. 4.2) have hollow-sphere and honeycombed structures, as discussed by Lightman and Street (1968) and by Street et al. (1969), and are highly porous. The specific surface (surface area:mass ratio) of the coke residue is usually between 10^4 and 8×10^4 m^2 kg^{-1} (Raask and Bhaskar, 1975).

The coke particles react with oxygen both at the outside surface, in a chemical rate-controlled process, and inside the pores, in a diffusion-controlled process. For the given external and internal surfaces of coke particles, the overall rate of combustion is proportional to oxygen concentration outside the stagnant diffusion layer surrounding the particles. Thus, the combustion rate depends on the degree of mixing of the burning coke particles and combustion air in the flame. The chemical

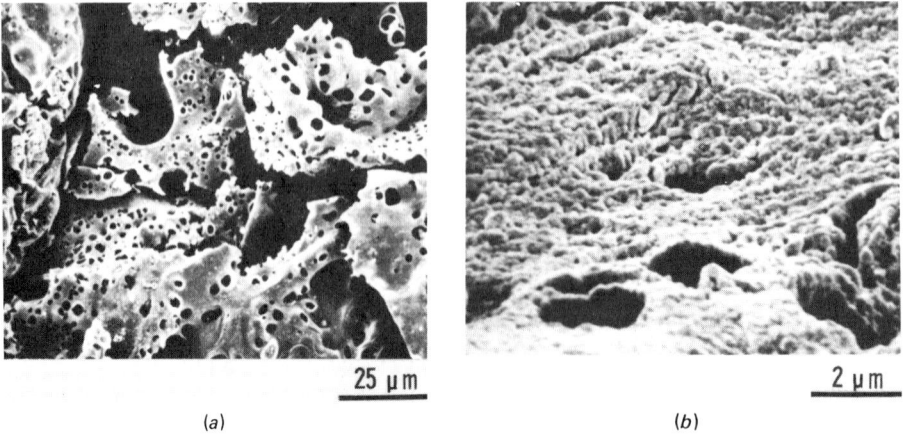

25 µm   2 µm

(a)   (b)

Figure 4.2 Coke residue in pulverized coal ash. (a) Large pores. (b) Micropores.

reaction rate is much more sensitive to temperature changes compared with that of diffusion controlled combustion. Field (1970) has shown that with anthracite particles below 78 µm in diameter the rate of chemical reaction rate is low at temperatures below 2200 K, but for low-rank coals this threshold temperature is much lower, around 1600 K. As a result, the burn-out time of the high volatile coal particles is short, 1 to 2 s, which is within the residence time of fuel clouds in large combustion chambers. Thus, the amount of unburnt coke residue in pulverized fuel ash of power station boilers should not exceed 2 or 3 percent by weight of ash, which represents a loss of 0.3 to 0.5 percent of the original coal energy. Higher amounts of carbon in ash are usually a sign of a shortcoming in boiler design or operation, e.g., a coarse milling, a lack of combustion air, or a poor mixing of pulverized coal and air.

With high-rank fuels—these include low-volatile bituminous coals (type 4 in Table 4.1) and in particular the anthracite fuels—it is more difficult to ensure that the loss of energy in the form of unburnt carbon in ash is kept below 1 percent of the total coal energy. The Central Electricity Generating Board (CEGB, 1971) recommends that with high-rank coals (anthracites) the fineness of milled fuel should be such that 80 to 85 percent by weight is below 75 µm in size, whereas for most other coals it is sufficient to have 70 to 75 percent of pulverized fuel below this particle size. Dolezal (1967) has suggested the following milling size requirements for different rank coals:

1. High-rank coals, 92 percent by weight below 90 µm
2. Low-rank bituminous coal, 65 to 72 percent by weight below 90 µm
3. Lignites, 40 to 58 percent by weight below 90 µm

Other boiler design and operation measures to reduce unburnt carbon losses include a good mixing of pulverized fuel and combustion air, and design of boilers with a long flame path to increase the residence time of burning coal particles above 1200

K (Sur, 1979). The relationship between the unburnt carbon loss and the combustion conditions in an anthracite-fired boiler has been investigated by Cutress and Peirce (1965) and by Peirce (1966). A theoretical relationship between the rate of combustion of a cloud of anthracite fuel and its particle size was established by Leesley and Hedley (1972). They concluded that the results were consistent with a combustion process in which the rate was controlled by the chemical reactivity of the surface layer of anthracite particles.

4.2 INHERENT AND ADVENTITIOUS MINERAL MATTER IN COAL

Gumz et al. (1958) have classified mineral matter introduced during the first stage of the coalification process as syngenetic, and that introduced after the formation of coal deposits as epigenetic. In research studies of the origin of the mineral species in coal this classification is useful, but for practical users it is more helpful to classify the mineral matter in coal as (1) inherent mineral matter, which represents the impurity material that is too closely associated with the coal substance to be readily separated from it by physical methods, and (2) adventitious mineral matter, which represents the material that is less intimately associated with the coal substance and thus more readily removable on coal cleaning.

The inherent mineral matter is generally found to be of syngenetic origin, and it is rich in iron, phosphorus, calcium, potassium, and magnesium. These are the elements known to be essential for plant life. Among other elements that appear to be less essential for sustenance of plant life but that are found in plants are silicon, aluminum, manganese, and sodium. The initial stages of coal formation frequently occurred under swamp conditions when the inherent mineral elements were incorporated in the carbonaceous phase. Additional mineral matter found in coal seams had their origin in dissolved salts and clay suspensions in the swamp or flood water.

Adventitious coal mineral matter is not homogeneously distributed and is frequently found to be present as deposits in cleavage and fracture cracks, and often appears as dirt bands in coal seams. That is, the amount of adventitious mineral matter associated with coal deposits has been largely determined by the geological environment under which the initial coal plant residues were consolidated into the stratified seams. During mining, some floor and roof material adjacent to the fuel seams is included in "run-of-mine" coals; consequently, the amount of adventitious mineral matter in untreated fuel can be high. Table 4.3 shows the ash (mineral matter) content of coal seams and that of the associated impurity bands in typical U.K. East Midlands mines (Dixon et al., 1964).

The data of Table 4.3 show that the inherent ash content of coals was between 6.1 and 7.5 percent, whereas the total ash content of run-of-mine smalls varied from 20 to 45 percent. High-ash coals are usually subjected to cleaning by washing in a high-density dispersion medium in order to reduce the amount of the adventitious mineral matter in the fuel delivered to power stations. The results of analysis of sulfur and chlorine in Table 4.3 show that the impurity species of these elements are closely associated with the coal seam and the intraseam dirt bands. It follows therefore that the sulfur and chlorine impurities in "as mined" fuel are not readily removed by the conventional methods of coal cleaning.

4.3 DISTRIBUTION OF PARTICULATE MINERAL IMPURITIES IN PULVERIZED COAL

As discussed in Section 4.1, it is necessary to reduce the coal combustible matter in size by milling to ensure sufficiently rapid combustion of the airborne particles in pulverized-fuel-fired boilers. The size distribution of pulverized coal can be expressed in the form of the Rosin–Rammler equation (Field et al., 1967):

$$d_w = 100 \exp\left(-\frac{d}{d_1}\right)^n \qquad (4.10)$$

where d_w is the weight percentage of particles greater than d in size, and d_1 and n are adjustable constants. The constant d_1 is a measure of the overall fineness, and n is a measure of the spread of sizes. A high value of n indicates a closely sized powder; for pulverized coals typical values of n are low, around 1.2, indicating a wide size spectrum. Values of d_1 are usually between 30 and 70 μm (Field et al., 1967).

It is to be expected that the particulate mineral matter in typical pulverized coal would have a distribution spectrum weighted towards smaller sizes compared with that of the coal substance particles. That is, the small size fractions of pulverized coal would contain relatively more mineral matter for two reasons. First, coals usually contain a large proportion of the total mineral matter in the form of fine clay sediment particles of 0.1 to 10 μm in size, as discussed in Chap. 3. Second, the density (specific gravity) of coal substance is between 1200 and 1300 kg m^{-3}, whereas the density of mineral species is between 2 and 4 times greater, i.e., between 2500 and 5000 kg m^{-3}. Transport of pulverized coal particles from the mill to boiler burners is effected by air elutriation, where the aerodynamic (Stokes) diameter is the important parameter. The relationship between the aerodynamic diameter d_a and the static (microscopically measurable) diameter d_s is given by

$$d_a = D^{1/2} d_s \qquad (4.11)$$

where D denotes the density of the particle material. Thus, for a given upper aerodynamic diameter limit—say 100 μm equivalent—the coal particles would be in the size range of 87 and 91 μm, whereas the mineral having the same Stokes diameter would be in the size range of 42 to 62 μm.

Rayner and Marskell (1963) and Littlejohn (1966) have reported the results of extensive studies on the distribution of mineral matter in pulverized fuel samples collected from coal milling plant of power station and industrial boilers. Fuel samples were size fractionated by sieving and by means of an aerodynamic classifier

Table 4.3 Ash, sulfur, carbonate, and chlorine contents of typical British coal seams associated strata

Colliery and seam	Sample	Percent by weight of air-dried samples			
		Ash	Sulfur	Carbon dioxide	Chlorine
Linby, High Main	Roof	90.5	0.06	4.01	0.10
	Intraseam dirt	52.1	0.31	0.04	0.16
	Floor	90.7	0.07	0.14	0.17
	Coal	6.2	0.65	0.87	0.46
	Run-of-mine smalls	25.7	0.53	1.91	0.37
Thoresby, Top Hard	Roof	78.7	0.30	8.86	0.35
	Intraseam dirt	75.1	0.79	0.04	0.34
	Floor	94.3	0.05	0.02	0.18
	Coal	6.1	1.22	0.88	1.05
	Run-of-mine smalls	20.2	0.95	1.65	0.94
Moorgreen, 2nd Waterloo	Roof	93.8	0.14	0.04	0.07
	Intraseam dirt	90.7	0.10	0.02	0.06
	Floor	91.3	0.10	0.02	0.07
	Coal	6.7	1.46	1.15	0.35
	Run-of-mine smalls	44.4	0.99	1.39	0.27
Westhorpe, Flockton	Roof	87.9	0.14	3.17	0.10
	Intraseam dirt	45.3	2.58	0.37	0.26
	Floor	94.0	0.06	0.22	0.09
	Coal, excluding cannel	6.9	3.54	0.46	0.16
	Cannel	24.7	3.54	0.46	0.16
	Run-of-mine smalls	36.5	2.12	0.37	0.26
Bentinck, Tupton	Roof	80.6	0.29	4.15	0.13
	Intraseam dirt	70.0	0.87	3.25	0.13
	Floor	83.2	0.08	0.05	0.13
	Coal, excluding cannel	7.5	0.96	0.27	0.53
	Cannel	8.4	2.49	0.26	0.28
	Run-of-mine smalls	28.2	0.98	0.55	0.48
Donisthorpe, Stockings	Roof	94.1	0.10	0.24	0.05
	Intraseam dirt	38.0	0.90	0.24	0.14
	Floor	92.3	0.23	0.02	0.07
	Coal	6.1	1.30	1.61	0.21
	Run-of-mine smalls	21.6	1.46	0.98	0.13

(Godridge et al., 1962), and the ash content and chemical composition of ash of different size fractions were determined. In each case the ash content of coal fractions increased as the particle size decreased from over 200 μm to below 10 μm. Some of the results published by Littlejohn (1966) have been used in Fig. 4.3 to illustrate the relationship between the ash content and particle-size fractions of pulverized coal. His results and more recent work on the distribution of mineral matter in milled coals by Raask and Bhaskar (1981, unpublished results) can be summarized as follows:

A. The relationship between the ash content and coal particle size:
   1. The ash content of milled coal in the particle size fraction from 200 μm down to 76 μm is lower than that of the corresponding ungraded fuel by a factor of 0.5 to 0.8.
   2. The middle size range fraction, 20 to 76 μm, which accounts for 40 to 60 percent by weight of the total coal, has an ash content close to that of ungraded fuel.
   3. The 10- to 20-μm fraction on average has an ash content that is 1.3 times greater than that of ungraded fuel, and the ash content of the fraction below 10 μm is between 1.5 and 2.5 times greater than that of ungraded fuel.
B. Distribution of different mineral species in size fractionated pulverized coal:
   1. Aluminosilicate clay minerals, which constitute the bulk of noncombustible impurities in most bituminous coals, occur to a large extent as a dispersion of sediment of 0.1- to 10-μm particles concentrated in small size fractions, i.e., below 20 μm.
   2. Silica in the form of quartz, and carbonates, chiefly calcite, ankerite, and siderite (Chap. 3), show an enrichment of the size cut in the upper middle range, i.e., between 20 and 75 μm.
   3. Pyrites, chiefly FeS_2, tend to concentrate in the lower middle range, i.e., between 10 and 45 μm. The pyrite mineral has a high density, above 3000 kg m^{-3}, and for this reason the impurity, when it occurs in coal or in the associated mineral strata in the form of large FeS_2 nodules, is ground to fine sizes before the particles are entrained in the stream of elutriating air. Also, some pyrite impurity occurs in the form of fine particles together with clay silt in coal deposits. The distribution of pyrite in U.S. high-sulfur coals is discussed in Chap. 15.
   4. Chlorides, which are significant impurity species in some of the British coals, are dispersed uniformly within the size spectrum of pulverized coal.

Figure 4.4 shows the relative distribution of the major coal impurity species in different size fractions. The total mineral content of ungraded pulverized coal was 16 percent by weight, and the amounts of different minerals in the ungraded fuel (weight percent of coal) were as follows: aluminosilicates, 9 percent; quartz, 2 percent; carbonates, 1.5 percent; pyrites, 2.5 percent; and chlorides, 1.0 percent.

The above composition breakdown of the major mineral impurities was based on the analytical data of East Midlands (U.K.) coals supplied to local power stations (Dixon et al., 1964, 1970a, b). The mode of occurrence of sodium, calcium, and iron species in U.S. low- and high-rank coals and the impurity-related problems in boiler operation are discussed in Chap. 15.

The inherent ash in the coarse and fine (below 5 μm diameter) size fractions consists largely of small particles, as shown in Fig. 4.5. Both fractions contained 0.1- to 5-μm particles, and it is therefore evident that the inherent ash in pulverized coal, irrespective of the particle size, consists largely of the mineral species of 0.1 to 5 μm in diameter. The adventitious mineral matter, separated from pulverized coal by flotation, consists chiefly of clay sediment particles in small size fractions (Fig.

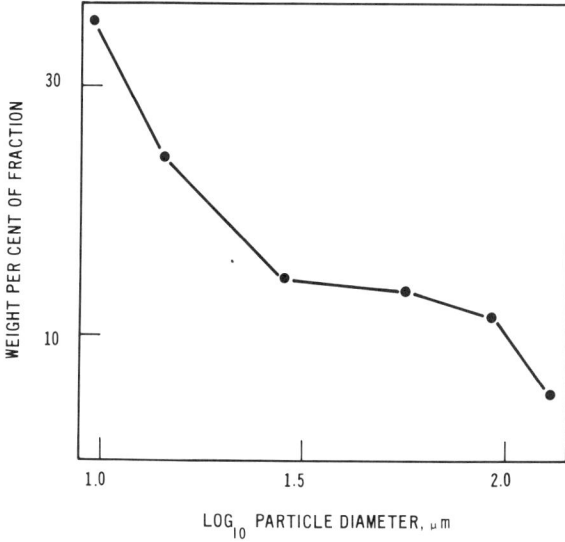

Figure 4.3 Ash content of different-size fractions of pulverized coal, average ash content 16.2 percent.

4.6a) and of ground rock in large fractions (Fig. 4.6b). Littlejohn (1966) has suggested that the particles of pulverized coals that float on a liquid of 1300 kg m^{-3} density could be classified as "pure" coal, and the particles that sink in a liquid of 2000 kg m^{-3} density can be classified as "pure" ash. The remainder—i.e., the middle fraction that sink in the 1300 kg m^{-3} density liquid but float on the 2000 kg m^{-3} density liquid—consists of mixed particles of coal and mineral matter, i.e., "middlings." The relative amounts of the density-separated fractions of typical pulverized bituminous coals are shown in Table 4.4.

Clean coal fractions in Table 4.4 contained 2–5 percent mineral impurities, and

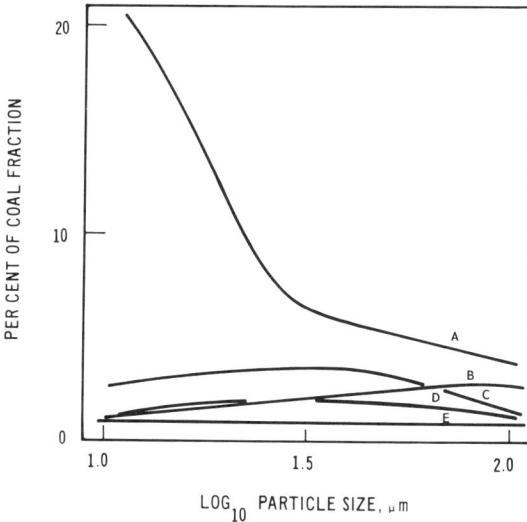

Figure 4.4 Distribution of different mineral species in size-fractionated coal: A, aluminosilicates; B, quartz; C, pyrites; D, carbonates; E, chlorides.

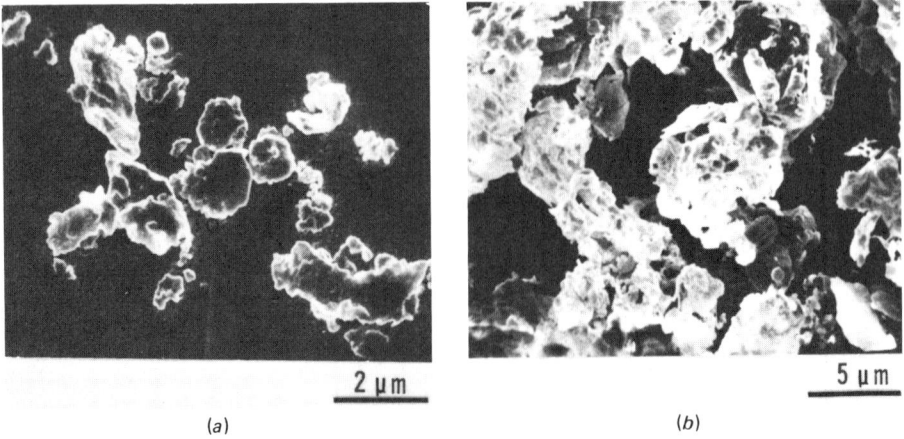

Figure 4.5 Inherent mineral matter in pulverized coal. (a) Particle size <5 μm. (b) Particle size >5 μm.

the dense fractions sinking in a liquid of density 2000 kg m^{-3} contained about 10 percent combustible matter. As expected, there was a close relationship between the ash content, determined by the usual combustion method, and the dense mineral rich fractions. The latter accounted for 90 percent of total mineral matter in coal, the remainder being present chiefly in the middling fractions.

Computerized techniques of analysis of single particles of pulverized coal are now available (Moza et al., 1979) for use as a valuable precursory in coal-cleaning studies and in investigations of the behavior of particulate mineral species in boiler flames. Combustion characteristics and the behavior of mineral matter in ash-rich coal particles are of particular interest. For example, the rate of oxidation of pyrite impurity dispersed in coal particles is likely to be significantly different from that of carbon-free pyrite particles. Also, the particles that contain both coal substance

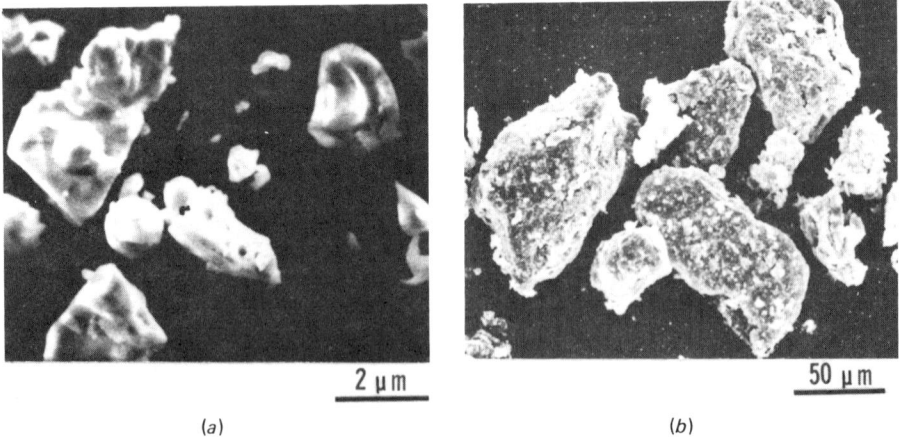

Figure 4.6 Adventitious mineral matter in pulverized coal. (a) Small particles. (b) Large particles.

Table 4.4 Clean coal, middlings, and mineral fractions of pulverized coals of differing ash content

Coal number	Percent by weight of ungraded coal			Ash content of ungraded coal
	Clean coal	Middlings	Mineral fraction	
1	65.8	30.7	3.5	5.4
2	67.9	25.6	6.5	7.0
3	50.3	33.2	16.5	16.2
4	56.4	27.5	16.1	16.4
5	55.8	27.9	16.3	17.2
6	14.1	69.9	16.0	17.3
7	49.3	33.5	17.2	18.3
8	54.0	27.1	18.9	18.6
9	38.2	42.0	19.8	18.7
10	48.1	34.1	17.8	19.1
11	64.0	16.4	19.6	22.9
12	48.0	31.9	20.1	23.5
13	47.9	29.7	22.4	23.7
14	3.9	66.6	29.5	32.6

and mineral matter are of interest in the study of coal/slag interface phenomena and in the formation of ash cenospheres as described in Chap. 6.

4.4 RELATIONSHIP BETWEEN THE MINERAL MATTER CONTENT OF COAL AND ITS ASH RESIDUE

The ash content of coal is usually determined by burning pulverized fuel spread in a 1- to 2-mm layer in a laboratory furnace with air circulation and kept at 1050 K. The ash residue, however, does not represent accurately the original mineral matter content of coal because of loss of SO_2, SO_3, CO_2, chlorine, and water from hydrated silicates, which is not balanced by the uptake of oxygen in oxidation of pyrites to iron oxides. To correct these losses, several formulas have been proposed for estimating the mineral matter in coal. Among the earliest was that of Parr (1928), which is still in use, derived from the ash and sulfur contents of coal as follows:

$$M = 1.08A + 0.55S \qquad (4.12)$$

where M = mineral matter of the coal

A = ash yield on combustion

S = sulfur content of the coal

Among other well-known formulas is that of King et al. (1936), which was subsequently modified by Millot (1958) to allow for the loss of water from clay minerals on heating:

$$M = 1.09A + 0.8\ CO_2 + 0.5\ S_{pyr} - 1.1\ SO_{3(ash)} + SO_{3(coal)} + 0.5\ Cl \quad (4.13)$$

where A, CO_2, S_{pyr}, $SO_{3(coal)}$, and Cl denote respectively the ash, CO_2, pyrite, sulfate, and chlorine contents of coal, and $SO_{3(ash)}$ is the sulfate content of ash.

A statistical formula has been derived by Brown et al. (1952) and can be used for most coals:

$$M = 1.06A + 0.53 \text{ S} + 0.74 \text{ CO}_2 - 0.32 \qquad (4.14)$$

As an alternative to the use of these formulas, several attempts have been made to determine the mineral matter content of coal by dissolution methods and by low-temperature combustion. Radmacher and Mohrhauer (1955) have described a method that involves removing the bulk of the mineral matter with hydrochloric and hydrofluoric acids and determining the loss in weight and the residual ash. The acid extraction procedure removes all minerals except pyrite, which must be determined separately. A modification of this method has been described by Bishop and Ward (1958) for coals with high concentrations of carbonates. The method developed by Brown et al. (1959) involves oxidation of coal at 650 K and weighing the incombustible residue, which is only slightly changed from the original mineral matter. Further advances in low-temperature ashing techniques have been reported by Gleit and Holland (1962) and Gluskoter (1965), as referred to in Chap. 3.

4.5 LABORATORY–PREPARED ASH CONTENT AND THE AMOUNT OF ASH, CLINKER, AND SLAG IN COAL–FIRED BOILERS

Instead of the correlation between the mineral matter content of coal and laboratory-prepared ash, power-station engineers are more interested in knowing how closely the laboratory-prepared ash content of coal represents the amount of ash produced in large boilers. The main difference between the ashing conditions in a laboratory muffle furnace and those in the boiler plant is the coal combustion temperature. The laboratory ashing furnace is usually kept below 1100 K, whereas the boiler flame temperature can reach 2000 K. The principal chemical reactions that affect the ash yield at different temperatures are:

High-temperature combustion   Low-temperature combustion

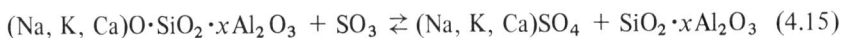

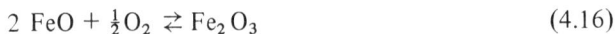

$$(\text{Na, K, Ca})\text{O} \cdot \text{SiO}_2 \cdot x\text{Al}_2\text{O}_3 + \text{SO}_3 \rightleftarrows (\text{Na, K, Ca})\text{SO}_4 + \text{SiO}_2 \cdot x\text{Al}_2\text{O}_3 \quad (4.15)$$

$$2 \text{ FeO} + \tfrac{1}{2}\text{O}_2 \rightleftarrows \text{Fe}_2\text{O}_3 \qquad (4.16)$$

The boiler ash cools rapidly at a rate of about 200 K s^{-1} through the temperature range from 1200 to 500 K (Fig. 6.1), and during this short time interval there is only a limited degree of sulfation [Eq. (4.15)] and oxidation [Eq. (4.16)] taking place. Thus the ash prepared in a laboratory furnace at 1100 K has a higher weight than that formed in the high-temperature boiler flame, due to the absorption of SO_3 in sulfate and additional oxygen in the ferric oxide. The relationship between the boiler ash A_b and laboratory-prepared ash A can be expressed in the form:

$$A_b = A - A \left[(SO_{3_l} - SO_{3_b}) + 0.1(FeO_l - FeO_b)\right] + C_b \qquad (4.17)$$

where SO_{3_l} and SO_{3_b} denote the sulfate content of laboratory prepared and boiler ashes respectively, and FeO_l and FeO_b denote the ferrous iron oxide content of the ashes. The factor 0.1 refers to the smaller oxygen content of FeO compared to that of Fe_2O_3:

$$1 - \frac{\text{molecular weight of FeO}}{\text{molecular weight of } Fe_2O_3} \cong 0.1 \qquad (4.18)$$

In Eq. 4.17, C_b denotes the unburnt carbon content of boiler ash; the laboratory prepared ash has zero carbon content. Further, the laboratory ash is usually prepared under conditions where iron oxide is oxidized to Fe_2O_3 state, so Eq. (4.17) reduces to

$$A_b = A - A \left[(SO_{3_l} - SO_{3_b}) + 0.1\, FeO_l\right] + C_b \qquad (4.19)$$

The sulfate content of laboratory prepared ash depends on its calcium, magnesium, and sodium content as shown in Fig. 4.7. The sulfate content of boiler ashes from bituminous coals shows a similar alkali dependence relationship, but on average the amount of sulfate in boiler ashes is approximately 0.5 of that of the laboratory prepared ashes of the same coal (Raask, 1982a). Further, the FeO content of boiler ashes is approximately equal to that of Fe_2O_3 (Raask, 1959, unpublished work); thus Eq. (4.19) reduces to

$$A_b = A \left(1 - 0.5\, SO_{3_l} - 0.05\, Fe_2O_{3_l}\right) + C_b \qquad (4.20)$$

where the notations are as for Eq. (4.17). Equation (4.20) does not apply to the sintered ash and slag discharged from the combustion chamber through the furnace bottom exit. The ash clinker material contains less sulfate and Fe_2O_3 than the

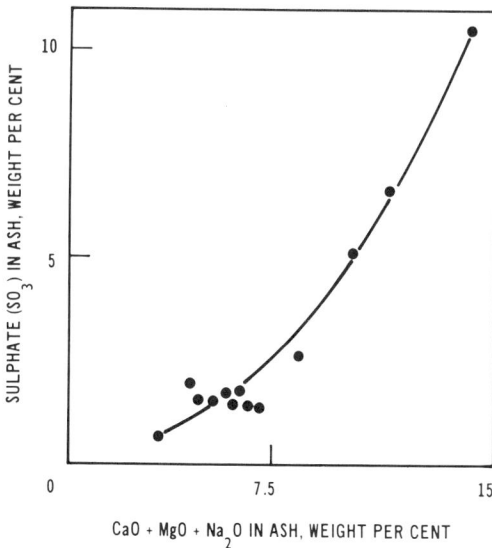

Figure 4.7 The sulfate content of laboratory-prepared coal ash.

corresponding precipitator ash, and the unburnt carbon content is practically zero. Thus for calculating the amount of boiler clinker ash (A_{bc}), Eq. (4.20) becomes

$$A_{bc} = A \, (1 - 0.2 \, SO_{3_l} - 0.05 \, Fe_2O_{3_l}) \qquad (4.21)$$

If the weight fraction of precipitated ash is x and that of clinker is $1 - x$, then the total ash products (A_b) are given by:

$$A_b = A \, [x(1 - 0.5 \, SO_{3_l} - 0.05 \, Fe_2O_{3_l} + C_b)$$
$$+ (1 - x)(1 - 0.2 \, SO_{3_l} - 0.05 \, Fe_2O_{3_l})] \qquad (4.22)$$

where A_b, A = weight percent of boiler ash and laboratory-prepared ash of coal, respectively

SO_{3_l}, $Fe_2O_{3_l}$ = sulphate and iron oxide concentrations of laboratory ash
$\quad C_b$ = (unburnt carbon) concentration of boiler ash are given in weight fractions of the respective ashes

As an example, coal may contain 16 percent laboratory prepared ash, and the ash may contain 2.0 percent sulfate and 10 percent iron oxide. Further, it is assumed that the boiler ash captured in the electrical precipitators amounts to 80 percent of the total ash input, and that the precipitator ash contains 3.0 percent of carbon residue. Thus, the amount of ash produced from 1 t coal under laboratory ashing conditions is 160 kg, and the following quantities of precipitator ash and sintered material (furnace-bottom ash) would be produced:

Precipitated ash yield $= 160 \times 0.8(1 - 0.5 \times 0.02 - 0.05 \times 0.1 - 0.03)$

$$= 129.9 \text{ kg}$$

and $\qquad$ Clinker yield $= 160 \times 0.2(1 - 0.2 \times 0.02 - 0.05 \times 0.1) = 31.6$ kg

Thus, the total of precipitator ash and furnace clinker amounts to 161.5 kg, which is only slightly more than that expected from the laboratory ashing result. That is, when the carbon content of precipitator ash is between 2 and 3 percent it compensates the loss of sulfate in the boiler ash, and the ash content of coal determined in laboratory represents accurately, within 1 ± 2 percent, the amount of ash and clinker produced in a pulverized-coal-fired boiler. If the unburnt carbon content is as high as 15 percent, as may be the case with low-volatile anthracite coals, then the total of boiler ash and clinker would be 177.3 kg, which is over 10 percent higher than predicted by the result of laboratory ashing.

By contrast, the total amount of ash products, precipitated ash, and slag produced in boilers fired with sub-bituminous coals can be significantly less than that predicted by the results of laboratory ashing. The data of Lissner (1963), as depicted in Fig. 4.8, show that the ash yield on burning sub-bituminous coals in a laboratory furnace depends markedly on the temperature. The ash yield drops rapidly above 1100 K. This is because calcium sulfate ($CaSO_4$), the principal ash constituent in low-temperature (laboratory) ash is unstable in the presence of silicates when the temperature exceeds 1100 K. The curves in Fig. 4.8 reflect the weight loss due to SO_3 evolution at higher temperatures.

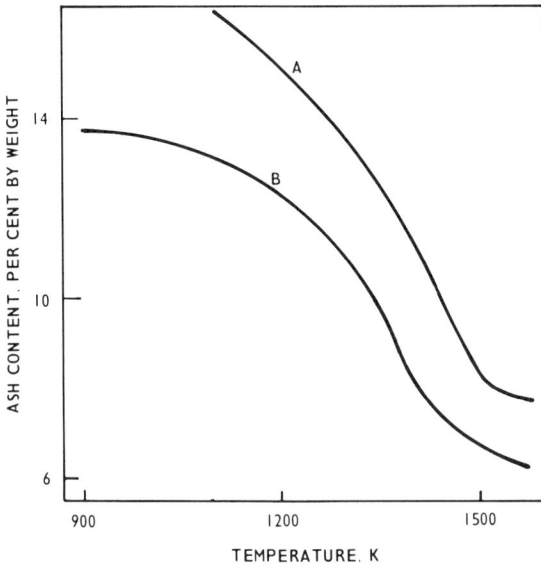

Figure 4.8 Ash content of high-calcium non-bituminous coals at different ashing temperatures: A, high-ash coal; B, medium-ash coal.

All systems of coal firing produce some fine ash, below 100 μm in diameter, which is entrained in the flue gas and has to be captured in the electrical precipitators. The gas-borne ash fraction varies from 85 percent of the total in pulverized-coal-fired boilers to about 15 percent in cyclone-fired boilers. The remainder is in the form of clinker and slag, or large-particle unsintered ash in the fluidized bed combustion. Table 4.5 shows the amounts of clinker, slag, and bed ash discharged from different systems. The estimates are based on the data published by Grunert et al. (1947) and on information supplied by boiler operation engineers.

The characteristics of slag and sintered ash deposits are discussed in Chaps. 9–14; the properties of fluidized-bed ash are compared with those of pulverized-fuel ash in Chaps. 18 and 20.

Table 4.5 Clinker, slag, and bed ash discharge from different combustion systems

Mode of firing	Nature of material	Percent of total ash
Pulverized coal with "dry" ash discharge		
Non-slagging coal ashes	Sintered clinker with variable	15–20
Moderately slagging coal ash	amounts of slag	20–25
Slagging coal ashes		25–30
Slag tap	Molten slag	40–60
Cyclone	Molten slag	75–85
Stoker	Sintered clinker	40–60
Fluidized bed	Unsintered particles	60–70

NOMENCLATURE

A	ash content of coal (laboratory prepared ash)
A_b	boiler ash
A_{ba}	boiler clinker
C	carbon content of coal substance
C_b	carbon content of boiler ash
CO_2	carbon dioxide content of coal carbonate
Cl	chloride content of coal
C_x	atomic ratio of carbon in coal substance
D	particle density
FeO	ferrous iron oxide content of ash
FeO_b	ferrous iron oxide content of boiler ash
FeO_l	ferrous iron oxide content of laboratory ash
Fe_2O_3	ferric iron oxide of ash
$Fe_2O_{3_l}$	ferric iron oxide content of laboratory ash
H_y	atomic ratio of hydrogen in coal substance
H_2	hydrogen content of coal substance
ΔH_{CO_2}	heat of formation of carbon dioxide
ΔH_l	latent heat of evaporation of water
ΔH_{H_2O}	heat of formation of water
M	mineral matter content of coal
N_2	nitrogen content of coal substance
O_2	oxygen content of coal substance
Q_g	gross calorific value of coal
Q_{ga}	calorific value of "as received" coal
Q_{go}	calorific value of dry and as free coal
Q_n	net calorific value of coal
S	sulfur content of coal
S_{pyr}	pyritic sulfur content of coal
$SO_{3(ash)}$	sulfur trioxide content of ash
SO_{3_b}	sulfur trioxide content of boiler ash
SO_{3_l}	sulfur trioxide content of laboratory prepared ash
$SO_{3(coal)}$	sulfur trioxide content of coal
V_m	volatile matter content of coal
W	water content of coal
W_t	water vapor content of flue gas
d	particle diameter
d_a	aerodynamic diameter of particles
d_s	microscopically measured diameter of particles
d_w	weight percentage of particles larger than d
d_1	adjustable constant in particle size equation
m	ratio factor in coal ballast/calorific value equation
n	power index in particle size equation

COAL GRINDING, ABRASIVE FUEL MINERALS, AND PLANT WEAR

5.1 COAL GRINDABILITY

Grindability of coals is usually assessed by the test devised by Hardgrove (1932), which has been incorporated in the ASTM (1971) procedure. The test machine is a miniature vertical-spindle ball mill in which usually a 50-g sample is ground by turning the spindle through 60 revolutions. The sample is then removed and screened, and the amount passing through a 200-mesh sieve is used to assign the Hardgrove grindability index G to the coal:

$$G = 6.93W + 13 \qquad (5.1)$$

where W is the weight (g) of the portion of 50-g sample passing through a 200-mesh sieve, i.e., below 76 μm in particle diameter. A critical appraisal of the method has been carried out by Edwards et al. (1980).

For a given moisture content, the Hardgrove grindability index depends on the hardness of coal substance, and the latter in turn is influenced by the rank of coal, i.e., its carbon content. Van Krevelen (1953) measured the Vickers micro-indentation hardness of seven coals of different rank. His results showed that the hardness increased from 28 to 38 kg mm^{-2} when the carbon content of coals, on dry ash-free basis, changed from 70 to 80 percent. It reached a maximum of 42 kg mm^{-2} for 84 percent carbon coal and then decreased to 28 kg mm^{-2} when the

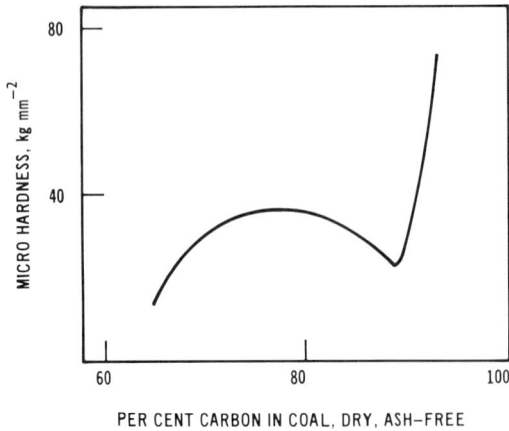

Figure 5.1 Vickers hardness and carbon content of coals.

carbon content reached 90 percent. Above 90 percent, the hardness increased again until the coal became so elastic that no indentation measurements could be carried out. Honda and Sanada (1957) found a similar relationship between the hardness and rank of different coals, as shown in Fig. 5.1.

British coals usually have a Hardgrove grindability index within the range of 45–60, which are the typical values for bituminous coals of high and medium volatile-matter content utilized in electricity utility boilers in many other countries. It is common practice to design the pulverizing plant to give full output for a coal of the grindability index of 50 and to provide a curve to show increase or decrease in output with softer and harder coals. A typical curve is given in Fig. 5.2 (CEGB, 1971), which shows that the output of a mill designed for bituminous coals is reduced by 20–30 percent when an anthracite coal of a low Hardgrove grindability index is fed to the mill.

Sub-bituminous coals have a high Hardgrove grindability index, but a problem may arise because of the high water content. Coals having a total moisture content

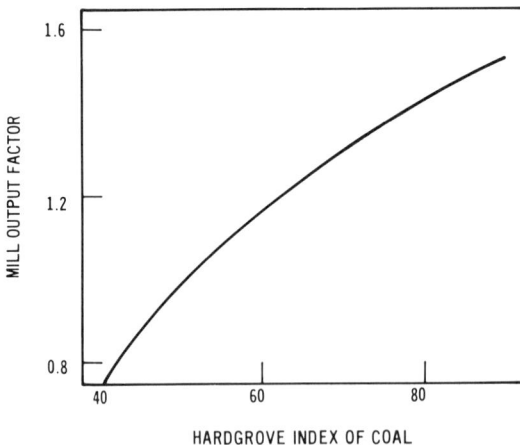

Figure 5.2 Mill output and grindability of coal. Design load = 1.0 for the index of 50. Coal is classified by Hardgrove index: 40–45, anthracite; 45–60, bituminous; and 60–90, subbituminous.

of 12–15 percent by weight can be handled in large mills, where all of the free moisture and up to half the inherent moisture is extracted by the passage of hot air, which carries pulverized fuel to the burners. An excess of residual water in coal can severely limit mill output, as shown in Fig. 5.3, in particular with tube ball mills where the grinding performance falls off rapidly when the water content of coal is above 4 percent. For high-moisture coals, it is necessary to provide the mills with additional air heaters. High-temperature flue gas for the air heaters is withdrawn from the superheat unit zone or from the bottom of the furnace and then is discharged back into the main flue gas stream after the conventional air heaters.

Another parameter that can limit mill output is the size of raw coal fed to the mill. The larger the coal size, the greater will be the energy required to break it down to pulverized fuel fineness; for constant energy input, therefore, the mill output will vary accordingly. It is customary in mill design to assume that raw coal will be below 20 mm in size, and Fig. 5.4 shows the departure in mill output when the size of raw coal increases. Figure 5.5 shows the relationship between the mill capacity and the required fineness of pulverized coal.

When a coal mill is correctly designed for a given supply of fuel, taking into account its grindability index, moisture content, and the size of raw coal, then the consistency in particle size and output of milled product depends largely on the amount and nature of the mineral matter in coal. The impurity fraction may include highly abrasive minerals, which can cause a rapid wear of mill grinding components, with the corresponding loss in plant performance as shown in Fig. 5.6. Wear-causing characteristics of different coal minerals and the measures to combat the mill plant and pulverized coal pipe wear are discussed in the following sections of this chapter.

5.2 ABRASIVE MINERALS IN COAL

The rate of wear in the body-contact sliding abrasion and that of particle impaction erosion is influenced by the abrasive characteristics of the wear-causing bodies or

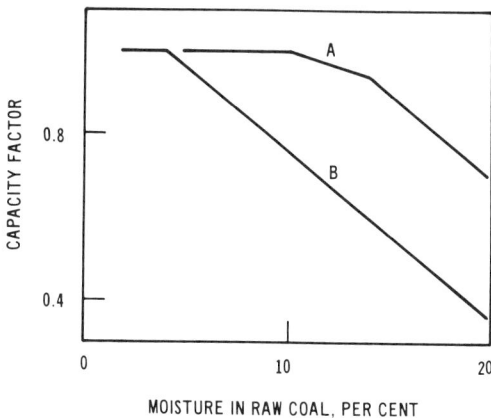

Figure 5.3 Variation in mill output with moisture content of coal: A, vertical-spindle mill; B, tube ball mill.

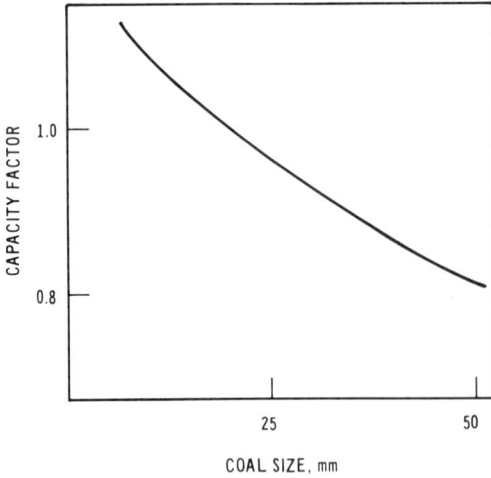

Figure 5.4 Variation in mill output with raw coal size.

particles. Hardness index given in the Mohs scale or as measured by the methods of Brinell, Vickers, or Rockwell (O'Neill, 1937), is a useful criterion that allows solids to be placed in a "pecking" order with respect to their potential abrasion- or erosion-causing property.

As discussed in Sect. 5.1, coal substance is not a hard material; it has a Vickers hardness value between 10 and 70 (Fig. 5.1), and thus as ash-free coal would not cause a significant abrasion or erosion wear in coal plant. The wear damage is caused chiefly by the mineral impurities in coal, and Table 5.1 gives the concentration and hardness of different mineral species in a typical bituminous coal. Cooling (1965) has collected the hardness data available in the literature and has discussed the relevance of the data to coal mill wear.

The hardness data in Table 5.1 show that kaolin (Fig. 5.7a) and the

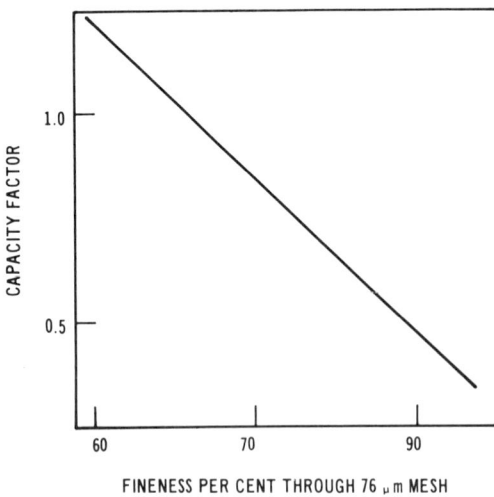

Figure 5.5 Variation in mill output with fineness of product.

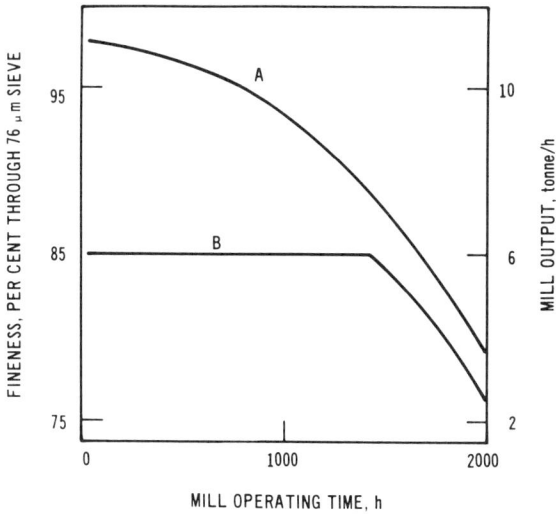

Figure 5.6 The effect of mill wear on grinding performance: A, mill output; B, fineness of pulverized coal.

plate-structured potassium silicates (Fig. 5.7b), which constitute the bulk of the mineral matter in coal, are soft materials, having a Vickers hardness below 70. It is therefore evident that the coal substance and the kaolin and mica minerals do not cause any significant erosion wear in the coal milling plant or in the pulverized-coal conveying pipes.

Experience in power stations has shown that the impact erosion wear of the coal plant components was particularly severe with coals of high quartz and pyrite

Table 5.1 Hardness of coal substance and mineral species in typical bituminous coal

Constituent	Approximate weight percent of coal	Mohs hardness number	Vickers hardness (kg mm^{-2})
Coal substance	85	1.5–2.5	10–70
Quartz	1.6	7	1200–1300
Pyrite	1.5	6–7	1100–1300
Silicates			
Kaolin	5	2–2.5	30–40
Illite	3	2–2.5	20–35
Muscovite	3	2–2.5	40–80
Orthoclase	<0.1	6	700–800
Kyanite	<0.1	6–7	500–2150
Topaz	<0.1	8	1500–1700
Carbonates			
Calcite	0.5	3	130–170
Magnesite	0.1	4	370–520
Siderite	0.2	4	370–430
Alumina	Rare	9	1200

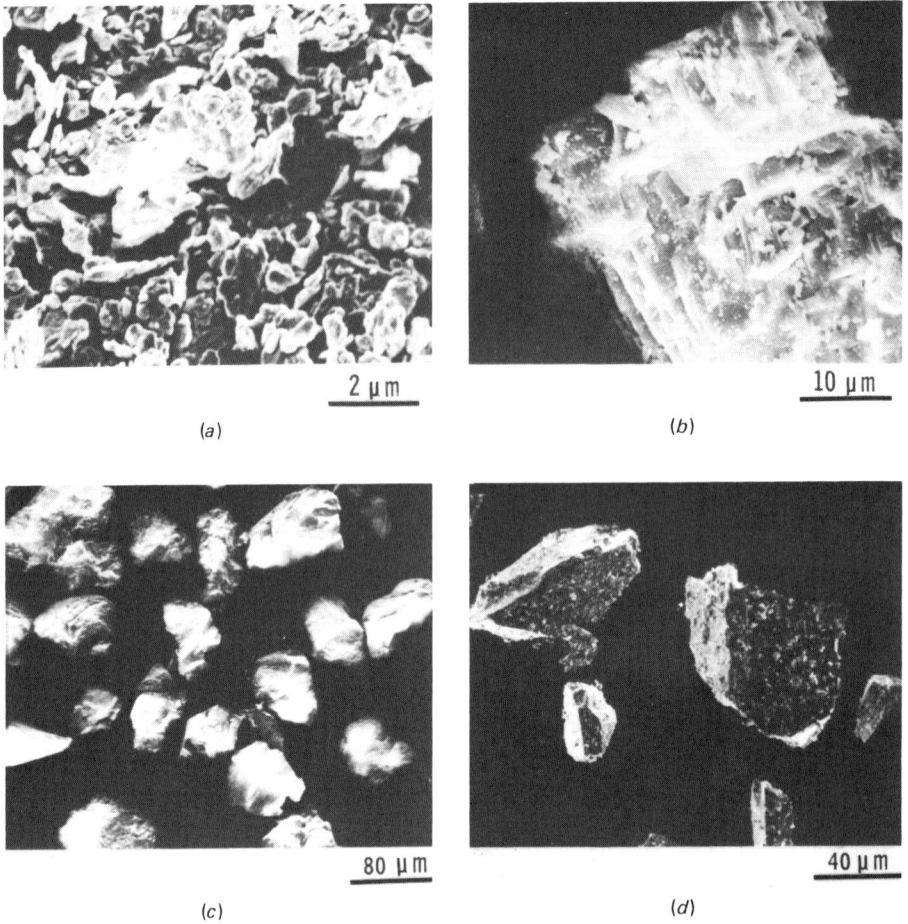

Figure 5.7 Nonabrasive and abrasive minerals in coal. (*a*) Soft kaolin. (*b*) Soft mica. (*c*) Hard quartz. (*d*) Hard pyrite.

contents. This is consistent with the hardness data given in Table 5.1. These minerals, in particular quartz, have a high Vickers hardness number, and sharp-edged fragments are produced on milling as shown in Fig. 5.7, *c* and *d*. The other hard minerals—orthoclase, kyanite, topaz, and alumina—are usually present in coals only in trace quantities, and it is unlikely therefore that these minerals cause any significant erosion wear.

Quartz and pyrite in coal can be determined by the x-ray diffraction analyses and by other techniques (Chap. 3), but these mineral species are not usually analyzed on a routine basis. In the absence of quartz and pyrite data, the abrasive property of coal can be estimated from the ash and sulfur content. British coals delivered to the power stations usually contain between 10 and 25 percent by weight of ash. The inherent ash of the coal—that is, the mineral matter intimately mixed with the coal substance—is usually about 5 percent, and this fraction contains very little quartz. The intraseam dirt bands contain between 3 and 10 percent quartz, but the highest concentrations of quartz are found in the floor and roof material, where it is between 25 and 30 percent (Table 3.5). An approximate relationship between the quartz content of coal and its ash content can be expressed in the form

$$C_q = b(A - a) \qquad (5.2)$$

where C_q is the weight percent of quartz in coal, A is the weight percent of ash in coal, and the values of a and b depend on the proportions of inherent ash, intraseam dirt, and floor and roof material in the coals delivered to the power stations. The approximate values of a and b for British coals are 5.0 and 0.15, respectively. These values are based on the mineralogical data published by Dixon et at. (1970a, b).

More accurate estimates of the quartz content of coals can be made when the results of chemical analysis of laboratory prepared ashes are available. As discussed in Chap. 3, the quartz content of a typical bituminous coal ash is given by Eq. (3.3), reiterated here:

$$[SiO_2]_q = [SiO_2]_t - 1.5[Al_2O_3] \qquad (3.3)$$

where the concentrations of quartz, total SiO_2, and Al_2O_3 are given in weight percent of ash. Equation (3.3) gives a useful estimate of the quartz content of most coals for the purpose of assessing their abrasive characteristics. Exceptions are the coals where the silica content of ash is outside the range of 35 to 55 percent by weight and the alumina content is outside the 15 to 30 percent range. In these cases direct determinations of quartz in coals are required by one of the methods discussed in Chap. 3.

Approximate values of the pyrite content of coals can be deduced by the result of sulfur analysis. Wandless (1959) has shown that the organic sulfur fraction of British coals is around 0.8 percent, and the difference between the total sulfur and 0.8 represents approximately the pyritic sulfur content of the coal. It is therefore possible to estimate the pyrite (FeS_2) content of coal using the expression

$$C_p = 1.9(S - 0.8) \qquad (5.3)$$

where C_p and S denote the weight percent of pyrite and sulfur in coal, respectively. A different formula for estimating pyrite content of the mineral matter in U.S. coals from sulfur analysis is given in Chap. 15.

Although the pyrite minerals can have nearly as high hardness number as that of quartz, as shown in Table 5.1, the abrasion- and erosion-wear damage caused by pyrite minerals in coal is likely to be significantly less than that caused by the same quantity of quartz. This is because quartz occurs in coal in the form of comparatively large particles of "free" mineral matter, whereas much of pyrite is dispersed in coal substance and clay sediment as described in Chaps. 3 and 4. The pyrite particles thus encapsulated in soft clay and coal matrices are likely to cause little abrasion- and erosion-wear damage in the coal milling plant and transport pipes. The results of particle-stream impact erosion tests with the quartz- and pyrite-rich fractions of coal mineral matter using the equipment described in Sect. 5.4 (Raask, 1969a) showed that the abrasiveness of pyrites was approximately 0.3 times that of quartz. This relative abrasiveness of the two mineral species will vary with different coals, but it should be within the limits of

$$I_p = (0.2 \text{ to } 0.5)I_q \qquad (5.4)$$

where I_p and I_q denote the relative abrasiveness of coal pyrites and quartz, respectively.

A relative abrasion index of coal I_c could be defined as

$$I_c = (C_q + xC_p)I_q \qquad (5.5)$$

where C_q and C_p denote the weight percent of quartz and pyrites in coal, respectively, the value of x should lie in the range of 0.2 to 0.5 as discussed above.

The abrasion and erosion wear of coal mill plant is usually directly proportional to the rate of fuel throughput. This in turn is dependent on the calorific value (net) of coal for a given boiler output. Thus, the abrasion and erosion wear index W_a of a coal could be expressed by

$$W_a = \frac{(C_q + xC_p)I_q}{Q_n} \qquad (5.6)$$

where Q_n denotes the net calorific value of coal, and C_q, x, C_p, and I_q have already been defined. The mill plant wear rate W_r for a given boiler load requirement with variable-quality fuel can be estimated using the expression

$$W_r = W_o \frac{(C_q + C_p)I_q}{Q_n} \qquad (5.7)$$

where W_o denotes the wear-rate proportionality "constant" of a mill plant operating at "standard" conditions. That is, once the rates of abrasion and erosion wear have been established for one particular type of fuel, Eq. 5.7 can be used to estimate the mill wear with changes in the coal quality.

Table 5.2 gives the estimated quartz and pyrite contents of typical bituminous coals using Eqs. 5.2 and 5.3, and the data were then used to compute the abrasive index I_c in Eq. (5.5) and the wear index W_a, in Eq. (5.6). The abrasive index of quartz of size 75–105 μm was taken to be 1.0. Analytical data for British and U.S. coals, respectively, were taken from publications of CEGB (1977) and O'Gorman and Walker (1972). The estimated calorific value of the U.S. coals was based on the fuel rank and ash content, as discussed in Chap. 4.

Based on the results of assessment of abrasiveness of typical power-station coals given in Table 5.2 and that of the exceptionally high quartz coals listed in Table 3.6, all solid fuels can be placed in four categories according to their abrasion- and erosion-wear characteristics:

1. Minimally abrasive coals, having an abrasive index below 0.01 and the corresponding wear index below 3×10^{-4}.

2. Moderately abrasive coals, having an abrasive index in the range of 0.01 to 0.025 and a wear index between 3×10^{-4} and 10^{-3}.

3. Highly abrasive coals, having an abrasive index between 0.025 and 0.05 and wear index between 10^{-3} and 2.5×10^{-3}.

4. Exceptionally abrasive coals, having an abrasive index above 0.05 and a wear index above 2.5×10^{-3}.

The corresponding abrasive indices of pulverized-coal ashes are discussed in Chap. 18.

5.3 CHARACTERISTICS OF SLIDING ABRASION WEAR

The type of abrasive wear most frequently encountered in coal milling is classified as high-stress grinding abrasion. In this situation, the abradant is interposed between two loaded surfaces where the applied stress exceeds the crushing strength of the abrasive and therefore comminution takes place. Wear of the crushing surface is the result of concentrated compressive stress at the point of abrasive contact, causing plastic deformation and fatiguing of ductile materials (Lipson and Colwell, 1961). Materials subjected to this form of abrasion show short deep scratches, grooves, and indentations. In order to resist this form of abrasion, therefore, the yield strength of the crushing surface must significantly exceed the crushing strength of the abrasive.

Table 5.2 The abrasive index and wear index of some British and U.S. coals

Coal type	Ash	Sulfur	Quartz	Pyrites	Calorific value (MJ kg^{-1})	Abrasive index	Wear index
		Percent by weight of coal					
			British coals				
Low ash	7.4	1.9	0.4	2.1	31.0	0.01	3.2×10^{-4}
	8.4	1.3	0.5	1.0	28.2	0.008	2.8×10^{-4}
Medium ash	15.2	1.6	1.5	1.5	24.3	0.02	8.2×10^{-4}
	17.8	1.8	1.9	1.9	22.8	0.025	1.1×10^{-3}
High ash	27.7	1.6	3.4	1.5	21.2	0.039	1.8×10^{-3}
	29.7	1.8	3.7	1.9	21.2	0.043	2.0×10^{-3}
			U.S. coals				
Low ash	7.2	1.6	1.0	1.5	32.0	0.015	4.7×10^{-4}
	7.8	1.1	0.1	0.6	30.5	0.003	1.0×10^{-4}
Medium ash	11.6	4.0	1.9	6.1	26.5	0.039	1.5×10^{-3}
	14.4	2.6	2.4	3.4	25.0	0.035	1.4×10^{-3}
High ash	24.4	1.3	0.7	1.0	23.0	0.01	4.3×10^{-4}
	28.0	2.2	4.5	2.7	21.5	0.054	2.5×10^{-3}

In the process of low-stress sliding abrasion, the abradant is not crushed and it acts like a cutting tool. Khruschov (1974) has summarized the significant parameters in the nonimpacting abrasive wear as follows.

1. Linear wear Δl is directly proportional to the stress P and the friction path Δs:

$$\Delta l = CP \, \Delta s \tag{5.8}$$

   where C is the wear coefficient depending on physical properties of the abrading and abraded materials.

2. At low friction velocities the rate of wear does not depend on the relative velocity of the abrading particles versus the abraded surface.

3. The wear rate rises approximately in direct proportion to the size of the abrasive grains to an optimum size.

4. The relative wear depends on the correlation between the hardness of the abrading H_a and that of the abraded material H_m as follows: if $H_a > H_m$, the wear rate does not depend on H_a, and if $H_a = \kappa H_m$ there is no significant wear when $\kappa < 0.7$ to 1.0.

5. The crushing strength, shape, size, and sharpness of the cutting edges of an abrading particle are the other important parameters that determine the rate of wear.

6. A significant influence on the rate of wear may be exercised by the interaction between the abrasive grains and the surface of abraded material. The wear could be reduced as the result of impregnating the surface of abraded material by debris of crushed grains.

7. Annealed steels show a direct proportionality between the relative resistance e_o and the pyramid hardness H_o—e.g., Vickers number—as follows:

$$e_o = b_o H_o \tag{5.9}$$

   where b_o is the coefficient of proportionality.

8. Heat-treated steels show a linear dependence of the relative wear resistance e on the hardness H_m:

$$e = e_o + b_1 (H_m - H_o) \tag{5.10}$$

   where b_1 is the coefficient of proportionality for different grades of steels, and e_o and H_o denote the relative wear resistance and the hardness, respectively, of the steels in the annealed state.

9. The relative wear resistance of cold-work-hardened alloys does not depend on the hardness when tested with an abrasive material that is harder than the abraded surface.

10. The relative wear resistance (e_0) of metals and alloys in an annealed state correlates with the modulus of elasticity E as follows:

$$e_0 = C_1 E^{1.3} \tag{5.11}$$

   where C_1 is the coefficient of proportionality. This expression does not apply to heat-treated steels.

Principal areas where abrasion wear occurs in the power-station coal handling plant are the walls of chutes and bunkers. Abrasion wear occurs also in coal mills, where the nature of metal wastage depends on the grinding process. A hammer mill, for instance, operates by beating the coal particles with rotating hammers, and impact forces are large. In a tube mill, impact occurs as the balls tumble over one another, and on the liner, and in a vertical-spindle mill, the balls or rolls are loaded on to a table either directly or through a coal bed and there should not be much impaction, although the balls tend to rebound. Hiorns and Parish (1966) have investigated the grinding wear in an experimental crusher designed to simulate the action of a vertical-spindle mill using a radioactive tracer technique. They found that the wear rate was proportional to the applied load, which is consistent with a wear mechanism by abrasion.

The quartz-rich nonbituminous coals mined in West Germany are highly abrasive and can cause severe damage to coal milling plant, as discussed by Burger (1961), Frenske (1961), and Beer (1961). The mineral matter content of coals burned in the local power stations is around 8 percent, varying from 3 to 30 percent. The adventitious minerals matter has a high quartz sand content, and thus the abrasiveness of coal is directly proportional to the ash content above 4 percent. An ash content below 4 percent represents the nonabrasive inherent mineral matter in coal.

5.4 PARTICLE IMPACTION EROSION

De Haller (1939), Bitter (1963), Sheldon and Finnie (1966), and Engel (1978) have shown that the impaction of solids on a surface gives rise to two types of wear. In deformation wear, repeated collisions deform a surface layer of impacted material and eventually a piece of material breaks loose. Pieces of the material are removed also by a cutting action of the moving particle. In practice both types of wear occur simultaneously. With brittle materials the deformation wear is more prominent, whereas ductile metals are eroded mainly by the cutting mechanism. The relative importance of deformation to cutting wear increases toward the latter as the angle of impaction changes from $\pi/2$ rad to zero, i.e., when the ratio of perpendicular to parallel velocity components decreases.

A number of attempts have been made to develop an equation to define the loss of metal in an erosion process. Bitter (1963) derived an expression for the deformation wear V_d (volume of metal removed):

$$V_d = \tfrac{1}{2} \frac{M(u \sin \theta - K_1)^2}{\epsilon_1} \tag{5.12}$$

and two equations for the cutting wear (V_c' and V_c''), the first being applicable when the velocity component parallel to the surface after the impaction is not zero and the second when it is zero (i.e., when particles slide along the surface after impaction).

$$V_c' = \frac{2MC(u \sin \theta - K_1)^2}{(u \sin \theta)^{1/2}} \left[u \cos \theta - \frac{(u \sin \theta - K_1)^2}{(u \sin \theta)^{1/2}} \epsilon_2 \right] \tag{5.13}$$

$$V_c'' = \frac{M\left[u^2 \cos^2 \theta - K_2(u \sin \theta - K_1)^{3/2}\right]}{2\epsilon_2} \tag{5.14}$$

where

$$C = 0.029\left(\frac{\rho}{y^5}\right)^{1/4} \tag{5.15}$$

$$K_1 = 0.016\left(\frac{y^5}{\rho}\right)^{1/2}\left(\frac{1 - q_1^2}{E_1} + \frac{1 - q_2^2}{E_2}\right)^2 \tag{5.16}$$

$$K_2 = 0.0082\left(\frac{y^9}{\rho}\right)^{1/4}\left(\frac{1 - q_1^2}{E_1} + \frac{1 - q_2^2}{E_2}\right)^2 \tag{5.17}$$

The symbols in metric units are M, the mass of particles; u, the velocity of particles; ρ, the density of particles; θ, the angle of impaction; E_1 and E_2, Young's moduli of the eroding material and impacting particles, respectively; q_1 and q_2, the respective Poisson's ratios; y, the elastic load limit of the impacted material; and ϵ_1 and ϵ_2, respectively, the deformation and cutting-wear factors.

The drawback is that the wear factors (ϵ_1 and ϵ_2) have to be determined experimentally. Further, Eqs. (5.12), (5.13), and (5.14) apply to erosion wear by spherical particles of radius r, and for nonspherical particles a term of "enhanced density" ρ_1 is necessary:

$$\rho_1 = \rho\left(\frac{r}{r_1}\right)^3 \tag{5.18}$$

where ρ is the density of particles and r_1 is equal to the radius of curvature of the contact points at the surface after impaction.

This treatise on erosion wear makes it clear that in practice it would be difficult to predict without any experimental work. Raask (1969a) used a laboratory assembly for measuring the erosion wear rates of different materials. Briefly, a powder feeder was used to introduce the impacting particles, 75- to 150-μm quartz grains, into the air stream. The particles were accelerated through a 2-mm-diameter tube, 500 mm long, and then allowed to strike the test specimen. The specimen and the impacting material were weighed before and after the test. Some measurements were made with glass spheres, 45 to 75 μm in diameter, and Fig. 5.8a shows the impacting and rebound streams, both at an angle of $\pi/4$ to the impacted surface of mild steel. It is evident that the stream of glass spheres had not dispersed on impact, whereas the rebound particles of quartz grains came off the target at various angles between 0 to $\pi/4$ rad to the surface, as shown in Fig. 5.8b. The angular particles of quartz had ploughed the surface for a distance of a few microns and escaped at a reduced velocity (Raask, 1969a).

Figure 5.9a shows typical cutting grooves in the surface of mild steel caused by quartz particles of 100 μm diameter impacting with a velocity of 20.1 m s^{-1} and at an angle of $\pi/4$ rad. By contrast, impaction of glass spheres with the same velocity and at the same impacting angle produced surface deformation but no cutting grooves. It is therefore evident that ductile or brittle behavior of a metal during the process of erosion wear depends on the abrasive characteristics of the impacting

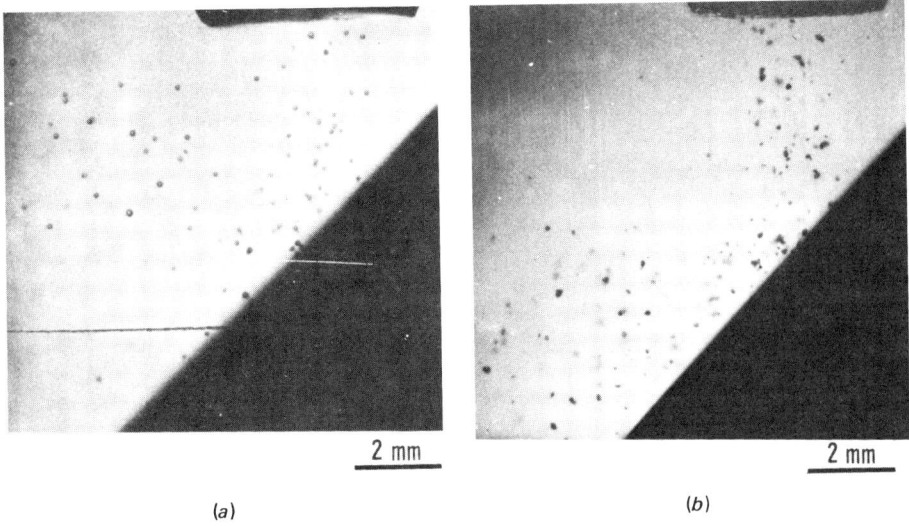

Figure 5.8 Particle impaction and rebound on steel target. (a) Impaction of glass spheres.
(b) Impaction of quartz sand.

particles. Highly abrasive, hard, and angular quartz grains result in a typical cutting erosion, whereas the action of glass spheres is to produce surface deformation first, leading subsequently to a brittle failure.

The erosion wear rates of mild steel were determined in the impact velocity range of 10 m s^{-1} to 30 m s^{-1}, and the angle of impaction was varied from $\pi/12$ rad to $\pi/2$ rad. Figure 5.9b shows a typical erosion cavity in mild steel after 0.5 kg quartz dust had been impacted at an angle of $\pi/4$ rad with a velocity of 27.5 m s^{-1}, which is a typical velocity for transporting pulverized coal from the coal mills to

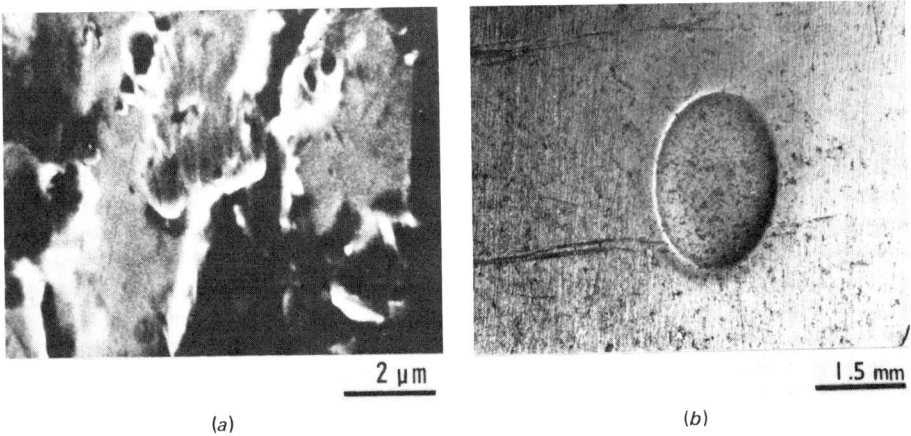

Figure 5.9 Erosion wear of mild steel on quartz particle impaction. (a) Initial scratch groove.
(b) Erosion cavity.

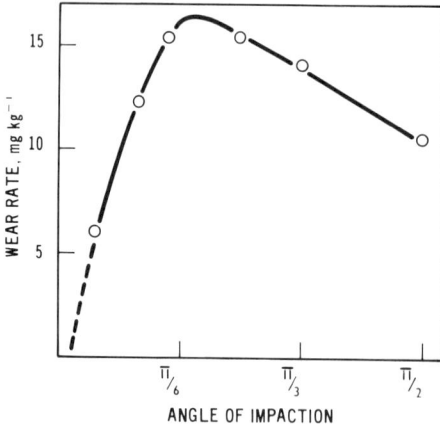

Figure 5.10 Erosion wear of mild steel at differing impaction angles. Impact material 125–150 μm quartz; impact velocity 27.5 m s^{-1}.

burners. The results in Fig. 5.10 show that the erosion wear rates fell off rapidly from the peak value at an impact angle of $\pi/5$ rad when the angle of impaction was reduced.

At a given angle of impaction, the relationship between the erosion wear and the impact velocity can be given by the expression

$$W_c = A_o W_m u^n \tag{5.19}$$

where W_c = the weight of metal loss
    W_m = the weight of impacting particles
    u = the impact velocity
    A_o = the erosion wear coefficient related to the abrasive characteristics of impacting particles and to the wear property of eroded material.

The velocity power index (n) was found to be 2.5 by Raask (1969a), but the experimental results reported by Goodwin et al. (1969) gave a slightly lower value of 2.3.

For a given concentration C_a of the abrasive particles in the gas stream, the weight of the impacting materials is proportional to the stream velocity. Therefore, the rate R of metal loss is given by:

$$R = A_1 C_a u^{n+1} \tag{5.20}$$

where A_1 is an erosion wear coefficient. Equation (5.20) shows that a comparatively small change in the impact velocity can have a marked effect on the rate of erosion wear. For example, a 10 percent increase in the velocity would result in doubling the wear rate.

A selection of commercially available and some experimental materials were tested for their erosion wear (Raask, 1979a). The results plotted in Fig. 5.11 show that there was a nonlinearly inverse relationship between the erosion wear rate and the hardness number of different steels. The erosion wear rate of 15.5 units for mild steel (hardness number 190) decreased to 4.2 units for Nihard, which is a cast-iron alloy with added nickel (hardness number 800). In order to further reduce the erosion wear it would be necessary to have refractory materials—e.g., good quality alumina, silicon, and tungsten carbide, as shown in Table 5.3.

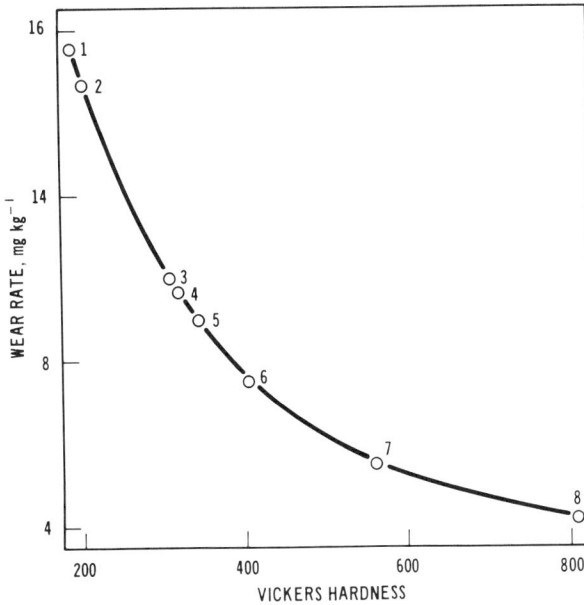

Figure 5.11 Erosion wear of steel alloys of differing hardness: impact material 125–150 μm quartz; impact velocity 27.5 m s^{-1}; and impact angle $\pi/4$ rad. 1, Mild steel; 2, austenitic steel; 3, 4, 5, 6, carbon-hardened steels; 7, Cr–boride-hardened steel; and 8, Ni-hardened steel.

Table 5.3 Erosion wear of mild steel and refactory materials[a]

Material	Erosion wear (mg kg^{-1})	Angle of impaction (rad)
Mild steel	15.5	$\pi/4$
Nihard (nickel cast iron)	4.2	$\pi/4$
Pyroceram (semicrystallized ceramic material)	14.8	$\pi/2$
Glassceram (semicrystallized glass)	12.5	$\pi/2$
Slagceram (semicrystallized slag)	7.5	$\pi/2$
Epoxy and refractory composite	7.0	$\pi/2$
Silceram (semicrystallized silicate)	4.0	$\pi/2$
Alumina, 75% Al_2O_3	5.3	$\pi/2$
Alumina, 97.5% Al_2O_3	0.6	$\pi/2$
Silicon carbide	1.0	$\pi/2$
Tungsten carbide	0.5	$\pi/2$

[a]Impacting material: quartz grains, 75–150 μm in size. Impacting velocity: 27.5 m s^{-1}.

5.5 MEASURES TO COMBAT ABRASION AND EROSION WEAR IN THE COAL PLANT

In current operating practice, abrasion- and erosion-resistant materials are extensively used to combat wear in the coal handling plant. The materials used and the methods of their applications vary for different component parts of the fuel plant, as described next.

Coal Chutes and Bunkers

The major problem area of coal chutes is related to impact sites where coal falls onto the surface of the chute from a conveyor. The magnitude of the problem is governed by the size and velocity of the impacting pieces of coal and rock, i.e., the kinetic energy and its impact angle with the chute surface. Coal chutes are usually fabricated from mild steel, and the practice at small stations has been to weld patches over thinned or perforated sections. This practice is no longer expedient for large stations, where coal throughputs of 20,000 t per day are required, which makes it necessary to have large chutes with more severe wear conditions.

A variety of wear-resistant materials have been tested in power-station coal chutes. These include alumina, cast basalt and quarry tiles, annealed aluminum alloys, hardened carbon steels, nickel cast iron (Nihard), and high-chromium alloys. A high-chromium alloy (type BF 253) and sintered alumina tiles have given satisfactory abrasion resistance performance in a number of applications. Hocke (1972) has evaluated a selection of materials for possible use to combat the sliding abrasion wear in the coke and sintered iron oxide chutes of steel works. His assessment was based on the relative rate of abrasion wear and the cost of materials (Table 5.4).

A major problem with bunkers is to ensure movement of coal under the action of gravity. Coal stocked in rectangular bunkers tends to stick first in corners and later on walls, so the usable bunker capacity is much reduced, often to half or less. It is therefore customary to line the bunker walls with antisticking materials. There is a wide range of bunker-lining materials available, the commonest being high-density (high-alumina) concrete and gunite (a sprayed mortar). These suffer from the disadvantage that in service the surface will become rough and, due to the high coefficient of friction, movement of coal is retarded. Glass tiles possess an attractive property of low friction and they have been used in this application, but they are expensive and suffer from breakage when the coal feed falls from a great height directly on the bunker walls. Other materials that have been used include stainless-steel and aluminum plates, and high-density polyethylene sheets. Composite materials made with bauxite filler and epoxy resin and those made with PTFE filler have proved successful in this application.

Mill Components

The grinding elements, usually in the form of rotating balls, are subjected to high-stress abrasion wear as discussed in Sect. 5.3. It is therefore essential that the

Table 5.4 Abrasion wear index and cost effectiveness index of different materials (cost index = wear index × unit cost[a])

Material	Wear index	Unit cost	Cost index
Ni–Cr martensitic cast iron	0.065	3.00	0.195
High-chrome cast iron	0.115	2.80	0.322
Nodular graphite cast iron	0.137	2.67	0.366
Fusion cast alumina special	0.172	2.67	0.459
Cr–Mo–W hard facing	0.079	6.50 (5 mm layer)	0.513
Cast basalt	0.846	0.66	0.564
Sintered alumina	0.173	4.85	0.840
En8	1.000	1.00	1.000
Tungsten carbide hard facing	0.192	10.50 (5 mm layer)	2.000
Concrete	18.000	0.20	3.600

[a]Unit cost was taken as that of a plate 305 × 305 × 30 mm in size unless otherwise stated.

Source: Hocke (1972).

ball material possesses a sufficiently high hardness to withstand abrasion and toughness to resist the high stresses. The cast iron modified by addition of nickel (Nihard) possesses a combination of the hardness and toughness requirements, and the alloys are extensively used to manufacture the coal grinding elements.

Avery and Chapin (1969) and Fairhurst and Rohrig (1976) have reported that in the arduous conditions of ore crushing operation, some high-chromium alloys had better performances than did Nihard materials. These martensitic alloys have a matrix that supports a massive and hard carbide phase resistant to abrasive wear (Norman, 1959; Boyes, 1969). The two alloys most frequently used are types 15Cr–3Mo and 19Cr–1.2Mo–0.7Ni; the latter is known as Paraboloy.

There has been a general move to use the Nihard and high-chromium alloys for manufacture of the other mill components, such as mill tables and tires, dam rings, tie bars, and wear cowls. Some components, such as access doors, mill bodies, and side plates, can be covered with alumina tiles in order to increase the life of the mill plant.

Pulverized-Coal Mill Exhauster Fans

Severe erosion wear of pulverized-coal exhauster fans is a well-known feature of suction mills, where the life of the fan blades can be exceptionally short, 500 to 600 h of operation, with abrasive coals of high quartz content. Traditionally the fan blades were made from mild steel, but more recently a wide range of materials and processes have been tested in order to increase the life of rapidly eroding blades.

The wear resistance of mild steel blades can be increased by surface hardening. One technique involves spreading of chromium boride paste on the blade surface and diffusing the boride into the metal during heat treatment with a welding torch.

Mild steel thus hardened has a Vickers hardness number of about 600, and its erosion wear rate is reduced to about one-third of that of untreated mild steel, as shown in Fig. 5.11 (sample 7). The hardening treatment has been successfully used to increase the life of fan blades and of other mill components—e.g., classifier plates—by a factor of 2 to 3. However, there is a problem of distortion resulting from the brutal heat treatment required to melt the steel surface to a depth of several millimeters.

Another method of protecting the fan components is to braze sintered tungsten carbide tiles to the plates. The results in Table 5.4 show that the erosion wear rate of tungsten carbide is about 30 times less than that of mild steel: thus the protected blades have a markedly enhanced life in a pulverized coal stream. However, the blade fastenings tend to erode, and tungsten carbide is an expensive material.

The internal surfaces of mill exhauster bodies also suffer erosion resulting in the side plates and segments wearing away. It is normal practice to restore the profile of worn components using abrasion-resistant mortars, or a composite material made with bauxite and epoxy resin. The materials are easy to use and relatively hard-wearing and are frequently used for quick repairs.

Pulverized-Coal Pipes

Erosion wear of pulverized-coal pipes remains a serious problem in many power stations. Previously, cast iron pipes were used because of their lower wear rate compared with that of mild steel pipes. However, brittle cast iron materials are no longer recommended for manufacture of pulverized-coal pipes because of the fire and explosion hazards in the mills and propagation of shock waves in the pipe lines.

The velocity of the pulverized-fuel stream may be significantly above 25 m s^{-1}, the minimum necessary to prevent settlement of coal dust in the pipeline. It should be noted that even a modest decrease, 5 to 10 percent, would result in a significant reduction of the wear rate, as shown by Eq. (5.20). Also, the radius of curvature of pipe bends, hence the angle of impaction of the coal stream inside the pipe, is an important factor, as shown by the results in Fig. 5.10. It is therefore imperative that the radius of curvature of pipe bends should be as large as practicable in given situations. Hoffmann et al. (1980) state that the erosion wear rate has a maximum when the D/d ratio is approximately 5, where D is the radius of pipe curvature and d is the pipe diameter.

A variety of lining materials and the mode of attachment have been considered to protect the coal-pipe bends and elbows against erosion wear. One technique utilizes a composite material made with an epoxy resin and erosion-resistant refractory aggregates. The composite is cast directly on the inside wall of mild steel bends and elbows, and the erosion-protective layer can be built to any required thickness; usually it is between 20 and 30 mm. The advantage of the method is that both new pipe bends and partially worn bends can be lined with the erosion-wear protecting material. Also, the lining of the pipe sections can be completed within a few days in an emergency.

A choice of abrasion-resistant alloys for manufacture of protective linings or sleeves inside the mild steel pipes can be made on the basis of the results given in Fig. 5.11. For example, best quality Nihard lining will last about four times as long as mild steel pipe of the same thickness, and this is probably the limit of the degree of protection to pulverized coal pipes that can be achieved by the use of hardened metal alloys. If a higher degree of protection is required, wear-resistant refractory material—e.g., tiles or sleeves made from 97.5 percent purity alumina—must be used. The results in Table 5.3 show that a high-grade alumina has an erosion wear rate approximately 25 times less than that of mild steel on the weight loss basis, or about 10 times less on the wall thickness loss basis.

Particle impaction erosion wear of boiler tubes will be discussed in Chap. 18 after changes in the abrasive characteristics of coal minerals in boiler flame have been evaluated.

NOMENCLATURE

A	ash content of coal
A_o	erosion wear coefficient
A_1	modified erosion wear coefficient
$[Al_2O_3]$	aluminium oxide content of ash
C	abrasive wear coefficient
C_a	particle concentration of gas stream
C_p	pyrite content of coal
C_q	quartz content of coal
C_1	proportionality coefficient
E	modulus of elasticity
E_1	Young's modulus of eroded material
E_2	Young's modulus of impacting particles
G	coal grindability index
H_a	hardness of abrading particles
H_m	hardness of abraded material
H_o	pyramid hardness of annealed steels
I_c	relative abrasiveness of coal
I_p	relative abrasiveness of pyrite
I_q	relative abrasiveness of quartz
M	mass of particles
P	stress acting on particles
Q_n	net calorific value of coal
R	erosion wear by weight
S	sulfur content of coal
$[SiO_2]_q$	quartz content of ash
$[SiO_2]_t$	total silica content of ash
Δs	friction path
V_d	volumetric erosion wear

V_c''	modified cutting wear
W	weight fraction of ground coal
W_a	wear index of coal
W_o	"standard" wear rate of coal mill
W_c	weight loss of eroded metal
W_m	weight of impacting particles
W_r	mill plant wear rate
a	adjustable constant
b	proportionality index
b_o	wear proportionality coefficient for annealed steels
b_1	wear proportionality coefficient for hardened steels
e_0	relative wear resistance of annealed steels
e	relative wear resistance of hardened steel
Δl	linear wear rate
n	power index
q_1	Poisson's ratio of eroded material
q_2	Poisson's ratio of impacting particles
r	radius of particles
u	velocity of impacting particles
x	abrasiveness ratio of pyrite to quartz
y	elastic load limit of impacted material
ϵ_1	deformation wear factor
ϵ_2	cutting wear factor
θ	angle of impaction
κ	hardness proportionality factor
ρ	particle density
ρ_1	particle enhanced density

PARTICULATE SILICATE MINERALS
IN BOILER FLAME

6.1 SHAPE TRANSFORMATION OF FLAME-BORNE
SILICATE-ASH PARTICLES

Silicate minerals, together with silica (quartz), constitute the major impurity fraction in most coals, as discussed in Chaps. 3 and 4. It is therefore essential to consider the physical changes and chemical reactions when particulate coal minerals are subjected to rapid heating in boiler flames. Figure 6.1 shows the temperature-time plot for the ash particles carried through the flame and subsequently through the heat-exchange sections of a 500-MW pulverized-coal-fired boiler. Small particles, below 1.0 μm in diameter, attain the temperature of the flame, whereas large particles will have a temperature as much as 200 K lower (Losel and Schmucker, 1977). This is because of time limitation in the heat transfer by radiation to the surface of ash particles of comparatively low emissivity (Chap. 13). In contrast, the ash particles trapped in burning coke residue (Fig. 8.4) may cool less rapidly than the rate depicted in Fig. 6.1.

Hydrated clay minerals in coal undergo endothermic dehydration reactions that commence at about 300 K with a loss of surface-adsorbed and interlattice water. Dehydroxylation of "chemically" bonded water in the minerals, chiefly kaolinite, montmorillonite, and illite, commences at 650 K (Fig. 16.1); at 1100 K the endothermic lattice destruction and the exothermic phase transformation will

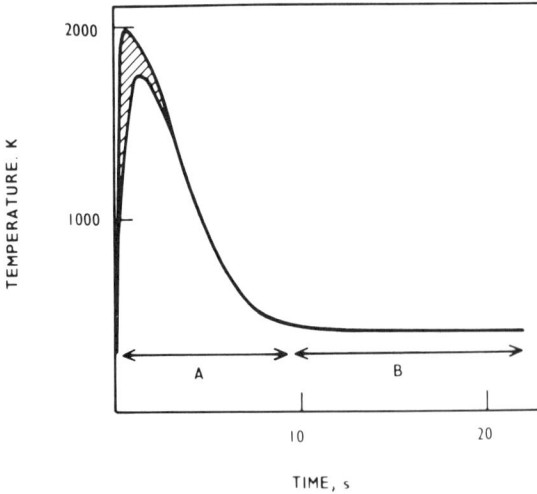

Figure 6.1 Temperature/time plot for ash particles in a 500-MW pulverized-coal-fired boiler, 0.1 μm (top curve) to 100 μm (lower curve) sizes. Section A: combustion and heat-exchange chambers; section B: electrical precipitators and chimney.

commence. The dehydration, dehydroxylation, lattice degradation, and phase-transformation reactions have usually been studied by differential thermal analysis (DTA) techniques where the heating rates are low, i.e., between 5 and 20 K/min (Mackenzie, 1957). In contrast, the heating rate of particulate silicates in a coal flame is higher by several orders of magnitude (Fig. 6.1). It is therefore highly likely that the water-loss and lattice-transforamtion reactions are not fully completed before the change of shape of silicate particles commences.

The change in shape of silicate particles on heating is the result of the action of surface tension on the sharp edges of small irregularly shaped particles, forcing these to take a spherical form. The rate of transformation is dependent on viscous relaxation, and thus the surface tension and viscosity of the particulate matter govern the change at a given temperature. The magnitude of the acting stress f depends on the surface tension γ and on the radius of curvature ρ of the surface:

$$f = \frac{2\gamma}{\rho} \tag{6.1}$$

The stress acting on the sharp edges of small particles can be high, as shown by the calculated values for different radii of curvature given in Fig. 6.2; surface tension was taken to be 0.32 N m^{-1} (Sect. 6.2). Frenkel (1945) has shown that the time t of transformation of an irregularly shaped particle to a sphere is given to a first approximation by:

$$r = r_o e^{-t/z} \tag{6.2}$$

where $z = 4\pi\eta/\gamma$ (6.3)

    r = the distance of a point on the original surface from the center of a sphere of equivalent volume having radius r_o

    η = viscosity

    γ = surface tension

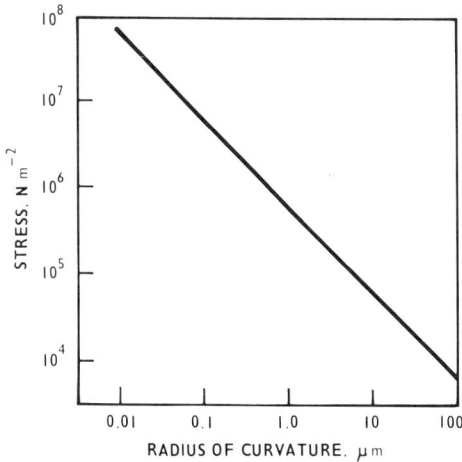

Figure 6.2 Magnitude of stress on particle edges of different radius of curvature. Surface tension, 0.32 N m^{-1}.

Equation (6.2) permits us to calculate the approximate time required for a particle to assume a spherical shape when the surface tension, viscosity, and size of particle are known. Alternatively, viscosity values can be calculated for the change in shape during a given time, e.g., 1 s for the results in Table 6.1. It was assumed that the thickness of surface layer reshaped as a result of the action of surface tension was 10 percent of the particle radius.

It is evident from Table 6.1 that the small irregularly shaped particles should transform to spheres in a coal flame when the viscosity of silicate glassy material is several orders higher than that required for bulk flow under gravity, which is about 25 N s m^{-2}. Raask (1969b) has determined experimentally the minimum temperature required to transform the irregularly shaped particles of silicate glasses and coal minerals 10 to 200 μm in diameter dispersed in a gas stream passing through a vertical furnace. The furnace temperature was varied from 1175 to 2025 K, and it was monitored by a radiation pyrometer and by thermocouples attached to the furnace tube. The particle temperature was calculated, assuming that the heat transfer took place chiefly by radiation with a possible error of ±25 K. The residence time of particles in the furnace after they had reached a steady temperature was calculated from Stokes' law, taking account of the velocity, viscosity and density of the gas stream. The residence time was 0.5 s for small particles, whereas for large particles it was 0.2 s. A mixture of nitrogen and oxygen, 5 percent O_2 in N_2 by volume, was used as the carrier gas. Particles were collected after heating at the bottom of the furnace and were examined by the optical and scanning electron microscope, and the furnace temperature was increased so that all particles acquired the spherical shape.

Table 6.1 Maximum viscosity for shape deformation of silicate-ash particles of differing sizes

Particle radius, μm	0.01	0.1	1.0	10	100
Viscosity, N s m^{-2}	2.5×10^7	2.5×10^6	2.5×10^5	2.5×10^4	2.5×10^3

Figure 6.3*a* illustrates a surface-fused particle of ground silicate (glass A in Table 6.2) after being heated to 1270 K, and Fig. 6.3*b* shows a particle of the same glass spheroidized at 1300 K. Figure 6.3, *c* and *d*, show typical shapes of coal mineral particles before heating and high temperature treated ash respectively. Table 6.2 gives the composition of silicate glass, mineral, and ash samples used in the laboratory work.

Figure 6.4 shows the relationship between the spheroidization temperature and

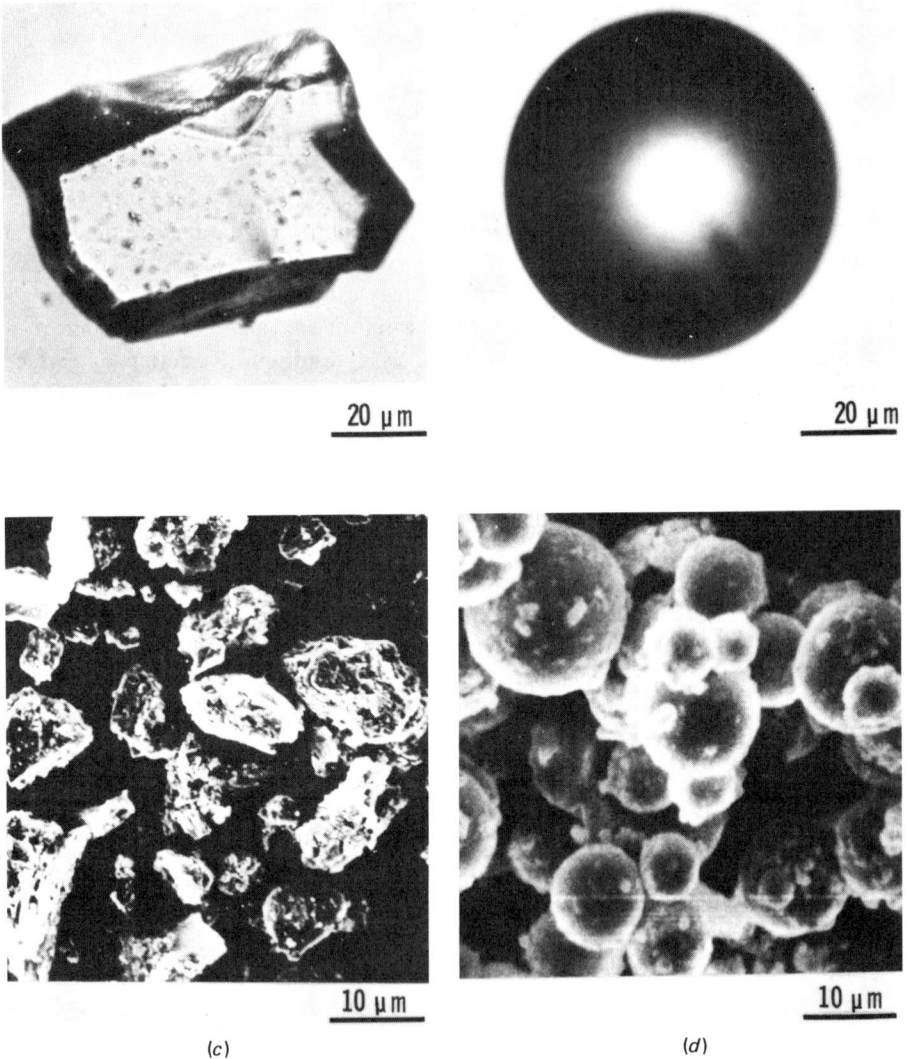

20 μm   20 μm

10 μm   10 μm

(c)   (d)

Figure 6.3 Gas-borne particles: (*a*) and (*b*) were heated in laboratory furnace and (*c*) and (*d*) in the boiler flame. (*a*) Surface-fused silicate glass. (*b*) Spheroidized silicate glass. (*c*) Coal minerals. (*d*) Spherical ash particles.

Table 6.2 Spheroidizing experiment materials

| Material | Constituent oxides, weight percent | | | | | | |
	SiO_2	Al_2O_3	Fe_2O_3	CaO	MgO	K_2O	Na_2O
Glass A	52.3	29.6	<0.1	16.3	<0.1	<0.1	1.8
Glass B	53.3	30.1	<0.1	16.6	<0.1	<0.1	<0.1
Glass C	51.8	29.3	<0.1	16.1	<0.1	2.7	<0.1
Illite	61.0	22.2	9.2	0.7	3.5	6.1	0.5
Muscovite	46.5	34.0	5.4	0.5	0.4	10.0	0.4
Native quartz	99.5	0.1	0.02	0.003	0.004	<0.001	<0.001
Chlorite	46.6	17.6	24.6	0.2	20.5	0.1	0.1
Coal ash	40.0	25.5	18.2	9.2	5.0	1.8	1.2

particle size of glasses A, B, and C in Table 6.2. Similar plots of the shape-change temperature against particle size for different silicate minerals are given in Fig. 6.5. The behavior of chlorite mineral illustrates the case where two components separate on heating and fuse at different temperatures (curves D_1 and D_2). The high-temperature phase (D_2) has the same fusion characteristics as native quartz. Curve C shows that particles of native quartz change their shape in the temperature range of 1775 to 2025 K. For this to occur, the viscosity of silica must be in the range of 10^2 to 10^4 N s m^{-2}, whereas the viscosity of pure silica is between 10^5 and 10^7 N s m^{-2} at these temperatures. The probable explanation here is that the sample of native quartz contained impurities (Table 6.2) that sharply reduced its viscosity (Bacon et al., 1960).

The particle-shape change of the mineral matter in pulverized coal when passed through the laboratory furnace is shown in Fig. 6.6, where curve A shows the percentage of rounded or fused particles in the total at different temperatures. About 10 percent of particles fused below 1575 K, the bulk of the irregular

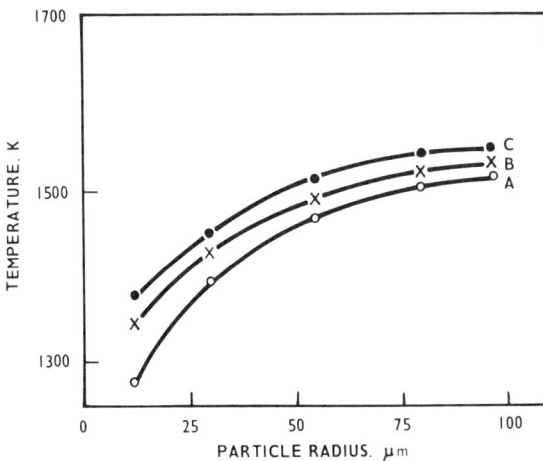

Figure 6.4 Fusion of particles of synthetic glasses: A, $3SiO_2 \cdot Al_2O_3 \cdot CaO \cdot 0.2 Na_2O$. B, $3SiO_2 \cdot Al_2O_3 \cdot CaO$. C, $3SiO_2 \cdot Al_2O_3 \cdot CaO \cdot 0.2 K_2O$.

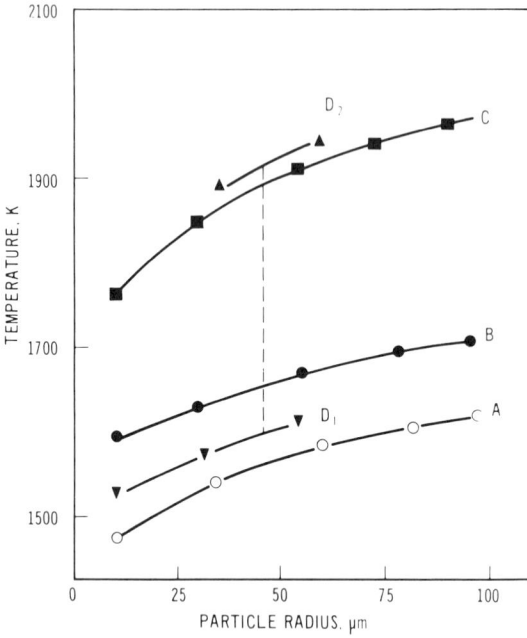

Figure 6.5 Spherical shape transformation of granular minerals in hot gas stream: A, illite; B, muscovite; C, native quartz; and D, chlorite.

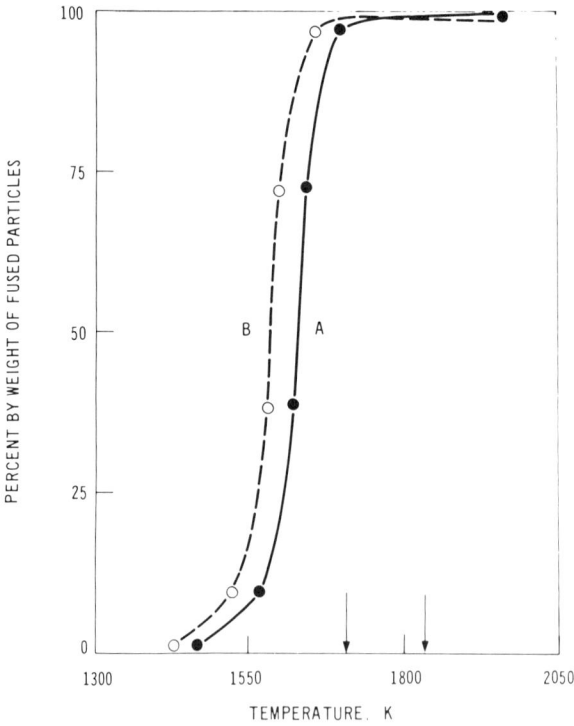

Figure 6.6 Fusion of particulate minerals in coal: A, in tube furnace, residence time 0.2 to 0.5 s; B, in boiler furnace, residence time 2 to 5 s.

particles changed to spheres between 1575 and 1725 K, and 5 percent of the residual particles changed their shape between 1725 and 1975 K. The particles which changed the shape below 1575 K were iron oxide formed from pyrite; the behavior of pyrite mineral on heating will be discussed in Chap. 7. The fraction which fused above 1725 K was found to be quartz.

The shape change of the particulate mineral matter in a large boiler is expected to be similar to that in the laboratory furnace. The only essential difference is that the residence time of particles in the boiler is up to 10 times longer: thus curve A in Fig. 6.6 would be displaced to curve B. This means that when the flame temperature of a pulverized-coal-fired boiler is between 1700 and 1850 K, as indicated in Fig. 6.1, the majority of particles emerging from the flame would be spherical, except for the large particles of quartz, which are surface-fused but retain their nonspherical shape as shown in Fig. 18.1b.

6.2 SLAG/COAL INTERFACE PHENOMENA: WETTING, SURFACE TENSION, AND COMBUSTION OF ASH–RICH COAL PARTICLES

Raask (1966a) has used a Leitz heating microscope to observe the wetting characteristics of coal ash slags and synthetic silicate melts on different surfaces. Figure 6.7 shows an ash pellet before heating (a), the slag on a wetting alumina surface (b), and on a nonwetting surface of coke (c). It was found that coal-ash slags and silicate melts wetted the refractory oxide materials silica, alumina, zirconia, and magnesium oxide, and also silicon carbide and platinum metal. Wetting of the surface of a refractory oxide by slag resulted from an interaction between the molten ash and solid surface that lowered the interfacial tension. The wetting behavior of molten slag on silicon carbide was influenced by the furnace atmosphere. The slag spread readily in an oxidizing atmosphere, as a result of the formation of SiO_2 at the surface.

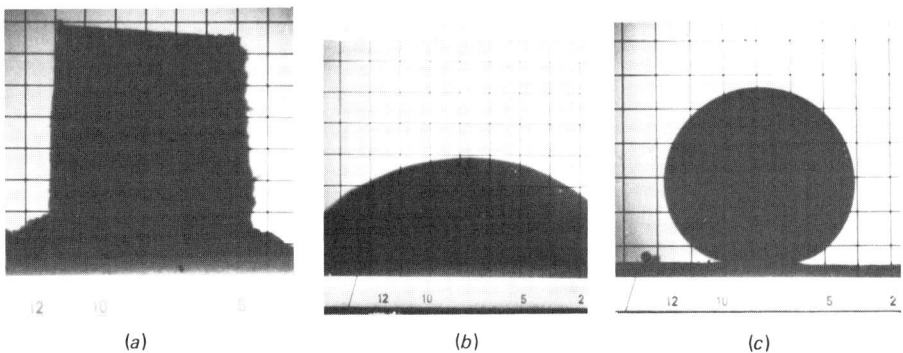

(a)	(b)	(c)

Figure 6.7 Ash fusion on wetting and nonwetting surface, 0.5 mm grid. (a) Ash pellet. (b) Slag on alumina. (c) Slag on carbon.

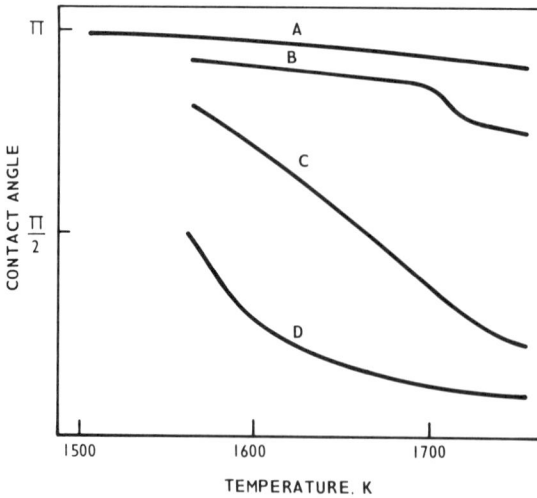

Figure 6.8 Contact angle of coal-ash slags on nonwetting and wetting surfaces: A, iron-free melt on carbon; B, coal-ash slag on carbon; C, coal-ash slag on carbon exposed to K_2SO_4 vapor; and D, coal-ash slag on alumina.

In the absence of a solid oxide layer or adsorbed film on carbon or graphite at high temperatures, the nonoxide surface is incompatible with that of silicate melts and coal ash slags. As a result there is a large contact angle between the surface of graphite, coke, or coal residue and the molten slag globule, as shown on Fig. 6.7c. The contact angle between the graphite surface and a globule of iron-free slag remained high at temperatures up to 1750 K (Fig. 6.8), but with a globule of coal ash slag which contained 17.8 percent by weight iron there was a significant decrease in the contact angle at about 1700 K. This was probably because of the formation of iron carbide at the graphite/slag interface as discussed in the following section.

The nonwetting property of coal-ash slag on the surface of graphite permits measurements of the surface tension of molten slags by the sessile drop method (Bockris et al., 1959). The technique involves photographing a molten slag globule on a graphite plate and can be conveniently carried out in a Leitz heating microscope. From the enlarged photograph, the equatorial (horizontal) diameter and the height of apex above the equatorial plane are measured. The density of slag may be determined from the volume of the drop as measured from the photograph. The shape of the drop depends upon parameter β as follows:

$$\beta = \frac{gD\rho^2}{\gamma} \tag{6.4}$$

where g = the force of gravity
D = density of slag
ρ = radius of curvature at the drop
γ = surface tension

The geometric derivation parameter β and the numerical data required to calculate the surface tension from the measurements have been published by Bockris et al. (1959).

The surface-tension measurements were carried out with a number of slags in the temperature range of 1600 to 1750 K, and the results varied from 0.25 to 0.4 $N m^{-1}$ giving an average value of the surface tension of 0.32 $N m^{-1}$ at 1675 K. Evolution of gas bubbles inside the molten droplets of slag when resting on a graphite plate, as discussed in the following section, limits the accuracy of the sessile drop method of measuring the surface tension of coal-ash slags. The bubbles distort the shape of the droplet and give erroneous results of the density of the slag. It is therefore necessary to photograph the globule immediately after escape of the gas bubble.

The fact that coal-ash slag does not wet or spread on the coke residue particles is of fundamental importance in high temperature combustion. In particular, the nonwetting behavior is a significant factor in the process of combustion of coal "middlings" (Table 4.4), i.e., the ash-rich particles of pulverized coal. As combustion proceeds, a layer of ash residue is left on the surface of the burning coal particle. This layer of ash is broken up into small beads as shown in Fig. 8.4a when the temperature of burning particle is sufficiently high to fuse the ash, that is, when the slag viscosity is below about 10^4 $N s m^{-2}$. As a result the surface of the burning particle is exposed to allow an access of oxidizing gases and permits the escape of combustion products.

The noninterference effect of slag on the rate of combustion of coal particles has been demonstrated by Raask (1966a). Two small cylinders of carbon, one of which is coated with coal ash, were heated in nitrogen above the ash fusion temperature. Figure 6.9a shows that a number of small globules had formed, leaving most of the surface free. Similar but smaller slag globules can be found on coke residue particles in pulverized fuel ash, as shown on Fig. 8.4a. The next photograph (b) shows that the presence of these small globules of slag on the surface of the carbon had no appreciable effect on combustion when air was introduced, and the ash-coated and uncoated carbon particles burned at the same rate.

Partially burnt coal-residue particles may be buried inside slag globules or in the surface layer of molten slag on furnace walls, particularly in cyclone and slag-top boilers. In these cases the carbon particles emerge at the slag surface as a result of gas evolution, as demonstrated in Fig. 6.10. The slag globule shown in Fig. 6.10a

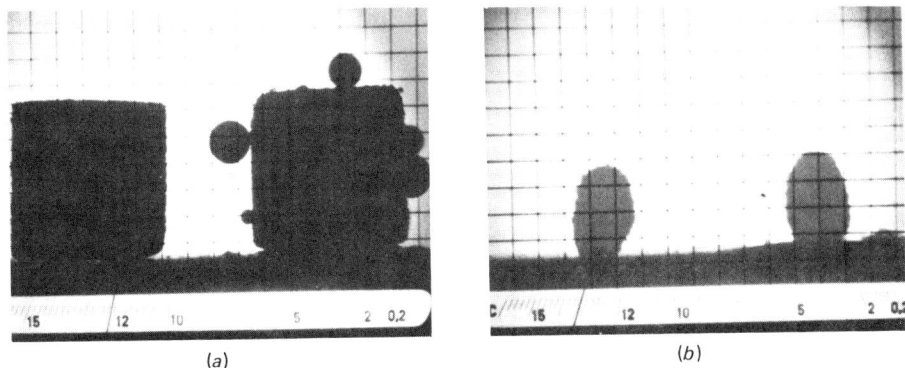

(a)   (b)

Figure 6.9 Combustion of carbon particle with slagged ash, 0.5 mm grid. (a) Slag globules on coal particle (righthand side). (b) Progress of combustion.

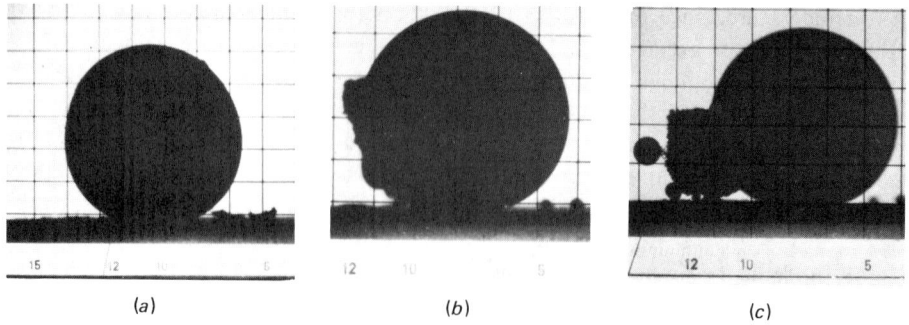

Figure 6.10 Ejection of coal particle from slag, 0.5 mm grid. (*a*) Ash globule with embedded coal particle. (*b*) Emergence of coal particle. (*c*) Ejection of coal particle.

has a carbon particle buried in it, and *b* and *c* show a partial and near-complete escape of the particle.

At lower, nonslagging temperatures, the presence of mineral matter in burning coal particles may enhance or it may retard the rate of combustion. Schmitt (1981) has suggested that gases penetrate into a coal particle most easily along the discontinuity channels associated with the presence of mineral-matter dispersions. It is therefore likely that the escape of combustible gases—e.g., hydrocarbons, CO, and H_2 generated inside a coal particle—will also be enhanced along the discontinuity routes. It could be thus argued that the inherent mineral matter in coal when present in the form of channel dispersions can enhance the rate of combustion in its early stages.

The presence of ash, however, can retard the rate of combustion of coal residue particles at ash-sintering temperatures of 1200 to 1400 K. When the temperature of the burning particles is sufficiently high to form a sintered layer of ash, but it is not high enough to slag the ash, the rate of combustion can be significantly reduced. Raask (1966a) has carried out a number of experiments in the Leitz heating microscope to assess the effect of mineral matter on combustion below the slagging temperature of ash. A layer of silica fume or a coating of unsintered ash had no marked effect on the combustion rates; there appeared to be no hindrance by the fume or ash deposit in its unsintered state on the diffusion of gases to and from the burning particle.

It is unlikely that the sintered ash from the inherent mineral matter constitutes a major barrier to combustion of coke particles formed from the high-volatile and medium-volatile coals in the pulverized-coal flame. The combustion is comparatively rapid, and the ash is present in the form of small globules at the surface of coke residue and also as a sintered matrix, as shown in Fig. 8.4, *a* and *b*, respectively. The presence of sintered ash in coke particles from a low-volatile coal may cause a slight but significant reduction in the rate of combustion and the high inherent ash content, above 5 percent of low-volatile coal may contribute toward the excessive amounts of unburned carbon in pulverized-fuel ash, as discussed in Chap. 5.

MacDonald and Murray (1952) have shown that there was a significant decrease in the combustion efficiency of stoker-fired boilers with increase in the ash content of coal. In particular, the increase in the inherent ash had a marked effect, whereas the adventitious ash had a less marked influence on combustion. It was observed that the loss in combustion efficiency was high with ashes of a low fusion temperature. This suggests that unburned coke particles were embedded in the sintered matrix, which markedly reduced the rate of combustion.

6.3 FORMATION OF ASH CENOSPHERES

It has been known for a number of years that a small portion of the particles in pulverized coal ash consists of thin-walled hollow spheres (cenospheres). The apparent density of these ash particles is less than that of water, so they separate from the dense ash in settlement lagoons. Amounts of the lightweight ash produced at different power stations vary: in some cases the quantity of ash cenospheres is negligible, whereas some boilers discharge sufficient amounts to form a thick layer of floating ash on lagoons. When the thickness of the floating layer exceeds 0.2 m, drying winds will entrain the lightweight ash from the surface of lagoons, thus causing a local nuisance.

Raask (1968a) has investigated the mode of formation of ash cenospheres in a Leitz heating microscope. Figure 6.11 illustrates the sequence of events leading to the formation of a cenosphere when a pellet of ash on carbon plate is heated in the microscope furnace. Coal-ash slag does not wet carbon or coke surface, as discussed in the previous section, and thus a spherical globule is formed from a cylindrical ash pellet (Fig. 6.11a) at 1600 K. Using the ash viscosity data published by the BCURA (1963), it was estimated that the viscosity of the slag globule at the time of its formation was about 1000 N s m^{-2}. By increasing the temperature by 75 K, the slag globule expanded (Fig. 6.11, b and c) to about double its original diameter before bursting. The process of gas generation inside the slag droplet was continuous, and the gas bubbles discharged at intervals of about 50 s. The viscosity of globule slag at the temperature of cenosphere formation was approximately 100 N s m^{-2}.

Figure 6.12 illustrates the effect of addition of iron powder on the gas

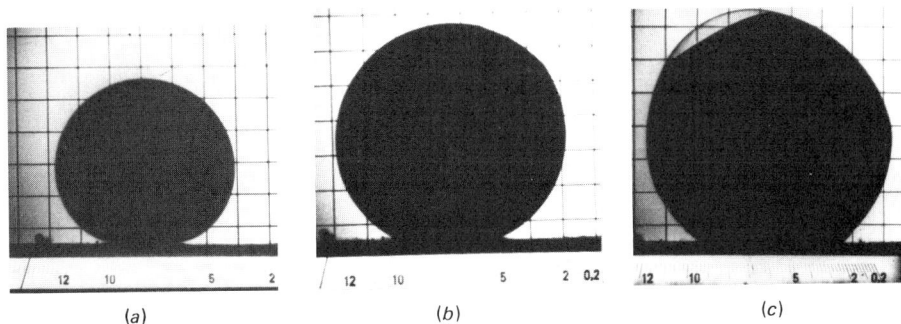

Figure 6.11 Formation of ash cenosphere in heating microscope, 0.5 mm grid. (a) Slag globule on carbon plate. (b) Half-stage inflation. (c) Cenosphere.

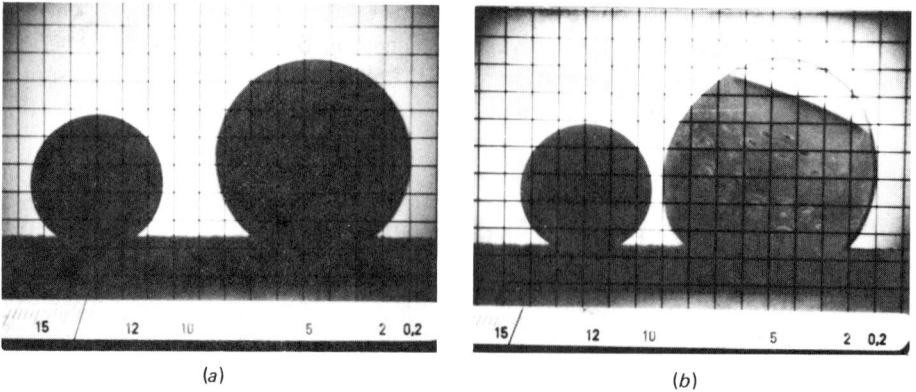

Figure 6.12 The effect of iron on gas evolution for cenosphere formation. (*a*) Partial expansion on right (from iron-free slag on left to iron added on right). (*b*) Full expansion on right (from iron free slag on left to iron added on right).

evolution. The first photograph shows the two slag droplets after melting: the globule on the left is free from iron. The other two photographs show clearly that the slag globule free from the iron retains its initial size on further heating, but the other droplet increases in size due to gas generation. The results of these experiments showed that the evolution of gases (carbon monoxide and dioxide) from the slag globule was catalyzed by the presence of iron. The same effect was observed when iron was added to the silicate melt in the form of oxide or metal powder.

Raask (1966a) has suggested that the mechanism of cenosphere genesis commenced at the slag/carbon interface where iron carbide formed and then diffused into the molten slag. Inside the slag the carbide reacted with silica, resulting in the release of carbon monoxide:

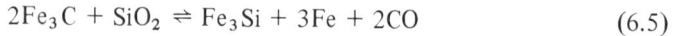

$$2Fe_3C + SiO_2 \rightleftharpoons Fe_3Si + 3Fe + 2CO \qquad (6.5)$$

According to the thermodynamic data published by Good (1962) and by Elliott and Gleiser (1960), the equilibrium pressure of CO produced in the reaction of Eq. (6.5) reaches 1 atm at 1675 K. This agrees with the experimental evidence that the gas evolution in the molten slag globule commenced at this temperature. The equilibrium pressure of carbon monoxide in Eq. (6.5) reaches 5 atm at 1825 K; thus the formation of ash cenospheres should be suppressed in a pressurized slagging combustor or gasifier.

An alternative but less likely explanation for the gas formation was that carbon had diffused into the slag, where it had reduced iron oxide or silica:

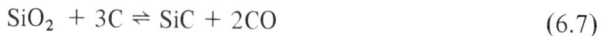

$$FeO + C \rightleftharpoons Fe + CO \qquad (6.6)$$

$$SiO_2 + 3C \rightleftharpoons SiC + 2CO \qquad (6.7)$$

The reaction in Eq. (6.6) is at equilibrium, with $p_{CO} = 1$ atm at 975 K, and the reaction of Eq. (6.7) is at equilibrium with the same pressure at 1825 K. Further, solubility of elemental carbon in molten slag is low; it is therefore unlikely that the reactions described by Eqs. (6.6) and (6.7) are relevant in the formation of ash cenospheres. It appears that gas evolution depends on the rate of carbon dissolved in molten ash in the form of iron carbide.

Time required for a cenosphere particle to expand to its final size is governed by the rate of gas evolution and by the rate of viscous flow of ash slag. Goodier (1936) has derived a relationship governing the rate of change of the radius of a hollow sphere, dr/dt, due to viscous relaxation,

$$\frac{dr}{dt} = \frac{r^2 \, \Delta p}{4\eta s} \tag{6.8}$$

where r = radius of the sphere
  Δp = difference between the internal pressure p and the equilibrium pressure
  η = viscosity of coal ash slag
  s = shell thickness
The value of Δp is given by the equation

$$\Delta p = p - p_a - 2\gamma \left(\frac{1}{r} + \frac{1}{r_1} \right) \tag{6.9}$$

where p_a = external pressure
  γ = surface tension
  r, r_1 = external radius and the internal radius, respectively
The cenosphere must now be assumed to contain a definite amount of gas throughout the expansion, so that

$$p = kr_1^{-3} \tag{6.10}$$

By combining Eqs. (6.8), (6.9), and (6.10), an expression for the rate of growth of cenospheres can be obtained:

$$\frac{dr}{dt} = \frac{1}{4\eta(r - r_1)} \frac{kr^2}{r_1^3} - p_a r^2 - 2\gamma \frac{r}{r_1} (r + r_1) \tag{6.11}$$

Equation (6.11) can be used to calculate the rate of change of radius of the particle as it expands to its final size. The relationship between r and r_1 is given by

$$r^3 - r_1^3 = k_1 = (\tfrac{3}{4})\pi \times \text{(volume of shell)} \tag{6.12}$$

The time required for the formation of different-sized cenospheres can be calculated from the area under the curve obtained by plotting the inverse of (dr/dt) against the radius. For example, a cenosphere of 50 μm diameter would require only about 0.3 ms for its formation at 1675 K; the surface tension and the viscosity of coal ash were taken to be 0.32 N m^{-1} and 100 N s m^{-2}, respectively.

The quantity of ash cenospheres formed in pulverized-fuel-fired boilers burning British coals varied from zero to 4.8 percent by weight; the usual quantity was between 0.3 and 1.5 percent (Raask, 1968a). The plot in Fig. 6.13 shows that the amount of cenospheres formed was negligible when the ash contained less than 5.0

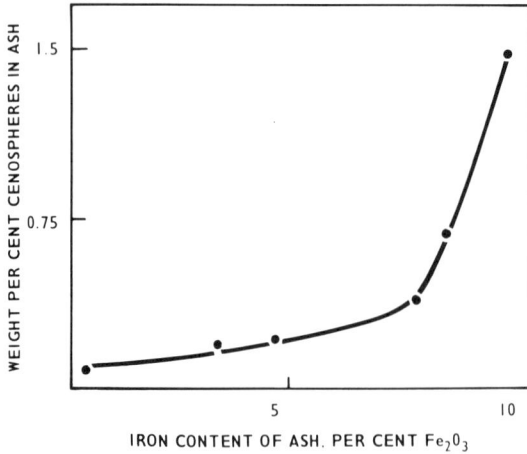

Figure 6.13 Dependence of ceno-sphere formation on iron in ash.

percent of iron oxide. This is in accordance with the postulated mechanism of formation of ash cenospheres as already described and summarized by Eq. (6.5): the presence of iron is essential for the gas bubbles to form inside the molten ash particles.

Photographs taken under the optical (Fig. 6.14a) and scanning electron microscope (Fig. 6.14, b, c, and d) show that ash cenospheres are colorless glassy particles of diameter ranging from 20 to 300 μm. There was a noticeable absence of small particles (less than 20 μm in diameter), except for a small amount of debris from the broken glass spheres as shown in Fig. 6.14c. Occasionally a blister could be seen on a cenosphere (Fig. 6.14d) but the shell-penetrating pores were absent; thus the hollow particles of ash are impervious to liquid and gases at the ambient temperature.

The cenospheres have a shell-wall thickness of 2 to 5 μm and have an apparent particle density of 200 to 500 kg m^{-3}. In addition, the floating material on ash lagoons contains particles that have an apparent density of 500 to 1000 kg m^{-3}, and in the nonfloating ash there are particles of density between 1000 and 2000 kg m^{-3}. These ash particles have a complex spheres-inside-spheres structure, shown in Fig. 6.15, a, b, and c; these configurations are sometimes referred to as plerospheres or plurospheres (Fisher et al., 1976). The ash particles of apparent density of 1000 to 2000 kg m^{-3} sometimes have a honeycomb structure encapsulated in a spherical envelope, as shown in Fig. 6.15d.

Several years after the original work by Raask (1968a), ash cenospheres were "discovered" in a big way by researchers in the United States (Fisher et al., 1976; Sarofim et al., 1977), with headlines proclaiming "mystery of the spheres." In fact, the mode of formation of ash cenospheres and multisphere formations, plerospheres, in boiler flames can be explained in terms of the surface-tension force, the nonwetting property of fused slag when in contact with coal or coke surface, and chemical reactions that produce gas bubbles inside slagged ash globules. The presence of carbon either inside or at the surface of ash slag particles is essential. It provides the "fuel" for gas generation and prevents the coalescence of the slag particles inside the external shell of cenosphere.

For the formation of plerospheres, an essential requirement is that coal substance and mineral matter be intimately dispersed. This can be found in the coal "middlings," i.e., pulverized-fuel particles that have a density of 1300 to 2000 kg m^{-3} (Table 4.4). When injected into the flame the particles acquire a fused silicate envelope, inside which the carbon residue prevents coalescence of encapsulated ash particles. It is therefore essential that by the time carbon residue is consumed the plerospheres must have cooled below about 1650 K to retain the structure as shown in Fig. 6.15, b and c. At higher temperatures the carbon-free plerospheres coalesce to dense spheres.

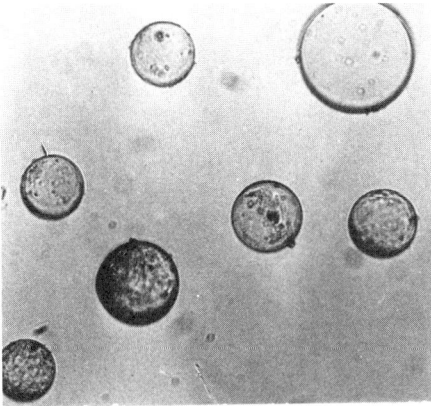

200 μm

(a)

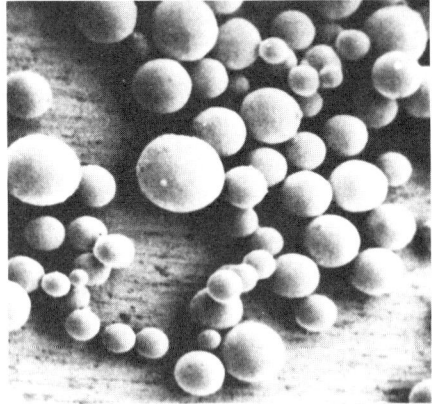

200 μm

(b)

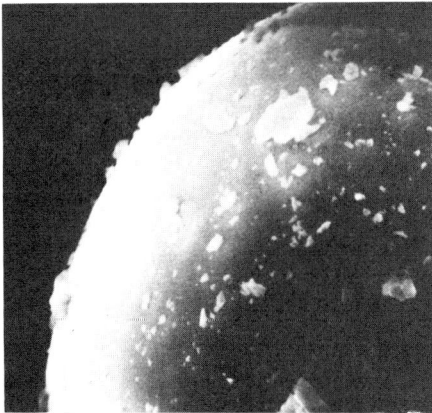

25 μm

(c)

25 μm

(d)

Figure 6.14 Ash cenospheres in pulverized fuel ash. (a) In optical microscope. (b) In electron microscope. (c) Close up. (d) Nipple growth.

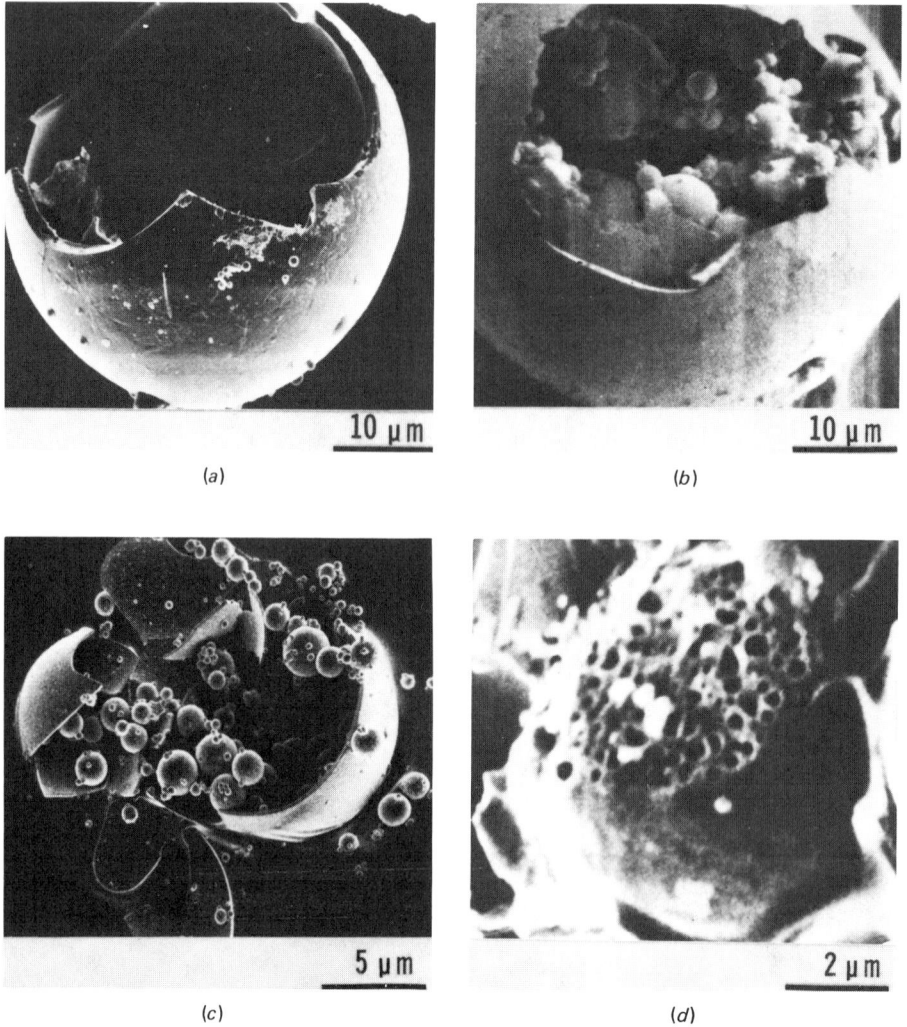

Figure 6.15 Plerosphere in pulverized-coal ash. (*a*) Impact damaged cenosphere. (*b*) Plerosphere. (*c*) Fractured plerosphere. (*d*) Honeycombed structure.

Favorable conditions for the formation of ash cenospheres and plerospheres in pulverized coal flames can be summarized as follows.

Cenospheres. 1. A minimum of 1 percent by weight of carbon should be incorporated or attached to fused silicate particles. It is therefore likely that a high fraction of the middlings particles of coal and ash (Table 4.4) in pulverized-coal ash will enhance the formation of ash cenospheres.

2. The presence of iron appears essential for catalysis of the reactions producing CO and CO_2 inside the fused ash particles. Most bituminous coal ashes

require above 7 percent iron oxide for the formation of a significant quantity of ash cenospheres. For some sub-bituminous coal ashes, 3 to 4 percent Fe_2O_3 is sufficient, since there is a high degree of intergrowth between the mineral matter and coal substance. Ashes with an exceptionally high amount of Fe_2O_3, above 20 percent, do not produce a significant quantity of cenospheres, as shown by the data published by Lauf (1981). This is probably because the iron-rich ashes have a low viscosity and the gas evolution is too rapid for the formation of stable cenospheres.

3. The residence time of the ash particles in the flame should not be excessive. For this reason the amounts of cenospheres formed in small- and medium-size pulverized-coal-fired boilers—i.e., 20- to 200-MW output units—are likely to be higher than those produced in 200- to 1000-MW boiler units. No significant amounts of ash cenospheres are likely to form in cyclone-fired boilers because of the high flame temperature and also because large mineral particles are thrown to furnace walls, where they coalesce to a flowing mass of molten slag.

Plerospheres. The conditions that favor the formation of ash cenospheres also enhance the creation of plerospheres. An additional requirement is that the coal substance and the mineral particles should be intimately dispersed inside the burning coke particle. It is therefore likely that pulverized coals that have high amounts of mixed particles of carbon and mineral matter with the initial density around 2000 kg m^{-3} will produce significant quantities of plerospheres.

6.4 RELEASE OF POTASSIUM FROM ALUMINOSILICATE CLAY MINERALS

Potassium minerals in coal are present chiefly as aluminosilicates (Chap. 3). The muscovite minerals contain around 11 percent by weight potassium, expressed as K_2O, and have an approximate chemical formula of $K_2O \cdot 2Al_2O_3 \cdot 6SiO_2 \cdot H_2O$. Illite minerals contain between 5 and 6 percent K_2O, and their chemical composition can be expressed approximately as $2K_2O \cdot 3(Ca,Mg,Fe)0.8Al_2O_3 \cdot 24SiO_2 \cdot 12H_2O$ (Gumz et al., 1958; Dixon et al., 1964; Kirsch, 1964, 1965).

The potassium mineral particles vitrify in pulverized-coal flame and change their shape to spherical forms, as described in Sect. 6.1. The silicate potassium is largely nonvolatile and remains dissolved in the glassy material, but a fraction of the metal escapes into the vapor phase and is subsequently converted to sulfate, as discussed in Chap. 7. The results of analysis of corrosive boiler deposits of alkali metal sulfates (Adams and Raask, 1963) gave rise to a suggestion that volatilization of potassium from silicates was enhanced in the presence of sodium chloride vapor in the coal flame. Jackson and Duffin (1963) demonstrated an exchange reaction of potassium for sodium in an aluminosilicate ceramic material in their experiments where sodium chloride was injected into propane flame constrained in a mullite tube. The exchange of potassium for sodium took place in the glazed surface layer of the ceramic material at 1675 K, and chloride deposit on a cooled target contained up to 0.22 mole fraction KCl.

Raask (1968b) has studied the release of silicate potassium in a closed combustion vessel, namely in a steel calorimeter where a small quantity of coal, usually 1 g, is burned in an oxygen atmosphere at 2.5 MN m^{-2} (British Standard, 1970). Coal samples were ground to the same fineness as that of the fuel used in pulverized-coal-fired boilers, namely 70 percent passing through a 76-μm sieve. Figure 6.16 shows the temperature profile of the mineral matter measured by inserting a platinum 13% rhodium–platinum thermocouple dipped in the calorimeter fuel sample. The results of the temperature measurements showed that the calorimeter combustion environment approximately resembled that in a large pulverized-coal-fired boiler. The combustion experiments were made with coals and with synthetic mixtures of aluminosilicates, sodium chloride, and benzoic acid fuel, to study the release of potassium from different minerals.

The amount of potassium volatilized from silicate minerals during combustion can be readily determined by a simple analytical technique. Volatilized potassium is present after combustion in the form of chloride with synthetic mixtures and sulfate when coal samples were burned. Alkali-metal chlorides and sulfates are readily soluble in water, whereas the rate of leaching of potassium from silicate ash is low. Thus, the difference between the total potassium content and that of the water-soluble fraction represents the amount combined in silicates. The same applies for the distribution of sodium in the water-soluble chloride and sulfate fraction and nonsoluble silicates. Table 6.3 gives the soluble and nonsoluble fractions of the two alkali metals in coals before combustion and in the calorimeter residue after combustion (Raask, 1968b).

The data in Table 6.3 show that with the exception of one coal (sample 4) there was a transfer of alkalis to fused-silicate residue on combustion. The alkali capture is expressed in terms of equivalent weight of sodium in the last column of Table 6.3 and, on average, 50 percent of water-soluble and potentially volatile sodium was captured by the silicate residue. Figure 6.17 shows the effect of sodium

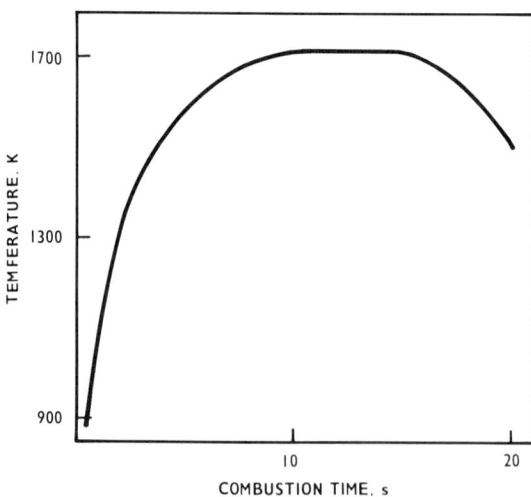

Figure 6.16 Temperature profile of coal combustion in bomb calorimeter.

Table 6.3 Silicate and nonsilicate alkalis in coals and combustion residues (weight percent of coal)

| Coal sample | Before combustion | | | | After combustion | | | | |
| | Silicate | | Nonsilicate | | Silicate | | Nonsilicate | | Alkali capture, Na-equiv. |
	Na	K	Na	K	Na	K	Na	K	
1	0.105	0.480	0.119	0.004	0.155	0.412	0.069	0.072	0.010
2	0.016	0.395	0.192	0.003	0.125	0.306	0.083	0.092	0.058
3	0.123	0.197	0.197	0.003	0.227	0.120	0.093	0.080	0.059
4	0.010	0.426	0.207	0.002	0.032	0.314	0.185	0.114	−0.044[a]
5	0.453	0.646	0.139	0.06	0.555	0.600	0.037	0.052	0.076
6	0.744	0.619	0.244	0.008	0.730	0.512	0.058	0.115	0.124
7	0.125	0.438	0.340	0.006	0.284	0.272	0.181	0.172	0.062
8	0.192	0.478	0.294	0.008	0.317	0.309	0.169	0.177	0.026

[a]Net release of alkalis.

chloride on the release of potassium from the synthetic mixes and from coal silicate during the bomb combustion. On average, 0.3 mole equivalent of potassium to sodium (0.5 weight equivalent) was released from the coal silicates and lesser amounts from the synthetic mixes. It is evident that with each mineral a limited amount of potassium was released, after which further addition of sodium chloride did not have any effect. More potassium was released from coal silicates than from the synthetic mixes, probably because the coal-ash residues had a lower viscosity and consequently diffusion of potassium from the interior to particle surface was enhanced.

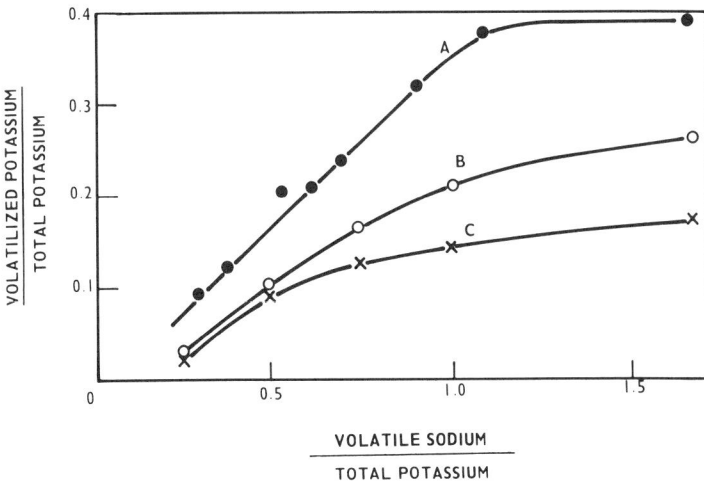

Figure 6.17 The effect of sodium chloride on release of potassium from silicates: A, coal silicates; B, muscovite; and C, orthoclase.

According to the thermodynamic data on silicates published by Kelley (1962) and those on chlorides published by Villa (1950), the alkali-metal exchange reactions have a net (positive) free energy (ΔG) value of 25 KJ mol^{-1} at 1675 K:

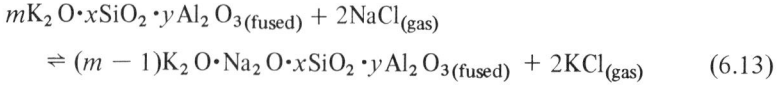

$$mK_2 O \cdot x SiO_2 \cdot y Al_2 O_{3(fused)} + 2NaCl_{(gas)}$$

$$\rightleftharpoons (m - 1)K_2 O \cdot Na_2 O \cdot x SiO_2 \cdot y Al_2 O_{3(fused)} + 2KCl_{(gas)} \qquad (6.13)$$

$$\Delta G = -RT \ln \frac{p_{(KCl)_2}}{p_{(NaCl)_2}} \qquad (6.14)$$

A positive ΔG value means that after the high-temperature exchange reactions the molar concentration of unreacted NaCl vapor should be higher than that of KCl. This is in accordance with data given in Fig. 6.17, which shows that for 1 mol NaCl, 0.15 to 0.30 mole equivalent potassium was released from the silicates.

In addition to the sodium chloride vapor, other coal mineral species can influence the release of potassium from aluminosilicate particles in pulverized-coal flame, as reported by Gibb and Angus (1983), who found that a high basic oxide content ($CaO + MgO + Fe_2 O_3$) of ash enhanced the potassium release. In contrast, a high total ash content of coal reduced the amount of potassium released, probably as a result of recapture of the volatile species by the vitrified silicate ash particles.

6.5 VOLATILIZATION OF SILICA IN COAL GASIFICATION AND COMBUSTION SYSTEMS

Hoy et al. (1962) have reported that volatilization of silica during coal gasification with oxygen and steam in a fixed-bed experimental assembly depended on the reaction zone temperature, as shown in Fig. 6.18. The degree of silica volatilization was estimated by comparing the composition of ash in coal fed to the gasifier with

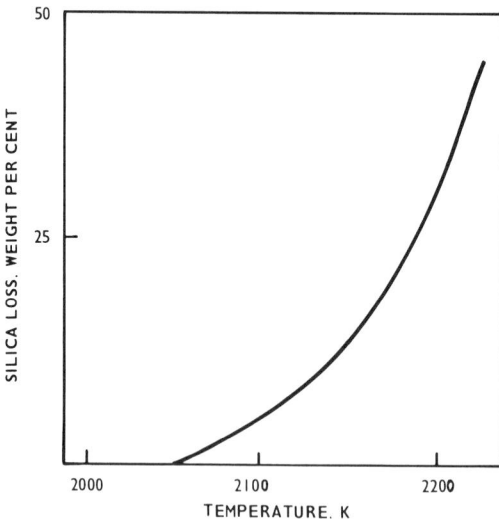

Figure 6.18 Silica volatilization in coal gasification with steam and oxygen.

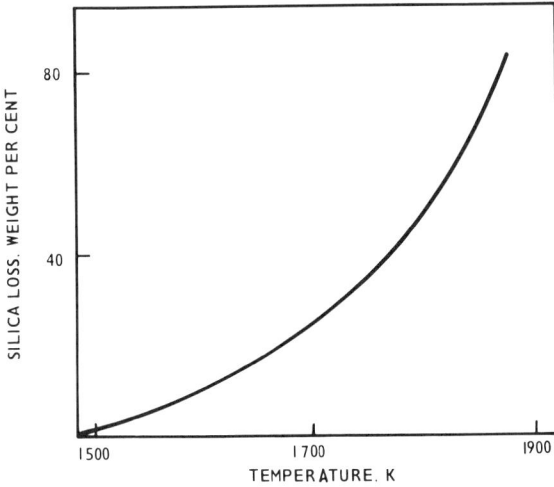

Figure 6.19 Silica loss from carbon/silica mixes in argon.

that of the slag discharged. It was assumed that no volatilization of Al_2O_3 or CaO occurred from the slag, and thus the change in SiO_2 to Al_2O_3 and SiO_2 to CaO ratios was taken to calculate the silica loss. Silica losses thus estimated from the slag and ash analysis were in agreement with visual observations that a fume emission commenced at about 2100 K and became more copious as the temperature of the gasifier reaction bed was increased.

Raask and Wilkins (1965) have made further studies on silica volatilization under oxidizing, neutral, and reducing atmospheres. In a series of laboratory experiments, mixtures of finely ground silica and excess carbon in a graphite crucible were heated in argon. Figure 6.19 shows the loss of silica in the temperature range of 1500 to 1875 K after the samples were heated for 1 h. By expressing the results on an Arrhenius plot of log loss against the reciprocal of the temperature, the energy of activation was found to be 280 kJ mol^{-1}.

The temperature at which significant volatilization of silica—i.e., formation of silicon monoxide (SiO)—commences in a coal combustion or gasification system depends on the locality where the reduction of SiO_2 takes place. Reduction of well dispersed silica in an excess of carbon, as in the above laboratory experiments, is governed by the vapor pressure of CO formed inside the coal matrix:

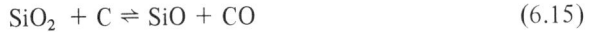

$$SiO_2 + C \rightleftharpoons SiO + CO \qquad (6.15)$$

that is,
$$p_{SiO} = \frac{k_o}{p_{CO}} \qquad (6.16)$$

where p_{SiO} and p_{CO} are the partial pressures of silicon monoxide and carbon monoxide, respectively, and k_o is the equilibrium constant.

When silica is exposed to the atmosphere in a gasifier, the following six reactions need to be considered:

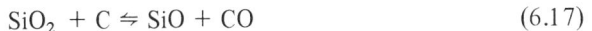

$$SiO_2 + C \rightleftharpoons SiO + CO \qquad (6.17)$$

$$SiO_2 + H_2 \leftrightharpoons SiO + H_2O \tag{6.18}$$

$$C + O_2 \leftrightharpoons CO_2 \tag{6.19}$$

$$C + \tfrac{1}{2}O_2 \leftrightharpoons CO \tag{6.20}$$

$$C + H_2O \leftrightharpoons H_2 + CO \tag{6.21}$$

$$H_2 + \tfrac{1}{2}O_2 \leftrightharpoons H_2O \tag{6.22}$$

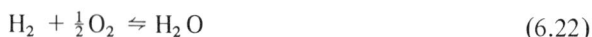

The vapor pressure of silicon monoxide can then be expressed in terms of the equilibrium constants k_1 to k_6 and the partial pressures of CO, CO_2, H_2, and H_2O:

$$(p_{SiO})^2 = \frac{k_1 k_2 k_3}{k_4 k_5 k_6} \cdot \frac{p_{CO}}{p_{CO_2}} \cdot \frac{p_{H_2}}{p_{H_2O}} \tag{6.23}$$

The equilibrium constants for Eq. (6.23) are given in the paper of Raask and Wilkins (1965) and were calculated from the relevant free energy data published by Kubaschewski et al. (1967), Elliott and Gleiser (1960), and Good (1962). Figure 6.20 (curve A) shows the partial pressure of SiO calculated from Eq. (6.23) and that compared from the amount of silica volatilization (curve B) in an experimental gasifier (Hoy et al., 1962). Considering the possible errors, both in the thermodynamic data and in the experimental measurements, the agreement is satisfactory.

Raask and Wilkins (1965) arrived at the following conclusions based on their laboratory studies and plant investigations.

1. The extent of silica volatilization and the subsequent formation of the fume is governed by the formation of silicon monoxide when silica or silicates react with carbon and hydrogen. The reduction of silica in clay minerals (Chap. 3) can be reduced to silicon monoxide by carbon at high temperatures.
2. During gasification with steam and oxygen, significant losses of silica would not be expected at atmospheric pressure, unless the reaction-zone temperature exceeds 2075 K; the corresponding temperature for gasification with preheated air would be about 2175 K. In a coal-fired MHD (magneto–hydro–dynamic) system operating at 0.5 MN m^{-2} (5 atm) pressure, as described in the book edited by Heywood and Womack (1969), silica volatilization may become significant above 2475 K.

The degree of silica volatilization in pulverized coal and cyclone fired boilers should be minimal under normal operating conditions. The CO and H_2 concentrations of the flue gas are low, so some SiO is formed only when silica is dispersed in a coal matrix. Assuming that 10 percent of the ash silicate is intimately mixed with coal substance, it is estimated that about 0.01 of the weight of the total silica is volatilized in the pulverized-fuel flame of a large (500-MW unit) boiler. This figure was obtained by taking into account the temperature and residence time of particles in the flame (Fig. 6.1) and the rate of SiO formation from the reaction between silica and carbon at different temperatures (Fig. 6.19).

In a cyclone-fired boiler, the amount of silica volatilized could be more than an

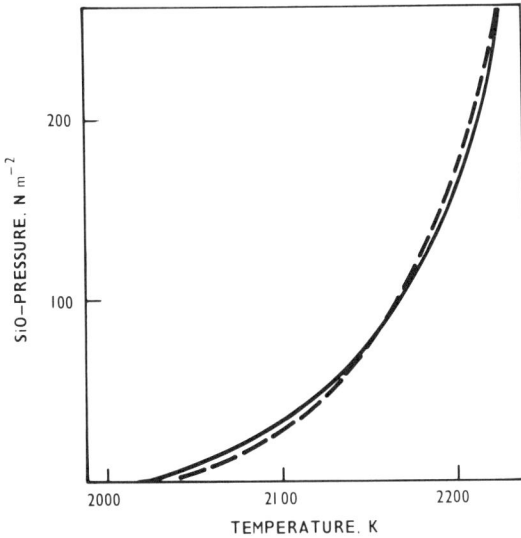

Figure 6.20 Vapor pressure of silicon monoxide in coal gasification with steam and oxygen: - - - - calculated from Eq. (6.24); —— calculated from silica losses.

order of magnitude higher than that in a "dry" ash boiler—that is, between 0.1 and 0.2 percent of the total ash silica could be volatized. This is because a coarsely ground coal is fed to cyclone-fired boilers, so that more silica is in intimate contact in coal, and also the boiler flame temperature is usually 100 to 200 K higher than that of a pulverized-coal-fired boiler. The mode of condensation of volatilized silica and the characteristics of the fume thus formed are discussed in Chap. 8.

NOMENCLATURE

D	slag density
dr/dt	rate of particle growth
ΔG	free energy change
g	gravitation force
k	constant mass of gas
k_0 to k_6	thermodynamic equilibrium constants
m	mole fraction of potassium oxide in aluminosilicates
P_{CO}	equilibrium pressure of carbon monoxide
p_{KCl}	vapor pressure of potassium chloride
p_{NaCl}	vapor pressure of sodium chloride
p_{SiO}	equilibrium pressure of silicon monoxide
p	equilibrium (internal) gas pressure
p_a	external pressure
Δp	excess pressure
R	thermodynamic gas constant
r	radius (external) of particle
r_1	internal radius of particle of cenosphere

r_0	radius of equivalent sphere
s	shell thickness
T	temperature
t	time
β	slag globule shape factor
γ	surface tension coefficient
η	viscosity coefficient
ρ	radius of curvature

REACTIONS OF NONSILICATE IMPURITIES IN COAL FLAMES

7.1 DISSOCIATION AND OXIDATION OF PYRITES

Pyrite minerals constitute in many coals a major source of the reactive impurity species that can cause a number of boiler operation and emission problems, namely slagging, fireside corrosion of boiler tubes, and chimney emission. Thermal decomposition of pyrite and marcasite minerals of the same chemical composition (FeS_2) has been extensively studied by numerous investigators, contributing to the technology of metallurgical chemistry of pyrite roasting for sulfuric-acid manufacture and to knowledge of the behavior of coal pyrites in boiler flames. Watt (1968) has made a comprehensive literature survey on thermal decomposition of pyrite and marcasite in different atmospheres.

When pyrite is heated in the absence of air it dissociates to form pyrrhotite (FeS) and sulfur gas, and at higher temperatures the pyrrhotite decomposes into sulfur and iron. The two-step reaction can be summarized as follows:

$$2FeS_2 \rightarrow 2FeS + S_2 \qquad \Delta G_T = 310.2 - 0.2925T \qquad (7.1)$$

$$2FeS \rightarrow 2Fe + S_2 \qquad \Delta G_T = 310.0 - 0.1017T \qquad (7.2)$$

where ΔG_T is the standard free energy change (KJ) and T is the temperature (K). The ΔG_T values were calculated from the thermodynamic data published by

JANAF (1963), Kelley (1960), and Kelley and King (1961), and are applicable in the temperature range of 500 to 1500 K. The corresponding equilibrium vapor-pressure curves of sulfur gas are shown in Fig. 7.1.

The solid phase formed on decomposition of FeS_2 does not have a definite composition, as discussed by Jensen (1942), and there is a further complication that sulfur vapor polymerizes to S_2 with some S_4 and S_6 species (Berkowitz and Marquart, 1963). Schwab and Philinis (1947) have found that decomposition of FeS_2 was a first-order reaction, having an activation energy of 121 kJ mol^{-1} in the initial stage rising to 138 kJ mol^{-1} at higher temperatures. They observed that there was a break in the rate curve after 20 to 30 percent of sulfur had been volatilized in an inert gas or in air. They concluded that the process of decomposition slowed down when pyrrhotite (FeS) first appeared as a separate phase.

Melting behavior of the residue formed on heating FeS_2 shows a complicated pattern. Asanti and Kohlmeyer (1951) have published a phase diagram of the FeO-FeS system, which has a minimum liquidus temperature of 1213 K (Fig. 7.2), but Bryers (1976) has claimed that a melting phase in the system can occur as low as 1075 K. When coal pyrite grains, 0.5 mm in size, were heated in a Leitz heating microscope purged with nitrogen, the FeS residue particles began to lose their sharp edges at 1300 K and they melted at 1350 K (Raask and Bhaskar, 1980, unpublished results). Pyrrhotite (FeS) mineral grains behaved in a similar manner on heating, except that they required slightly higher temperatures for the initial deformation at 1350 K and for melting at 1425 K. Wetting characteristics of molten sulfide on alumina, platinum, carbon, and slag surfaces were examined in a heating microscope, and the observations showed that the FeS melt was a highly mobile liquid that spread rapidly on all surfaces.

Halstead and Raask (1969) have studied the decomposition and oxidation of iron pyrite particles when passed through a laboratory furnace in the form of air-borne cloud. The furnace temperature was kept between 1375 and 1775 K for different runs, and the residence time of the particles in the hot zone was 0.5 s.

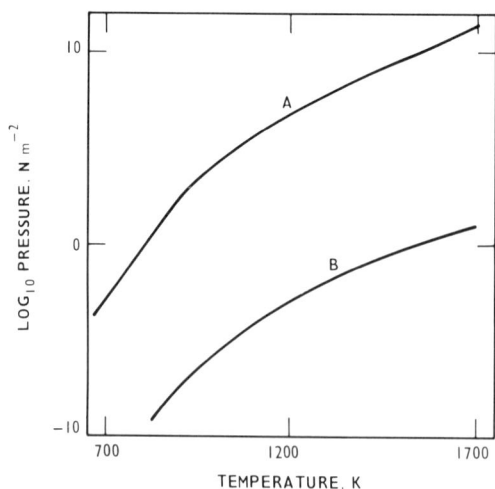

Figure 7.1 Sulfur vapor pressure on dissociation of iron sulfide: A, $2FeS_2 \rightarrow 2FeS + S_2$; B, $2FeS \rightarrow 2Fe + S_2$.

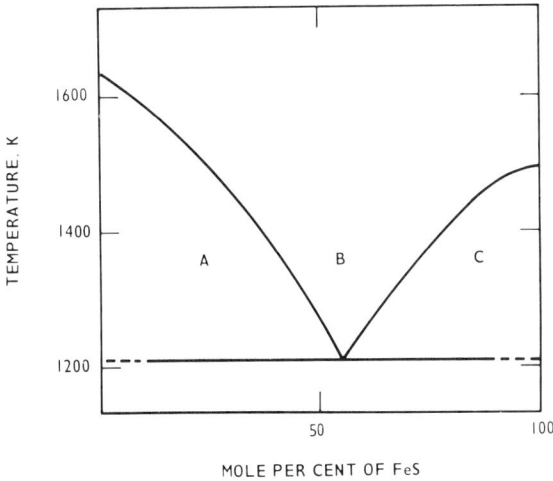

Figure 7.2 Liquid and solid phases of iron oxide/sulfide system: A, FeO + melt; B, melt; C, FeS + melt.

Figure 7.3a shows FeS residue before melting, and it is evident that a vigorous evolution of sulfur gas on rapid heating had created pores, channels, and occasionally a rupture in the particles. Subsequently, FeS particles melted at 1300 K and acquired a rounded shape, as shown in Fig. 7.3b. Molten FeS droplets will readily form couplets and chains, as shown in Fig. 8.5, a and b, when touching, and neck-bridged particles of iron oxide can be occasionally found in pulverized-coal ash (Fig. 8.5e). The duration of liquid phase must have been short, less than 1 ms, otherwise a complete coalescence would have occurred, as discussed in Chap. 10.

Iron oxide particles found in pulverized fuel ash are spherical in shape, and this can be taken as a proof that a liquidus phase occurs before oxidation of FeS to FeO:

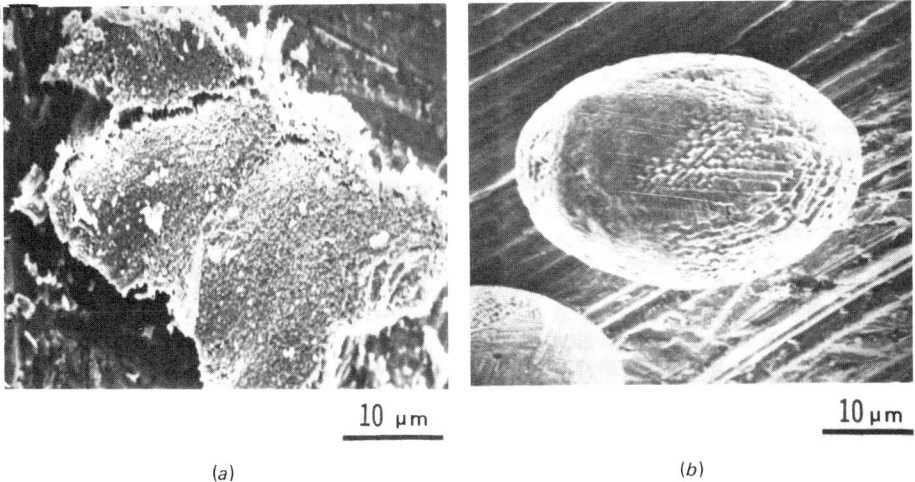

(a)

(b)

Figure 7.3 Dissociation products of particulate pyrite. (a) FeS residue of FeS_2 before melting. (b) Molten FeS residue.

$$FeS_{(molten)} + \tfrac{3}{2}O_2 \rightarrow FeO_{(solid)} + SO_2 \qquad (7.3)$$

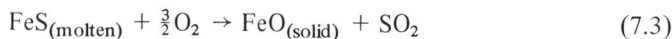

The only debatable point here is duration of the liquidus phase when the gas-borne pyrite particles dissociate and oxidize in pulverized-coal flame. Table 7.1 shows the amount of sulfur in the residue when FeS_2 particles, 100 μm in diameter, were passed through the vertical furnace kept at different temperatures. The residence time of the moving particles in the furnace hot zone was approximately 0.5 s.

Figure 7.4 shows log $(S_o - S_x)/S_o$ values plotted against the inverse of temperature, where S_o is the sulfur fraction in FeS and S_x is the fraction in FeS_xO_y residue. The slope of the Arrhenius plot gives a temperature coefficient of 100 kJ K^{-1} mol^{-1}, which is of the same magnitude as the temperature coefficient of partial pressure of sulfur vapor on dissociation of FeS (Fig. 7.1). It is therefore likely that the overall rate of oxidation of FeS depends on the diffusion of sulfur vapor generated inside the particles and that of oxygen into the residue. Further, it appears that when a solid layer of FeO and Fe_3O_4 forms at the surface of the particles it does not offer much resistance to gaseous diffusion because of the porous nature of the solid phase, as is evident from Fig. 7.3b. This is consistent with the fact that no significant quantities of residual sulfur in the form of sulfide have been found in the magnetite-ash particles captured in the electrostatic precipitators of large coal-fired boilers.

Returning to the question of the duration of liquidus phase in the flame-borne pyrite particles, it is likely to be between 0.5 and 5 ms, depending on the particle size and on the oxygen concentration of the flue gas. This time range for a liquid phase of FeS particles in coal flame was arrived at by taking into account the results of laboratory experiments described above, and the phase diagram of the iron–iron sulfide system published by Jensen (1942). In a fuel-rich zone of the flame, a liquidus phase will persist over a longer period and FeS particles may dissociate to a solid iron core surrounded by a molten layer of sulfide. The possible key role of molten FeS residue in early stages of the formation of furnace wall deposits in pulverized-coal-fired boilers has been pointed out by Reid (1971), Raask (1973), Bryers (1976), Dik et al. (1977), and Borio and Narciso (1978). This topic will be discussed further in Chaps. 10, 11, 12, and 15.

Table 7.1 Sulfur content of pyrite-particle residues heated in a laboratory furnace for 0.5 s

Temperature, K	Percent sulfur by weight
(Unheated)	53.1
1375	28.5
1475	22.7
1575	17.8
1675	15.1
1775	10.1

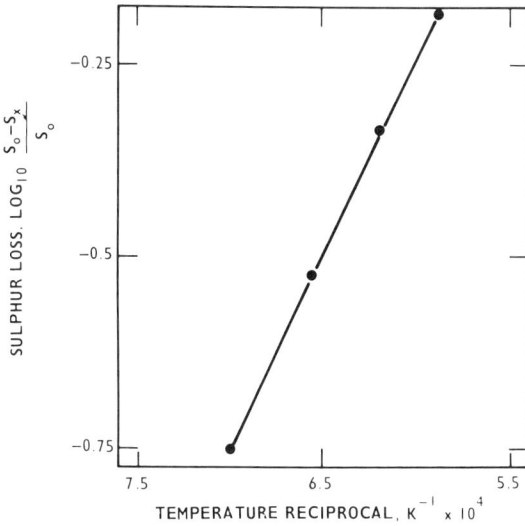

Figure 7.4 Sulfur loss $(S_O - S_x)$ from pyrite residue (FeS) on heating $(S_O =$ sulfur content of FeS).

Significant volatilization of iron in the form of metal vapor or via formation of volatile iron carbonyl is unlikely to occur in pulverized-coal-fired boilers. This could occur in a high-temperature process of coal gasification where the reducing conditions favor the formation of highly volatile carbonyl. In the coal-fired MHD (magneto-hydrodynamic) system, some volatilization of iron metal may occur as a result of the high flame temperature.

7.2 VOLATILIZATION AND SULFATION OF CHLORIDES

When coal is burned below 1000 K, chlorine is given off in the form of HCl before evaporation of NaCl takes place, as shown by results of Schoen (1956), who studied the loss of chlorine from coals of different chlorine content when heated in air. He found that evolution of HCl began at about 500 K, continued at a sharply increased rate in the temperature range of 575 to 675 K, and was completed at 875 K. There was no significant loss of alkalis below 1100 K. Similar results were obtained by Edgcombe (1956), who found HCl evolution took place when coal was heated in air at 475 K but none was lost when the same coal was heated in nitrogen. He concluded that the HCl evolution under these conditions represented a loss of the part of chlorine that was held by ion-exchange linkages in the coal substance, whereas the chloride salts, chiefly NaCl and $CaCl_2$, do not volatilize below 475 K.

Daybell and Pringle (1958) found that when a coal was impregnated with a solution of sodium chloride, dried, and heated in air at 475 K for 24 h, it lost nearly half its chlorine without a corresponding loss of sodium. Since NaCl on its own did not volatilize at such a low temperature, they concluded that the decomposition was brought about by reaction with some compound in the coal, either originally present or formed during heating. They found that sodium sulfate was formed at 475 K by oxidation of pyrites and organic sulfur compounds. Dunderdale et al. (1963) have shown that metallic sodium can form when sodium salts are heated in the absence of air. Volatilization of sodium in the form of metal vapor may be of particular interest with some sub-bituminous coals where the alkali metal is present as organometal compounds.

As discussed in Chap. 3, practically all chlorine compounds in coal are soluble in water, and power-station coals usually contain over 10 percent by weight of water. It is therefore to be expected that the chlorine compounds, chiefly sodium and calcium chlorides, are present in a saline solution of raw coal water. On drying, first in coal mills and subsequently in the flame, a chloride residue in the form of small crystals is likely to be precipitated on the surface and in the pores of coal and mineral particles.

Evaporation of the discrete chloride particle depends on the rate of heat transfer to the particle and of diffusion of the vapor from the surface. The radiative heat transfer dQ/dt from the flame to the evaporating particles is given by

$$\frac{dQ}{dt} = 4\pi r^2 \sigma \epsilon (T_o^4 - T^4) \tag{7.4}$$

where r = particle radius
  σ = radiation constant (5.67×10^{-8} W m^{-2} K^{-4})
  ϵ = particle emissivity
  T_o = temperature of heat source
  T = particle temperature

Frossling (1938) has shown that the rate of evaporation dm/dt from a liquid droplet is given by the equation

$$\frac{dm}{dt} = \frac{4\pi r D_v M(p_o - p)}{RT} [1 + k(\mathrm{Re})^{1/2}] \tag{7.5}$$

where r = particle radius
  D_v = diffusivity of NaCl vapor
  M = molecular weight
  p_o = saturation vapor pressure of NaCl
  p = vapor pressure of NaCl in flue gas
  R = gas constant
  T = particle temperature
  k = constant
  Re = the Reynolds number

Equation (7.5) can be simplified when the relative velocity between the particles and the carrier gas is small, when $p \gg p_o$, and Re → 0:

$$\frac{dm}{dt} = \frac{4\pi r D_v M p}{RT} \tag{7.6}$$

To sustain evaporation at the rate of dm/dt, the rate of heat transfer must be

$$\frac{dQ}{dt} = \frac{dm}{dt}\left(\frac{L}{M}\right) \tag{7.7}$$

where L is the latent heat of evaporation/mol NaCl.

From Eq. (7.7), allowing for convective heat transfer, the temperature of a steady-state evaporation may be estimated. When the boiler-flame temperature is between 1800 and 1900 K, the evaporating particles of 1 to 100 μm have a temperature between 1300 and 1500 K. The particle size of crystalline chloride that precipitates on evaporation of saline water is likely to be in the range of 0.1 to 10 μm. When the chloride microcrystals are carried into the flame at the surface of coal and mineral particles, their evaporation is complete within 0.1 s after the particles have attained a temperature above 1300 K. Chlorides trapped inside the pores of coal particles are likely to be released with coal volatile matter, which is usually completed within 0.5 s in a pulverized-fuel flame.

Concentrations of sulfur and chlorine compounds in the flue gas can be calculated from the results of analysis of coal for carbon sulfur and from the CO_2 content of the flue gas. The latter is usually monitored for the purpose of combustion control. The formula for calculating the concentration of sulfur gases, chiefly SO_2, is

$$\frac{V_{SO_2}}{V_{CO_2}} = \frac{M_C W_C}{M_S W_S} \tag{7.8}$$

where V_{SO_2}, V_{CO_2} = volume concentration of SO_2 and CO_2
$\qquad M_C, M_S$ = atomic weight of carbon and sulfur
$\qquad W_S, W_C$ = weight percentage of sulfur and carbon in coal
Introducing molecular weights of the two elements, Eq. (7.8) becomes

$$V_{SO_2} = 3.75 \times 10^3 \, V_{CO_2}\left(\frac{W_S}{W_C}\right) \tag{7.9}$$

The factor of 10^4 has been introduced in the equation so that it now gives the SO_2 concentration in ppm by volume of the flue gas when the CO_2 concentration is given in percent by volume. The SO_2 concentration thus calculated refers to the maximum concentration of the gaseous sulfur compound in the flue gas before deposition of alkali-metal and calcium sulfates takes place. The sulfate deposition reduces SO_2 concentration of the flue gas by the corresponding amount, as discussed by Raask (1982a). The corresponding formula for calculating the concentration of chlorine compounds in the flue gas is

$$V_{Cl} = 3.38 \times 10^3 \, V_{CO_2}\left(\frac{W_{Cl}}{W_C}\right) \tag{7.10}$$

where V_{Cl} is the concentration of volatilized chlorine compounds in ppm by volume, V_{CO_2} is the CO_2 concentration in percent by volume of the flue gas, and W_{Cl} and W_C are the chlorine and carbon contents of coal by weight per unit.

Concentration of volatile alkaline metal compounds—e.g., chlorides, oxides, hydroxides, and sulfates—cannot be estimated on the basis of the total amount of sodium and potassium in coal. In the boiler flame there is a release of potassium from silicates and a capture of sodium by the fused ash particles, as discussed in Chap. 6 [chiefly Eq. (6.13)]. Approximate concentrations of the flame-volatile sodium and potassium can be estimated as follows. Water-soluble sodium represents approximately the amount of volatilized sodium and is reduced by the adsorption in fused silicate ash by about 50 percent. This amount then releases approximately the same quantity by weight of potassium originally present in silicates (Table 6.3). The approximate formulas for calculating concentrations of volatilized sodium and potassium in the flue gas are

$$V_{Na} = 2.6 \times 10^3 \ V_{CO_2} \left(\frac{W_{Na-sol}}{W_C} \right) \tag{7.11}$$

and

$$V_K = 10^3 \ V_{CO_2} \left(\frac{W_{Na-sol}}{W_C} \right) \tag{7.12}$$

where V_{Na}, V_K = volume concentration (ppm) of volatilized sodium and potassium if present as chloride or hydroxide, or half of these values if present as oxide or sulfate in the flame

W_{Na-sol} = amount of water-soluble sodium in coal by weight percent

W_C = weight percent of carbon in coal

V_{CO_2} = volume percent of CO_2 in the flue gas

Table 7.2 gives the concentrations of sulfur, chlorine, and vaporized alkali-metal species in coal flame calculated according to Eqs. (7.9)–(7.12). It was taken that coals contain 60 percent carbon by weight and that CO_2 content of the flue gas was 15 percent by volume.

Volatilized chlorides are thermodynamically unstable in the presence of sulfur gases, oxygen, and water vapor in coal flame. Boll and Patel (1960) and Halstead and Raask (1969) have considered the sulfation of sodium chloride based on the available thermodynamic data. The conversion proceeds through a series of reactions summarized as

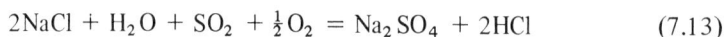

$$2NaCl + H_2O + SO_2 + \tfrac{1}{2}O_2 = Na_2SO_4 + 2HCl \tag{7.13}$$

where all reactants and products are in the vapor state. Table 7.3 gives the standard

Table 7.2 Calculated sulfur, chlorine, and volatilized alkali-metal concentrations of pulverized-coal flames

Percent by weight of coal			Parts per million (ppm) by volume of flue gas			
Sulfur	Chlorine	Sodium (water-soluble)	$SO_2 + SO_3$	Cl	Na	K
0.8	0.8	0.35	750	680	230	130
1.2	0.4	0.2	1100	340	130	75
1.8	0.15	0.07	1700	120	45	30
3.5	0.07	0.03	3300	65	20	12

Table 7.3 Standard free energy changes ΔG_T° of sulfur and sodium compounds in the temperature range of 500 to 1500 K

	$G_T^\circ = A + BT$ (kJ)	
Reaction[a]	A	$B \times 10^2$
$2FeS_2\,(s) \rightleftharpoons 2FeS(s) + S_2\,(g)$	+310.24	−29.25
$2FeS(s) \rightleftharpoons 2Fe(s) + S_2\,(g)$	+301.04	−10.17
$S_2\,(g) + 2O_2 \rightleftharpoons 2SO_2\,(g)$	−724.88	+14.60
$2Fe(s) + O_2\,(g) \rightleftharpoons 2FeO(s)$	−525.72	+13.35
$2NaCl(l) + SiO_2\,(s) + H_2O(g) \rightleftharpoons 2HCl(g) + Na_2\,SiO_3\,(l)$	+215.06	−8.87
$NaCl(g) + H_2O(g) \rightleftharpoons NaOH(g) + HCl(g)$	+96.44	+1.05
$2NaCl(g) + H_2O(g) + SO_2\,(g) \frac{1}{2}O_2\,(g) \rightleftharpoons Na_2\,SO_4\,(g) + 2HCl(g)$	−409.82	+30.25
$2NaOH(g) + SO_2\,(g) + \frac{1}{2}O_2\,(g) \rightleftharpoons Na_2\,SO_4\,(g) + H_2O(g)$	−582.62	+26.82

[a]Use of (s), (l), and (g) refers to the solid, liquid, and gaseous phases, respectively.

free energy changes ΔG_T° of a number of the possible reactions involving sulfur and sodium compounds in boiler flame. The calculated ΔG_T° values are based on the thermodynamic data published by JANAF (1963, 1971), Kelley (1960), and Kelley and King (1961).

Figure 7.5 shows the saturation vapor pressure of sodium chloride and sodium sulfate (Zimm and Mayer, 1944; Jackson and Duffin, 1963; Halstead and Raask, 1969), with the equilibrium concentration of the chloride and sulfate in an oxidizing coal flame. The NaCl and $Na_2\,SO_4$ concentration curves show that below 1200 K sulfate is the dominant species. Further, the $Na_2\,SO_4$ concentration curve crosses the corresponding saturation vapor pressure at 1200 K, which is the "dewpoint" temperature for condensation of sulfate.

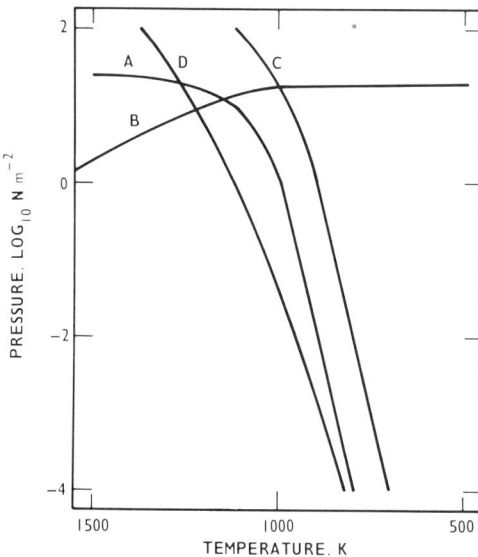

Figure 7.5 Equilibrium concentration of sodium chloride (A) and sulfate (B) in boiler-flue gas and saturation pressure of NaCl (C) and $Na_2\,SO_4$ (D) (0.4% Cl and 1.2% S in coal, 5% CO in flue gas.

It is difficult to estimate the time required for the conversion of volatilized sodium chloride to sulfate in boiler flame by simulated laboratory experiments or by calculations, because of the complexity of the sulfation reactions and the lack of relevant kinetic data. One method to check the degree of conversion of chloride to sulfate is to insert cooled probes into boiler flue gas and to analyze the deposited material. The results of deposition probe measurements in coal-fired boilers (Raask, 1962) showed that the conversion of alkali-metal chlorides to sulfate was virtually complete by the time the flue gas of a 200-MW boiler reached the superheater tube banks. The gas temperature at the sampling point was 1200 K, and the estimated residence time—that is, the time available for evaporation of sodium chloride to its conversion to sulfate—was about 1 s (Fig. 6.1). Deposition of chloride occurs when the residence time is exceptionally short (Raask, 1962; Halstead and Raask, 1969) or when there is a lack of oxygen in the boiler flue gas, as discussed in Chap. 12.

Some coals contain substantial quantities of water soluble calcium in the form of chloride (Table 3.7). It is therefore likely that this fraction of calcium will volatilize in the boiler flame and calcium chloride will be converted to sulfate in a manner similar to that of the sulfation of volatilized sodium chloride. However, the principal calcium mineral, that of carbonate, does not volatilize; its sulfation is discussed in the next section.

7.3 DISSOCIATION AND SULFATION OF CARBONATES

The principal carbonate minerals in coal are those of calcium, magnesium, and iron (Chap. 3). Magnesium carbonate is usually found in association with calcium in a form of double carbonate, $CaCO_3 \cdot xMgCO_3$, which is known as dolomite when $x = 1$ or dolomitic limestones when $x < 1$. Iron carbonates in the form of siderite, $FeCO_3$, and ankerite, $CaCO_3 \cdot xMgCO_3 \cdot yFeCO_3$, are frequently found in significant quantities in different coals, as discussed by Kemezys and Taylor (1964), Dixon et al. (1970b), and O'Gorman and Walker (1972).

On heating, carbonates dissociate to a solid residue of oxide and CO_2,

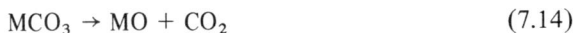

$$MCO_3 \rightarrow MO + CO_2 \tag{7.14}$$

and each species has its characteristic threshold temperature where thermal decomposition will commence. Differential thermal analysis (DTA) is the usual technique for studying the behavior of carbonates on heating, as exemplified by the results published by Webb and Heystek (1957), Lehmann and Mueller (1960), Powell (1965), French and Rosenberg (1965), Warne (1965, 1970, 1975), and O'Gorman and Walker (1973). The curves in Fig. 7.6 show the decomposition temperatures of different coal carbonates, where siderite is the first to dissociate to FeO and CO_2 at 800 K (curve A):

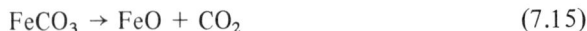

$$FeCO_3 \rightarrow FeO + CO_2 \tag{7.15}$$

Powell (1965) has concluded from his investigations that decomposition of siderite

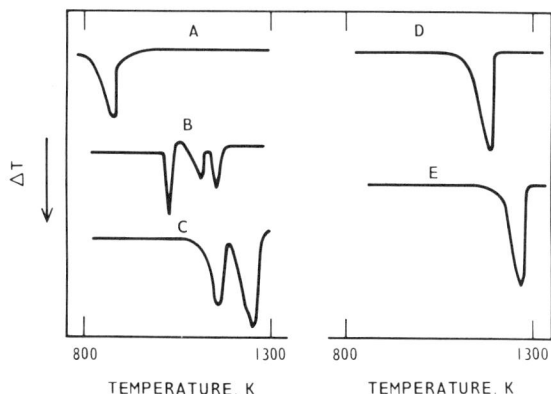

Figure 7.6 Dissociation of coal carbonates by differential thermal analysis: A, siderite; B, ankerite; C, dolomite; D, aragonite; E, calcite.

is a great deal more complex than it appears at first sight. FeO (wustite) produced is unstable below 845 K and oxidizes to Fe_3O_4 (magnetite) or Fe_2O_3 (hematite) when decomposition takes place in a DTA furnace where the rate of heating is comparatively low, usually between 5 and 20 K/min.

Ankerites, $CaCO_3 \cdot xMgCO \cdot yFeCO_3$, as might be expected from the variability of their composition and from the fact that they contain three cationic elements, show a complicated pattern of CO_2 evolution on heating. Ankerite is usually characterized by three endothermic peaks, as shown by curve B in Fig. 7.6. The origin of these peaks in terms of chemical and structural changes has been discussed by Beck (1950) and Schwob (1950). Kulp et al. (1951) have found that the characteristic temperature of the first endothermic decomposition peak decreased linearly with increasing iron oxide content of the ankerite.

In pulverized-coal-fired boilers where the rate of heating of flame-borne particles can be as high as 1000 K/s (Fig. 6.1), siderite and ankerite particles fragment into 0.1- to 1.0-μm FeO particles as a result of the rapid CO evolution, as discussed in Chap. 8. Iron oxide thus formed in boiler flame is highly reactive and is readily dissolved in fused silicate particles or deposits. Iron sulfate is not formed in the flame because of its thermal instability, but some iron oxide will dissolve in molten alkali-metal sulfate formed on boiler tubes in the temperature range of 800 to 900 K, as discussed in Chap. 12.

Dissociation of magnesite, unlike that of calcite, is not reversible, and MgO does not react readily with CO_2 gas other than at high pressures. Decomposition of dolomite takes place as a single-stage reaction when CO_2 given off is swept away by a purging gas stream:

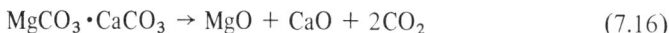

$$MgCO_3 \cdot CaCO_3 \rightarrow MgO + CaO + 2CO_2 \tag{7.16}$$

In the presence of CO_2 at a partial pressure higher than the dissociation pressure of the $CaCO_3$, the decomposition becomes a multistage reaction. That is, the initial decomposition into oxides is followed by recarbonation of the calcium oxide (Raask, 1974), as shown on curve C in Fig. 7.6:

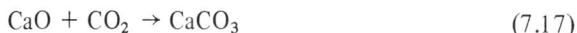

$$CaO + CO_2 \rightarrow CaCO_3 \tag{7.17}$$

Decomposition of magnesium-free calcium carbonate depends chiefly on the crystal structure: aragonite starts to decompose at 1075 K (curve D in Fig. 7.6), which is some 100 K lower than that of calcite (curve E). The latter has a well ordered crystalline structure and it thus requires a higher temperature for dissociation.

Decomposition of a crystalline material such as calcium carbonate is said to proceed through four stages: nucleation, reaction, recrystallization, and sintering. Gregg (1953) calculated that the nucleation period required to initiate decomposition of a calcium carbonate particle is about 0.3 s, which is about the same as that required to heat the carbonate particles in coal flame. After nucleation, which takes place at the crystal surface, the interface of dissociation reaction moves at a uniform rate towards the center of the particle. That is, the rate of decomposition of $CaCO_3$, dm/dt, is proportional to the surface area S:

$$\frac{dm}{dt} = k_1 S = k_2 m^{2/3} \tag{7.18}$$

where m is the mass of the decomposing particle and k and k_1 are the constants.

The initial reaction product is an oxide in which the structural units are still arranged according to the lattice of original carbonate as discussed by Ingraham and Marier (1963). This is a "pseudo-lattice" and appears as an amorphous material on the X-ray diffraction analysis. Subsequently, the primary structure rearranges itself to a large number of small crystallites. The specific surface (ratio of surface area to mass) of the recrystallized material is high, and in this form CaO reacts readily with sulfur gases to form sulfate:

$$CaO + SO_3 \rightarrow CaSO_4 \tag{7.19}$$

$$CaO + SO_2 + \tfrac{1}{2}O_2 \rightarrow CaSO_4 \tag{7.20}$$

Above 1200 K the nonsulfated CaO crystallites sinter to larger crystals, and this change is accompanied by a decrease in surface area and hence in a reduction of chemical reactivity of the oxide. For this reason, lime (CaO) and magnesium oxide (MgO) formed at high temperature from originally large particles of carbonate are much less reactive than those formed below 1200 K, where crystal sintering does not take place. Consequently the basic materials, when injected in the form of carbonate particles into boiler flame, are not effective additives for the capture of SO_2 or SO_3.

Bituminous coals usually contain less than 1.0 percent carbonates, above 1.0 percent sulfur, and above 5 percent silica. There is therefore an excess of "acidic" constituents in ash, and calcium and magnesium carbonates are in part sulfated; the remainder of CaO and MgO is captured by silicates, as discussed in the next section. Unreacted CaO and MgO are usually not detected in pulverized-fuel ash. However, some lignite and sub-bituminous coals that are rich in calcium and sodium but low in sulfur and silica (Sondreal et al., 1968) leave a highly alkaline ash. That is, the ash contains a sufficient quantity of unreacted calcium oxide that can give a saturated solution of $Ca(OH)_2$ of high pH value when dispersed in water.

7.4 DISTRIBUTION OF ALKALIS IN IMMISCIBLE SULFATE AND SILICATE PHASES

Bituminous coals usually leave a highly silicious ash on combustion. That is, fused aluminosilicates constitute an acidic media at high temperatures, capable of absorbing large quantities of basic metals in the form of oxides, e.g., those of sodium, calcium, and magnesium. At lower temperatures the corresponding sulfates are thermodynamically more stable in the presence of sulfur gases. The equilibrium distribution of alkaline oxides between molten sulfates and fused silicates at different temperatures can be calculated from the appropriate thermodynamic data. The behavior of potassium sulfate and silicates at high temperatures has been studied by Raask (1965, 1966b) because of the importance of potassium as an ionizable metal in the high-temperature flame of the MHD system (Heywood and Womack, 1969). Figure 7.7 shows the equilibrium plot for the potassium sulfate/ silicate system in an oxidizing atmosphere in the presence of SO_2:

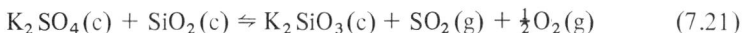

$$K_2SO_4(c) + SiO_2(c) \rightleftharpoons K_2SiO_3(c) + SO_2(g) + \tfrac{1}{2}O_2(g) \qquad (7.21)$$

where c and g denote the condensed and gaseous states, respectively. The relevant data were taken from those published by Briner et al. (1948), Kubaschewski et al. (1967), Elliott and Gleiser (1960), and Richardson et al. (1960). Figure 7.7 gives an upper temperature limit of 1450 K for thermodynamic stability of K_2SO_4 when in contact with molten silicates in the flue gas, which contains 4 percent oxygen and 0.1 percent SO_2. The temperature limits of thermal stability of Na_2SO_4, $CaSO_4$, and $MgSO_4$ in the same atmosphere are approximately 1300 K, 1150 K, and 1000 K, respectively.

Fused silicate particles in a pulverized-coal flame absorb volatilized alkalis in a

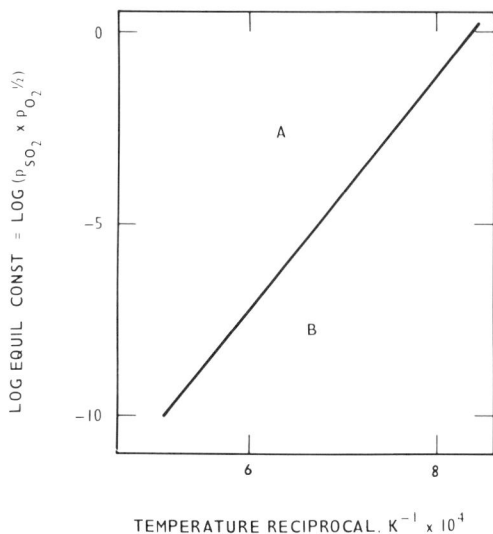

Figure 7.7 Thermodynamic equilibrium of system $K_2SO_4 + SiO_2 \rightleftharpoons K_2SiO_3 + SO_2 + \tfrac{1}{2}O_2$: A, sulfate phase; B, silicate phase.

surface layer to a depth of about 0.05 μm (Hosegood and Raask, 1964), but a significant fraction of alkali metals and calcium are present as sulfate (Raask, 1982a). The distribution of sodium in the silicate and sulfate phases can be expressed in a general form:

$$m_{sul} = m_o - k_3 w^{2/3} \tag{7.22}$$

where
$$m_{sil} + m_{sul} = m_o \tag{7.23}$$

and where m_{sil}, m_{sul}, m_o = amount of sodium in silicate and sulfate fractions, and the total sodium in ash, respectively
k_3 = constant
w = ash content of coal

When the sodium-to-ash ratio is below 1 to 100, the bulk of sodium is captured by the silicate particles and Eq. (7.23) reduces, according to the results of Raask (1968b), to

$$m_{sil} \simeq m_o \tag{7.24}$$

and consequently the amount of sodium available for the formation of sulfate is small. Hosegood and Raask (1964) have suggested that the surface layer of semimolten silicate ash particles in pulverized coal flames is capable of absorbing sodium in concentrations up to 50 mg Na/m² ash surface. Subsequent to the absorption of alkali metals in the fused layer of silicate ash, submicron-size particles of sulfate will form at the surface, as discussed in Chap. 8.

Kordes et al. (1951) and Pearce and Beisler (1965) have found that the melt of composition $Na_2SO_4-Na_2O-SiO_2$ has one liquid phase at 1475 K, but as the proportion of silica increases, the melt separates into two layers. The change from the miscible to immiscible phase of the system has been explained by alterations in the silicate structure as the ratio of Na_2O to SiO_2 decreases. In more basic, less viscous melts, the silicate ions exist in the form of SiO_4^{4-} tetrahedrons that have the same mobility as sulfate ions, and thus homogeneity of the system is to be expected. As the silica content is increased, the complexity of the silicate structure reaches a point where the silica anions become relatively immobile for a separation of sulfate from silica to take place.

Raask and Jessop (1966) have studied the miscibility of the corresponding potassium sulfate–silicate system by the usual crucible method, as well as by a technique of a hanging droplet that has been used by Nettleton and Raask (1967) to determine the rate of evaporation of potassium sulfate in the temperature range of 1400 to 1800 K. Droplets of potassium sulfate–silicate mixtures, 3 mm in diameter, were suspended from 0.5 mm platinum wire which had a semispherical head 1.5 mm in diameter. Separation of the silicate (internal) and sulfate phases in the droplets can be observed directly in the Leitz heating microscope, which is used in its conventional mode of operation to assess the fusion characteristics of coal ashes (Radmacher, 1949; DIN, 1976). Figure 7.8a shows the two-phase separation of $2K_2SO_4-K_2O-2.1SiO_2$ system at 1575 K. The outside envelope is the transparent sulfate phase, through which the platinum wire heat (top) and a globule of

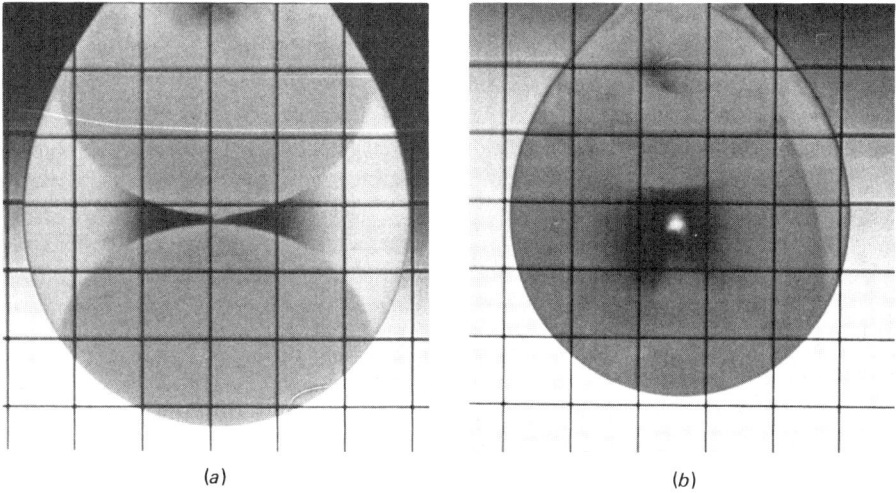

(a)   (b)

Figure 7.8 $2K_2SO_4-K_2O-2.1SiO_2$ droplet in heating microscope—0.5 mm grid. (a) Two phases at 1575 K. (b) Single phase at 1725 K.

molten silicate (bottom) can be seen. As the temperature was increased to 1725 K, the two phases became miscible (Fig. 7.8b) because of the increased solubility of sulfate in the silicate melt at the higher temperature.

The $K_2SO_4-K_2O-SiO_2$ phase diagram is depicted in Fig. 7.9, which shows that the system is miscible at 1575 K when the molar ratio of K_2O to SiO_2 is above 0.5. Less basic melts separate into two immiscible liquids. This is the case with most coal-ash slags where the molar ratio of basic oxides (sum of Na_2O, K_2O, CaO, and MgO) to SiO_2 is well below 0.5. Exceptions to this are the sodium- and calcium-rich ashes of some lignite and nonbituminous coals, which can have sufficient amounts of alkalis to form a single-phase melt of miscible sulfates and silicates, but there

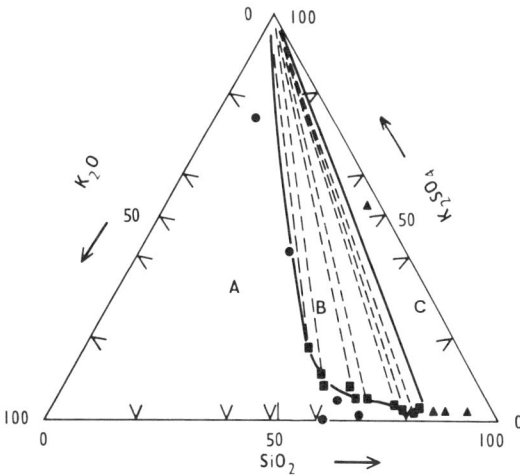

Figure 7.9 Miscibility gap in $K_2SO_4-K_2O-SiO_2$ system at 1575 K: A, single phase; B, two phases; C, liquid + solid SiO_2.

have been no reported cases where such boiler deposits have been found. This, however, does not rule out the possibility of the formation of a molten layer of sodium sulfate and silicate at the surface of ash particles. The surface layer of low fusion temperature would enhance the rate of sintering, as discussed in Chap. 10, and it would lead to a rapid build-up of boiler deposits with nonbituminous coals of high sodium content mined in the midwestern states of the United States, as discussed in Chap. 15.

The relevance of the flame-imprinted characteristics of the silicate and non-silicate species to boiler operation problems has been discussed previously by Raask (1981b), and is also discussed in Chaps. 10 to 14. Chapter 8 deals with the formation, coalescence, and properties of the flame-borne ash and fume particles.

Analytical Note

The sodium, potassium, and calcium contents of the sulfate phase in fly ash and boiler deposits can be determined by chemical analysis. The author has used a method where ground sample (100 mg) was dispersed in 100 ml of 0.02 molar $HClO_4$, boiled for 10 min, cooled, and filtered, and the filtrate was analyzed for the required elements. A weakly acidic solution was used to prevent hydrolysis of aluminum and iron sulfates (chiefly corrosion ingredients or products, Chap. 17). Perchloric acid solution was chosen as the extraction medium, as it does not interfere with sulfate and chloride determinations.

The extraction medium of dilute acid leaches out small amounts of alkali metals and calcium present in the silicate glassy material of the flame-heated fly ash and the initial deposit on boiler tubes (Chap. 12). The soluble silicate fraction can be determined by analysis of the filtrate for silicon and aluminum or by acid/alkali titration, as discussed by Raask (1982a).

NOMENCLATURE

A	thermodynamic constant
B	thermodynamic constant
D_v	vapor diffusivity
ΔG	standard free energy
L	latent heat of evaporation
M	molecular weight
M_C	atomic weight of carbon
M_S	atomic weight of sulfur
m	mass of decomposing particle
m_o	total sodium in ash
m_{sil}	sodium in silicates
m_{sul}	sodium in sulfates
dm/dt	rate of evaporation of decomposition
p	vapor pressure

p_o	saturation vapor pressure
dQ/dt	rate of heat transfer
R	thermodynamic gas content
r	radius of particle
Re	Reynolds number
S	surface area
T	temperature
T_o	temperature of heat source
V_{Cl}	volumetric Cl (chloride) concentration
V_{CO_2}	volumetric CO_2 concentration
V_K	volumetric potassium vapor concentration
V_{Na}	volumetric sodium vapor concentration
V_{SO_2}	volumetric SO_2 concentration
W_C	carbon content of coal
W_{Cl}	chlorine content of coal
W_{Na-sol}	water-soluble (evaporable) sodium content of coal
W_S	sulfur content of coal
W	ash content of coal
ϵ	emissivity of particle
σ	radiation constant
κ	constant
κ_1	constant
κ_2	constant
κ_3	constant

EIGHT

CREATION, CAPTURE, AND COALESCENCE OF PARTICULATE ASH IN BOILER FLAME

8.1 FRAGMENTATION BY THERMAL SHOCK AND BY GAS EVOLUTION

Mineral matter in pulverized coal is introduced into the boiler flame in different degrees of association with the combustible coal substance, as discussed in Chap. 4. That is, the distribution spectrum covers the range where the coal particles contain a few percent by weight of the inherent mineral matter; the middlings, which have substantial amounts of both coal and mineral matter; and the mineral particles with small amounts of coal substance.

Fracture disintegration of different coal minerals, free from carbon, on rapid heating has been studied by passing ground minerals through a vertical furnace and in a Leitz heating microscope (Raask, 1984a). It was observed that the silicate minerals, quartz, illite, and muscovite did not fracture when subjected to rapid heating in the laboratory furnaces. An exception was the behavior of chlorite mineral, which dissociated to high-melting quartz particles and a lower melting fraction (Fig. 6.5). Based on the results of these tests, it was concluded that no significant number of submicron-size particles were generated on rapid heating of the silicate minerals in the pulverized coal flame at temperatures up to 2000 K.

On combustion of the devolatilized coke particles, a shell of fused ash may be formed, encapsulating a large number of small particles of coal inherent ash in the form of plerospheres (Fig. 6.15). A gas evolution takes place inside the ash shell as a result of the reaction between hot carbon and the encapsulated ash particles. It was observed in the slag–coal interface studies (Chap. 6) that as a result of the gas eruption the thin shell of ash (Fig. 6.12) was ruptured leading to ejection of some particles. A similar occurrence of particle ejection was observed by Ramsden (1969). It was concluded from these experiments that the process of gas evolution is unlikely to create many new ash particles in pulverized-coal flame. Sarofim et al. (1977) have observed that on average only three to five ash particles were formed from the mineral matter present in each coal particle. Much of the inherent ash in coal is present in the form of fine particles of 0.1 to 0.5 μm in diameter, and several hundreds of these in "clean" coal and several thousands in the "middling" particles produced only a few ash particles, as discussed in Sect. 8.5. In contrast, Smith et al. (1979) and Smith (1980) have suggested that bubble bursting leads to the formation of fume particles found in pulverized-coal ash. It would be difficult to obtain quantitative experimental evidence for or against these conflicting claims, but it is worth noting here that even at high temperatures coal silicate melts have a much higher viscosity than that of water at room temperature, and viscous liquids are not easily disintegrated by bubble-bursting phenomena.

The nonsilicate materials–chiefly pyrites and carbonates, in particular iron carbonate–fracture and disintegrate when the particles are rapidly heated in laboratory furnace and also in the pulverized-coal flame, as discussed by Raask (1984a). Dissociation of particulate pyrites has been discussed in Chap. 7, and Fig. 7.3a shows a partial rupture of FeS residue as a result of vigorous evolution of sulfur gas before melting. On melting the fragments will readily coalesce, as will be illustrated in Fig. 8.5 (Sect. 8.5). Molten pyrite residue (FeS) is delineated on rapid cooling to segments of 0.5 to 3 μm in size, as shown in Fig. 8.1a. The fume particles, 0.1 to 0.5 μm in diameter, escaping on dissociation of FeS_2 were captured on a cooled target and are shown in Fig. 8.1b.

Iron carbonate minerals, siderite and ankerite, disintegrated extensively on rapid heating. Figure 8.1c shows a sintered residue of siderite after it had been heated to 1675 K. The residue material constituted about 20 percent by weight of the original material; the remainder was entrained in the carrier gas (nitrogen) in the form of a fume. Figure 8.1d shows that the deposited fume particles consisted of nonspherical particles 0.1 to 1.0 μm, together with a number of plate-shaped particles with a length-to-width ratio between 2 to 1 and 10 to 1. Their shape suggests that the particles had not gone through a melting phase. The submicron-size particles of iron oxide thus formed from carbonates and pyrites constitute a reactive fluxing material that dissolves in fused silicate ash, resulting in lowering of the viscosity of slag deposit. Further, the formation of fume particles of iron oxide explains the presence of magnetically separable particles, chiefly Fe_3O_4, in fine fractions of chimney emission solids (Raask and Goetz, 1981).

The residue formed on dissociation of calcite ($CaCO_3$) on rapid heating in a Leitz heating microscope is illustrated in Fig. 8.1e. It shows that particles 50 μm in diameter were fractured to segments of 10 to 20 μm across, separated by channels 1 to 2 μm wide. The segments appeared to consist of 0.2- to 0.5-μm particles

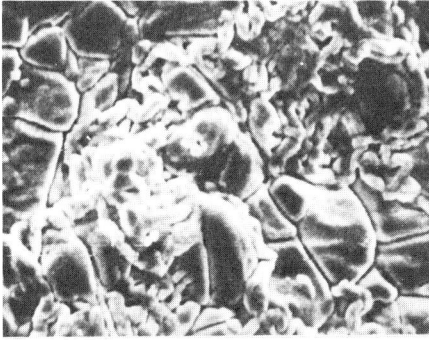

2 μm

(a)

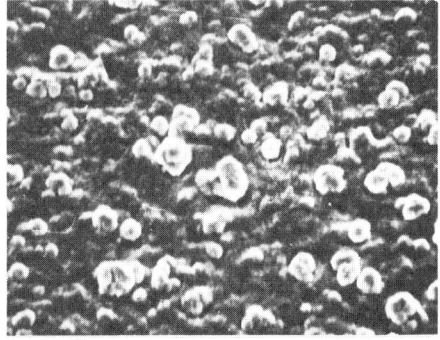

2 μm

(b)

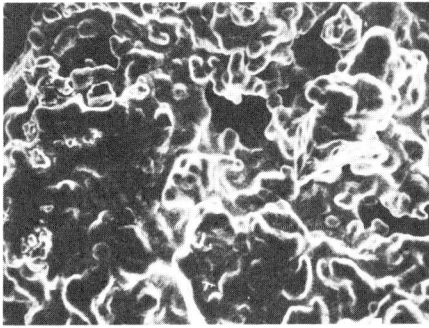

5 μm

(c)

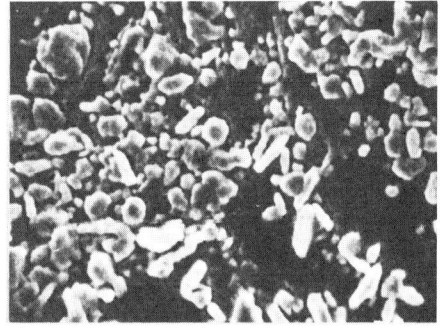

2 μm

(d)

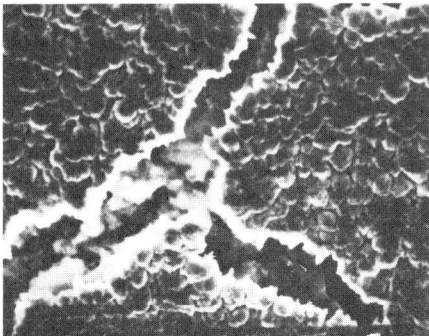

2 μm

(e)

2 μm

(f)

Figure 8.1 Particle fracture and fume formation on rapid heating of nonsilicate coal minerals. (*a*) Pyrite residue. (*b*) Pyrite fume. (*c*) Siderite residue. (*d*) Siderite fume. (*e*) Calcite residue. (*f*) Calcite fume.

sinter-bonded to form a matrix. Some of the submicron particles were entrained in the nitrogen carrier gas before sintering and were subsequently captured on a cooled target. Figure 8.1f shows a selection of the captured CaO particles, 0.05 to 0.5 μm in diameter. In the presence of sulfur gases—e.g., in boiler flue gas—these particles are sulfated, and Raask and Goetz (1981) have shown that calcium sulfate ($CaSO_4$) is present in high concentrations in the 0.1- to 0.5-μm-size fraction of the ash escaping the electrical precipitators.

8.2 FORMATION OF SILICA FUME FROM SILICON MONOXIDE VAPOR

Flagan and Friedlander (1976), commenting on the narrow size distribution of submicron-sized particles in pulverized-coal ash, considered that a substantial number of new particles had been produced by condensation. Desrosiers et al. (1978) have calculated that up to 0.05 percent of silicate ash constituents are volatilized in pulverized coal-fired boilers when the maximum flame temperature is 1960 K. In cyclone-fired boilers where the flame temperature can reach 2200 K, up to 0.8 percent of silicate material can be transferred into the vapor phase for subsequent fume formation. The above estimates of silica volatilization in the coal-fired boiler are higher than those cited in Chap. 6 by Raask and Wilkins (1965). Estimates based on their work suggest that only about 0.01 percent of silica was volatilized in pulverized-coal-fired boilers, and up to 0.1 percent in the high-temperature cyclone-fired boilers.

On the other hand, Gumz et al. (1958) and Ulrich et al. (1977) claim that substantial quantities of silica fume are produced in coal fired boilers. These different estimates are difficult to substantiate experimentally, for the parameters which govern the silica volatilization cannot be defined accurately. These include the degree of intermixing of carbon and silicate minerals in the pulverized coal particles, the degree of mixing of fuel and combustion air, and the residence time of mixed particles of carbon and ash at high temperatures. Raask and Wilkins (1965) have reported that a visible silica fume was emitted from an experimental gasifier when the reaction zone temperature exceeded 2000 K. Silica-fume emission from a stoker boiler has been reported by Murphy et al. (1957), but there have not been any observed occurrences of visible silica fume being emitted from pulverized-coal- or cyclone-fired boilers.

There has been a great deal of controversy about the structure of the material resulting from deposition of gaseous silicon monoxide. A widely accepted view is that silicon monoxide as a solid is unstable at all temperatures (Brewer and Green, 1957). On cooling in a neutral atmosphere, the monoxide decomposes into silica and silicon,

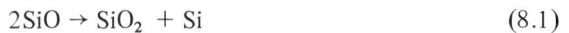

$$2SiO \rightarrow SiO_2 + Si \tag{8.1}$$

and in the presence of oxygen, water vapor, or CO_2, it oxidizes to silica:

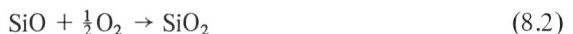

$$SiO + \tfrac{1}{2}O_2 \rightarrow SiO_2 \tag{8.2}$$

Raask and Wilkins (1965) showed that when pure silica was volatilized, the condensation product consisted of a white deposit of silica or a grey deposit of silica and silicon, but when coal-ash slag was volatilized, the silica fume was contaminated with aluminum, iron, calcium, and magnesium oxides. Coal-ash slag was heated in a graphite crucible in argon atmosphere containing a trace amount of oxygen, and the condensation deposit was collected from a cooler part of the furnace tube. The results of analysis of the deposit material are given in Table 8.1.

The high alumina content of sample B suggests that a substantial amount of $Al_2 O_3$ was volatilized at 2175 K and subsequently condensed or sublimed at about 1600 K. It is likely that volatile sub-oxides of aluminum were formed at high temperatures and condensed and reoxidized on cooling:

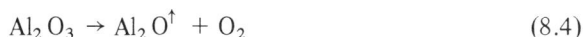

$$Al_2 O_3 \rightarrow Al_2 O_2^{\uparrow} + \tfrac{1}{2} O_2 \tag{8.3}$$

$$Al_2 O_3 \rightarrow Al_2 O^{\uparrow} + O_2 \tag{8.4}$$

The particles of silica fume formed on oxidation of silicon monoxide vapor [Eq. (8.2)] were spherical in shape and had a diameter between 0.015 and 0.15 μm (150–1500 Å units). The "primary" particles had a tendency to form chains and rings, as shown in Fig. 8.2a. SiO vapor in the experiments was produced by heating a silica glass rod in a carbon arc, and it was subsequently oxidized in air. When SiO vapor was oxidized in a graphite crucible as a result of slow ingress of argon containing 100 ppm oxygen, whiskers of 1 to 2 μm thick grew from the container wall, as shown in Fig. 8.2b.

Although volatilization of silicon monoxide and subsequent formation of silica fume may not be extensive in pulverized-coal-fired boilers, nevertheless, silicate minerals constitute the principal source for volatilized potassium (Chap. 6). Formation of the noncaptive submicron-size particles of potassium, sodium, and calcium sulfates and the captive species on silicate ash is discussed the next two sections.

Table 8.1 Composition of fume deposit from coal-ash slag–graphite reaction

Condition	Experiment	
	A	B
Temperature of slag in graphite crucible (K)	1875	2175
Deposit temperature (K)	1375–1475	1575–1675
Deposit constituents (weight percent)		
SiO_2	95	28
$Al_2 O_3$	<2	43
FeO	2.5	4
CaO	<1	2.5
MgO	4	4.5

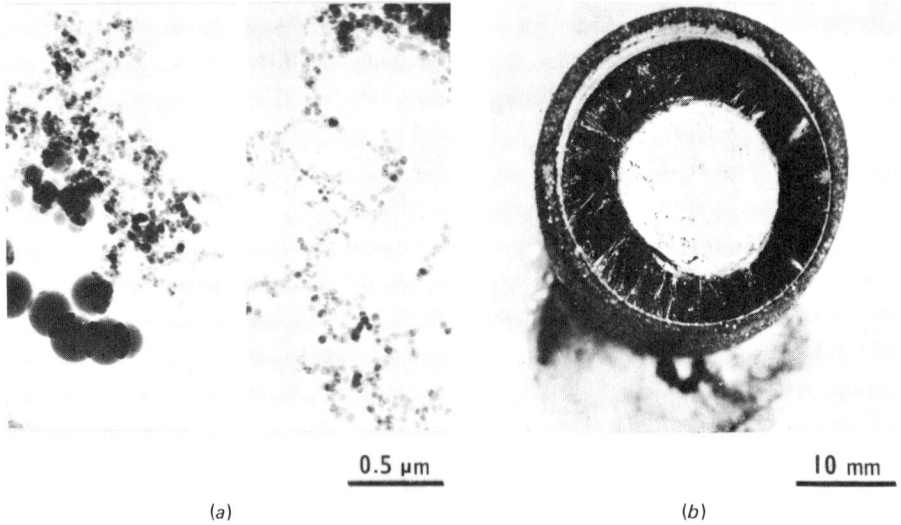

0.5 μm   10 mm

(a)   (b)

Figure 8.2 Oxidation products of silicon monoxide. (a) Silica fume. (b) Silica whiskers.

8.3 FORMATION OF SULFATE FUME
IN BOILER–FLUE GAS

Principal constituents of the fume formed as a result of reactions of the volatilized species in coal flames are sodium and potassium sulfates (Na_2SO_4 and K_2SO_4). These sulfates, together with calcium sulfate ($CaSO_4$), account for the majority of "new" submicron-size particles generated in pulverized-coal-fired boilers (Raask and Goetz, 1981). The concentration of sulfate can be estimated in terms of the weight percent of coal or that of ash, and also in terms of the number of submicron-size particles in a unit volume of the flue gas. The flue gas volume (V_g) produced at the atmospheric pressure from a unit weight of coal is given by

$$V_g = \frac{22.4 C_w T}{12 \times 273 \; V_{CO_2}} = 6.84 \times 10^{-3} \; \frac{C_w T}{V_{CO_2}} \tag{8.5}$$

where C_w is the weight percent of carbon in coal and V_{CO_2} is the volume percent of CO_2 in the flue gas at temperature T. The numerical values of 12 and 22.4 refer to the molecular weight of carbon and to the molecular volume of a gas at standard pressure (atmospheric) and temperature (273 K), respectively. The weight of a spherical particle (W) is given by

$$W = \frac{4\pi}{3} \left(\frac{d}{2}\right)^3 \rho_p = 0.523 \; d^3 \rho_p \tag{8.6}$$

where d and ρ_p are the diameter and density of the particle, respectively. Thus, the number N of sulfate particles produced from a unit weight of coal is given by

$$N = \frac{0.01 W_{sul}}{0.523 d_1^3 \rho_p} = \frac{0.019 W_{sul}}{d_1^3 \rho_p} \qquad (8.7)$$

where W_{sul} is the weight percent of sulfate produced on combustion and d_1 is the weighted mean diameter of the particles. The number of particles produced in a unit volume of the flue gas (N_V) is obtained by combining Eqs. (8.5) and (8.7):

$$N_V = \frac{0.019 W_{sul} V_{CO_2}}{(6.84 \times 10^{-3}) d_1^3 \rho_p C_w T} = \frac{2.78 W_{sul} V_{CO_2}}{d_1^3 \rho_p C_w T} \qquad (8.8)$$

A typical value of C_w, the weight percent of carbon in coal, is 65, and the concentration of CO_2 in the flue gas by volume (V_{CO_2}) of pulverized-coal-fired boilers is usually around 13 percent. The density ρ_p of the sulfate can be taken to be as 2700 kg m^{-3}. This value represents the density of $Na_2 SO_4$ and $K_2 SO_4$ at room temperature, but this is admissible when measurements of the diameter of sulfate particles are carried out at the same temperature. Equation (8.8) thus becomes

$$N_v = \frac{(2.06 \times 10^{-4}) W_{sul}}{d_1^3 T} \qquad (8.9)$$

where d_1 is the diameter of particles in meters and N_v is the number of particles in cubic meter of the flue gas. It is more convenient to have the diameter in μm units, so Eq. (8.9) becomes

$$N_v = \frac{(2.06 \times 10^{14}) W_{sul}}{d_2^3 T} \qquad (8.10)$$

A typical value of W_{sul} —i.e., the amount of sulfate fume ($Na_2 SO_4 + K_2 SO_4$) produced per weight percent of the original coal—may be taken as 0.4, and the temperature at which the fume particles are formed is around 1200 K. Further, a typical value of the mean diameter of the sulfate fume particles may be taken to be 0.12 μm from the measurements reported by Raask (1981a), and from Eq. (8.10) it is possible to calculate the number of particles: i.e., $4.0 \times 10^{13}/m^3$ flue gas. Table 8.2 shows the estimated number of submicron-size particles of sulfate formed in pulverized-coal-fired boilers when fueled with coals of different volatile alkali-metal contents (Table 6.3).

It should be noted that the estimates made in Table 8.2 of the number of alkali-metal sulfate fume particles refer to the maximum numbers that would have formed in the absence of silicate ash. There is some evidence to show that a significant number of the sulfate particles are formed directly on larger ash particles (Raask and Goetz, 1981) and do not therefore have an independent aerodynamic existence. Also, a significant amount of alkali-metal sulfate is deposited on cooled furnace wall and superheater tubes as the flue gas traverses the boiler system. The reduction in the number of alkali-metal fume particles as a result of capture by silicate ash and by cooled boiler tubes is partly balanced by the submicron-size $CaSO_4$ particles formed from the fragments of calcium carbonate (Fig. 8.1e). The calculated weight quantities of submicron-size fume (Table 8.2) are consistent with

Table 8.2 Formation of alkali-metal sulfate fume particles in coal flame

Coal[a]	Weight percent of coal		Sulfate particle concentration in flue gas at 1200 K (number/m^{-3}) for diameter (d) of particles		
	Volatilized sodium	Volatilized potassium	$d = 0.05\ \mu m$	$d = 0.1\ \mu m$	$d = 0.25\ \mu m$
1	0.069	0.072	5.1×10^{14}	6.4×10^{13}	4.1×10^{12}
2	0.083	0.092	6.3×10^{14}	7.9×10^{13}	5.1×10^{12}
3	0.093	0.080	6.3×10^{14}	7.9×10^{13}	5.1×10^{12}
4	0.185	0.114	1.1×10^{15}	1.4×10^{14}	9.1×10^{12}
5	0.037	0.052	3.2×10^{14}	4.0×10^{13}	2.5×10^{12}
6	0.058	0.115	6.0×10^{14}	7.6×10^{13}	4.8×10^{12}
7	0.181	0.117	1.1×10^{15}	1.4×10^{14}	9.0×10^{12}
8	0.169	0.177	1.3×10^{15}	1.6×10^{14}	1.0×10^{13}

[a]Coals 1, 2, 3, 5, and 6 are typical British coals of medium and high content of volatilizable alkali metals. Coals 4, 7, and 8 have exceptionally high amounts of the volatile alkalis.

the results of measurements of McElroy et al. (1982). They found that between 0.2 and 2.2 weight percent of the total ash was present in the form of submicron-size particles with a peak in the distribution curve at a particle diameter near 0.1 μm.

There have been several attempts made to predict the number and particle size of the aerosol particles formed in coal combustion flame. Notable among these is that of Flagan and Friedlander (1976), who used the self-preserving theory for a free molecular aerosol, developed earlier by Lai et al. (1972). It is described by the equation

$$\left(\frac{dN_s}{dt}\right)_{coag} = -\frac{1}{2}\left(\frac{3}{4\pi}\right)^{1/6}\left(\frac{6kT}{\rho_p}\right)^{1/2}\left(\frac{P}{RT}\right)I_1\, V_s^{1/6} N_s^{1/6} \qquad (8.11)$$

where N_s, V_s = number and volume of new particles, respectively, per unit mass of combustion products

ρ_p = density of particles

T, P = temperature and pressure of the system, respectively

k = Boltzmann gas constant

R = thermodynamic gas constant

I_1 = dimensionless coagulation integral with the value here of is 6.67.

The authors assumed that the coal combustor was a plug flow reactor with uniform composition and temperature at any position along its length. The flame temperature was taken to remain constant for 0.5 s, then decreased to 1400 K in 0.5 s, and subsequently to 425 K in 2 s. The authors further assumed that the condensible material accounted for 1.0 percent of the total ash, and that the fume particles were formed in an homogeneous process early in the combustion. They predicted that there would be 2.3×10^{14} particles/m^3 with an average diameter of 0.05 μm. Flagan (1978) has discussed the reasons for discrepancies that may arise between the theoretically arrived figures and those obtained by measurements.

8.4 CAPTIVE FORMATION OF SULFATE
ON THE SURFACE OF FLUE–GAS–BORNE ASH

Ash surface characterization studies by Raask and Goetz (1981), and Raask (1981a) showed that pulverized-coal ash captured in the electrical precipitators and the corresponding chimney-stack solids consisted chiefly of spherical or rounded particles of aluminosilicates 0.5 to 50 μm in diameter, and that these particles appear to carry a large number of submicron-size particles. The silica particles both in the electrical precipitator ash and in the chimney-stack solids seemed to constitute a platform for anchorage for the submicron-size captive particles, as shown in Fig. 8.3a. Natusch (1979) has shown that ash particles in pulverized-coal-

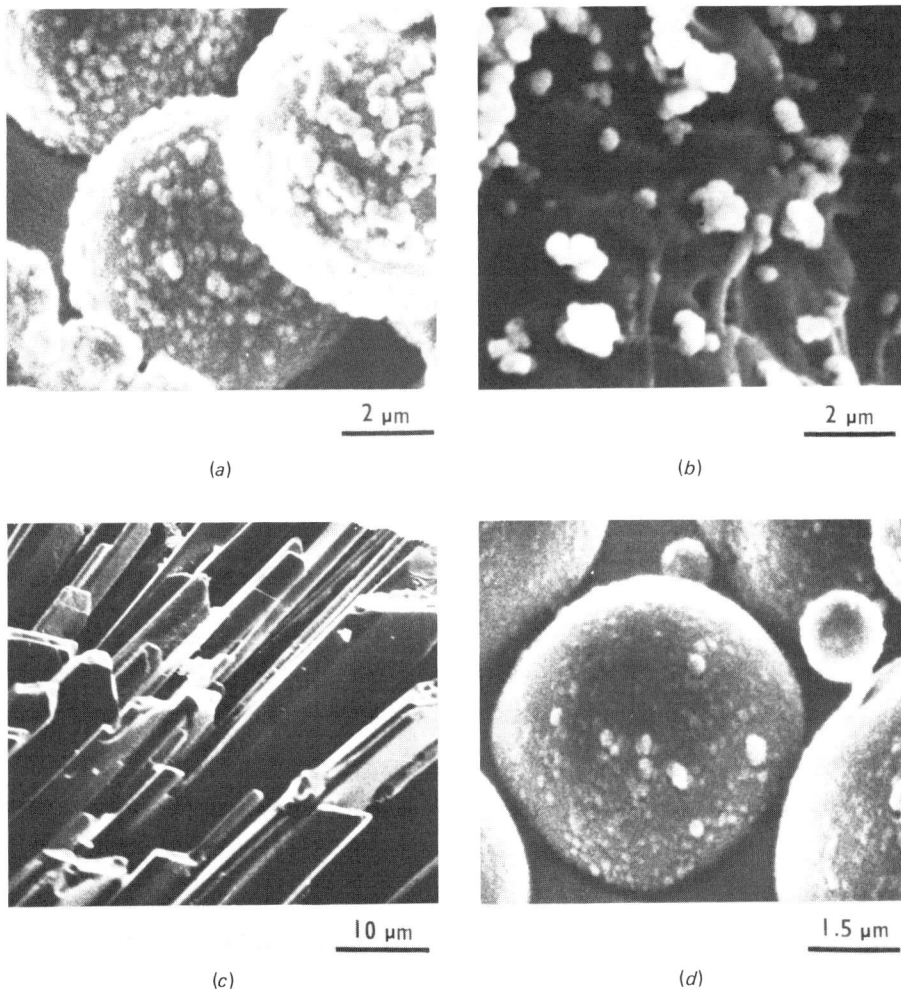

2 µm

(a)

2 µm

(b)

10 µm

(c)

1.5 µm

(d)

Figure 8.3 Sulfate fume in pulverized coal ash. (a) Alkali-metal sulfate on silicate ash. (b) Calcium sulfate-rich fume. (c) Calcium sulfate recrystallized from water. (d) Dilute acid-leached ash.

fired boilers acquire a sulfate-rich surface layer, and Raask and Goetz (1981) have established that K_2SO_4 and Na_2SO_4 were the dominant sulfate species at the surface of silicate ash particles. The surface layer of particles of pulverized-coal ash and chimney solids were analyzed to a depth of about 2 μm by the EDAX technique (Chap. 3). The results of analysis are given in Table 8.3, together with those of fume material (Fig. 8.3b) and the evaporation residue of water-soluble matter in ash (Fig. 8.3c).

The results in Table 8.3 show that concentrations of calcium in the ash surface layer was low, but in contrast, $CaSO_4$ was the dominant sulfate species in the chimney-fume material shown in Fig. 8.3b. Further, hydrated calcium sulfate shown in Fig. 8.3c was the principal ingredient of the residue left on evaporation of ash water leachate. Figure 8.3d shows that water and dilute acid (0.1 M HCl) "cleaned" the surface of ash particles—that is the submicron-size particles on the surface of ash as shown in Fig. 8.3a had been dissolved in the etching media.

A possible mechanism for the formation of sodium and potassium sulfate at the surface of ash particles is that vaporized alkali metals (Chap. 7) were first captured at the surface of fused silicate particles originally low in alkalis, e.g., kaolin particles. Subsequently, the alkalis in the form of oxides were sulfated by SO_2 and SO_3 in the boiler-flue gas. In classical chemistry notations, the sulfation reactions can be summarized as follows:

$$SiO_2 + 2NaCl + H_2O \rightarrow SiO_2 \cdot Na_2O + 2HCl \tag{8.12}$$

$$SiO_2 \cdot Na_2O + SO_3 \rightarrow SiO_2 + Na_2SO_4 \tag{8.13}$$

The fact that molten silicates and sulfates are immiscible (Figs. 7.8 and 7.9) means that the sulfation of alkalis at the surface of fused silica particles can take place at higher temperatures than would be predicted by the silicate–sulfate equilibrium curve (Fig. 7.7). The requirement is that the rate of transport of alkali-metal vapor, sulfur gases, and oxygen in the flue gas is greater than the rate of diffusion of alkalis from the surface to interior of the host silicate particles.

There may be some sulfation of nonvolatilized alkalis taking place at the

Table 8.3 Analysis of surface layer, fume, and soluble fractions of pulverized-fuel ash (weight percent)

| | Surface layer | | | |
| | Precipitator | Chimney | Fume material | Recrystallized soluble |
Constituent	ash	solid	of chimney solid	matter in ash
SiO_2	48.3	47.9	20.2	0.9
Al_2O_3	25.2	26.2	8.8	1.8
Fe_3O_4	2.2	6.5	7.8	0.1
CaO	0.4	1.1	12.8	26.8
K_2O	3.0	2.6	1.4	0.7
Na_2O	1.1	1.5	0.9	2.4
SO_3	10.0	14.3	43.5	61.0

surface of silicate particles originally rich in alkalis, e.g., illites. The sulfate reaction leads to a formation of a separate sulfate phase, and denudation of alkalis in the silicate surface layer is replenished by diffusion from the interior. This is in accordance with findings of Stinespring and Stewart (1981), who have studied the surface concentration of potassium and sulfate in heated illite particles by the Auger microprobe technique. They have established that heating of illite samples at 1475 K for 12 h resulted in a 10-fold enrichment of potassium and sulfate in a 0.01-μm-thick surface layer when compared with that of the interior.

The direct sulfation of nonvolatile alkalis is unlikely to be extensive in pulverized-coal-fired boilers, where the residence time of flame-borne particles is short (Fig. 6.1), but some K_2SO_4 could be formed by this route because practically all potassium is usually present in coal minerals as potassium aluminosilicates. In contrast, a significant fraction of coal sodium is volatilized rapidly in the flame and subsequently captured by silicate ash (Chap. 7). It is therefore to be expected that the ratio of sodium sulfate to potassium sulfate at the surface of different particles can vary markedly according to their former original composition, either alkali-free silicates or potassium aluminosilicates.

Microscopic counts of the submicron-size particles at the surface of silicate ash (Fig. 8.3a) gave an average concentration of 10 particles/μm^2 (10^{13} particles/m^2 ash surface) of the mean diameter of 0.12 μm. Estimates of the concentration of ash in the flue gas and that of the geometric surface (ratio of surface area to weight) gave values of 0.0044 kg m^{-3} and 200 m^2 kg^{-1}, respectively. This would give a value of $10^{13} \times 0.0044 \times 200 = 8.8 \times 10^{12}/m^3$ as the number of the submicron-size sulfate particles carried by the silicate ash in a cubic meter of the flue gas. The coal burned in the pulverized-fuel-fired boiler was type 2 in Table 8.2, and thus it estimated that a total of 3.7×10^{13} particles/m^3 flue gas at 1200 K was generated in the flame. Thus the fraction of sulfate fume carried by the host particles amounts to about 24 percent of the total generated. It appears that the remainder of alkali-metal sulfates and practically all calcium sulfate particles were formed independently from the captive and catalyzing surface of ash in boiler flame. The submicron-size particles (Fig. 8.3b) thus formed are not subsequently captured by the silicate ash. This is probably because in the absence of any significant thermophoresic and electrophoresic effects, there is no enhancing influence for transport of a fume to the surface of large particles. These effects can play a significant role in deposition of the submicron-size particles on cooled surfaces in the boiler-flue gas, and a preferential capture of small particles results in a marked change of the size distribution of the flue-gas-borne ash leaving the high-temperature boiler zone, as discussed in Chap. 12.

8.5 COALESCENCE OF THE INHERENT ASH IN BURNING COAL PARTICLES

Work published by Raask (1968a), Ramsden (1969), Fisher et al. (1976), and Sarofim et al. (1977) show that the mineral matter dispersed in coal particles—i.e., the inherent mineral matter in coal and "middling" particles—undergoes an extensive degree of coalescence on combustion of pulverized fuel. A coal particle of 50 μm diameter may contain 5 to 50 percent of the mineral matter (Chap. 4) in the form of a nonvolatile clay dispersion made up of several thousand particles 0.1 to 5 μm in size (Fig. 4.5). From these particles, a cenosphere or plerosphere of a diameter between 25 and 100 μm may form. Inside plerospheres there may be a large number of small particles, as shown in Fig. 6.15, b and c, but in aerodynamic behavior these constitute single particles. Also, there must have been a large number of ash plerospheres formed that coalesce subsequently to a single dense particle. This occurs when the combustible coal substance in an envelope of fused ash, which keeps the trapped ash particles apart, is all consumed before the plerosphere particle "freezes."

Pulverized coal ash usually contains 2 to 5 percent by weight of unburnt carbon in the form of large particles of coke residue, 10 to 100 μm in diameter (Fig. 4.2). The coke residue contains ash in two different forms. First, there are the spherical globules of slag at the surface, as shown in Fig. 8.4a, that had been prevented from coalescence by the carbon substance. Second, there is a lace network of fine ash inside the coke particles, revealed as a skeleton (Fig. 8.4b) when the combustible matter is burned off in air at 900 K. It is therefore evident that the coke residue particles found in pulverized-fuel ash had largely retained their inherent mineral matter, partly as an internal lace structure of sintered ash and partly as small slag globules attached to the surface of coke.

There is no evidence to show that a significant degree of agglomeration takes place as a result of collision of the flame-borne particles of silica ash and sulfate fume in pulverized-coal-fired boilers. Molten residue of pyrites appeared to be the only species of coal mineral impurities that had an observable tendency to particle-to-particle bridging when heated in the stream of nitrogen passing through a laboratory furnace. Figure 8.5a shows typical samples of bridged particles found in the residue, and occasionally remarkable chain formation of bridged particles could

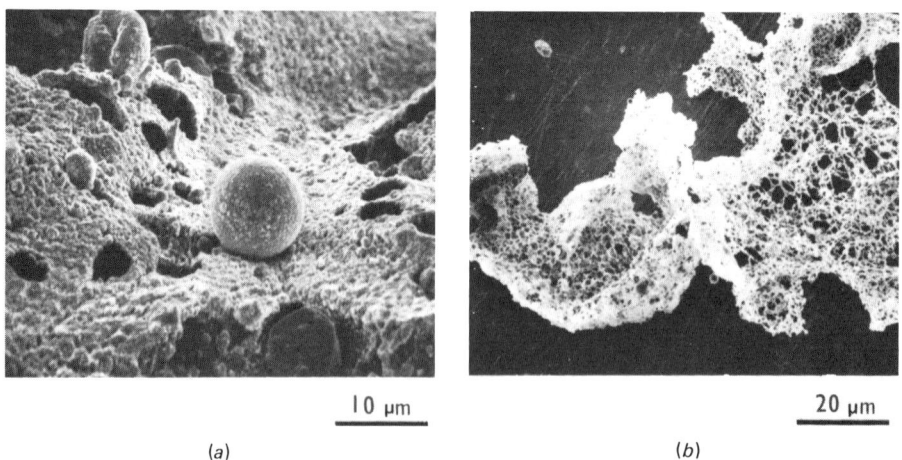

10 μm	20 μm
(a)	(b)

Figure 8.4 Slag beads and sintered ash in coke-residue particles. (a) Ash at the surface of coke particles. (b) Lace skeleton of sintered ash in coke.

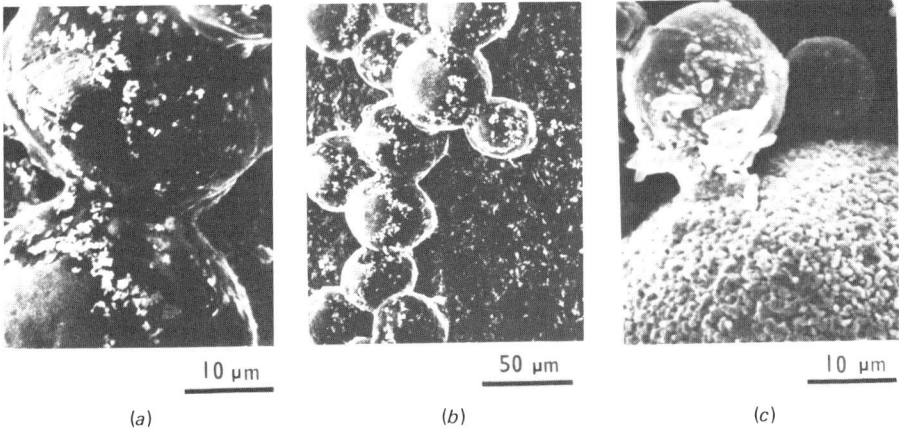

Figure 8.5 Sinter bridging of FeS particles when molten. (*a*) Couplet of FeS. (*b*) Catena of FeS. (*c*) Bridged iron oxide particles in ash.

be observed in the laboratory-heated material, as shown in Fig. 8.5*b*. Doublets of iron oxide, originating from pyrite particles, are occasionally found in pulverized coal ash, as shown in Fig. 8.5*c*. Finite time required for the formation of an effective particle-to-particle bonding by viscous flow as a result of action of surface tension is defined by the Frenkel equation given in Chap. 10 [Eq. (10.2)]. Molten droplets of pyrite residue will have a low viscosity, probably below 10 N s m^{-2}; according to the Frenkel equation, the time thus required for the formation of the degree of bonding shown in Fig. 8.5 would be less than 1 ms.

Regarding the behavior of main constituents of the flame-borne ash, namely silicates and sulfates, it was shown in Sect. 8.3 that the average distance between ash and fume particles in boiler flue gas is likely to be between 10 and 100 μm. It appears therefore remarkable that the particles are unable to traverse such short distances to agglomerate and to coalesce to larger sizes at high temperatures. One of the reasons could be that the particles acquire electrostatic charge of the same polarity, preventing collisions of the low-mass particles. Further, it has been shown (Chap. 7) that many of the sulfate fume particles—chiefly $CaSO_4$, $Na_2 SO_4$, and $K_2 SO$—are formed in the boiler-flue gas at temperatures below their melting points. It is therefore to be expected that when particle collisions occur these do not result in an effective bond being formed.

Table 8.4 summarizes the formation of "new" ash particles, and capture and coalescence of the existing and flame-created particles in pulverized-coal- and cyclone-fired boilers.

8.6 SIZE CHARACTERISTICS OF GAS–BORNE ASH IN PULVERIZED–COAL– AND CYCLONE–FIRED BOILERS

Coalescence of ash and the flame-created fume particles occurs subsequent to deposition on furnace wall and high-temperature superheater tubes, in both

Table 8.4 Creation, capture, and coalescence of ash particles in boiler flames

	Pulverized-coal firing	Cyclone firing
Fuel and firing regime		
Coal milling	Fine, below 100 μm in size	Coarse, up to 500 μm in size
Maximum flame temperature, K	2000	2150
Furnace-slag surface temperature, K	1650	1800
New particle creation		
Particle fracture on heating	Some disintegration of carbonate and pyrites; no significant fracture of silicate particles	As in pulverized-coal-fired boilers
Silicate-melt bubble bursting	Some ejection of molten globules on gas evolution, but number of particles thus created is not significant	The gas evolution at coal–slag interface takes place at the surface of furnace-wall slag
Volatilization and condensation	Extensive sulfate fume formation related to the amount of volatilized sodium; no significant silica fume formation	Sulfate fume formation as in pulverized-coal-fired boilers; some silica volatilization under malcombustion (reducing) conditions
Capture and coalescence		
Coalescence of flame-borne particles and formation of ceno-spheres and plero-spheres	Extensive coalescence of inherent ash of 0.1 to 5 μm in size to form cenospheres, plerospheres, dense particles, and laced skeletons	Coalescence of flame-borne particles less extensive; it takes place mainly at the slag surface on boiler walls
Amount of ash capture on wall and super-heater tubes, weight percent	15–30	75–85
Selectivity of ash deposition and capture	Particles of 0.5 to 5.0 are preferentially retained on deposition; there is an enrichment in sulfates in the deposit material	Large particles captured by molten slag on furnace walls; ash and sulfate deposition on superheater tubes as in pulverized-coal-fired boilers
Ash characteristics	Contains 0.1 to 4 percent by weight of large cenosphere particles and fume particles 0.05 to 0.25 μm in size	Concentration of cenospheres and plerospheres is low, but fume particle concentration is high

pulverized-coal- and cyclone-fired boilers. Table 4.5 showed that 20 to 80 percent of weight of total ash is deposited on boiler tubes, where it coalesces and is subsequently discharged from the combustion chamber as sintered clinker or molten slag. The degree of agglomeration may be even higher than that shown by the above percentage figures. Soot blowing with high-velocity steam or air jets breaks up lightly sintered deposits on boiler tubes and the debris, particles up to a few millimeters in diameter, are entrained in the flue gas. Subsequently, these large

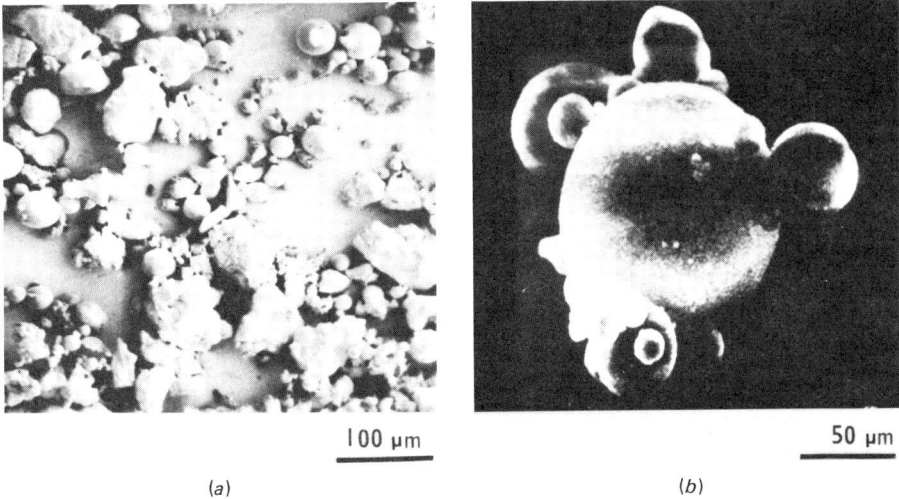

100 μm

(a)

50 μm

(b)

Figure 8.6 Coarse ash particles in coal-fired boilers. (*a*) Boiler grit. (*b*) Segment of boiler deposit dislodged by soot blowing.

particles, shown in Fig. 8.6, *a* and *b*, settle in boiler passages and in the first stage ash collectors: the material is known as boiler grit.

Not only does the formation of boiler deposits significantly reduce the total ash burden of the flue gas, but it also can markedly change the size distribution spectrum of the particles. This is because deposition and retention of different-size particles on cooled boiler tubes can be highly selective. As discussed in Chap. 12, in pulverized-coal-fired, "nonslagging" boilers, the ash particles in the size range of 0.5 to 5.0 μm are preferentially retained in the deposited ash, whereas large particles have sufficient rebound kinetic energy to escape the noncaptive surface. Further, a significant fraction of alkali-metal sulfates is deposited from the flue gas on cooled boiler tubes, resulting in the formation of sulfate-rich deposits. Correspondingly, the number of sulfate particles formed in the flame is subsequently reduced during passage of the flue gas through the boiler-tube banks. There remain, however, sufficient quantities of uncaptured fume to give the particle concentration numbers per unit volume of the flue gas at 400 K similar to, or higher than, those given in Table 8.2. The latter data were calculated for the flue gas temperature at 1200 K, and by the time the gas reaches the electrical precipitator it is contracted in volume by a factor of three.

In cyclone-fired boilers, the large mineral and coal particles are captured by the layer of molten slag on the furnace-wall tubes. In this system, the secondary combustion air is fed into the furnace tangentially and the large particles—coal and mineral matter—are thrown by the centrifugal force to the highly captive surface of molten slag on the walls of the combustion chamber. Smaller ash particles below 10 μm in diameter do not have sufficient kinetic energy to reach the slag surface and will remain entrained in the flue gas. The high flame and slag temperatures in the

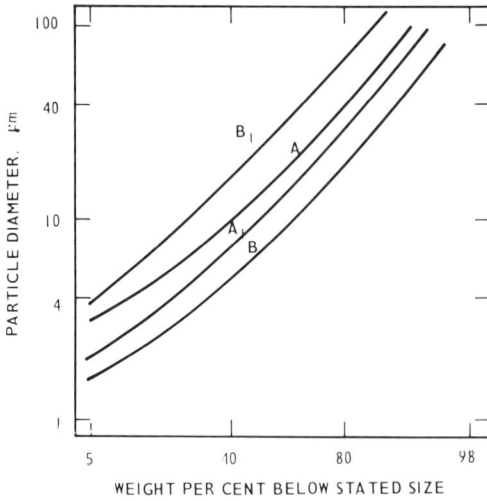

Figure 8.7 Particle size of coal ashes: A, pulverized-coal boiler ash; A_1, pulverized-coal laboratory ash; B, cyclone boiler ash; and B_1, cyclone coal laboratory ash.

cyclone-fired boiler result in a more extensive volatilization of alkali-metal and silica species. Subsequently the fume particles below 1 μm in diameter will be formed from the volatile species. It is therefore to be expected that the flue-gas-borne ash reaching the electrostatic precipitators has a significantly smaller mean particle size when compared with that of pulverized-fuel ash.

Curves A and B in Fig. 8.7 show the particle size distribution of typical pulverized-coal and cyclone ashes, respectively. The size distributions of ashes produced from pulverized coal and the coarser fuel of cyclone boiler when these were burned in a laboratory furnace at 1000 K are shown by curves A_1 and B_1, respectively. The difference in the two pairs of curves A and A_1, B and B_1, demonstrates the effect of high-temperature flame reactions and selective deposition on the particle size of ash remaining entrained in the flue gas. The relevance of flame-imprinted characteristics of the gas-borne ash to boiler slagging, corrosion, and erosion has been discussed by Raask (1981c).

NOMENCLATURE

C_W	carbon content of coal
d	diameter of particle
d_1	mean diameter of particles in meter
d_2	mean diameter of particles in μm
I_1	dimensionless integral
k	Boltzmann gas constant
N	number of particles
N_s	number of particles per unit mass of combustion products
N_v	number of particles in unit volume

P	gas pressure
R	thermodynamic gas constant
T	temperature
V_{CO_2}	CO_2 content of flue gas
V_g	gas volume
W	weight of particle
W_{sul}	flame-formed sulfate, weight percent of coal
ρ_p	particle density

NINE

SLAG VISCOSITY

9.1 METHODS OF VISCOSITY MEASUREMENTS

There are many high-temperature processes—e.g., manufacture of ceramic, glass, and refractory products, enamel coatings, and formation of metallurgical process slags and coal combustion slags—where the viscosity parameter is a useful criterion when assessing the sintering, slagging, and flow properties of these materials. The measurable viscosity of glassy materials can extend over a wide range, more than 10 orders of magnitude. For example, the viscosity of a standard soda–lime–silica glass (U.S. Bureau of Standards) has been determined in the range of 10 to 10^{14} N s m^{-2} (Napolitano and Hawkins, 1964). In the work concerning coal-ash sintering, slagging, and slag flow, it may not be necessary to cover such a wide viscosity range, but discernible effects can be observed within 10 orders of magnitude, as discussed in Sect. 9.4.

There is no single method capable of measuring the viscosity of coal-ash slags over the entire range of 10 orders of magnitude. Studies of metallurgical slags by Dantuma (1928), Endell et al. (1936), and Bockris and Lowe (1953), and of glasses by Lillie (1929) and by Robinson and Peterson (1944) have demonstrated the usefulness of a rotating crucible viscometer for obtaining data on the flow characteristics of high temperature melts. Bockris et al. (1959) describe the construction of a rotating-crucible viscometer where the crucible containing the melt is rotated at constant angular velocity Ω. The torque τ produced on the concentric inner cylinder, suspended via a partly immersed connecting spindle and torsion wire, is given by

$$\tau = 4\pi\eta\Omega \frac{a^2 b^2}{b^2 - a^2} (L + L_1 + L_2)$$ (9.1)

where η = viscosity

a = radius of the inner cylinder

b = radius of the outer crucible

L = depth of immersion of the inner cylinder

L_1, L_2 = end corrections due to the finite length of the inner cylinder and the immersed connecting spindle

The torque τ in a torsion wire equals $k\theta$, where θ is the angular displacement, and the equation for the viscosity η is

$$\eta = \frac{k_1 \theta}{\Omega}$$ (9.2)

where k_1 is the apparatus constant, which is usually obtained by calibration with oils of known viscosity.

Reid and Cohen (1944) have described a rotating viscometer using a stationary crucible and rotating cylinder both made from platinum–rhodium alloys. With the apparatus viscosities in the range of 10^{-2} to 10^3 N s m^{-2} could be determined both in oxidizing and reducing atmospheres, but with coal ash slags in conventional boilers, viscosities below about 5 N s m^{-2} are not usually measured. Shaw (1961) and Watt and Fereday (1969) have described a rotating viscometer used to determine the viscosity characteristics of British coal ashes. They used a molybdenum crucible (rotating) and cylinder (stationary), and the upper temperature limit of the apparatus was 2100 K, whereas the top temperature of a platinum–rhodium measuring assembly is around 1900 K. However, the molybdenum viscometer and furnace assembly need to be protected against oxidation by a purge of forming gas (10 percent hydrogen by volume, balance nitrogen), and thus measurements cannot be made in an oxidizing atmosphere.

The viscosity of less fluid slags in the range of 10^3 to 10^8 N s m^{-2} can be measured by a rod penetration viscometer. The technique was developed by Kelley et al. (1964) for measuring the viscosity of glasses, and the method does not require precalibration. In this sense it is an absolute method where the viscosity of glass can be calculated directly from the measurement:

$$\eta = \frac{F(\ln L/a - 0.72)}{2\pi v L}$$ (9.3)

where F = force that drives the rod into the melt

L = depth of rod penetration

a = radius of the rod

v = rate of penetration

The penetrator rod usually takes the form of a flat-bottom needle, 1.0 to 1.5 mm in diameter, made from an alloy of platinum and rhodium, or from alumina, mounted in an alumina rod carrier. Compared with the rotating cylinder viscometer, the rod penetrator instrument is less cumbersome, and it is easier to construct and operate. Boow (1969), Kiss et al. (1972), and Gibb (1981) have used the rod penetration method for measuring the viscosity of Australian and British coal ashes.

The viscosity measurements with the rod penetrameter become slow above 10^8 $N s^{-1} m^{-2}$ and cannot be speeded up by placing heavier loads on the small penetrating rod. In glass technology, methods of fiber elongation for measurements of high viscosities have been developed by Lillie (1931), Norton (1935), Boow and Turner (1942), Robinson and Peterson (1944), and Poole (1949). The technique involves stretching glass fibers, 0.5–1.0 mm in diameter and about 100 mm long, under loads ranging from 0.05 to 1.0 kg. The equation for determining the viscosity η of glass fibers of length L has been derived by Lewis et al. (1942):

$$\eta = \frac{FL}{3A(dL/dt)} \tag{9.4}$$

where F = stretching load

A = cross-sectional area of fiber

dL/dt = rate of stretching

The fiber stretching method can be used to determine the viscosity of glasses in the range 10^8 to 10^{14} $N s^{-1} m^{-2}$, but so far the technique has not been applied for measurements of the viscosity of coal-ash slag. It is difficult to prepare suitable fibers of coal ash slag and would require specialized high-temperature fiber-drawing equipment.

Another, probably more useful, method of measurement of high viscosities of glasses and coal-ash slags is one that utilizes the basic equation of particle-to-particle sintering by viscous flow, derived by Frenkel (1945):

$$\eta = \frac{1.5r\gamma t}{x^2} \tag{9.5}$$

where η = viscosity

x = radius of the neck grown between two spherical particles of radius r after time t

γ = surface tension

Kuczynski (1949a, b) has used the sintering technique for measuring the viscosity of glasses. Glass spheres, 50 μm in diameter, were placed on a glass plate of the same composition and heated in a furnace for periods up to 4 days. The diameter of neck between the glass sphere and plate was measured by a microscope after cooling to room temperature. Kiss et al. (1972) and Raask (1973) have used a similar technique for measuring the rate of sintering of coal-ash slag. The technique and the method of preparing spherical particles of slag are described in Chap. 10. The surface tension of coal-ash slag required to calculate viscosity from the results of sintering measurements was obtained by the sessile drop method, as described in Chap. 6.

Viscosities that can be measured by the droplet technique extend from 10^7 to 10^{11} $N s m^{-2}$ and cover the range where the particle-to-particle sintering by viscous flow commences as an initial stage in the formation of boiler deposits. Further, the results of viscosity and surface-tension measurements with spherical slag particles can be related directly to the rate of sintering in the Frenkel equation, as discussed in Sect. 9.4 and in Chap. 10.

9.2 THE NEWTONIAN AND NON-NEWTONIAN FLOW CHARACTERISTICS OF VISCOUS MELTS

The viscosity η of glasses decreases exponentially with increase of temperature T in the form

$$\eta = \frac{A_1 \exp E_\eta}{RT} \tag{9.6}$$

where A_1 is a preexponential constant and R is the thermodynamic (gas) constant. E_η is referred to as the activation energy for viscous flow and may be thought of as an energy barrier that must be overcome by molecular clusters or chains of the liquid moving relative to one another. It is usual to express Eq. (9.6) in its logarithmic form, and for many glasses the viscosity data fit the equation derived by Fulcher (1925):

$$\log_{10} \eta = A_2 + \frac{B}{T - T_o} \tag{9.7}$$

where A_2 is a constant, B is the temperature coefficient (energy of activation), and T_o is a temperature correction required to fit the data in a straight-line plot of log of the viscosity against the inverse of temperature (T). For soda–lime–silica glasses, the temperature correction is usually small; it was 7 K in the viscosity measurements made by Napolitano and Hawkins (1964).

By contrast, coal-ash slags usually show distinctly non-Newtonian flow characteristics when the viscosity exceeds about 10 N s m^{-2}, as shown on Fig. 9.1 (BCURA, 1963). There are, however, some ash slags where the viscosity of the Newtonian flow range exceeds 1000 N s m^{-2} (Nicholls and Reid, 1940). The temperature at which a sudden transition in viscosity occurs, as shown by curve A in Fig. 9.1, is called the temperature of critical viscosity. The rapid increase in viscosity on cooling below this temperature is a result of crystallization. Figure 9.2 shows the clusters of mullite needles dispersed in a porous slag; dark circles are gas-bubble holes. Figure 9.2b shows that the needles were 2 μm in diameter, with the ratio of length to diameter between 1 to 1 and 1 to 10.

Once the crystalline material is formed in a coal-ash melt, its rate of dissolution on reheating is sluggish and a viscosity–temperature plot shows a hysteresis effect on thermal cycling (curve B, Fig. 9.1). Similarly, when a solid material, soluble in slag, is added in the form of fine powder, the melt requires over an hour before the viscosity acquires a steady value. Figure 9.3a shows that the viscosity of the melt increased over a period of 20 min as bentonite powder was stirred into the melt. Subsequently the viscosity decreased exponentially to its final value, and this

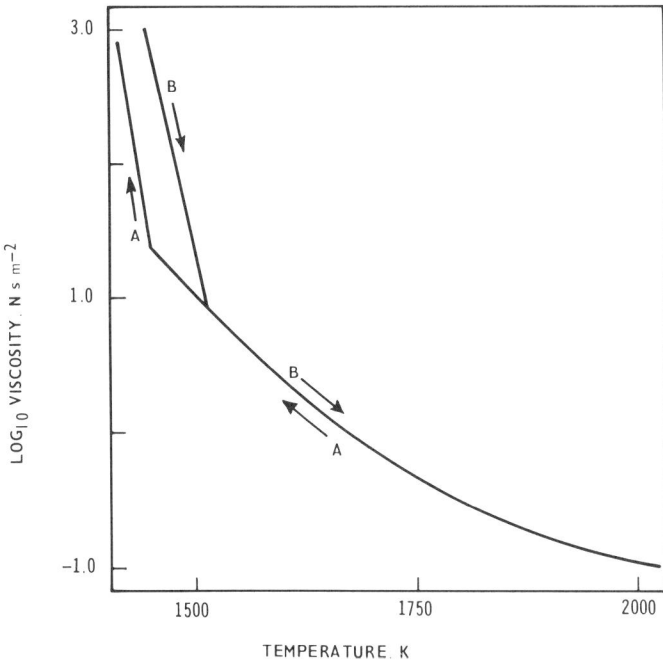

Figure 9.1 Viscosity of a typical coal-ash slag: A, on cooling; B, on reheating.

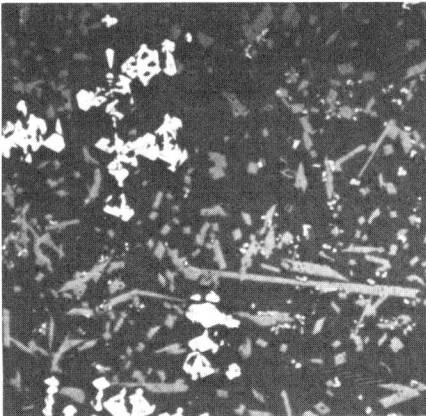

(a)

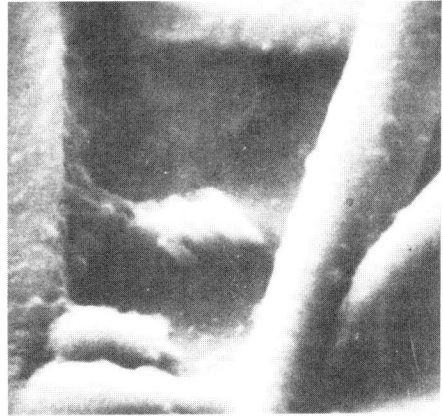

(b)

Figure 9.2 Crystallization of mullite in coal-ash slag. (a) Mullite needle clusters (white) in slag. (b) Detailed view of mullite needles.

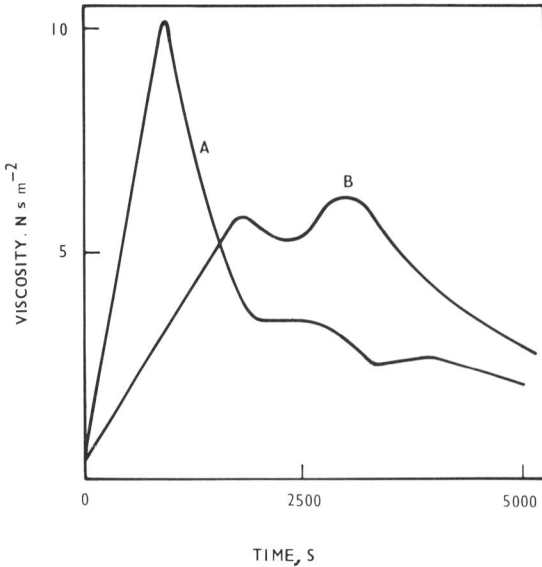

Figure 9.3 Change in viscosity on dissolution of additives in coal-ash slag: A, 21% by weight bentonite at 1745 K; B, 21% by weight muscovite at 1725 K.

suggested that the dissolution of solids in coal ash slags is a diffusion-controlled process. Figure 9.3b shows that muscovite powder dissolved more sluggishly in the melt, and that the final viscosity was higher than that with bentonite powder. This was to be expected because muscovite contains potassium and addition of potassium to aluminosilicate melts may increase the viscosity, as discussed in Sect. 9.3.

At higher viscosities there are limited ranges where some coal ash slags appear to have Newtonian flow characteristics as judged by a near-linear plot of the logarithm of viscosity against inverse of temperature, as shown on Fig. 9.4 (Kiss et al., 1972). This is likely, however, to be a fortuitous occurrence, and the results depend on the method of measurements. The results given in Fig. 9.4 were obtained with a coal ash of high iron content (slagging ash) and the measurements were made by a platinum-needle penetrometer, as described in Sect. 9.1. It appears that where the degree of crystallization is not excessive, the glassy layer at the penetrating rod–slag interface behaves as a Newtonian liquid. However, when a massive crystallization occurs in slag melts, the rod penetration method cannot be used to measure viscosity (Boow, 1969).

Figure 9.5 shows the results of determination of the viscosity of two slags of typical British coals by measuring the rate of neck growth between spherical particles (Raask, 1973). The lower viscosity values obtained with coal B (curve B) reflected correctly its higher deposit-forming propensity compared with that of coal A when burned in a pulverized-coal-fired boiler. The fact that the same results were obtained when the viscosity measurements were made in air or in hydrogen (curve B) suggests that at these high viscosities the rate of sintering was not influenced by the external atmosphere.

The laboratory measurements of slag viscosity are carried out under isothermal conditions, but in boiler plant there is a temperature gradient across the layer of deposits on the heat-exchange tubes. As a result of the temperature gradient, there is also a viscosity gradient across the ash deposits as discussed in Sect. 9.4.

9.3 THE RELATIONSHIP BETWEEN SLAG VISCOSITY AND COAL-ASH COMPOSITION

In order to explain the relationship between the viscosity of silicate slags and their composition, it has been suggested that silicate anions constitute a network where cations such as Na^+, K^+, Ca^{2+}, Mg^{2+}, and Fe^{2+} are distributed between the anion units (Mackenzie, 1957; Gumz et al., 1958; Kozakevitch, 1959; Turkdogan and Bills, 1960). Bockris et al. (1955) have suggested that the anion units are present in the form of mixtures of chains and rings depending on the nature and concentration of cations. A miscellany of rings and cages in high-silica and low-alkali slags results in an increase of viscosity. Addition of alkalis and other cations will reduce the viscosity by breaking Si–O–Si bonds and thus producing smaller silica units.

Where alumina is present, as in all coal-ash slags, it can play an amphoteric role (Kozakevitch, 1959; Turkdogan and Bills, 1960). In highly silicious systems it markedly reduces the viscosity; the species present are Al^{3+} ions. In alkaline melts a part of alumina adds on to the silicate anions as AlO^- units, which are isomorphous with SiO_2. The length of the silicate-ion chains and the size of the rings are thus increased resulting in higher viscosities. Addition of the alkalis or alkaline-earth oxides should decrease the viscosity of aluminosilicate melts, and this is usually the case (Eitel, 1954; Mackenzie, 1957). However, the effect of potassium oxide can be an exception to this rule. Endell and Zauleck (1950) and Glover (1969) have shown that potassium in aluminosilicate slags increases the viscosity rather than decreases

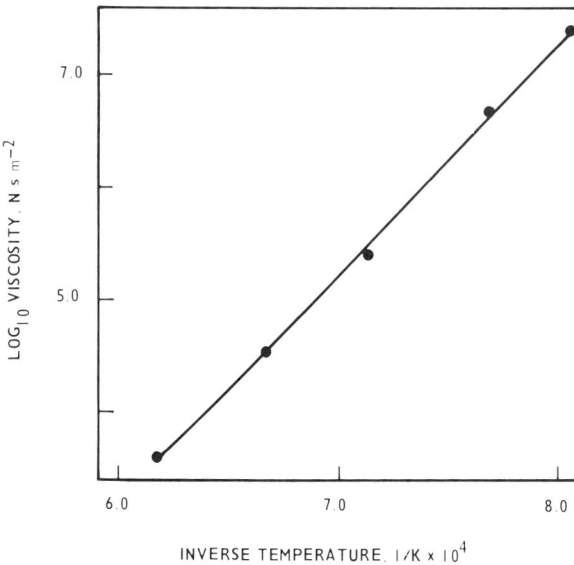

Figure 9.4 Viscosity of coal-ash slag measured by needle penetrometer.

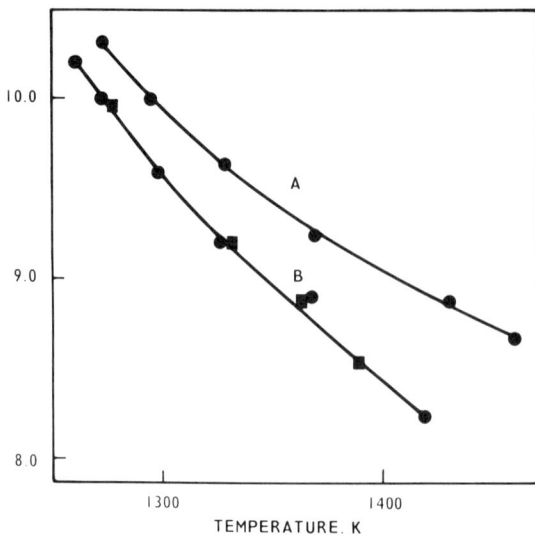

Figure 9.5 Viscosity by particle sinter-bond measurements with ashes of low (A) and high (B) deposit-forming propensity: —•— in air, —■— in reducing gas.

it. That is, instead of breaking the silicate chains or rings to smaller units potassium increases the complexity of aluminosilicate ions by changing the concentration ratio of AlO_2^- ions (network extenders) to Al^{3+} ions (network modifiers):

$$2K_2O + Al^{3+} \rightarrow AlO_2^- + 4K^+ \qquad (9.8)$$

When large amounts of potassium oxide, up to 20 percent by weight, were added to a typical bituminous coal ash slag, there was a marked increase in the viscosity as discussed by Glover (1969). His experimental results were relevant to the MHD system (Heywood and Womack, 1969), where potassium is needed as a "seed" material in order to increase the electrical conductance of the high temperature flame gases. For the conventional coal-fired boilers it is considered that the typical amounts of potassium, 0.5 to 3 percent K_2O of ash, do not have a significant effect either in increasing or decreasing the slag viscosity. This may not be the case, in particular with nonbituminous coal and lignite ashes, which have a high sodium content and a low silica-to-alumina ratio as shown by the data published by Duzy (1965), O'Gorman and Walker (1972), and Moore and Ehrler (1973).

Iron oxides can have a predominant influence on the viscosity and flow characteristics of coal-ash slags. The iron content of coal ash, expressed as Fe_2O_3, can vary from a few percent by weight to over 30 percent, and iron in coal minerals is present in many different species, chiefly in pyrites and carbonates, but also in silicates as a substitution element as discussed by Malden and Meads (1967); in nonbituminous coals a significant quantity of iron may be present in organometal compounds. Uniform distribution of iron in boiler slags is facilitated by formation of the reactive intermediate products on decomposition of pyrite and carbonate, namely a mobile liquid phase of FeS residue and fine particle size of FeO from carbonate (Chaps. 7 and 8).

Iron oxides in coal ash slags constitute a "fluxing" agent that can be highly sensitive to the external atmosphere. In a neutral or in mildly reducing atmosphere at high temperatures, iron in the form of FeO is a strong fluxing agent; the effect on the viscosity is comparable with that of CaO. In an oxidizing atmosphere the ferric iron may behave as a silicate network extender in a manner similar to that of Al_2O_3. The slag viscosity will also increase as a result of crystallization of iron spinels and iron-containing silicates.

In a strongly reducing atmosphere at high temperatures, iron can separate out as liquid metal, usually alloyed with silicon, and this occasionally occurs in cyclone-fired boilers under conditions of malcombustion (Gumz et al., 1958). The behavior of iron in coal-ash slags can be summarized as follows:

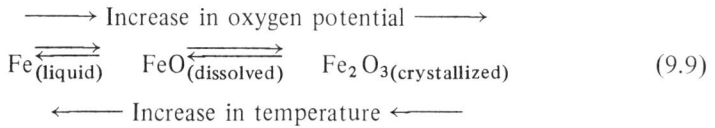

$$\longrightarrow \text{Increase in oxygen potential} \longrightarrow$$

$$\text{Fe}\overrightarrow{\underset{(\text{liquid})}{\longleftarrow}} \quad \text{FeO}\overrightarrow{\underset{(\text{dissolved})}{\longleftarrow}} \quad \text{Fe}_2\text{O}_{3(\text{crystallized})} \qquad (9.9)$$

$$\longleftarrow \text{Increase in temperature} \longleftarrow$$

The effect of atmosphere on the viscosity of coal-ash slags has been demonstrated in the results of Nicholls and Reid (1940). The authors have shown that when the measurements were carried in the viscometer purged with air, high concentrations of iron in the ferric state induced crystallization in slag, resulting in a rapid increase in viscosity below the critical temperature, as shown by curve A in Fig. 9.6. Curve B shows that the critical temperature, which is also defined as slag-flow temperature, was decreased from 1670 to 1415 K.

Numerous empirical expressions have been derived that relate the viscosity of coal-ash slags to composition; the best known of these is by Reid and Cohen (1944). They published their findings in the form of a monogram relating viscosity,

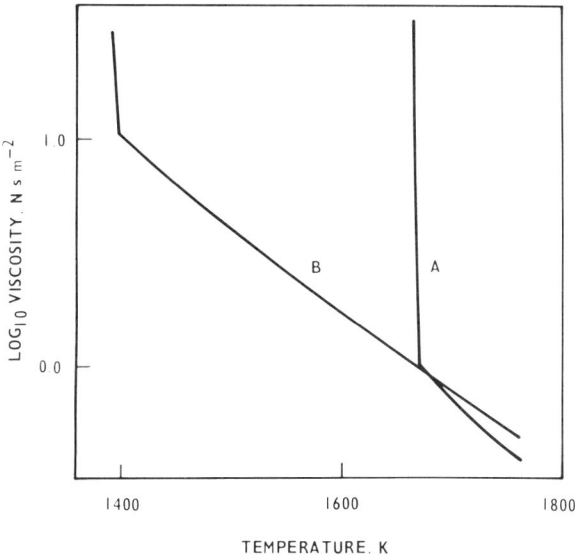

Figure 9.6 Effect of atmosphere on viscosity of iron-rich slag: A, in air; B, in nitrogen.

temperature, and composition of slags of a number of U.S. coals. The slag composition was given in terms of the silica ratio (SR):

$$SR = \frac{100 \; SiO_2}{SiO_2 + equiv. \; Fe_2O_3 + CaO + MgO} \tag{9.10}$$

where in terms of weight percentages,

$$SiO_2 + equiv. \; Fe_2O_3 + CaO + MgO = 100 \tag{9.11}$$

According to Reid and Cohen (1944), the relationship between viscosity and temperature can be expressed by an exponential equation:

$$\eta - 0.161 = (1.72 \times 10^{-4})T - c_1 \tag{9.12}$$

where η = viscosity in N s m^{-2}

T = temperature, K

c_1 = constant, to be computed from the viscosity measurements at given temperatures or from the viscosity monogram provided by the authors

Hoy et al. (1965), after a large number of slag viscosity measurements, have developed an empirical equation taking into account the ash composition in terms of silica ratio (SR) as defined earlier by Reid and Cohen (1944):

$$\log_{10} \eta = 4.468 \left(\frac{SR}{100}\right)^2 + 1.265 \left(\frac{104}{T}\right) - 8.44 \tag{9.13}$$

where the units of η and T have already been specified. Later Watt and Fereday (1969), using the same experimental data, derived another expression by using regression analysis:

$$\log \eta = \frac{10^7 \; m}{(T + 123)^2} + c_2 \tag{9.14}$$

η (N s m^{-2}) and T (K) have already been defined, and m and c_2 are "constants" calculated from composition of slags, where

$$m = 0.00835 \; SiO_2 + 0.00601 \; Al_2O_3 - 0.109 \tag{9.15}$$

and $\quad SiO_2 + Al_2O_3 + equiv. \; Fe_2O_3 + CaO + MgO = 100$

The value of c_2 is calculated from the expression

$$c_2 = 0.0415 \; SiO_2 + 0.0192 \; Al_2O_3 + 0.0276 \; equiv. \; Fe_2O_3 + 0.0160 \; CaO - 4.92 \tag{9.16}$$

The authors have considered the effect of alumina (Al_2O_3) on viscosity, which previous workers had left out in the formulations of predicting the slag flow characteristics. The viscosity measurements were carried out in a reducing atmosphere, but the results are likely to be appropriate for the cyclone-furnace combustion conditions where many of the burning fuel particles are thrown on the surface of hot slag.

Hoy et al. (1965) were able to define the critical temperature (T_c) in terms of coal-ash composition:

$$T_c = 3263 - 1470 \left(\frac{SiO_2}{Al_2O_3} \right) + 360 \left(\frac{SiO_2}{Al_2O_3} \right)^2$$

$$- 14.7 \, (Fe_2O_3 + CaO + MgO) + 0.15 \, (Fe_2O_3 + CaO + MgO)^2 \quad (9.17)$$

where T_c is in degrees K and, as in Eq. (9.15),

$$SiO_2 + Al_2O_3 + equiv. \, Fe_2O_3 + CaO + MgO = 100$$

Equation (9.17) shows that for a given flux content $(Fe_2O_3 + CaO + MgO)$, the critical temperature depends on the silica-to-alumina ratio, and that a slag with SiO_2-to-Al_2O_3 ratio of 2 to 1 by weight, or 4 to 1 approximately by molar ratio, would have the lowest critical temperature.

The main uncertainty in calculating the viscosity from the composition of ash arises from the variable oxidation state of iron oxide dissolved in the slag. The equilibrium ratio of ferrous to ferric iron is slow to respond to changes in the temperature and atmosphere, as depicted by Eq. (9.9). More research is required to establish the effect of the oxidation state of iron on slag viscosity, and hence on the rate of sintering and slagging and on deposit adhesion as discussed in Chaps. 10 to 12. The Mossbauer spectroscopy and magnetic susceptibility techniques (Bancroft, 1973; Allen et al., 1981) can be used to determine the ferrous-to-ferric ratio of iron in the surface layer of particulate ash before sintering. Similar measurements can be made to establish the ferrous-to-ferric ratio of iron in different layers of slag deposit.

9.4 VISCOSITY AS A RATE–CONTROLLING PARAMETER IN FORMATION OF SINTERED DEPOSIT AND MOLTEN SLAG

Kuczynski (1949b) obtained measurable sinter bridges between glass particles as a result of viscous flow when the viscosity of the glassy material exceeded 10^{11} $N \, s \, m^{-2}$. He used comparatively large particles, 200 μm in radius, and the time required to produce slight sintering was comparatively long, over 100 h. The particle diameter of ash deposited on boiler tubes is likely to be smaller by 1 to 3 orders of magnitude—i.e., in the range of 0.1 and 10 μm (Chap. 12)—and correspondingly the time of initial sintering is reduced to a few hours in the temperature range where the ash viscosity is 10^{11} $N \, s \, m^{-2}$.

Table 9.1 lists the events involving ash particles in the pulverized-coal-fired boiler in chronological order from the moment when the particulate minerals are injected into the flame, on a viscosity scale that extends from 10^{11} $N \, s \, m^{-2}$ for the initial sintering down to 10 $N \, s \, m^{-2}$ for running slag flow in cyclone-fired boilers.

The approximate relationships between the rates of sintering and slagging, viscosity, and temperature given in Table 9.1 refer to the behavior of bituminous coal ashes, which cause occasional furnace-wall and superheater fouling in pulverized-coal-fired boilers. The terms onset of sintering, medium and rapid sintering, and slagging are arbitrary definitions, but when referred to the viscosity,

Table 9.1 Viscosity scale for ash deformation, cenosphere formation, sintering, and slagging: Typical bituminous coal ashes

Event	Viscosity range (N s m^{-2})	Temperature (K)	Comment
	In flame		
Deformation of flame-borne silicate particles	10^3–10^7	1450–1650	Discussed in Sect. 6.1
Cenosphere formation and coalescence of flame-borne particles	10^2–10^4	1600–1700	Discussed in Sects. 6.3 and 8.4
	On boiler tubes		
Onset of sintering	10^9–10^{11}	1200–1300	Discussed in Sect. 9.3 and Chap. 10
Medium-rate sintering	10^7–10^9	1300–1400	
Rapid sintering	10^5–10^7	1400–1550	
Formation of nonflowing slag	10^3–10^5	1550–1650	
Slow movement of slag due to gravity	10^2–10^3	1650–1700	Discussed in Sect. 9.2
Satisfactory flow from cyclone-fired boilers	<25	<1700	

which is an important rate-controlling parameter, these terms give an idea of the rates of coalescence of deposited ash particles leading to the formation of boiler deposits.

Ash deposits on cooled furnace-wall and superheater tubes constitute a heat barrier (Chap. 13) and, as a result of poor heat conductance of the silicate material in comparison to that of boiler-tube steels, a steep temperature gradient is set up across the deposit layer. It follows therefore that within an ash deposit layer the processes of sintering and slagging can occur at different rates, which may cover the entire range listed in Table 9.1. Figure 9.7a shows the temperature gradient across a deposit layer 20 mm thick on the furnace-wall tubes of a pulverized-coal-fired boiler, together with the corresponding viscosity gradient for a typical bituminous coal ash. Figure 9.7b depicts the temperature and viscosity gradients across a deposit of the same thickness on the high-temperature superheater tubes. In this case, the tube metal temperature is higher but the thermal gradient and the maximum deposit temperature are lower.

The presentation in Fig. 9.7 depicts the viscosity gradient for thick deposits of silicate ash where there is no low temperature phase of complex sulfates present (Chaps. 12 and 17). Viscosity of the inner layer of deposits—i.e., first 0 to 2 mm on cooled boiler tubes enriched in sulfate—can change with time, as depicted in Fig. 9.8. The ash particles on a clean boiler tube have a high viscosity; as a result, the rate of build-up of deposit is slow. Gradually a layer of molten sulfate may be formed that has a low viscosity (curve S), whereas the viscosity of the silicate phase (curve A) increases due to crystallization (Sect. 9.2). Subsequently, the sulfate phase may freeze as a result of the thermal shielding effect of the thick layer of silicate ash deposit.

Reid and Cohen (1944) and Reid (1971) have used the viscosity data—i.e., the temperature of critical viscosity [Fig. 9.1 and Eq. (9.17)]—to estimate the thickness of "frozen" slag on boiler tubes. Figure 9.9 shows the increase in relative thickness of deposits in a cyclone-fired boiler, with the change in slag flow properties as characterized by the temperature of critical viscosity. The authors claim that the viscosity measurements can be useful in explaining the cause of massive slag accumulations on combustion-chamber walls.

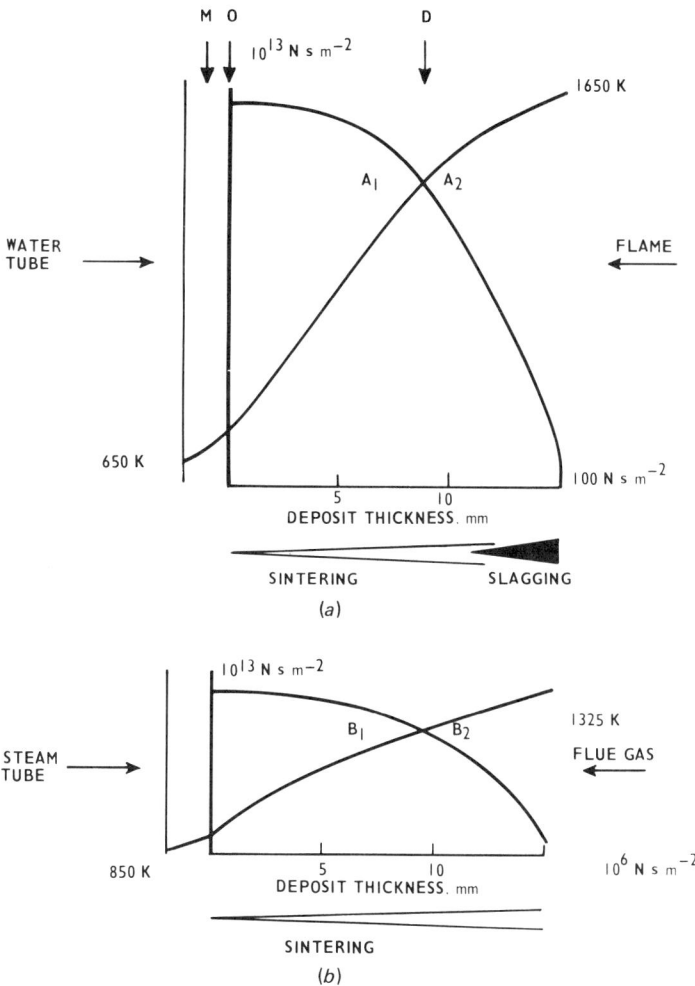

Figure 9.7 Temperature and viscosity gradients in (a) furnace wall and (b) superheater deposit of nonsegregated ash: M, metal; O, oxide; D, deposit.

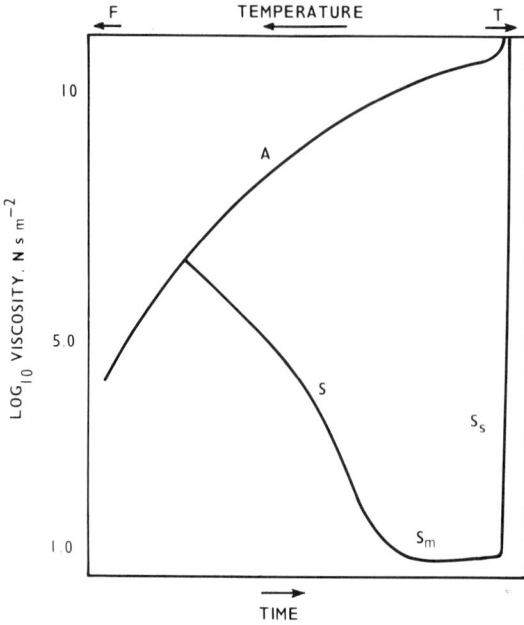

Figure 9.8 Viscosity change on formation of layer-structured phases from initial deposit material: A, silicate phase; S, formation of complex sulfate; S_m, molten sulfate; S_s, frozen sulfate; F, flame; and T, tube surface.

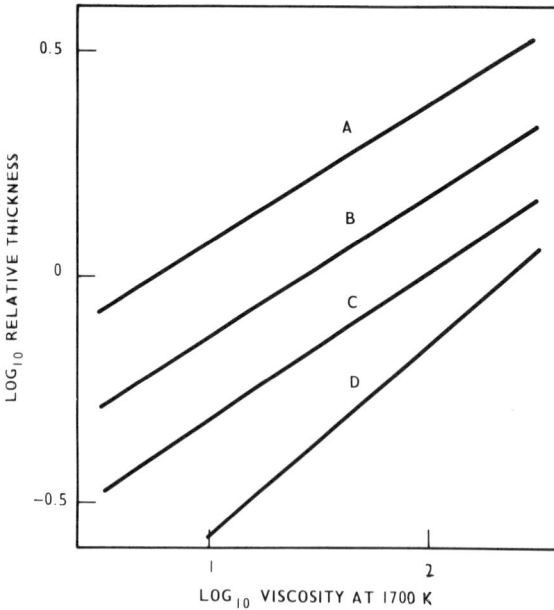

Figure 9.9 The relationship between slag-layer thickness and critical viscosity temperature (T_c): flame temperature 1865 K; tube surface temperature 920 K; A, $T_c = 1810$ K; B, $T_c = 1755$ K; C, $T_c = 1700$ K; D is Newtonian slag, no critical temperature.

NOMENCLATURE

A	cross-sectional area of glass fiber
A_1	preexponential constant
A_2	logarithmic constant
a	radius of cylinder or rod
B	temperature coefficient for viscous flow
b	radius of crucible
c_1 and c_2	constants
dL/dt	rate of fiber stretching
E_η	exponential coefficient for viscous flow
F	applied force; stretching load
k and k_1	constants
L	depth of penetration; length of glass fiber
L_1 and L_2	end corrections
m	slag composition parameter
R	thermodynamic gas content
r	radius of spherical ash particle
SR	silica ratio
T	temperature
T_c	critical temperature for viscous flow
T_o	temperature correction for viscous flow
t	time
v	velocity of penetration
x	radius of neck bridge between particles
Ω	angular viscosity
γ	surface-tension coefficient
η	viscosity coefficient
τ	torque

SINTERING, FUSION, AND SLAGGING PROPENSITIES OF COAL ASHES

10.1 SINTERING MODEL

Frenkel (1945) has developed a sintering model that describes the rate of coalescence of particles in terms of measurable parameters, viscosity, surface tension, and particle size. He assumed that sintering commenced with a deformation of particles under the influence of surface tension, which causes a viscous flow where the viscosity η and diffusion D_m of the material are related by the equation

$$\frac{1}{\eta} = \frac{(\Delta d)D_m}{kT} \tag{10.1}$$

where Δd = interatomic distance
k = Boltzman constant
T = temperature

Frenkel then derived an equation relating the growth of the interface between two spherical particles or a particle and a semiinfinite body:

$$x^2 = \frac{3r\gamma t}{2\eta} \tag{10.2}$$

where x = radius of the interface assumed to be circular
r = radius of the spherical particles
γ = surface tension
t = time

Rearranging Eq. (10.2) in terms of x/r and t, it becomes

$$\frac{x}{r} = 1.225 \left(\frac{\gamma}{\eta r}\right)^{1/2} t^{1/2} \tag{10.3}$$

and is applicable when $x/r < 0.3$.

Kuczynski (1949a, b, 1977) has examined four different mechanisms of sintering by which particle-to-particle bonding can take place: viscous flow, vapor condensation, diffusion, and surface tension. He showed that the sintering of glassy particles was governed by viscous flow [Eq. (10.2)]. Watt (1968) has suggested that the mechanism of vapor transport of alkali metals can play a significant role in the formation of sintered coal-ash deposits. Mackenzie and Shuttleworth (1949) and Roberts (1950) have described the mode of transport of volatile species from the convex to concave surfaces resulting in particle bridging at the contact points.

The alkali metals dissolved in flame-heated silicate-ash particles have a low vapor pressure (Boow, 1972), but they can significantly reduce the viscosity (Chap. 9). It is therefore evident that sintering by viscous flow has a dominant role in the formation of deposits in coal-fired boilers. The rate of sintering of different size particles for a given viscosity (10^8 N s m^{-2}) can be seen in Fig. 10.1, where the ratio x/r [Eq. (10.3)], is plotted against time on a logarithmic scale. The surface tension of fused ash was taken to be 0.32 N m^{-1}. Figure 10.2 shows plots where the ratio of x/r represents the degree of sintering of 5-μm-radius particles having different viscosities. Table 10.1 lists four arbitrary stages of sintering of ash deposit on boiler tubes, leading from the initial contact between the particles to the formation

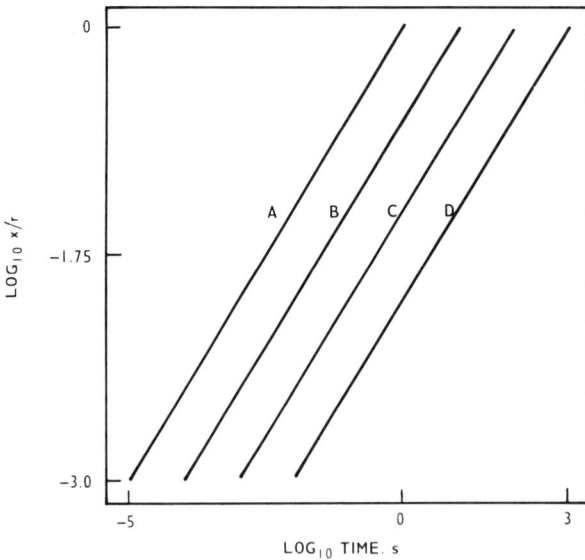

Figure 10.1 Log–log plot of degree of sintering (x/r) against time; viscosity 10^8 N s m^{-2}. Particle radius: A, 0.05 μm; B, 0.5 μm; C, 5.0 μm; D, 50 μm.

Table 10.1 Degree of sintering based on the ratio of neck bond radius to particle radius (x/r)

Value of x/r	Degree of sintering	Comment
0.001	Onset of sintering	The deposit of this degree of sintering on boiler tubes has no significant cohesive strength and would probably fall off under the action of gravity and boiler vibration.
0.01	Slightly sintered matrix	The deposit on boiler tubes would probably be removed by soot blowing.
0.1	Strongly sintered deposit	The deposit on boiler tubes would be difficult to remove by soot blowing.
>0.3	Slagging	The ash particles lose their original identity and the deposit on boiler tubes cannot be removed by soot blowing.

of fused slag where the shape of initial constituent particles is no longer distinguishable.

Rapid formation of sintered boiler deposits and slags is usually explained by the presence of a liquid phase or molten layer on the surface of ash particles. In high-temperature technology—i.e., in blast-furnace slag—a liquidus phase is considered to have a viscosity value below 10 N s m^{-2}. The plots in Figs. 10.1 and 10.2 show that with small ash particles it is not necessary to evoke the presence of a liquid phase for a rapid sintering. For example, particles 0.1 μm in diameter

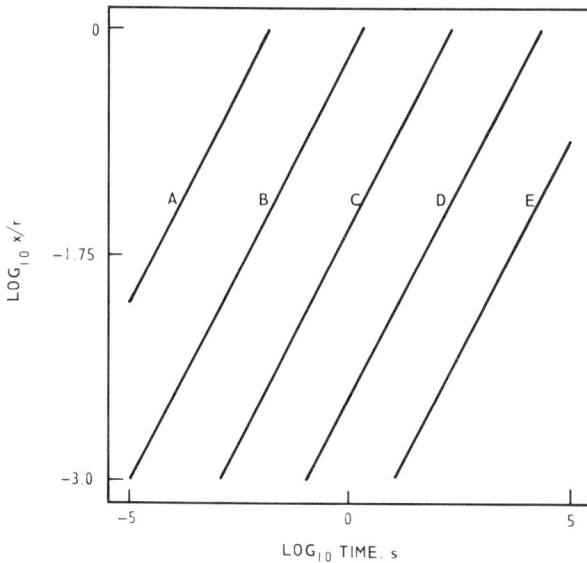

Figure 10.2 Log–log plot of degree of sintering (x/r) against time, particle radius 5 μm. Viscosity (N s m^{-2}): A, 10^4; B, 10^6; C, 10^8; D, 10^{10}; E, 10^{12}.

would require about 10 ms to form a substantial sinter bond when the viscosity has a high value of 10^8 N s m^{-2}. With the same viscosity, 10-μm particles would require about 10 s to achieve the same degree of bonding. The inverse relationship between the rate of sintering and the particle size has been confirmed by the results of measurements with size-separated ash fractions carried out by Dering et al. (1972). The rate of sintering of fine ashes is further enhanced by the high ratio of mass of the alkali-rich surface layer to the total mass of small particles, as discussed by Hosegood and Raask (1964).

The two principal parameters that govern the rate of sintering—the surface tension and the viscosity—both decrease with temperature, as shown in Fig. 10.3. However, the temperature coefficient of surface tension is small (curve A) and is approximately proportional to the inverse of the square root of temperature (Boni and Derge, 1956), whereas the viscosity changes exponentially with temperature, as shown by curve B. It is therefore evident that the rate of ash sintering will show an inverse relationship with the viscosity and will increase rapidly as the temperature of coalescing particles is raised.

10.2 SINTERING RATE MEASUREMENTS

The model for coal-ash sintering discussed in the previous section is based on the viscous deformation and flow at the contact points between spherical particles. It would therefore be logical to start a synopsis of different methods of determining the rate of sintering with a technique where the measurements are based directly on Frenkel's equation (10.2). Kuczynski (1949b) has carried out sintering measurements by placing spherical glassy particles on the surface of a glass slab of the same composition. Raask (1973) has reported some results of sintering of coal-ash particles based on a similar technique. The method requires spherical particles of ash, and these were prepared by passing ground fused slag through a vertical furnace

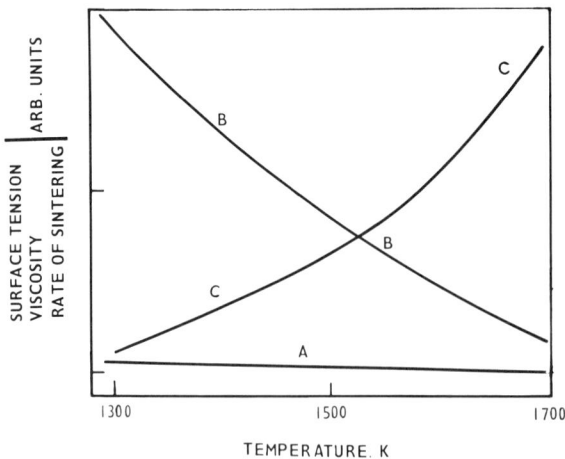

Figure 10.3 The effect of temperature on ash-sintering parameters: A, surface tension; B, viscosity; C, rate of sintering.

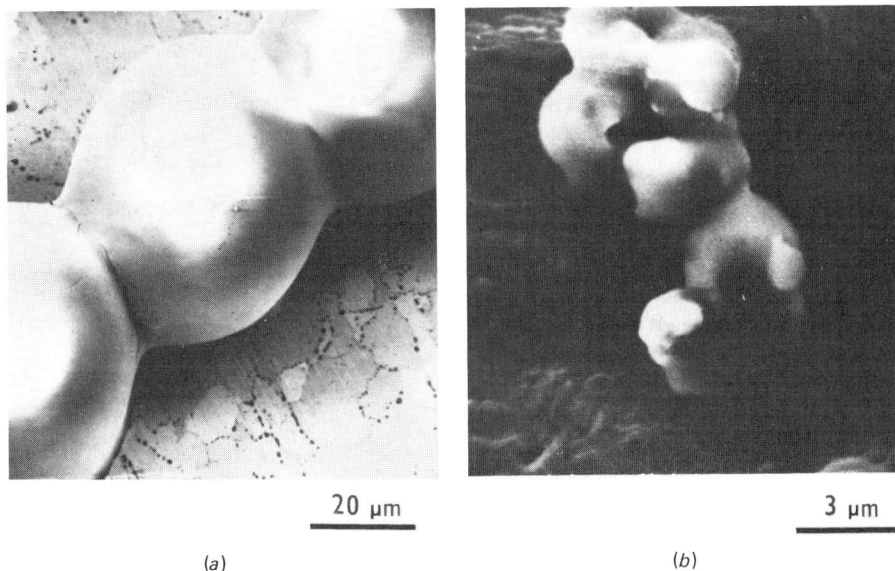

20 μm   3 μm

(a)   (b)

Figure 10.4 Ash sintering. (a) In laboratory furnace. (b) Strand of boiler deposit.

(Chap. 6). Subsequently the particles were placed in a narrow groove on a platinum foil, as shown in Fig. 10.4a. The particles were then introduced into a preheated furnace and kept at a constant temperature in air or in a gas mixture for a period of 5 min to several hours. The radius of the sinter neck between the particles (Fig. 10.4a) was measured microscopically at the ambient temperature. Figure 10.4b shows a sinter bridged segment of typical superheater deposit.

Figure 10.5 shows the rate of neck growth between spherical particles of slag, 60 μm in radius, when heated in air. The spherical particles were prepared from boiler slag from a typical British bituminous coal of composition similar to that in Table 6.2. The time for a firm degree of sintering ($x/r = 0.1$, Table 10.1) was 135 s and 70 s at 1375 and 1425 K, respectively. From these measurements the time required for a given degree of sintering can be calculated for the ash particles of different sizes. For example, the ash deposited on boiler tubes in pulverized-coal-fired boilers contains a large number of particles of 0.5 to 5 μm in diameter, and these particles require only a few seconds to form a strongly sintered deposit (Fig. 10.4b) in the same temperature range.

Particle-to-particle sinter bonding usually results in a shrinkage of the external dimensions of a powder compact, and the dilatometric shrinkage measurements have been extensively used to determine the rate of sintering of glass and refractory materials. Kuczynski and Zaplatynski (1955) have measured the shrinkage of glass capillaries; Oel (1960) has determined the rate of sintering of angular glass powder compacts and showed that the Frenkel model was applicable also for nonspherical particles.

Density D_o of a powder compact of spherical particles of radius r is given by

$$D_o = \frac{\kappa}{\beta r^3} \qquad (10.4)$$

where κ is a constant related to the specific gravity of material and β is a packing factor. Assuming that the κ and β values do not change significantly in the initial stage of sintering, then the density D of the sintered product is given by

$$D = \frac{\kappa}{\beta(r^2 - x^2)^{3/2}} \qquad (10.5)$$

where x is the radius of sinter neck as in Eq. (10.2), with the limitation that x/r is not greater than 0.3. It therefore follows that

$$\frac{D}{D_o} = \left(\frac{r^2}{r^2 - x^2}\right)^{3/2} \qquad (10.6)$$

and the degree of sintering in terms of x/r, as in Table 10.1, is given by

$$\frac{x}{r} = \left(\frac{D^{2/3} - D_o^{2/3}}{D^{2/3}}\right)^{1/2} \qquad (10.7)$$

Smith (1956) has used a dilatometric shrinkage technique to study the sintering characteristics of pulverized-fuel ash, and an intercept of the shrinkage curve on the temperature axis was taken to define the sinter point. These measurements give useful information, and the results can be related to different degrees of sintering as outlined in Table 10.1. With some coal ashes, however, anamalous results can be obtained where the shrinkage measurements show no change although a significant degree of sintering has taken place. This deviation in the sintering behavior from the Frenkel model makes it necessary to monitor another parameter related to the coalescence of ash particles.

Viscosity measurements by the rod penetration method have been applied by Boow (1972), Raask (1973), and Gibb (1981) to assess the slagging characteristics

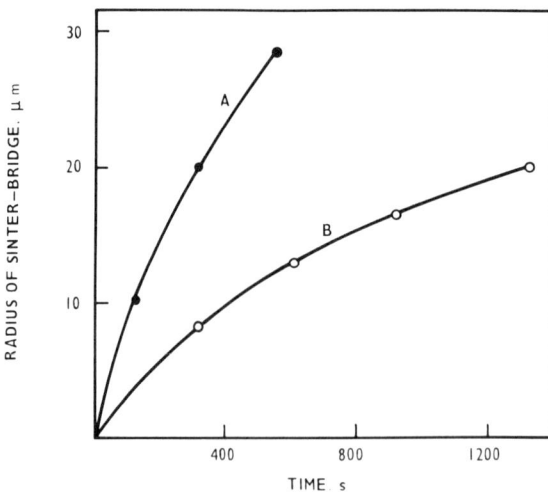

Figure 10.5 Growth of sinter bridge between particles, radius 60 μm: A, at 1425 K; B, at 1375 K.

of different coal ashes. However, the rate of initial sintering cannot be measured by this technique, and Raask (1979b) therefore considered the use of electrical conductance measurements for monitoring the rate of sintering of coal ashes. Previously, Ramanan and Chaklader (1975) used the same technique to study sintering of glass sphere and nickel powder compacts. The essential requirement is that the materials should have measurable electrical conductance at sintering temperatures. This is not a limitation when sinter-testing coal ashes by the conductance technique, because of the presence of sodium, potassium, and calcium species in the silicious material.

An ash compact before sintering would have a low conductance because of lack of particle-to-particle contacts. As the cross-sectional area of sinter bonds between the particles grows, the conductance path is increased correspondingly:

$$\Lambda = A \left(\frac{x^2}{r^2 - x^2} \right) \tag{10.8}$$

where Λ = the conductance

x = radius of neck growth

r = radius of particles as in the Frenkel model equation (10.2)

A = constant that varies with packing geometry

Combining Eqs. (10.6) and (10.8), the conductance Λ is given in terms of the density:

$$\Lambda = A_1 \left[\left(\frac{D}{D_o} \right)^{2/3} - 1 \right] \tag{10.9}$$

where D_o and D are the density of a powder compact before and after sintering, and A_1 is constant. Equation (10.9) applies to sintering at a constant temperature, and for continuously increasing temperature the exponential effect must be taken into account:

$$\Lambda = A_2 \left[\left(\frac{D}{D_o} \right)^{2/3} - 1 \right] \exp \left(\frac{-E}{RT} \right) \tag{10.10}$$

where Λ, D, and D_o have already been defined, E is the energy of activation of sintering, R is the thermodynamic (gas) constant, T is the temperature, and A_2 is a constant. When the degree of sintering does not change—e.g., on cooling after the process of particle coalescence has reached the stage of density D—Eq. (10.10) reduces to:

$$\Lambda = A_3 \exp \left(\frac{-E}{RT} \right) \tag{10.11}$$

Raask (1979b, 1982b) has devised a furnace assembly (Fig. 10.6) for simultaneous measurements of the electrical conductance and the dilatometric shrinkage of coal ash. An alumina crucible, 8 mm in internal diameter and 14 mm high, was used to contain the ash compact. Small holes were drilled through the base of the crucible and platinum wire was looped through the holes. A platinum disc 7.5

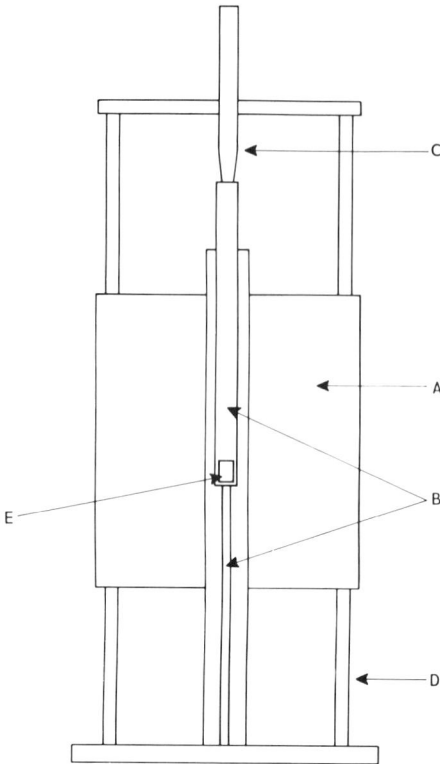

Figure 10.6 Schematic of furnace for simultaneous shrinkage and conductance measurements: A, furnace; B, conductance leads and thermocouple; C, displacement probe; D, furnace has balanced lowering and raising action; E, sample.

mm in diameter was placed on the wire contacts, and 400 mg of ash was lightly compacted to a height of 12 mm, giving a bulk density of 750 kg m^{-3}. Another platinum disc was placed on top of the sample, and the conductance and thermocouple connections were completed as shown in Fig. 10.6. A displacement transducer resting on an aluminum extension tube was used to record the linear movement of the sample to an accuracy of ±2 μm. An AC bridge recorder was used for the conductance measurements.

Figure 10.7a shows the furnace in its down position for exposure of the sample well; the sample crucible and three pellets of sintered ash from previous runs are shown at the well. Figure 10.7b shows the furnace in the operation position. Heating at a rate of 0.1 K s^{-1} (6 K min^{-1}) was chosen to be the same as that used in the ASTM (1968) ash fusion tests. The sinter tests can be carried out in air or in simulated flue gas. Care is needed to stop heating when the ash sample has decreased 30 percent in height to avoid slagging: once slag is formed, it is difficult to remove the frozen material from the crucible.

Initial experiments were made with a soda glass, ground below 100 μm in particle size, of known viscosity and temperature characteristics published by Napolitano and Hawkins (1964). This was done to establish the validity of the simultaneous dilatometric and conductance measurements for determining the rate of sintering of powder compacts. The results are shown in Fig. 10.8, where line A$_1$

(a)

(b)

Figure 10.7 Furnace assembly for ash-sintering measurements. (a) Down position for sample change. (b) Operation position.

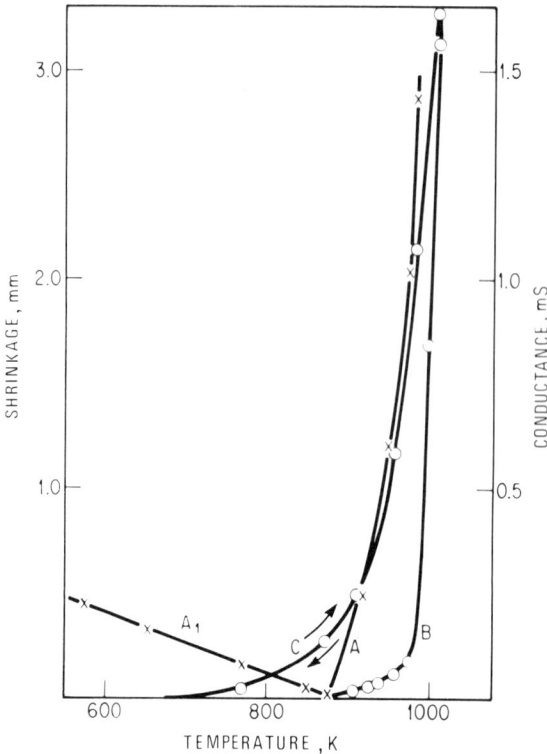

Figure 10.8 Simultaneous shrinkage and electrical conductance measurements of soda glass: A_1, thermal expansion; A, shrinkage; B, conductance on heating; C, conductance on cooling (left arrow) and reheating (right arrow).

depicts thermal expansion of the alumina support tubes and the sample, and curve A shows the linear shrinkage of a 10-mm-high sample of powdered glass. The intercept of curve A on the temperature axis at 875 K can be defined as the sinterpoint temperature. The plot of the results of conductance measurements, curve B, meets the temperature axis at the same point, and it is therefore evident that sintering of the powdered glass compact had proceeded according to the model as defined by Eq. (10.10).

The conductance results on cooling curve C show a large hysteresis effect: that is, these are significantly higher than the corresponding results on heating. On reheating, however, the conductance measurements fit closely to curve C. This behavior is in accord with the sintering model, and the measurements on first heating when sinter bonds are formed fit Eq. (10.10). Since on subsequent cooling and reheating the process of particle coalescence is "frozen," the conductance change is governed exponentially by temperature as defined by Eq. (10.11).

Figure 10.9 shows that there was an inverse relationship between the viscosity and electrical conductance of partially sintered glass. The conductance is dependent on the ionic mobility—i.e., on the rate of diffusion of sodium and calcium ions in the glass matrix—and thus an inverse relationship between viscosity and self-diffusion is established, as stipulated by the Frenkel model of sintering [Eq. (10.1)]. The results of conductance measurements plotted in Fig. 10.9 cover the viscosity range of 10^6 to 10^{10} N s m^{-2}, and earlier (Table 9.1) it was argued that this is a relevant viscosity range for the formation of sintered deposits in coal-fired boilers. That is, strong sinter bonds can form in this viscosity range within a few minutes or several days, depending on the particle size (Figs. 10.1 and 10.2).

The cell "constant" of the conductance assembly can be calculated from the diameter of platinum discs (7.5 mm) and the distance between the plates, which are first 12 mm apart, decreasing to 8 mm as the ash sample shrinks on sintering. Thus, in terms of the specific conductance (conductivity), measurements can be made in the range of 0.2 to 1000 mS m^{-1}.

A number of ashes from British bituminous coals were sinter tested, and Fig. 10.10 shows a typical shrinkage-temperature (curve A) and the conductance plot (curve B). The sintering behavior of this ash and of a number of other bituminous coal ashes was as predicted according to the sintering model defined by Eq. (10.10). That is, a measurable change in shrinkage and electrical conductance commenced at the same temperature, as shown in Fig. 10.10. Isothermal tests with the same coal ashes also gave results that were consistent with the sintering model. Figure 10.11 shows that the conductance at a given time was proportional to the reciprocal of pellet length raised to the power of 0.67, in accordance with Eq. (10.9).

A limited number of U.S. bituminous coal ashes also have been investigated for their sintering characteristics by the simultaneous shrinkage and conductance measurements. Figure 10.12 shows a typical shrinkage curve and the conductance

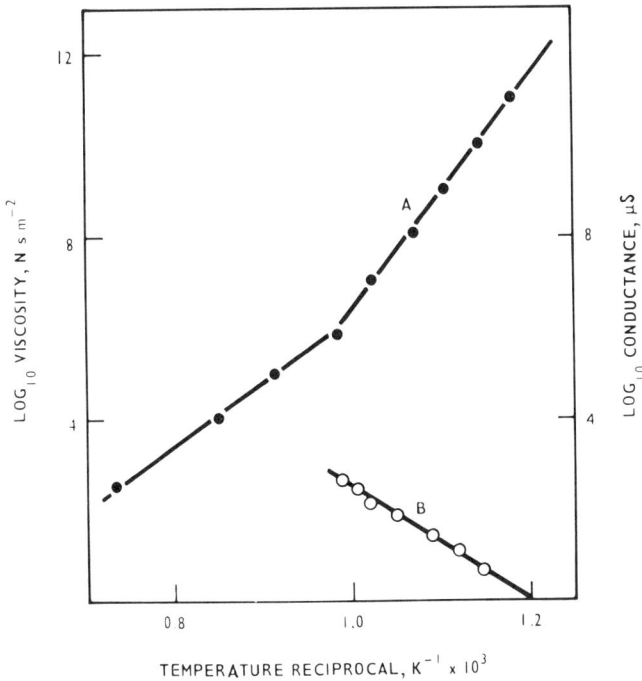

Figure 10.9 Arrhenius plots of viscosity (A) and conductance (B) of soda glass.

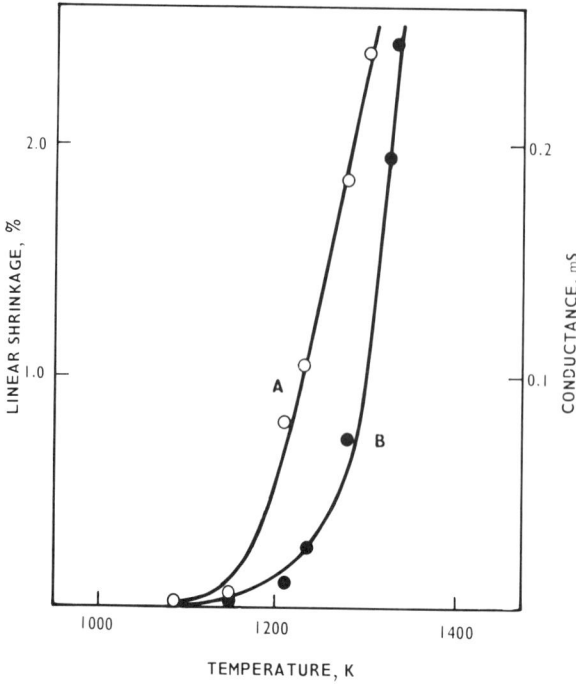

Figure 10.10 Simultaneous shrinkage (A) and conductance (B) measurements, Pye Hill ash.

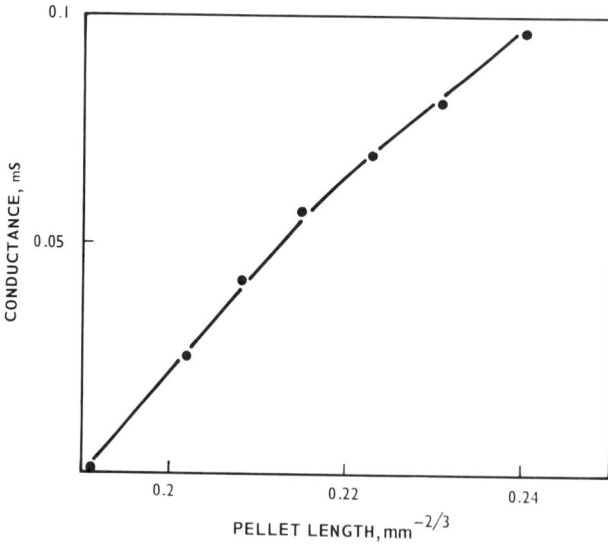

Figure 10.11 The relationship between shrinkage and conductance, Pye Hill ash at 1250 K.

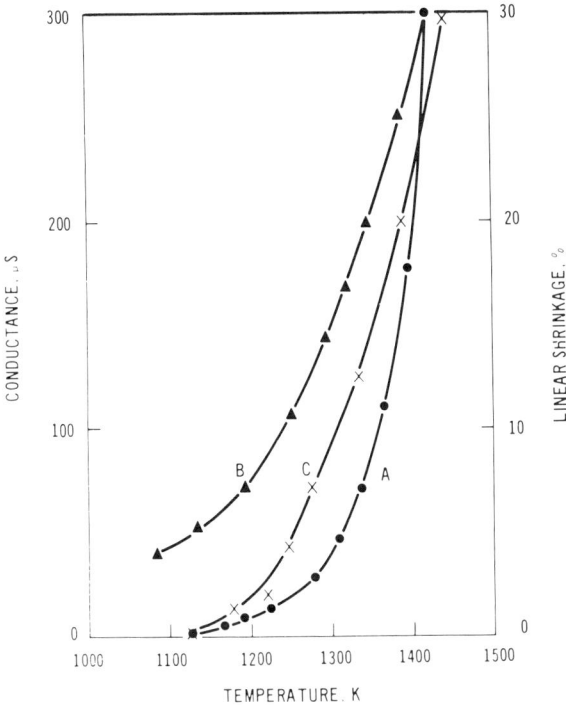

Figure 10.12 Conductance and shrinkage plots on sintering of Illinois coal ash: A, conductance on heating, 5 K min^{-1}; B, conductance on cooling, 5 K min^{-1}; C, shrinkage on heating, 5 K min^{-1}.

plots on heating and on cooling, obtained with an Illinois coal ash. It is evident that with this ash and other U.S. coal ashes tested, sintering proceeded according to Eq. (10.10). Cumming (1980) has pointed out that the conductance data can be examined in more detail on the plot of log Λ against $1/T$, as depicted in Fig. 10.13. The conductance plot on heating is nonlinear, as expected from Eq. (10.10), whereas on cooling the plot was linear in accord with Eq. (10.11).

Raask (1982b) has suggested that the conductance plot on heating can be divided into three sections, where the logarithmic conductance values in μS (microSiemens) are as follows: 0 to 1, 1 to 2, and 2 to 3, corresponding to the range of 1 to 1000 μS in the linear scale. Figure 10.14 shows schematically the conductance path and shrinkage in three different degrees of sintering. The strength of the sintered pellet to crushing was determined at room temperature at the end of sinter tests. The presentation shows that an increase of 155 K from the initial sinterpoint temperature of 1125 to 1180 K resulted in a high degree of sintering where the conductance readings were above 100 μS.

There are some coal ashes that exhibit in their sintering behavior a remarkable degree of divergence from the particle coalescence model as defined by Eq. (10.10). Figure 10.15 shows that a sub-bituminous coal ash (Leigh Creek, Australia) commenced sintering at 1100 K according to the conductance measurements (curve B). There was, however, no shrinkage of the ash pellet before the temperature reached 1350 K (curve A). That is, there was a gap of 250 K between the ash sinterpoint temperature indicated by the conductance measurements and that

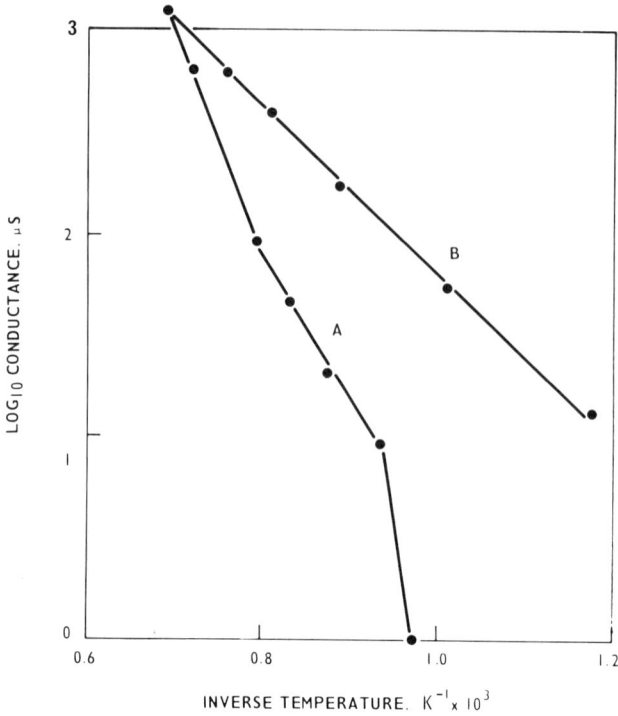

Figure 10.13 Conductance logarithm/inverse temperature plot on sintering of coalite breeze ash: A, heating, 5 K min^{-1}; B, cooling, 5 K min^{-1}.

deduced from the shrinkage measurements. The results of simultaneous measurements of the electrical conductance and shrinkage illustrate the sintering behavior of ashes that appear to consist of the infusible and low-temperature fusible constituents. The infusible particles (quartz) are sinter-bonded by a comparatively small amount (<10 percent) of low-temperature-melting (low viscosity) glassy material rich in sodium. This is a probable explanation for the formation of large masses of deposits ("sand bombs") with high-quartz ashes rich in sodium (Hein, 1976). The conventional ash fusion tests are insensitive for detecting the small amounts of low-viscosity (i.e., low-temperature-melting) phases, and thus the ashes could be classified as having a low deposit-forming propensity. In contrast, the electrical conductance is a measure of the ionic, chiefly sodium, mobility related to the viscosity: hence it is a sensitive method of monitoring the development of low-temperature sinter bonding.

The Leigh Creek ash had another unusual sintering feature in that after the initial rise in conductance with temperature, there was a minimum in the curve (Fig. 10.15) at 1325 K. This was probably because the ash contained unusually high amounts of sodium, 6.4 percent Na_2O by weight of ash, and it is possible that some sodium aluminate had crystallized in the temperature range of 1275 to 1325 K, resulting in a reduction of sodium ion concentration in the sintered matrix.

Laboratory-prepared ashes can be categorized according to the results of simultaneous measurements of shrinkage and the electrical conductance on sintering of an ash compact. Most coal ashes will coalesce and shrink according to the model defined by Eq. (10.10). That is, particle deformation and particle-to-particle bonding are accompanied by the external shrinkage and a change in shape of an ash pellet. With these ashes the shrinkage measurements and the conductance measurements give equally valid results. Further, the conventional ash fusion tests discussed in Sect. 10.5 usually give meaningful results with these coals. There are, however, some coal ashes rich in sodium that can give anamolous results when sinter-tested as exemplified by the curves in Fig. 10.15. That is, there may be a high degree of the internal particle-to-particle adhesion, resulting in the formation of ash compact or deposit of high strength without the usually noted corresponding external shrinkage or change in the shape of an ash pellet. With these ashes the electrical conductance measurements give more meaningful results in early stages of sintering than those of any other technique.

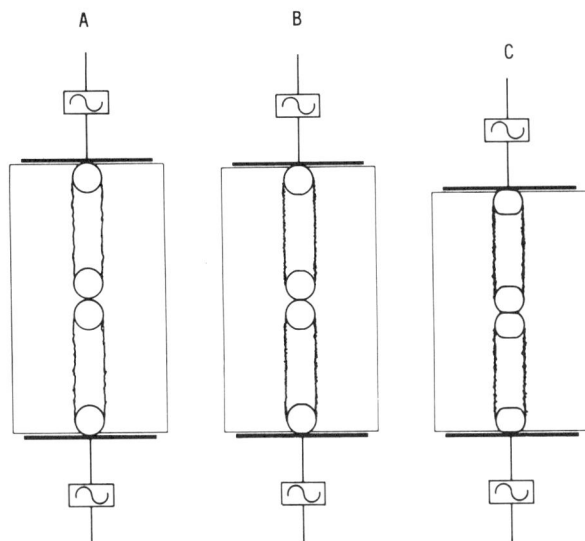

Stage	Temperature, K	Conductance, μS	Shrinkage, %	Strength, MN m^{-2}
A. First contact	1125	1.0	0.02	0
B. Initial sinter	1195	10	2.0	0.1
C. Rapid sinter	1280	100	10	1.0

Figure 10.14 Three stages in the sintering of Illinois coal ash: 12 mm high, 8 mm diameter pellet between platinum plates.

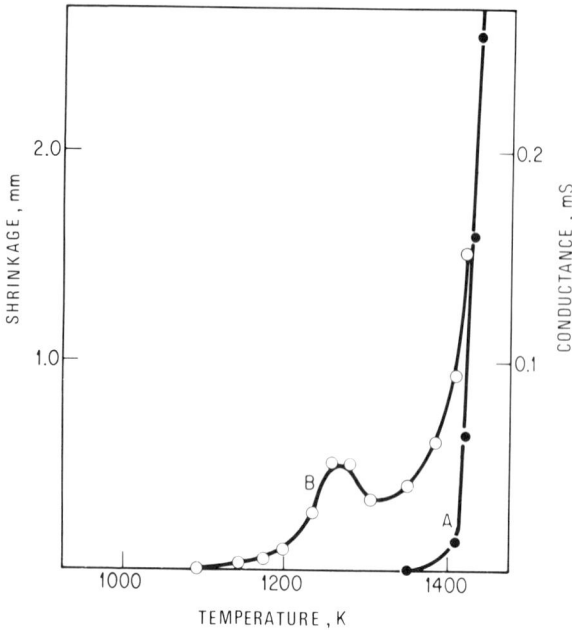

Figure 10.15 Simultaneous shrinkage (A) and conductance measurements (B) on Leigh Creek (Australia) coal ash.

10.3 SINTERED ASH STRENGTH MEASUREMENTS

Several investigators have used crushing strength measurements of laboratory-prepared specimens to assess the sintering characteristics of coal ashes. The best known of these is that employed by Barnhart and Williams (1956) and by Attig and Barnhart (1963). They argued that it is essential to have the ash prepared for sintering tests under conditions that simulate as far as possible those of full-scale combustion. Pulverized coal was burned in a 10-kg/h combustor with the flame temperature between 1650 and 2000 K, and the ash was collected in a cyclone separator and a cloth filter bag.

The combustor ash was reheated in air at 750 K to burn off the carbon residue and then reduced in particle size below 150 μm. Pellets of ash 19 mm high and 15 mm in diameter were made in a mould at a pressure of 1.0 MN m^{-2}, and from each ash 24 specimens were made. Six samples were heated at four different temperatures—1090, 1145, 1190, and 1255 K—for 15 h, and after cooling the crushing strength of the sintered ash pellets was measured. Some of the results of Attig and Barnhart (1963) are shown in Fig. 10.16.

The investigators claim that the results of their sintering tests were consistent with ash fouling and slagging characteristics in large coal fired boilers. That is, the sintering strength of slagging coal from Illinois (Fig. 10.16, curve A) in the temperature range of 1145 to 1245 K was markedly higher than that of less slagging Indiana coal ashes (curves B and C). The conventional ash fusion tests (Sect. 10.5) gave the same slagging characteristics for all three coal ashes. This was probably because the ashes had a high iron content, above 20 percent as Fe$_2$O$_3$, and in the presence of that amount of iron oxide "flux," the additional effect of fairly high amount of CaO, 8.8 percent in coal A was not discernible by the conventional ash fusion tests. An alternative method of sinter testing of coal ashes of a high CaO content is that described in Sect. 10.2, i.e., the simultaneous shrinkage and electrical conductance measurements.

Ash preparation for sintering tests by burning pulverized coal in an experimental combustor is an expensive method, and not many researchers have such a facility available to them. Gibb (1981) decided therefore to use a more conventional method of ash preparation by burning samples of pulverized coal in a muffle furnace at 775 K. Cylindrical pellets, 9.5 mm in diameter and 0.8 g in weight, were formed in a die where enough pressure was applied to give a length of 10.5 mm. The bulk density of the ash pellets thus was 1100 kg m^{-3}, and with an average value of the specific gravity of different ash constituents of 2400 kg m^{-3}, there was a porosity value of the ash pellets of 54 percent by volume. Batches of three ash pellets were heated for 1 h either in air or in a reducing atmosphere (a mixture of hydrogen and carbon dioxide, 5 percent by volume each in nitrogen). The temperature range for the sintering tests extended from 1050 to 1325 K, and it was about the same as that in the earlier sintering tests of Attig and Barnhart (1963).

The crushing strength of ash pellets was plotted against temperature, and from the graph the temperature T_s corresponding to a strength of 5 MN m^{-2} was taken to represent the sintering characteristics of different pellets. Gibb then compared his results of sintering tests with viscosity measurements by the rod penetrator technique (Sect. 9.1) and with conventional ash fusion tests (Sect. 10.5). From the viscosity-temperature plot he chose the temperature T_V corresponding to a value of 10^6 N s m^{-2} to represent the sintering characteristics of different ashes on a basis of the viscosity parameter as discussed in Sect. 9.4. Table 10.2 gives a comparison of the results of sintering tests, viscosity measurements, and ash fusion tests.

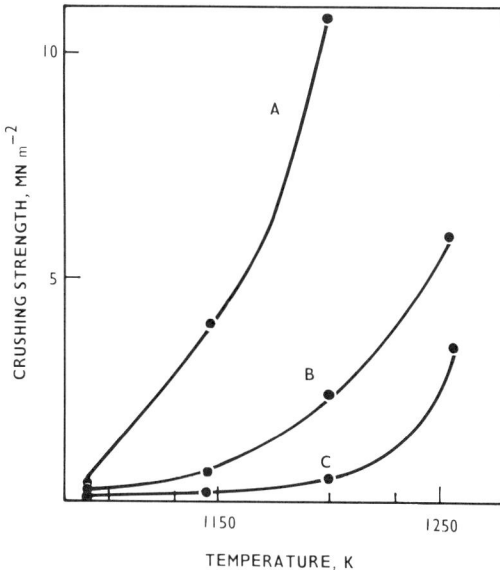

Figure 10.16 Sinter strength of coal ashes related to boiler slagging: A, severe; B, moderate; C, none.

Table 10.2 Sintering, viscosity, and ash fusion tests

Ash[a]	Temperature (K) for sintering strength 5 MN m^{-2}	Temperature (K) for viscosity 10^6 N s m^{-2}	Ash fusion		
			Initial deformation	Half sphere	Flow
		Oxidizing atmosphere (air)			
1	1275	1395	1515	1625	1675
2	1195	1235	1565	1600	1635
3	1205	1515	1575	1675	1675
4	1335	1220	1525	1675	1675
5	1255	1220	1375	1655	1675
6	1085	1205	1525	1605	1635
		Reducing atmosphere (5% H$_2$ + 5% CO$_2$ + 90% N$_2$)			
1	1165	1130	1375	1465	1675
2	1135	1210	1435	1495	1535
3	1125	1140	1365	1495	1565
4	1250	–	1355	1495	1595
5	1180	–	1385	1565	1675
6	1060	–	1455	1555	1605

[a]Ashes 1 to 5 are typical of those from British bituminous coals. Ash 2 has been known to cause boiler slagging when the coal is not mixed with those of nonslagging ashes. Coal ash 3 has given occasional slagging difficulties. Ash 6, of a sub-bituminous coal mined in Australia (Leigh Creek), has a high sodium content, over 6 percent Na$_2$O, and it has caused severe boiler slagging.

Table 10.2 shows that the sintering tests as described by Gibb (1981) gave the results (first temperature column) that were consistent with the slagging and fouling propensities of coals when burned in pulverized-coal-fired boilers. That is, coal ashes 2, 3, and 6, in particular the latter, have caused boiler fouling and slagging and have correspondingly low sintering temperatures. The results of viscosity measurements were less conclusive, and the conventional ash fusion tests gave misleading results with sample 6 where the temperature of initial deformation was high both in air and in a reducing atmosphere. Thus, there was no indication that this coal ash could form strongly sintered deposits at comparatively low temperatures of 1060–1085 K.

10.4 A RAPID TEST FOR INITIAL SINTERING

Stallmann and Neavel (1980) have reported a simple method to assess the sintering propensity of different ashes in the temperature range of 810–1360 K. They have found some sintered ash agglomerates in coal pyrolysis reactors at temperatures well below the initial deformation of coal ash according to the ASTM ash fusion tests (Sect. 10.5). They considered that a more sensitive method of assessing the sintering characteristics was required.

Ash for sintering tests was prepared by burning pulverized coal in air at 810 K, and there was no significant degree of ash sintering in the oxidizing atmosphere at this temperature. The ash was passed through a 100-mesh (150-μm) sieve, placed in a small boat with minimum of compaction, and heated for 30 min at temperatures ranging from 810 to 1365 K. The furnace was purged with a gas mixture of 40 percent CO by volume, balance CO_2, to simulate the gas atmosphere in their pyrolysis and gasification systems.

After cooling, samples were sieved to determine the weight percentage retained on a 100-mesh sieve. Since the ash initially passed through a 100-mesh sieve, the fraction retained after heating represented the ash agglomerates formed on sintering. Some of the results of agglomeration sintering tests of Stallmann and Neavel (1980) are given in Fig. 10.17, which show that with some coal ashes the initial bonding of particles to form aggregates over 150 μm in size commenced below 1000 K. The authors do not give chemical composition of the ashes they tested, but it is likely that the initial agglomeration occurred as a result of bonding by low-melting sulfides, chiefly FeS, CaS, and Na_2S. Hein (1979) has pointed out that sulfides may play a significant role in the formation of large masses of lightweight deposits ("sand bombs") in the superheater zone of boilers fired with nonbituminous coals.

The ash agglomeration–temperature plots in Fig. 10.17 seem to indicate that the results were consistent with the sintering model as set out in Sect. 10.1. It states that the rate of a sintering process governed by viscous flow increases rapidly with temperature. The model [i.e., Eq. (10.2)] defines that at a given temperature the contact area between the particles, hence the degree of agglomeration, should be directly proportional to time. However, Stallmann and Neavel (1980) found that when the heating time was varied from 0.5 to 8 h there was no significant difference in the degree of ash agglomeration. Clearly further work is necessary to

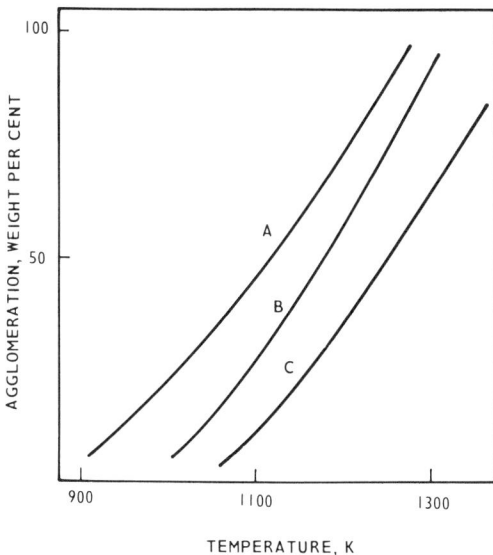

Figure 10.17 Ash sintering by agglomeration tests: A, Illinois coal; B, Texas lignite; C, Arkansas lignite.

elucidate the mechanism of coalescence of particulate ash in its early stages, in particular in reducing atmospheres where molten sulfides may be present.

Table 10.3 summarizes different sintering techniques that can be used to assess the deposit-forming propensities of coal ashes.

The sintering tests listed in Table 10.3 have occasionally been used to obtain empirical fouling indices of different coal ashes, but the deposit-forming–ash composition relationships are based largely on the conventional ash fusion tests discussed in Sect. 10.5.

10.5 CONVENTIONAL ASH FUSION TESTS

Ash fusion tests are based on the external change in shape—i.e., deformation, shrinkage, and flow—of a pyramidal or cylindrical pellet of ash when heated in a laboratory furnace. A pyramidal shape is often used because it is easier to observe rounding of the pointed tip of the specimen than rounding of the edge of a cylindrical pellet. The methods are empirical, and strict observance of the test conditions is necessary to obtain reproducible results; these are laid down in the

Table 10.3 Summary of sintering techniques

Measuring technique	Equipment	Comments
Neck-growth measurements between spherical particles (Kuczynski, 1949b; Raask, 1973); Sect. 10.2.	A furnace and a microscope.	This method is suitable for homogeneous material when available in the form of spherical particles. It is not suitable for routine sinter testing of coal ashes.
Simultaneous shrinkage and electrical conductance measurements (Raask, 1979b, 1982b); Sect. 10.2.	Needs a purpose-built furnace assembly for more accurate measurements; a simpler version uses platinum-wire electrodes as described by Raask (1979b) and by Cumming (1980).	Each ash could be provided with sintering-rate curves. The method needs to be tested and assessed with a wide selection of ashes.
Crushing strength measurements of sintered ash pellets (Atting and Barnhart, 1963; Gibb, 1981); Sect. 10.3.	A furnace and a crushing strength measuring device.	The method has been used by several researchers but there is no agreed-on procedure.
Ash agglomeration by sieving test (Stallmann and Neavel, 1980); Sect. 10.4.	A furnace and sieving equipment.	This is one of the simplest methods of testing for initial sintering, and it warrants more systematic tests.
Ash plug flow method (Bluckman et al., 1976; Basu and Sarka, 1983).	A tubular, silica or alumina, furnace assembly with perforated plate to support an ash plug.	This method is relevant to ask sintering in fluidized bed combustion (Chap. 2).

Table 10.4 Ash fusion tests by different national standards

British standard[a] (1970)	U.S. standard D1857 (1968)	German standard DIN 51730 (1976)	French standard M03-012 (1945)
Sample dimensions			
Trigonometric pyramids 8 to 13 mm high with equilateral base side 0.3 to 0.5 of height	Trigonometric pyramid 19.1 mm high with equilateral base 6.4 mm side	Cylindrical samples, 3 mm high, 3 mm in diameter	Trigonometric pyramid, 12 mm high and with equilateral 6 mm base and 2 mm top
Oxidizing atmosphere			
Air or CO_2	Air	Air	Air
Reducing atmosphere			
50% CO + 50% H_2	60% CO + 40% CO_2	70% CO + 30% CO_2	50% CO or H_2 + 50% water vapor
Heating rate			
5 K min^{-1}	6 K min^{-1}	10 K min^{-1}	2.5 or 10 K min^{-1}
Observed stages in ash fusion temperatures			
A—Initial deformation, rounding of tip of pyramid B—Hemisphere temperature, height = 0.5 base C—Flow temperature, height = 0.3 base	A—Initial deformation, rounding of tip of pyramid B—Softening temperature, height = width C—Hemisphere temperature, height = 0.5 width D—Fluid temperature, height = 1.6 mm	A—Initial softening B—Hemisphere temperature, height = 0.5 base diameter C—Fluid temperature, all changes are recorded photographically	Changes in height viewed with 10× magnification and plotted against temperature; no recommended stages

[a]British Standard includes an alternative method of ash fusion tests that are similar to the German Standard.

U.S. ASTM (1968), British Standard (1970), German (DIN, 1976), and French (Norme Française, 1945) procedures.

The French standard stipulates that the change in specimen height should be plotted against temperature, and the curve characterizes the fusion property of ash as exemplified by the work of Kent and Champion (1964). The German standard specifies the use of a cylindrical specimen, and the change in the shape and shrinkage are recorded photographically using a heating microscope as originally developed by Radmacher (1949). A summary of different methods of the ash fusability tests is given in Table 10.4.

The methods of ash fusion tests summarized in Table 10.4 do not include recommendations for correlating the laboratory results with the rate of build-up of sintered ash and slag deposits in coal-fired boilers. Reid (1971), Corey (1974), and Winegartner (1974) have offered some advice for interpreting the results of ash fusion tests when carried out according to the ASTM method as summarized below.

The *initial deformation temperature* is the temperature at which a standard pyramid of ash just begins to fuse or show evidence of deformation at the top as it is being heated. This would correspond to the temperature in an operating furnace at which the particles of ash, in transit through the furnace, have been cooled to a point where they still retain a slight tendency to stick together and slowly form deposit on heat-exchange surfaces. When the ash particles entrained in the flue gas have cooled to a temperature below the initial deformation temperature, the deposit on cooled tubes tends to accumulate as a "dry" product.

The *softening temperature*, usually called the fusion temperature of an ash, corresponds to an observed condition between initial deformation and fluidity where the height of ash cone is equal to the width of sample.

The *hemisphere temperature* is read when the cone has further deformed to the point where it assumes a hemispherical shape and its height equals half its width. Both softening and hemisphere temperatures are related to conditions at which the ash shows a greatly accelerated tendency to mass together and stick in large quantities to heat-exchange surfaces.

The *fluid temperature* is the temperature at which the test cone flattens out to a pancake shape. The fluid temperature is related to the temperature at which the coal ash may be expected to flow in drips or streams from furnace walls.

On the viscosity parameter scale given in Table 9.1, the four stages observed in the ASTM ash fusion tests have the approximate viscosity ranges shown in Table 10.5.

Experience has shown that for pulverized-coal-fired boilers ("dry ash" boilers) where a formation of excessive deposits is highly undesirable, the temperature of initial deformation and softening observed in ash fusion tests is an important criterion in boiler layout regarding the heat transfer and flue gas temperatures. Design engineers usually aim to limit the temperature of flue gas entering the convection superheater section to a value below the initial deformation temperature noted in ash fusion tests. When the flue gas temperature and consequently that of the ash particles deposited on high-temperature steam tubes is significantly higher

Table 10.5 Viscosity ranges for ASTM ash fusion tests

Observed stage	Viscosity range $(N\ s\ m^{-2})$
Initial deformation	10^7-10^9
Softening	10^5-10^7
Hemisphere	10^3-10^5
Flow	$<10^3$

Table 10.6 Coal-ash slagging and temperature

Slagging index (F_s) temperature (K)	Boiler slagging
1505–1615	Medium
1325–1505	High
<1325	Severe

than the initial deformation temperature of a laboratory-prepared ash sample, it may result in a rapid formation of sintered deposits.

A debatable point here is whether the initial deformation temperature observed in an oxidizing atmosphere or that observed in a reducing atmosphere should be taken as a guiding criterion in boiler design. With some ashes, such as sample 3 in Table 10.2, the deformation temperature in reducing atmosphere can be lower by some 200 K than that in oxidizing atmosphere. Pulverized-coal-fired boilers usually operate with an excess of oxygen, 3 to 5 percent by volume of the flue gas, and combustion should be completed when the flue gas reaches the superheater zone; consequently an oxidizing atmosphere should prevail. In practice, however, intermittent streams of reducing gases reach this section of the boiler, and it is therefore advisable for boiler-design engineers to take the initial deformation temperature of ash in a reducing atmosphere as a guidance criterion for prevention of severe superheater fouling.

Gray and Moore (1974) have proposed a slagging index (F_s) based on the initial deformation (IT) and the hemisphere temperature (HT) observed in ash fusion tests:

$$F_s = \frac{4\ IT + HT}{5} \tag{10.12}$$

The authors have offered guidance classification on coal-ash slagging characteristics based on the F_s temperature thus derived, as summarized in Table 10.6.

When testing coals for suitability for cyclone-fired boilers, the initial ash-deformation temperature is less important than that of the final flow. The viscosity of slag at this stage is about 25 N s m^{-2}, which is regarded as a maximum value for a satisfactory discharge of slag from cyclone boilers. Boiler-design engineers consider that the ash temperature T_{25} should not exceed 1700 K, otherwise the coals are not suitable for cyclone firing and would be better utilized in pulverized-coal ("dry ash") boilers. The slag flow temperatures obtained by the laboratory ash fusion tests are usually checked against T_{25} temperatures computed from the empirical formulas based on ash composition (Chap. 9).

It has been recognized by many researchers that although ash fusion tests can give useful information regarding the fouling and slagging propensities of coal ashes, there are serious shortcomings. First, the tests are based on subjective observations and not on precise scientific measurements, and they have a large margin of error. For example, the ASTM method, which is widely used in many countries, allows for a 55 K margin or reproducibility in the initial deformation, softening hemisphere temperatures in an oxidizing atmosphere, and a 70 K margin of uncertainty in determining the initial deformation temperature in a reducing atmosphere. Within that 70 K margin the viscosity can change by more than an order of magnitude (Chap. 9), and consequently the rate of ash sintering will change by the same factor.

Another shortcoming with the ash fusion method is that when testing some coal ashes there occurs an extensive degree of particle-to-particle bonding without any visible sign of deformation in the shape of pellet. This takes place in particular with high-sodium and high-calcium ashes from sub-bituminous coals, as discussed in Sect. 10.2. Briefly, the reason for the phenomenon is that there are infusible constituents, e.g. quartz, present in sufficient quantity to retain the original shape of an ash pellet when the particle-to-particle bonding is caused by a low-temperature phase of fused sodium and calcium silicates, sulfates, or sulfides. For coal ashes of such sintering characteristics there is a need to have a different method of assessing their fouling propensity. The sintering techniques listed in Table 10.3 could be used to give additional information to the conventional ash-fusion tests.

10.6 EMPIRICAL FOULING AND SLAGGING INDICES BASED ON ASH COMPOSITION

For the purpose of assessing their fouling and slagging propensities, laboratory-prepared coal ashes are analyzed for 10 most abundant constituent elements, other than oxygen: Si, Al, Fe, Ca, Mg, K, Na, Ti, S, and P. Titanium (Ti) is sometimes omitted from the analysis, since its concentration is usually small and consistent, between 0.5 and 1.0 percent of the total ash, expressed as TiO_2, and it is considered that TiO_2 does not have any specific role in the formation of boiler deposits. Phosphorus (P) is left out frequently because of its low concentration in ash, usually less than 0.5 percent expressed as P_2O_5.

Chemical analysis of coal ashes can be carried out by a variety of methods discussed in Chap. 3, but the procedures specified in the standard methods of coal analysis—e.g., British Standard (1970) and ASTM Standard (1969)—should be consulted. Important points to note are the method of preparing ash and the accuracy required. Ash residue prepared for the analysis by burning of a pulverized and homogeneously mixed coal sample in air at 1025 K should not contain any carbonaceous matter, carbonates, hydroxides, or water. The analytical results are expressed in the terms of weight percent of corresponding oxides, e.g., SiO_2, Al_2O_3, Fe_2O_3, SO_3. The sum of oxide percentages should be between 98 and 101.5. Less accurate results cannot be relied on to give any meaningful assessment of the fouling and slagging propensity of ashes. Analytical errors caused by mutual interference of elements present in coal ashes are usually the reason for the unacceptable data. Other causes for erroneous results may be incomplete combustion of coal substance, absorbed CO_2 and water, or the presence of an element above 1.0 percent quantity that is usually found in coal ashes in trace quantities.

The sulfate (SO_3) content of boiler slag is usually small, whereas the laboratory ash prepared at 1025 K can contain high amounts of sulfate, as discussed in Chap. 4. For this reason the results of chemical analysis are usually normalized to an SO_3-free basis for the purpose of assessing the fouling and slagging propensities of different coal ashes.

Before a discussion on various fouling and slagging indices of different coal ashes, it is expedient to explain the term "fouling" and "slagging" by ash deposits as understood by the boiler-design and operation engineers. By fouling is meant a build-up of bonded and sintered deposits on superheater and reheater tubes in the convection heat-transfer passages of a boiler. The gas temperature in this boiler section should not usually exceed 1350 K; consequently a slag formation here would be an unusual occurrence, indicating either a high gas temperature or an exceptionally low slagging temperature of the ash. Ashes rich in calcium and sodium, in particular, can cause severe superheater fouling.

The ash deposit on heat-exchange surfaces in the combustion chamber first undergoes the fouling (sintering) stage, but as the thickness increases, the outer layer of deposit is transformed to a tacky material having a viscosity value below $1000 \ N \ s \ m^{-2}$. Subsequently, large masses of semimolten slag can build up quickly, within hours, when the adhesive bond at the tube-surface interface is sufficiently strong to support the weight of large masses of slag. Severe slagging of pulverized-coal-fired boilers is usually associated with coal ashes rich in iron. However, large masses of fouling deposit can also build up in upper reaches of the combustion chamber with sodium-rich nonbituminous coal, as described by Hein (1976).

Several methods of calculating the flow temperature of slag in cyclone-fired boilers were given in Chap. 9. Further formulas for calculating viscosity from ash composition have been published by Sage and McIlroy (1960), Duzy (1965), Reid (1971), and Winegartner (1974). The silica ratio, as developed by Reid and Cohen (1944) and given by Eq. (9.10) has been used for assessing the slag flow characteristics in cyclone-fired boilers. The silica ratio has been used, together with iron oxide content of ash, by Raask (1973) to make an appraisal of the deposit forming propensity of British coal ashes as summarized in Table 10.7.

Coals in class A do not cause any fouling or slagging problems in pulverized-fuel-fired boilers, and those in class B result occasionally in some difficulties. Coals in class C can cause severe boiler slagging but can be burned when mixed with

Table 10.7 Sintering and slagging characteristics of British coal ashes

Classification of coal ash	Nonslagging	Some slagging	High slagging
Iron oxide of ash, weight percent Fe_2O_3	3–8	8–15	15–23
Silica ratio, $$\frac{SiO_2 \times 100}{SiO_2 + Fe_2O_3 + CaO + MgO}$$	72–80	65–72	50–65
Slow sintering temperature, K	1350–1450	1250–1350	1150–1250
Rapid sintering temperature, K	1450–1550	1350–1450	1250–1350
Slagging temperature, K	1550–1700	1450–1600	1350–1500

nonslagging coals, as discussed in Chap. 14. The sintering and slagging characterization as summarized in Table 10.7 applies reasonably well for coal ashes that do not contain sodium (Na_2O) at above 2.5 percent or calcium (CaO) at above 7.5 percent. For those coals, different fouling indices have been formulated.

Eight principal oxide constituents of sulfate-free ash can be divided into those that are acidic in the pyrochemical sense (SiO_2, Al_2O_3, and TiO_2) and those that are basic (Fe_2O_3, CaO, MgO, K_2O, and Na_2O). Several investigators have used the ratio of the sum of basic oxides to the sum of acidic oxides to assess the deposit- and slag-forming propensity of bituminous and nonbituminous coal ashes. The ratio of basic to acidic oxides in ash ($R_{b/a}$) is given by

$$R_{b/a} = \frac{Fe_2O_3 + CaO + MgO + K_2O + Na_2O}{SiO_2 + Al_2O + TiO_2} \tag{10.13}$$

where the constituent oxides are given as weight percent of the total.

Duzy (1965) and Duzy and Walker (1965) applied the relationship to U.S. coal ashes and showed that the lowest fusion temperature occurred when the oxide ratio was 0.55 (Fig. 10.18). Most bituminous coals have a pyrochemically acidic ash with the $R_{b/a}$ ratio between 0.2 and 0.4. With these coals, additional amounts of basic oxides, chiefly Fe_2O_3, CaO, and Na_2O, in ash would increase the deposit-forming propensity, and a maximum value of 0.5 for the basic/acidic oxide ratio is sometimes specified for pulverized-coal-fired boilers. For cyclone-fired boilers, the ratio values below 0.27 usually result in a slag that is too viscous to flow readily (Winegartner, 1974).

With sub-bituminous coal ashes of high CaO content, additional amounts of silica would enhance the ash deposit-forming characteristics, and this is reflected by the empirical fouling index F_x for the low-rank western U.S. coal ashes derived by Sondreal et al. (1978):

$$F_x = 0.38\ Na_2O + 0.006\ SiO_2 - 0.008\ CaO + 0.062\ Ash + 0.0037 \tag{10.14}$$

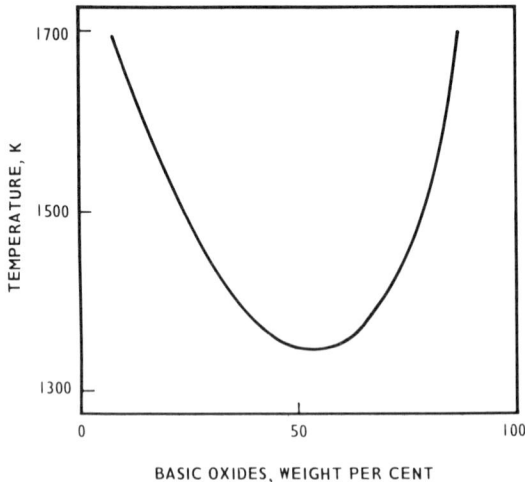

Figure 10.18 The relationship between ash fusion temperature and composition.

Table 10.8 Ash fouling index based on sodium and chlorine content of coals

Weight percent of coal		
Sodium (Na_2O) of western U.S. nonbituminous coal	Chlorine of British bituminous coals	Boiler fouling
<0.3 (<2.5% of ash)	<0.3	Slight
0.3–0.5 (2.5–4.0% of ash)	0.3–0.5	Moderate
>0.5 (>4.0% of ash)	>0.5	Severe

where the oxide constituents are expressed as percent of ash and the ash content is given as percent of dry coal.

The above expression shows a strong influence of sodium oxide content on ash fouling, which is also illustrated by the results of Sondreal et al. (1978), summarized in Table 10.8 with the findings of Jackson and Ward (1956). The latter have used the chlorine content—of British coals as an indirect index of deposit-forming tendency of the ashes resulting from reactions with sodium.

Sedor et al. (1960), Duzy and Walker (1965), and Gronhovd et al. (1970) have shown that when the sodium content of ashes is used to assess their deposit-forming propensities it is necessary to consider the bituminous coal ashes and lignitic fuel ashes separately (Table 10.9).

The ASME Research Committee on Boiler Fouling (Winegartner, 1974) has defined "bituminous type ash" as having an iron oxide content higher than that of calcium oxide, and "lignitic type ash" has more CaO than Fe_2O_3. Alternative definitions are:

1. For bituminous type ash: $SiO_2 > (Fe_2O_3 + CaO + Na_2O)$
2. For lignitic type ash: $SiO_2 < (Fe_2O_3 + CaO + Na_2O)$

These show that bituminous coal ashes are usually pyrochemically acidic, whereas lignitic fuel ashes can be highly alkaline. The fouling criteria in Table 10.9 suggest that an alkaline (lignitic) ash requires higher amounts of sodium than are required in an acidic (bituminous) ash to produce the same degree of boiler fouling. This is consistent with the relationship between the ratio of basic oxides to acidic oxides in

Table 10.9 Deposit-forming propensity of bituminous and lignitic coal ashes based on sodium content

Bituminous coal ash		Lignitic coal ash	
Na_2O in ash (weight percent)	Boiler fouling	Na_2O in ash (weight percent)	Boiler fouling
<0.5	Low	<2.0	Low
0.5–1.0	Medium	2–6	Medium
1.0–2.5	High	6–8	High
>2.5	Severe	>8	Severe

Table 10.10 Sodium equivalent criterion for boiler fouling with bituminous coal ashes

Na_2O_{eq} of coal (weight percent)	Boiler fouling
<0.3	Low
0.3–0.45	Medium
0.45–0.6	High
>0.6	Severe

ash and its fusion characteristics, as discussed by Duzy (1965) and by Duzy and Walker (1965). It appears, however, that the sodium-content criteria for boiler fouling with bituminous coal ashes, as set out in Table 10.9, are rather cautious. That is, there are many bituminous coals with sodium content of ash between 1.0 and 2.5 percent that cause minimal boiler fouling.

Sedor et al. (1960) and Griffin and Profita (1965) have used the total amount of alkalis, $Na_2O + K_2O$, in coal to assess its ash deposit-forming propensity:

$$Na_2O_{eq} \text{ of coal} = (Na_2O + 0.659\, K_2O)\frac{ash}{100} \qquad (10.15)$$

where Na_2O and K_2O are given in weight percent of ash and "ash" is given in weight percent of coal. The factor 0.659 before K_2O is the molecular weight ratio of Na_2O to K_2O. Table 10.10 shows the Na_2O_{eq} values that have been suggested as a guide for bituminous coal ashes. This classification assessment based on the total alkali metal content of coals does not apply to many sub-bituminous coal ashes (Winegartner, 1974).

Attig and Duzy (1969) modified the ratio of basic to acidic oxide [Eq. (10.13)] by introducing a multiplication factor, the Na_2O content of ash:

$$F_y = \frac{(Fe_2O_3 + CaO + MgO + K_2O + Na_2O)Na_2O}{SiO_2 + Al_2O_3 + TiO_2} = R_{b/a} \times Na_2O$$

$$\qquad (10.16)$$

and
$$F_{y'} = R_{b/a} \times (Na_2O)_{ws} \qquad (10.17)$$

where F_y and $F_{y'}$ are the fouling indices when Na_2O and $(Na_2O)_{ws}$ denote the total and water-soluble fraction of sodium in ash, respectively. Values of F_y given by the original authors and those of $F_{y'}$ inferred from their data by Winegartner (1974) are given in Table 10.11. The F_y relationship does not apply to lignitic type coal ashes. With these coals it is more appropriate to use the $F_{y'}$ formula [Eq. (10.17)].

Attig and Duzy (1969) have derived also an empirical slagging index $(F_{s'})$ of coal ashes based on the ratio of basic to acidic oxides in ash multiplied by the sulfur content of coal:

Table 10.11 Fouling propensity of coal ashes assessed from base–acid ratio and sodium content

F_y	$F_{y'}$	Boiler fouling
<0.2	<0.1	Low
0.2–0.5	0.1–0.25	Medium
0.5–1.0	0.25–0.7	High
>1.0	>0.7	Severe

$$F_{s'} = \frac{(Fe_2O_3 + CaO + MgO + K_2O + Na_2O)S}{SiO_2 + Al_2O_3 + TiO_2} = R_{b/a} \times S \quad (10.18)$$

where S denotes the weight percent of sulfur in dry coal. The slagging enhancement factor of sulfur was introduced on the basis that much of sulfur in slagging coals is present as pyrite and the pyrite residue can be an effective fluxing agent. The authors recommend an interpretation of the slagging index for bituminous coal ashes, as shown in Table 10.12. The guidelines of Table 10.12 do not apply to sub-bituminous coal ashes of high calcium content and low in silica content.

Bryers and Taylor (1976) and Bryers (1978) have suggested that the slagging propensity of coal ashes can be assessed by determining the amount and particle size of pyrite in pulverized coal. They advocate that coals of moderate and high sulfur content should be analyzed for the pyrite fraction above 7.5 μm in size in pulverized coal by size and density separation techniques. This is of particular importance in cases of lignite and sub-bituminous coals, which are not usually required to be ground to the same fineness as that of bituminous, less easily combustible coals, and much of pyrite may be present in large particles. When the pyrite mineral in pulverized coal occur in small sizes and dispersed with the silicious mineral matter, less boiler slagging would occur compared with that when the same amount of pyrites were present as large particles. This is because the residence time of the large particles of pyrite is not sufficient to oxidize the molten FeS residue to the solid FeO phase before the particles impact boiler tubes, as discussed in Chaps. 7 and 12.

Table 10.12 Slagging index of bituminous coal ashes from base–acid ratio and sulfur content

Slagging index, F_s	Boiler slagging
<0.6	Low
0.6–2.0	Medium
2.0–2.6	High
>2.6	Severe

The slagging index as expressed by Eq. (10.18) highlights the high fluxing efficiency of FeO, which is the oxidation product of FeS_2. However, this is not in accord with the results of Kent and Champion (1964), who have claimed that the fluxing power of FeO was only 0.55 times that of CaO. Winegartner and Rhodes (1975) also conclude, after their regressional analysis of ash fusibility and composition data, that CaO has more important influence on the ash-fusion temperature than it is usually assumed. CaO has a relatively low molecular weight in comparison with that of FeO, and calcium oxide is a strongly basic oxide having no tendency for an amphoteric behavior. By contrast, iron oxide can be present in a silicate melt as an "acidic" oxide—that is, it acts as a silicate network extender rather than a network modifier (chiefly Al_2O_3, Chap. 9).

Huffman et al. (1981) have discussed the relative effects of iron and calcium compounds in the formation of boiler deposits and have suggested that iron in the form of molten but transient constituent FeS may play an important role in early stages of deposit build-up, particularly in a reducing environment. However, compared with FeO the basic CaO has a more marked effect on reducing the viscosity of silicate glassy material in an oxidizing atmosphere. The authors have claimed that potassium also can be an important fluxing agent under oxidizing conditions. The amount of glassy phase formed below 1475 K was proportional to the potassium illite content of coal mineral matter. This is in accordance with the results of high-temperature experiments with the illite mineral published earlier by Raask (1968b) and discussed in Chap. 6. Figure 6.5 showed that the illite particles started to vitrify and change their shape below 1500 K (curve A), whereas the potassium muscovite particles required a temperature some 100 K higher for the change.

NOMENCLATURE

A	packing geometry constant
A_1, A_2, A_3	constants
D	density of sintered powder compact
D_o	original density of powder for electrical conductance
D_m	diffusion coefficient
Δd	interatomic distance
E	exponential coefficient
$F_s, F_{s'}$	slagging indices
$F_x, F_y, F_{y'}$	fouling indices
HT	hemisphere temperature
IT	initial deformation temperature
R	thermodynamic gas constant
$R_{b/a}$	ratio of basic to acidic oxides in slag
r	radius of spherical particles
T	temperature
t	time

x	radius of interface between spherical particles
β	packing factor
Λ	electrical conductance
γ	surface tension coefficient
η	viscosity coefficient
κ	constant

ELEVEN

ADHESION OF ASH DEPOSIT ON BOILER TUBES AND REFRACTORY MATERIALS

11.1 VAN DER WAALS, ELECTROSTATIC, AND SURFACE–TENSION FORCES

Particulate ash deposited on boiler tubes is initially held in place by surface forces, i.e., van der Waals, electrostatic, and surface-tension forces. Van der Waals forces become important when molecules or solid surfaces are brought close together without a chemical interaction taking place. The first coherent theory of van der Waals forces is that of London (1930), who showed that a neutral molecule has an instantaneous electrostatic dipole moment that interacts with neighboring molecules to produce an attraction force. To a first approximation, these forces are additive and thus they may be integrated over the whole volume of a solid to provide a value for the resultant force between it and the neighboring solid body. For a hemisphere of radius r held at a distance of nearest approach h from a plane, the resultant force F is given by

$$F = \frac{Ar}{6h^2} \tag{11.1}$$

where A is the Hamaker constant (Tabor, 1971).

Equation (11.1) applies only over short distances, and Casimir and Polder (1948) showed that such "retarded" van der Waals forces between a sphere and surface decay rapidly with increase in distance. Lifshitz (1956) has used another approach, based on the bulk dielectric properties of solids, to derive an equation for the retarded van der Waals forces F' between a sphere of radius r at the distance h from a flat surface:

$$F' = \frac{2B\pi r}{3h^3} \tag{11.2}$$

where B is the appropriate constant for the given material. Other pioneering studies in this field have been carried out by Derjaguin et al. (1956), Kitchener and Prosser (1957), Black et al. (1960), Sparnaay and Jochems (1960), Krupp (1967), and Israelachvili and Tabor (1973).

Tabor and Winterton (1968) and Tabor (1971) have measured the unretarded [Eq. (11.1)] and retarded [Eq. (11.2)] van der Waals forces between polished mica plates. They found that the changeover occurred at a distance of about 150 Å $(1.5 \times 10^{-8}$ m); the corresponding value of the Hamaker constant A in Eq. (11.1) was found to be 10^{-19} N (Newton), and that of the Lifshift constant B in Eq. (11.2) was 8.9×10^{-29} N m. These values have been used to compute the ratio of van der Waals forces to the gravitational force on small ash particles approaching a flat surface.

The gravitational force F_g on an ash particle of radius r and the density D is given by

$$F_g = \frac{4\pi r^3 Dg}{3} \tag{11.3}$$

where g is the gravity acceleration constant $(9.81$ m s$^{-2})$. Thus, the ratio (F_r) of the short-distance van der Waals forces F to the gravitational force F_g on a particle is

$$\frac{F}{F_g} = F_r = \frac{A}{8\pi Dgr^2 h^2} = \frac{1.62 \times 10^{-25}}{r^2 h^2} \tag{11.4}$$

where the value of D for ash was taken to be 2500 kg m^{-3} and when $h < 1.5 \times 10^{-8}$ m. The corresponding ratio $F_{r'}$ of the retarded van der Waals forces F' to the gravitational force is given by

$$\frac{F'}{F_g} = F_{r'} = \frac{2B}{Dgr^2 h^3} = \frac{8.15 \times 10^{-34}}{r^2 h^3} \tag{11.5}$$

where $h > 1.5 \times 10^{-8}$ m and, as in Eq. (11.4), the radius of the particle is in meters.

Figure 11.1 shows a comparison of van der Waals and gravitational forces on small ash particles as these approach a collecting surface. Curves A and B indicate that the submicron-size particles are readily held on a surface by van der Waals forces. The capture of small particles of ash on boiler tubes is further enhanced by surface irregularities of oxidized metal. Pfefferkorn and Vahl (1963) have shown that the oxide layer has an outward growth of spikes and needles, which are able to trap small particles. Steel and Brandes (1963) considered that the attachment of fine particles on the oxide whiskers is a result of electrostatic attraction between

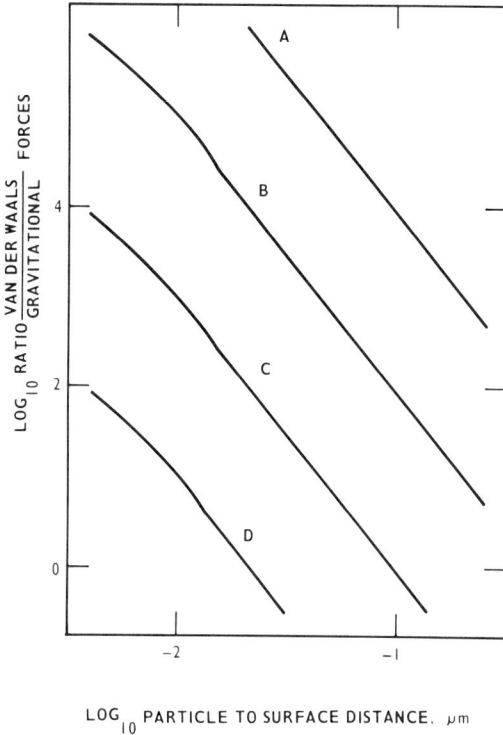

Figure 11.1 Comparison of van der Waals and gravitational forces on ash particles near collecting surfaces. Particle diameter, μm: A, 0.01; B, 0.1; C, 1.0; D, 10.0.

the particles and the spiked surface. They carried out some preliminary experiments that showed that when an electrically isolated steel probe in the flue gas was charged up to 3 kV, the steel surface remained free from oxide scale and deposit. However, when the polarity of charge on the probe was changed, small particles were retained at the surface.

Raask (1963a) has suggested that an electrostatic attraction force enhanced the transport and retention of submicron-size particles on a steel probe inserted in the flue gas of coal-fired boilers (Chap. 12). Penney and Klinger (1962) have found that a layer of electrically precipitated deposit of ash has a cohesive strength between 5 and 40 times higher than that formed by sedimentation. They suggested that the particles in an electric field have permanent dipole characteristics that led to these being orientated to form a cohesive layer of ash. Their experiments were carried out at low temperatures, and it is not certain that the observations are applicable for high-temperature ash deposition. It appears, however, that the combined effects of van der Waals and electrostatic forces of attraction and surface irregularities are sufficient to hold the submicron-size particles of ash deposit on boiler tubes until liquid-phase adhesion and chemical and mechanical bonds form at the ash deposit/metal oxide interface.

The requirement of a liquid phase in the formation of glass/metal seals and enamel coatings has been emphasized by King et al. (1959), Holland (1964), and Klomp (1979). The role of a liquid film is to provide the initial adhesion of solid particles as a result of surface tension. The work of adhesion W_a is given by

$$W_a = \pi + \gamma (1 + \cos \theta) \tag{11.6}$$

where γ is the surface tension of the liquid and θ is the contact angle at the solid/liquid interface. With perfect wetting—i.e., when θ equals zero—W_a has the highest value:

$$W_a = \pi + 2\gamma \tag{11.7}$$

The work of cohesion W_c of a liquid is given by

$$W_c = 2\gamma \tag{11.8}$$

With wetting liquids, therefore, W_a can be higher than W_c and failure will take place within the liquid layer, whereas with nonwetting liquids the rupture occurs at the solid–liquid interface.

Alkali-metal sulfates frequently constitute a liquid phase in ash deposits, and the molten sulfates readily wet and spread on the surface of boiler tubes. In a reducing atmosphere and when in contact with carbon, sulfates are reduced to sulfides, which wet and spread on any surface. The coefficient of surface tension of sulfates is fairly high, 0.20 N m^{-1} for Na_2SO_4 and 0.14 N m^{-1} for K_2SO_4 near their respective melting-point temperatures (International Critical Tables, 1928; Bertozzi and Soldani, 1967). Thus the work of cohesion of molten sulfate layer in boiler deposit is between 0.3 and 0.4 N m^{-1}, and the work of adhesion is higher because of a low contact angle at the sulfate/tube surface interface. It is therefore to be expected and it is observed in practice that when the deposit is removed—e.g., by sootblowing—there remains a film of sulfate adhering to boiler tubes.

The effect of furnace atmosphere on the adherence strength of vitreous coatings to an iron surface has been discussed by Eubanks and Moore (1955). An iron-free silicate melt does not spread on a metal surface in a reducing atmosphere, but it wets readily in an oxidizing environment, as shown by the results of King et al. (1959) depicted in Fig. 11.2. Iron-containing silicates, including coal-ash slags, wet

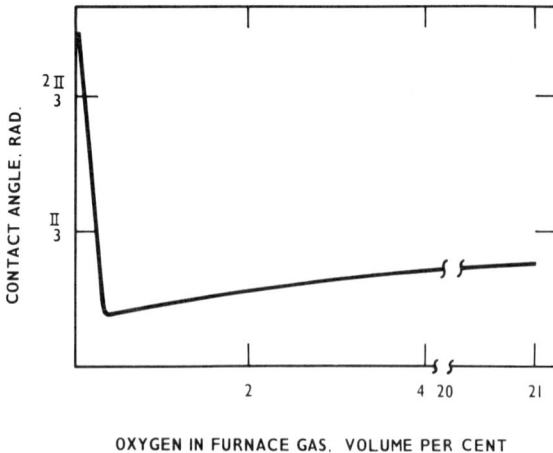

Figure 11.2 Effect of atmosphere on contact angle between iron and enamel at 980 K.

most surfaces in both oxidizing and reducing atmospheres with the exception of that of carbon (Chap. 6). The surface tension of coal ash slag determined by Raask (1966a) was 0.32 $N m^{-1}$, which is about twice that of sulfates; consequently the work of adhesion [Eq. (11.6)] and the cohesive bond strength [Eq. (11.8)] are also high.

Cross and Picknett (1963) have pointed out that for very fine particles only a small amount of liquid, about a hundred-molecule-thick layer, is sufficient for adhesion. In the case of a volatile liquid, the equilibrium thickness of the film, and thus the adhesion, varies with the partial pressure of vapor in the surrounding atmosphere. When evaporation from a liquid film occurs—i.e., as a result of increased temperature—the adhesion first rises to a maximum value due to the meniscus effect, but it breaks down as the film thickness is reduced to molecular dimensions. However, before the breakdown of the surface-tension adhesion occurs, chemical and mechanical bonds may develop between the deposited ash and boiler-tube surface, as discussed in the following sections.

11.2 THE TEMPERATURE ENVIRONMENT FOR ADHESION OF ASH DEPOSIT IN COAL COMBUSTION SYSTEMS

Benjamin and Weaver (1961) have discussed the adhesion of evaporated metal films on glass surfaces, whereas in coal-fired boilers the reverse adhesion phenomenon occurs. That is, the glassy silicate and sulfate constituents of ash deposit on boiler tubes (Chap. 12) and subsequently form a sinter or slag matrix strongly adhering to the metal surface. The main problem area is in the high-temperature section of boiler plant where the particulate ash and condensable sulfates form adhesive deposits on cooled surfaces. The middle section of the coal-fired boiler, namely the economizer (water preheater) and primary superheaters and reheaters, is usually free from adhesive deposits, although accumulations of compacted ash can occur. The performance of air heaters, which complete the process of extracting usable heat from the flue gas, can be impaired as a result of the formation of bonded deposits on the heat-exchange plates. The primary cause for deposit formation in this low-temperature section is usually condensation of sulfuric acid, which then reacts with airheater metal and deposited ash to form bonded deposits, as discussed in Chap. 19.

Table 11.1 summarizes the high-temperature environmental conditions for formation of adhesive ash deposit on cooled and uncooled surfaces. The latter occurs mainly in coal-gas-fired turbines and on refractory surfaces that are partially cooled in coal-fired boiler systems.

Refractory materials are not extensively used in pulverized-coal-fired boilers with "dry" ash discharge. Their use is limited chiefly to an area around the burner quarls in combustion chamber walls. The main function of burner quarls is to shape coal/air flow for an optimum combustion, and to protect the water tubes from excessive erosion by impaction of abrasive coal mineral particles. As a result of

Table 11.1 Temperature environment for adhesion of ash deposit in coal-fired systems

Locality	Deposit adhesion interface temperature (K)	Deposit exposure temperature (K)
Conventional boilers		
Water tubes in combustion chamber	650–750	1500–1750
Primary superheater tubes exposed to flame radiation	750–800	1300–1500
Secondary superheater and reheater tubes in nonradiative flue-gas passage	800–950	1100–1300
Refractory burner quarls in pulverized-coal-fired boilers	900–1000	1600–1750
Refractory lining of furnace walls of cyclone-fired boilers	900–1000	1700–1900
Fluidized-bed combustion and gasification	750–850	1000–1200
Coal-gas-fired turbines	800–1000	1000–1200

build-up of ash deposits on burner quarls, the flow of coal/air stream can be badly distorted, causing severe malcombustion.

Silicon carbide is the principal refractory material used for lining the burner quarls of pulverized-coal-fired boilers. The carbide refractories are also widely used as a lining material on the furnace-wall tubes of cyclone-fired boilers, where the high flame temperature makes it necessary to protect the boiler tubes against high-temperature corrosion.

The rate of development of adhesive bond R_a between a heat-exchange surface and ash deposit may be considered to be a function first of the interface temperature T_a and also of the temperature difference between the flame or flue gas and the deposit-collecting surface, ΔT:

$$R_a = f(T_a, \Delta T) \tag{11.9}$$

Table 11.1 shows that the temperature of the deposit-collecting surface of water tubes in the combustion chamber is comparatively low, but the tubes are exposed to flame radiation and consequently there is a large temperature difference, about 1000 K, between the exposure temperature and that of deposit/tube-surface interface. As a result, the ash particles arriving at the collecting surface have a high temperature, probably above 1500 K. Further, there is a steep temperature gradient, maybe as much as 400–500 K mm^{-1}, across a thin layer of deposited material. Within the porous deposit matrix, condensable ash constituents, chiefly sulfates and some chlorides (Chap. 7) migrate toward the tube surface, as discussed in Chap. 12.

As the temperature of the deposit-collecting surface increases to 900 K—i.e., that of the secondary (hottest) superheater steam tubes—the exposure temperature must be limited so that ΔT, as defined in Eq. (11.9), does not exceed 500 K. Otherwise, combined effects of the high-temperature deposit collecting surface, the "nonfrozen" ash particles, and a steep temperature gradient within the deposit would result in a rapid formation of sintered and bonded deposits.

11.3 THERMAL AND CHEMICAL COMPATABILITY OF ASH DEPOSITS WITH BOILER-TUBE SURFACES

The steel alloys that are most frequently used for construction of the high-temperature heat-exchange tubes in coal fired boilers may be placed in three categories:

1. Mild steels, annealed ("killed") low-alloy steels used to fabricate water preheater (economizer) tubes and steam generating tubes
2. Ferritic steels, e.g., 1CrO·5Mo, 2.25Cr·11Mo, 5 Cr, 9Cr, and 12Cr steels, used to construct superheater and reheater tubes
3. Austenitic steels, e.g., 18Cr·12NiNb, used in the steam-outlet section of super-heaters

The figures before the chemical symbols refer to weight percent of the alloying metals added to iron to obtain steels of required properties. The steel alloys are chosen for their metallurgical and fabrication properties, e.g., creep resistance, and for their resistance to high-temperature corrosion, as discussed in Chap. 17. Resistance to high-temperature corrosion of different steel alloys in boiler-flue gas depends largely on the chemical and structural composition of oxide layers formed on the tube metal, and on the strength of the adhesive bond between the protective oxide and parent metal.

The chemical composition and thermal behavior characteristics of the oxide layer on boiler-tube steels influence markedly the strength of adhesive bond formed between the oxide surface and ash deposits. It is therefore expedient to consider how far the metal oxide interface is chemically and thermally compatible with both its progenitory alloy and the ash deposit. Thermal compatibility, or lack of it, is probably easier to assess than a chemical match or mismatch. The research and experience in fabrication of glass/metal seals and in manufacture of enamel-coated products show that for strong seals or adhering coatings it is necessary to have the coefficient of thermal expansion of metal L_m compatible with that of coating material L_g, as discussed by Holland (1964). For an enamel coating on steel it is propitious to have an L_m value slightly higher than that of L_g (Hull et al., 1941):

$$L_g \cong 0.9 L_m \tag{11.10}$$

This is because steel is strong both in compression and in tension, whereas glass is strong in compression but weak in tension. On cooling, the glass phase of a metal/glass seal would remain in compression, and a failure would not take place as long as the residual unbalanced stresses do not exceed the interface bond strength.

Table 11.2 shows approximate values of the coefficient of thermal expansion of the ferritic and austenitic steels, metal oxides, and silicate ash deposit material. In the case of adherence of ash deposit on boiler tubes, we are looking for a mismatch in the thermal property so that the adhesive bond may weaken on thermal cycling.

The data in Table 11.2 show that the coefficient of thermal expansion of mild steel and ferritic steels is not greatly different from that of their oxides and the

Table 11.2 Coefficient of thermal expansion of boiler-tube steels, oxides, and silicate constituents of ash deposit[a]

Material	Thermal expansion, $\Delta m/m$ (K^{-1})
Steels	
Mild steel and ferritic steels	$11-12 \times 10^{-6}$
Austenitic steels	$16-18 \times 10^{-6}$
Oxides	
Tube metal oxides (Fe_3O_4, Cr_2O_3, NiO)	$8-10 \times 10^{-6}$
Deposit constituents	
Glassy material	$6-9 \times 10^{-6}$
Quartz (crystalline)	$5-8 \times 10^{-6}$
Silicates in fired brick	$7-8 \times 10^{-6}$

[a]From Hodgman (1962).

ash-deposit constituents. It is therefore evident that there is no gross incompatibility in the thermal expansion characteristics, and strongly bonded ash deposits once formed on mild-steel tubes are not easily dislodged on thermal cycling.

In contrast, the thermal expansion of austenitic steel is significantly higher than that of the oxides and deposit material. In the absence of boiler deposit, the oxide material in the form of thin film is able to absorb thermal stresses and the adhesive film remains intact on cooling. However, it appears that the oxide layer is unable to absorb thermal stresses in a similar manner when contaminated and constrained by bonded ash deposits. It is therefore a usual occurrence that ash deposits peel off the austentitic steel tubes on cooling, whereas the deposit formed on ferritic steels under the same conditions remain firmly attached to the tubes. This was particularly noticeable when a variety of ferritic and austenitic steels were tested in an experimental loop in the high-temperature superheated steam section of a coal-fired boiler, as discussed in Sect. 11.5. When the ash deposit peeled off the austenitic tubes on cooling, there remained a thin layer of black oxide attached to the deposit. However, it appears that this small loss of oxide did not significantly enhance the rate of corrosion of the austenitic steel tubes.

The poor adhesion of ash deposits on austenitic-steel tubes and strong bonding on ferritic-steel surfaces are only partly due to their thermal expansion properties. Chemical compatibility of the oxidized steel surface with silicate ash is also of great significance. King et al. (1959) have stated that in order to obtain good adherence of enamel coatings on metals, the enamel material at the interface must become saturated with an oxide of the metal, e.g., FeO on the surface of low-alloy steels. Coal-ash deposit on boiler tubes contains a substantial quantity of iron oxide, usually between 5 and 25 percent by weight, and thus the deposit interface can become saturated with FeO. An approximate solubility limit of FeO in silicate ashes of different composition can be calculated from the "solubility factors" given by King et al. (1959):

$$FeO_{SOL} = 2.23 \ P_2O_5 + 1.7 \ B_2O_3 + 1.2 \ SiO_2 \qquad (11.11)$$

$$- (0.63 \text{ Na}_2\text{O} + 0.45 \text{ K}_2\text{O} + 1.21 \text{ Al}_2\text{O}_3 + 0.75 \text{ ZrO}_2 \qquad (11.11)$$
$$+ 1.34 \text{ TiO}_2 + 0.4 \text{ CaO} + 0.11 \text{ BaO} + 0.92 \text{ MgO}) \qquad \text{(Cont.)}$$

where FeO_{SOL} denotes the solubility limit of iron oxide by weight percent of a silicate melt having the concentration of constituent oxides given also by weight percent.

Concentration of acidic oxides of P_2O_5 and B_2O_3, and those of basic oxides of BaO and ZrO_2 in typical coal ashes, are small. Thus, for the purpose of calculating an approximate solubility limit of FeO these oxides can be left out and Eq. (11.11) simplified to

$$\text{FeO}_{\text{SOL}} = 1.2(\text{SiO}_2 - \text{Al}_2\text{O}_3) - 0.6(\text{CaO} + \text{MgO} + \text{Na}_2\text{O} + \text{K}_2\text{O} + \text{TiO}_2)$$
$$(11.12)$$

The decreasing effects of basic oxides on the solubility of FeO was averaged to 0.6 in Eq. (11.12). The possible error thus introduced is not large when the main basic oxide constituents are Na_2O, K_2O, and CaO. For ashes of exceptionally high MgO and TiO_2 contents, Eq. (11.11) should be used to calculate the FeO solubility limit. Table 11.3 gives the calculated solubility limits of FeO in bituminous coal ashes of differing composition.

The data in Table 11.3 show that the FeO content in iron-rich ashes (samples 4 and 5) exceeds the limit of solution of FeO in a glassy phase of silicate ash, and it is therefore to be expected that with these deposits a strong adhesive bond would readily develop at the oxide interface. The difference between the FeO solubility limit FeO_{SOL} and the amount of iron oxide present in ash FeO_A can be taken as an adhesion index I_{AD},

$$I_{\text{AD}} = \text{FeO}_{\text{SOL}} - \text{FeO}_A \qquad (11.13)$$

where a negative or a small positive difference would indicate that the ash deposit has an explicit tendency to adhere on ferritic steel boiler tubes. The empirical adhesion index thus derived is related to the silica ratio (Nicholls and Reid, 1940)

Table 11.3 Calculated solubility limit of FeO in coal-ash slags[a]

Ash num-ber	Percent by weight									
								FeO		
	SiO_2	Al_2O_3	CaO	MgO	Na_2O	K_2O	TiO_2	Solubility limit	Amount present	Differ-ence
1	50.7	34.1	1.7	1.7	0.3	1.8	1.2	15.9	6.8	9.1
2	48.6	28.0	3.4	1.9	1.9	3.1	1.0	17.9	8.1	9.8
3	43.6	24.6	7.7	2.9	0.7	2.2	1.0	14.1	11.3	2.8
4	42.2	28.1	4.3	1.1	0.9	1.7	0.8	11.6	15.0	-3.6
5	37.2	21.4	8.1	1.6	0.6	1.4	0.8	11.5	21.4	-9.9

[a]Analytical data by Watt and Thorne (1965) and Raask (1959; unpublished results).

and to the viscosity index of Watt and Fereday (1969). Further, the adhesion index as defined by Eq. (11.13) is related to the sintering, fouling, and slagging indices discussed in Chap. 10. All these relationships are based on the premise that an increase in the amount of FeO, CaO, and Na_2O in ash enhances the deposit-forming propensity and adhesion. This does not apply for the sub-bituminous coal ashes, where aluminosilicates do not constitute a major fraction of the deposit material.

11.4 MECHANICAL AND CHEMICAL BONDING AT BOILER–TUBE/ASH–DEPOSIT INTERFACE

Ash deposits on boiler tubes can be keyed to the surface of metal oxide by mechanical and chemical bonds. Mechanical bonding is enhanced by extended surface at the interface, as shown in Fig. 11.3a. Boiler tubes are not polished and thus have an extended surface that is further increased by oxidation and chemical reactions between the oxide layer and ash deposits. It is therefore evident that a comparatively rough surface of boiler tubes constitutes an anchorage for keying ash deposits to the heat-exchange elements.

Dietzel (1934) and Staley (1934) have proposed that the chemical reactions at the enamel/metal interface can be discussed in terms of electrolytic cells set up

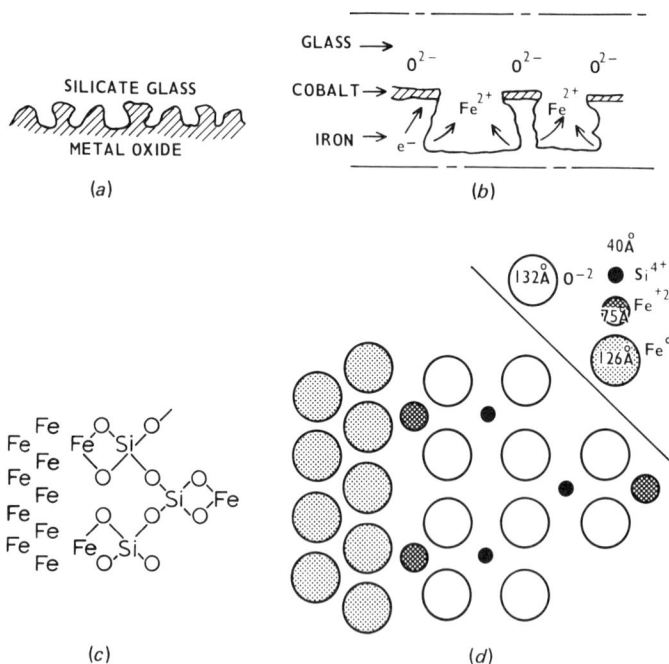

Figure 11.3 Schematic representation of mechanical and chemical bonds at boiler tube/ash deposit interface. (a) Mechanical bonding. (b) Galvanic corrosion cavities. (c) Hypothetical structure of ferrous silicate glass. (d) Chemical bond at iron/silicate interface.

Table 11.4 Free energy (ΔG, kJ mol^{-1} O$_2$) of formation of oxides of boiler-steel and coal-ash constituents

Oxide	$-\Delta G$	Oxide	$-\Delta G$
$2Ca + O_2 \rightarrow 2CaO$	1070	$W + O_2 \rightarrow WO_2$	393
$Zr + O_2 \rightarrow ZrO_2$	915	$Mo + O_2 \rightarrow MoO_2$	393
$\frac{4}{3}Al + O_2 \rightarrow \frac{2}{3}Al_2O_3$	907	$Sn + O_2 \rightarrow SnO_2$	393
$2Ti + O_2 \rightarrow 2TiO$	849	$\frac{4}{3}Fe + O_2 \rightarrow \frac{2}{3}Fe_2O_3$	376
$Ti + O_2 \rightarrow TiO_2$	761	$2Co + O_2 \rightarrow 2CoO$	334
$Si + O_2 \rightarrow SiO_2$	694	$\frac{2}{3}Mo + O_2 \rightarrow \frac{2}{3}MoO_3$	330
$\frac{4}{3}B + O_2 \rightarrow \frac{2}{3}B_2O_3$	677	$2Ni + O_2 \rightarrow 2NiO$	318
$2Mn + O_2 \rightarrow 2MnO$	619	$\frac{4}{3}As + O_2 \rightarrow \frac{2}{3}As_2O_3$	305
$4Na + O_2 \rightarrow 2Na_2O$	581	$\frac{4}{3}Sb + O_2 \rightarrow \frac{2}{3}Sb_3O_3$	301
$\frac{4}{3}Cr + O_2 \rightarrow \frac{2}{3}Cr_2O_3$	577	$2Pb + O_2 \rightarrow 2PbO$	263
$2Zn + O_2 \rightarrow 2ZnO$	497	$2Bi + O_2 \rightarrow 2BiO$	251
$\frac{4}{5}V + O_2 \rightarrow \frac{2}{5}V_2O_5$	460	$\frac{4}{3}Bi + O_2 \rightarrow \frac{2}{3}Bi_2O_3$	209
$\frac{4}{5}P + O_2 \rightarrow \frac{2}{5}P_2O_5$	426	$4Cu + O_2 \rightarrow 2Cu_2O$	201
$2Fe + O_2 \rightarrow 2FeO$	393	$2Cu + O_2 \rightarrow 2CuO$	146

Sources: King et al. (1959); Janaf (1967); Kubaschweski et al. (1967).

between the metals of different electrochemical potential. It was suggested that cobalt or nickel, precipitated in the enamel when in contact with the steel surface, forms short-circuited local cells in which iron is the anode. The current flows from iron though the melt to cobalt and back to iron. The result is that iron goes into solution, the surface becomes roughened, and the enamel material anchors itself into the cavities, as shown in Fig. 11.3b. The galvanic reactions where cobalt acts as an oxygen carrier could be summarized as follows:

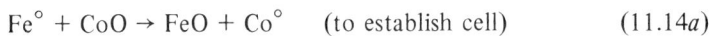

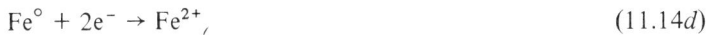

$$Fe^\circ + CoO \rightarrow FeO + Co^\circ \quad \text{(to establish cell)} \qquad (11.14a)$$

$$2Co^\circ + O_2 \rightarrow 2CO^{2+} + 2O^{2-} \qquad (11.14b)$$

$$CO^{2+} + 2e^- \rightarrow Co^\circ \qquad (11.14c)$$

$$Fe^\circ + 2e^- \rightarrow Fe^{2+}, \qquad (11.14d)$$

Harrison et al. (1952), using radioactive cobalt in a porcelain enamel, confirmed that cobalt was liberated from molten enamel and deposited during early stages of firing. The data on the free energy change of formation of metal oxides ΔG given in Table 11.4 show that nickel also can function as a cathode metal in a galvanic cell with iron.

The galvanic reactions will take place at a much faster rate in the low-viscosity phase of sulfates in boiler deposit than in highly viscous silicate glass. However, rapid reactions at the tube surface/deposit interface may not be necessary or appropriate for development of a strong bond between the ash deposit and boiler tubes. In metal/glass seal and metal/enamel coating technology, the adhesive bonds formed on heating have to be completed in a few hours, whereas those in boiler deposits can form over a period of days or weeks. Weyl (1953) has suggested that the adhesive bond between metal and glass can be high when there is a gradual rather than abrupt change in the glass-phase composition near the interface.

When the glassy material of ash deposits is brought into intimate contact with boiler tubes, either by the action of surface tension or by the galvanic reaction, the controlling parameter in mechanical bonding is the strength of the glassy phase at the narrowest cross-sectional area of contact cavities (Fig. 11.3b). The strength of a glass can vary, and an annealed glass may have a tensile strength of around 50 MN m^{-2}, giving a mechanical bond of the maximum strength of 35 MN m^{-2}. However, the glass at the interface may be stronger or weaker depending on whether the conditions in the keying cavities increase or decrease local flaws and resultant stresses.

Chemical bonds—ionic or covalent, as shown in Fig. 11.3, c and d—at the metal oxide/deposit interface are potentially strong, with theoretical values over 10^9 N m^{-2}. It is, however, impossible to estimate the number of sites and the size of contact areas at the interface where the chemical bonds may be effective. In any case, the cohesive strength of the deposit matrix is the limiting factor, since it is lower than that of chemical bonds by several orders of magnitude. In practice, this means that when a strongly adhering deposit is subjected to a destructive force—e.g., sootblower jet—failure occurs within the deposit matrix and there remains a residual layer of ash material firmly bonded to the tube surface.

11.5 BOND–STRENGTH MEASUREMENTS

A variety of techniques have been used to measure the strength of adhesive bonds formed on heating metals when in contact with glasses, enamels, and refractory material. It is usual practice to carry out the strength measurements after the seal formed at high temperatures has been allowed to cool to the room temperature. Hull et al. (1941) prepared a glass sandwich disc joint, 13 mm in diameter, between two metal rods for their tensile strength measurements of glass/metal bond. Klomp (1972) prepared the adhesion test pieces where two lengths of a tube, glass or alumina, were joined by a metal seal. The seal specimens then were tested in a four-point bending arrangement and the strength σ_b of the interface bond was calculated from the applied force F using the relation

$$\sigma_b = \frac{8F_1(L - 2L_1)d_1}{\pi(d_1^4 - d_2^4)} \tag{11.15}$$

where L = distance between the supports

L_1 = distance between the point of application of the loading force and the joint

d_1, d_2 = outside and inside diameter of the two tubes, respectively

King et al. (1959) measured the strength of adherence of enamel coating on metals by bonding the test pieces, 34-mm squares, with an epoxy resin to the adaptor blocks in a tensile testing machine. The method may have a limitation where the strength of the experimental joints exceeds that of the adhesive used to bond the joints in the testing assembly. For example, the tensile strength of an epoxy resin is around 40 MN m^{-2}, and the enamel coating on metals frequently has an adhesive strength above 50 MN m^{-2}. A method of overlapping joints can be used for testing metal/glass and metal/enamel bonds when the metals are available in the form of sheets or plates, as described by Matting and Witt (1967). They measured the strength of glass bonds between two pieces of metal to failure in tension and in shear.

Nicholas et al. (1968) and Nicholas (1968) have used an interesting method of measuring the adhesion strength of frozen metal droplets on alumina. Metal specimens 3 mm in diameter and 5 mm in height on alumina plaques were melted in a vacuum furnace. The interfacial strength was measured after cooling by placing the specimen in a shearing jig attached to a tensometer and noting the force needed to push the drop off the plaque. It was established that when the contact angle was greater than 1.9 rad (110°) the interface failed in tension and that the bond strength σ_b is related to the contact angle θ by the expression

$$\sigma_b = 0.25 \; \sigma_t \cos\left(\theta - \frac{\pi}{2}\right) \qquad (11.16)$$

where σ_t is the tensile strength of the interface. At lower contact angles, the bond strength equalled the shear strength of the interface and was independent of the contact angle. Of the nine metals tested—Ag, Al, Au, Co, Cu, Fe, Ni, Pd, and U—iron had the highest interfacial strength to alumina, up to 19 MN m^{-2}, and the cobalt/alumina bond broke on cooling below 500 K. This was explained by the fact that cobalt and also uranium undergo a martensitic transformation on cooling, resulting in severe shear stresses that destroyed the cobalt/alumina and the uranium/alumina bond.

Moza and Austin (1979), and Moza et al. (1980) have used a droplet technique to measure the adhesive bond of coal-ash slag on the surface of oxidized ferritic steel. An ash pellet 4 mm in diameter, 4 mm in height, and weighing about 50 mg, supported on a 0.3-mm platinum wire, was fused in a small gas-fired furnace. The slag droplet thus formed, about 3 mm in diameter, was then allowed to strike a steel disc after falling through a distance of 30 mm. The temperature of the slag droplet was between 1600 and 1700 K, and that of the steel surface was kept between 700 and 950 K. Subsequently, the adhesive bond strength was measured by shearing the droplet off the steel substrate with a push rod. Their results, shown in Fig. 11.4, were consistent with the chemical composition of ashes, chiefly calcium and iron oxide content (Table 11.5). The high CaO content, together with the fairly high amount of iron oxide (Fe_2O_3), in Wyoming sub-bituminous ash produced strongly adhering slag on the surface of mild steel at 850 K, as shown by curve A in Fig. 11.4.

The adhesion temperature of soda glass on a metal substrate was determined by Oel and Gottschalk (1966) by heating a glass disc sandwiched between two metal discs in a vertical furnace. Raask (1973) had adopted the technique for measuring

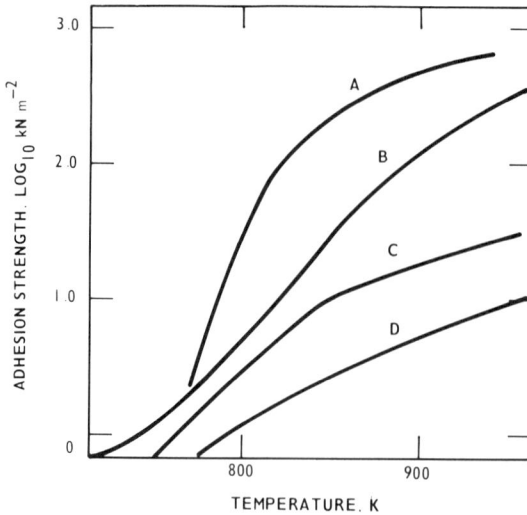

Figure 11.4 Adhesion of coal-ash slags on mild steel: A, Wyoming sub-bituminous; B, Texas lignite; C, Pennsylvania bituminous; D, Illinois bituminous.

the strength of the adhesive bond developed when a boiler deposit was sandwiched between two discs of ferritic or austenitic steels. The deposit material was taken immediately after boiler shut-down from the superheater tubes of a pulverized coal fired boiler fuelled with a mixture of East Midlands (U.K.) coals. The flue-gas temperature in the superheater prior to boiler shut-down was about 1300 K, and the tube metal temperature was 850 K. The deposit material consisted of 30 percent alkali-metal sulfates in weight ratio of 2 to 1 of Na_2SO_4 to K_2SO_4, the remainder being silicate ash.

A layer of deposit 3 mm thick was sandwiched between two metal discs (20 mm in diameter) made of boiler-tube steels, as shown in Fig. 11.5a, and introduced in the preheated, vertical furnace, similar to that shown in Fig. 10.7, used for the electric conductance measurements of ash sinter compacts. After a time interval

Table 11.5 Analysis (weight percent) of coal ashes used in adhesion Tests[a]

Ash constituents (wt. %)	Ash A (Wyoming sub-bituminous, ash 19.9 wt. %)	Ash B (Texas lignite, ash 15.0 wt. %)	Ash C (Illinois bituminous, ash 12.3 wt. %)	Ash D (Pennsylvania bituminous, ash 17.1 wt. %)
SiO_2	50.9	47.7	49.7	52.4
Al_2O_3	19.0	12.0	15.8	27.9
TiO_2	1.3	2.1	1.3	1.9
Fe_2O_3	7.7	4.7	13.0	9.5
MgO	2.2	2.9	0.9	1.0
CaO	11.8	15.0	6.5	2.0
Na_2O	0.3	0.6	1.1	0.4
K_2O	1.3	0.4	1.9	3.3
SO_3	5.5	14.6	7.0	1.7

[a]By Moza et al. (1980).

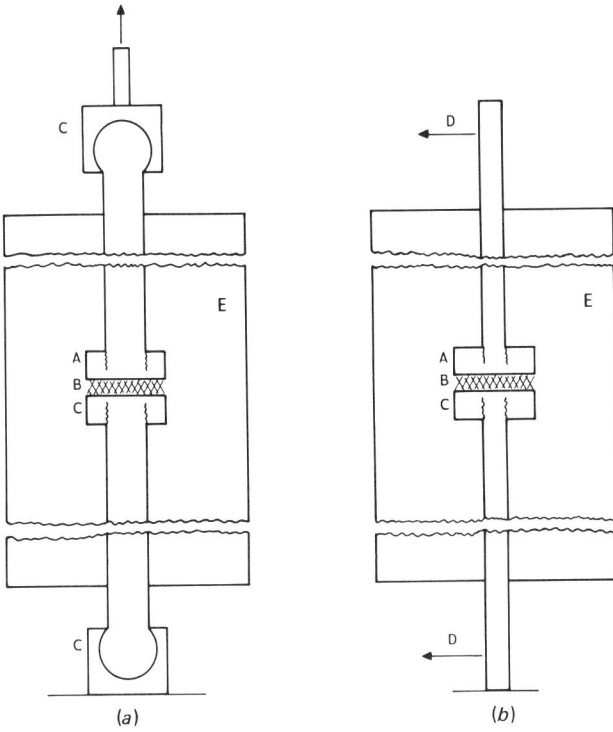

Figure 11.5 Furnace assembly for studying bond development between boiler-tube steels and ash deposits. A, boiler-steel discs; B, deposit sample; C, spherical joints; D, to A/C conductance bridge; E, furnace. (*a*) Bond strength measurement. (*b*) Electrical conductance.

lasting from 1 to 25 d, the bond was ruptured by applying a tensile force without prior cooling. The results in Fig. 11.6 show that there was an approximately linear relationship between the bond strength and time. Further, it is evident that the bond strength between boiler deposit and austentitic steel (type 18Cr·13Ni·Nb) was significantly lower than that with ferritic steel (9 percent Cr steel). This was to be expected as a result of the thermal and chemical composition incompatibility of the austenitic steel surface with boiler deposit, as discussed in Sects. 11.2 and 11.3. Figure 11.7 shows that within a limited temperature range of 775–900 K, the rate of formation of adhesive bond between ferritic steel and boiler deposit as determined by the tensile-strength measurements increased exponentially with temperature.

In relating the results of laboratory measurements of the bond strength to formation of boiler deposits, it is useful to note that an adhesive bond of 0.1 KN m^{-2} strength is able to support the weight of a lightly sintered ash deposit 10 mm thick, having a density of 1000 kg m^{-3}. In order to support a large mass of slag, 100 mm thick and of the density of 2000 kg m^{-3}, a bond strength of 2 KN m^{-2} is required.

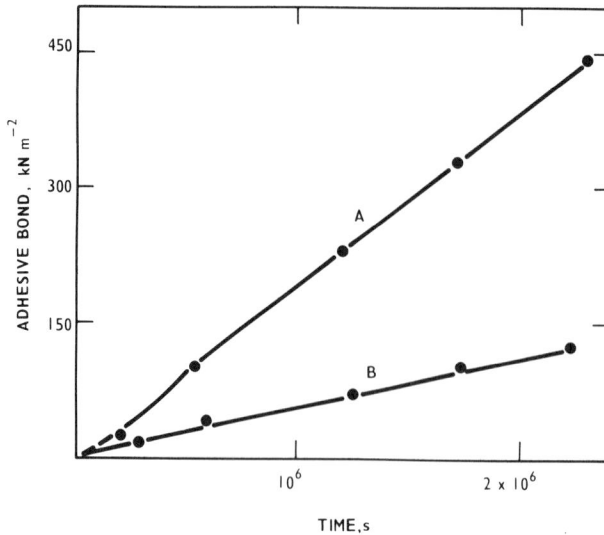

Figure 11.6 Bond strength between ash deposit and boiler-tube steels at 900 K: A, ferritic steel; B, austenitic steel.

The adhesive bond between boiler-ash deposit and the surface of ferritic steels can attain exceptionally high strengths. This was established by examining the deposits formed on different steel specimens tested in an experimental superheater loop described by Alexander and Raask (1959). Favorable conditions for the formation of firmly bonded deposits were as follows:

1. The iron oxide content of coal was above 20 percent expressed as Fe_2O_3, giving a high adhesive index according to Eq. 11.13.
2. The ash-collecting surface was a 5 percent chromium ferritic steel, which formed a strongly adhering oxide layer for a firm anchorage for the deposit.
3. The tube metal temperature was high, 950 K, which enhanced the rate of ash capture, as discussed in Chap. 12. The flue-gas temperature at that position was approximately 1250 K.
4. The ferritic tube in the experimental loop was sheltered from direct action of sootblower. Weak turbulence caused by the jet removed some of the unsintered silica ash, leaving iron-rich deposit firmly bonded to the tube.

The iron-rich deposit had grown in thickness to about 20 mm after 9 mo, and cohesive strength of the deposit material increased toward the surface of tube metal. There was a corresponding enrichment of iron along the deposit profile, as depicted by curve A in Fig. 11.8. Microscopic examinations showed that there was no discernable interface boundary between the ash deposit and metal oxide.

Formation of the adhesive bond between ash deposits and steel surfaces can be studied by monitoring the electrical conductance across a layer of deposit material

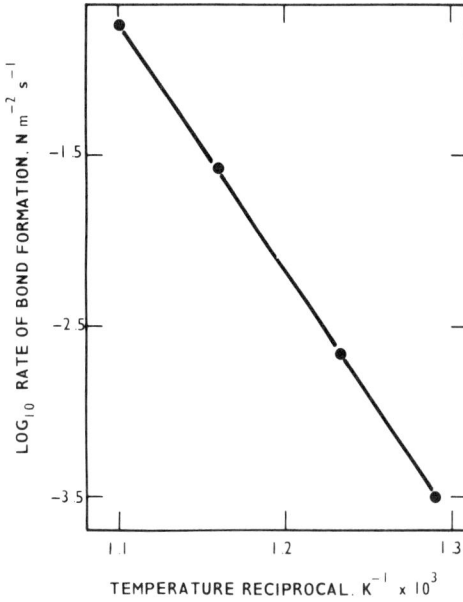

Figure 11.7 The effect of temperature on ash deposit/ferritic steel bond.

sandwiched between the metal discs, as shown in Fig. 11.5b. The furnace assembly and the method of conductance measurements are the same as those described in Chap. 10. The essential differences are that the alumina support tubes and platinum electrode contacts in the sintering measurements were replaced by steel disc electrodes screwed on the steel extension tubes. The electrode discs are replaceable so that the adhesion of ash deposit on different boiler steels can be investigated. Further, the electrode contact can be cooled by a gas stream in order to simulate

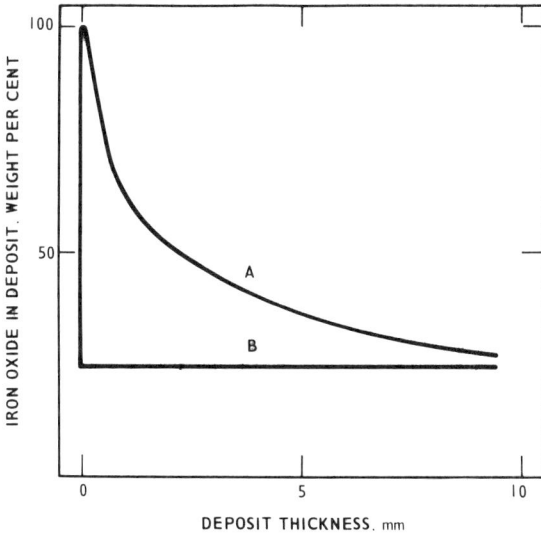

Figure 11.8 Iron oxide concentration profile of (A) firmly adhering deposit and (B) nonadhering deposit.

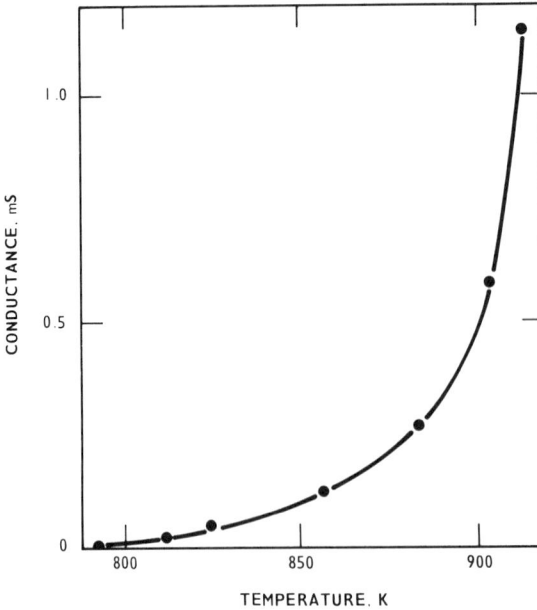

Figure 11.9 Electrical conductance of glass/steel interface at different temperatures.

approximately the temperature gradient under which layer-structured deposits form on boiler tubes in coal-fired boilers.

Initial measurements were made with soda glass (U.S. Bureau of Standard Glass, Chap. 10) where a 3-mm-thick layer of powdered glass was placed between the electrode discs made of 5 percent chromium steel. The results in Fig. 11.9 suggest that the metal/glass interface bond started to develop at 800 K and that it increased exponentially with temperature. Figure 17.6 shows a similar conductance-against-temperature plot for a boiler deposit that contained 10 percent alkali-metal sulfates of the molar ratio of $2 Na_2SO_4$ to K_2SO_4. The results indicate that a strong deposit bond started to develop at 900 K. So far there are insufficient experimental data to evaluate the usefulness of the conductance measurements for studying the phenomenon of adhesion of ash deposits on boiler tubes. However, these measurements should give some additional information of the activity of bond-forming constituents in boiler deposits when combined with bond-strength measurements.

NOMENCLATURE

A	Hamaker constant
B	constant
D	particle density
d_1	outside diameter of tube
d_2	inside diameter of tube
F	van der Waals force
F'	retarded van der Waals force

F_g	gravitational force on particle
F_r	ratio of van der Waals to gravitational forces
$F_{r'}$	ratio of retarded van der Waals to gravitational forces
F_1	applied force
FeO_A	iron oxide content of ash
FeO_{SOL}	solubility factor of iron oxide in slag
g	gravitation constant
h	distance between plane and hemisphere surfaces
I_{AD}	adhesion index
L	distance between supports
L_1	distance between bond and point of force application
L_g	thermal expansion of coating material
L_m	thermal expansion of metal
R_a	rate of adhsion bond development
r	radius of hemisphere
T_a	temperature of deposit-collecting surface
ΔT	temperature difference
W_a	work of adhesion
W_c	work of cohesion
θ	contact angle
γ	surface tension force
σ_b	shear strength of interface bond
σ_t	tensile strength of interface bond

TWELVE

DEPOSITION MECHANISMS, RATE MEASUREMENTS, AND THE MODE OF FORMATION OF BOILER DEPOSITS

12.1 DEPOSITION BY VAPOR AND SMALL–PARTICLE DIFFUSION

It was established in Chaps. 7 and 8 that alkali-metal sulfates, $Na_2 SO_4$ and $K_2 SO_4$, are the principal condensable salts formed in pulverized-coal flame. Although the sulfates are in part captured or formed directly on the surface of silicate-ash particles, as discussed by Raask and Goetz (1981) and Raask (1981a), there remains a substantial fraction, over 50 percent of the total, in the vapor phase when the flue-gas temperature is above 1200 K (Halstead and Raask, 1969).

Hedley et al. (1965) have shown that the rate of deposition of sodium sulfate (R_D) on a cooled target in flue gas can be calculated from the mass transfer equation

$$R_D = \frac{A(M_g - M_s)}{(Sc)^{2/3}(Pr)^{1/3}} (\mu\rho)^{1/2} \left(\frac{v}{D}\right)^{1/2} \tag{12.1}$$

where μ, ρ, v = viscosity, density, and velocity of the flue gas, respectively
$\quad\quad D$ = diameter of collecting target
$\quad M_g, M_s$ = concentration of $Na_2 SO_4$ vapor in flue gas and at the deposit surface, respectively
$\quad\quad A$ = constant; 0.5 for stream-lined gas flow, and higher for turbulent flow

$$Sc = \frac{viscosity}{density \times diffusivity} \quad (Schmidt \ number) \tag{12.1a}$$

$$Pr = \frac{specific \ heat \times viscosity}{thermal \ conductivity} \quad (Prandtl \ number) \tag{12.1b}$$

Equation (12.1) shows that the rate of deposition of volatilized species in the flue gas on a cooled target is proportional to the difference between the partial pressure of the species in the gas (M_g) and that at the deposit surface (M_s). The latter is usually small in comparison with the M_g value, and it can be omitted in Eq. (12.1).

Brown (1968) has suggested that sulfate particles form in the boundary layer of flue gas surrounding a cooled target, and that subsequent transport to the collecting surface is effected by solid-particle diffusion. He postulated this deposition model to explain comparatively high rates of sulfate deposition when there was a large difference between the temperature of target and that of sulfate dewpoint. The results of deposition measurements in coal- and oil-fired boilers (Sect. 12.5) showed, however, that the temperature of the cooled probe below sulfate dewpoint did not influence the deposition rate. These results suggest that nucleation and growth of sulfate particles had taken place both at the surface of the collecting target and on the ash particles.

The flue gas in coal-fired boilers is likely to contain a small weight fraction of silicate ash in the form of fume (Chap. 8) with the particle diameter below 0.1 μm. These particles can be transported to boiler tubes by their random (Brownian) movements; this phenomenon is known as particle diffusion deposition. From kinetic theory, Einstein (1906) has shown that distance λ covered by a moving particle is controlled by its diffusivity D_v:

$$\lambda^2 = 2D_v t \tag{12.2}$$

where t is the time and D_v is given by

$$D_v = \frac{kT}{3\pi\mu D} \tag{12.3}$$

where k is the Boltzmann constant, T is the temperature, μ is the absolute viscosity of flue gas, and D is the diameter of the collecting target. Diffusivity D_v can be calculated also from the Stokes–Einstein equation,

$$D_v = \frac{kT[1 + 0.86(\lambda/r)]}{6\pi\mu r} \tag{12.4}$$

where λ is the mean free path in molecular motion and r is the radius of particles; other notations are as above.

Stairmand (1950) has obtained a deposition efficiency e_d of a target by considering the time taken for particles to diffuse through a thin layer of gas onto the target surface:

$$e_d = \frac{(8D_v)^{1/2}}{vD} \tag{12.5}$$

where v is the velocity of flue gas. Johnstone and Roberts (1949) have considered particles diffusing through a gas layer of thickness s_g and derived an equation for the efficiency of deposition on a cylindrical target:

$$e_d = \frac{(4D_v)^{1/2}}{v s_g} \tag{12.6}$$

The value of s_g was obtained by analogy with heat transfer:

$$\frac{d}{s_g} = 2 + 0.557 \frac{(\rho v d)^{1/2}}{\mu} + \frac{\mu^{1/3}}{\rho D} \tag{12.7}$$

where d is the particle diameter and other notations are as above.

Hedley et al. (1965) have made an assessment of relative significance of the vapor-phase and particle diffusion mechanism in the mass of particles transported from the flue gas to boiler tubes. Their conclusion was that the amount of material deposited by the particle diffusion is likely to be less significant than that transported by the vapor-phase diffusion.

The surface roughness of oxidized steels and the deposited ash particles on boiler tubes will enhance condensation in the presence of a liquid film, e.g., molten sulfate. This is because the meniscus surface has a lower vapor pressure compared with that at the plane surface. It follows, therefore, that condensation will take place at the points of contact of ash particles on boiler tubes, and the resulting film grows to an equilibrium shape, which has the vapor pressure equal to the particle pressure of the condensing species.

12.2 THERMOPHORESIC AND ELECTROPHORESIC DEPOSITION

Gas-borne particles moving in a space where there is a temperature gradient are subjected to a thermophoresic force that causes diffusion toward the lower temperature region. Einstein (1924) suggested that this force resulted from a difference in heat absorption and subsequent radiation on the hot and cold sides of particles. This, according to Newton's third law, would result in the particle moving away from the heat source. Epstein (1929) has derived an expression for the thermal force F_t on the particle:

$$F_t = 4.5\pi \left(\frac{d\mu^2}{\rho T}\right) \left(\frac{\kappa_g}{2(\kappa_g + \kappa_p)}\right) \frac{dT}{dx} \tag{12.8}$$

where d = diameter of particle
μ, ρ, T = gas viscosity, density, and temperature, respectively
κ_g, κ_p = thermal conductivity of gas and particle, respectively
dT/dx = temperature gradient

Smith (1952) has equated the expression for thermal force derived earlier by Cawood (1936) to that of Stokes' law of viscous drag for particles within a boundary layer and obtained an expression:

$$v_\tau = \frac{R\rho_o\lambda_o}{12}\left(\frac{d}{D}\right)\frac{\text{Nu}}{\mu}(T_g - T_t) \qquad (12.9)$$

where v_τ = terminal velocity of particle
R = thermodynamic (gas) constant
ρ_o = density of gas stream at NTP
λ_o = molecular mean free path at NTP
d = diameter of the particle
D = diameter of collector
μ = absolute viscosity of gas stream
T_g, T_t = temperature of gas stream and collecting target, respectively
Nu is the Nusselt number and it is defined:

$$\text{Nu} = \frac{\text{heat transfer coefficient} \times \text{tube diameter}}{\text{thermal conductivity}} \qquad (12.9a)$$

Smith concluded that deposition by thermal diffusion becomes significant when the particle diameter was in the range of 0.5–5 μm. Figure 12.1 shows the effect of temperature gradient on terminal velocity of the particles moving towards a cooled target.

Rosenblatt and LaMer (1946) have derived a simplified equation for terminal velocity v_τ of the particles, in the form

$$v_\tau = \frac{k_1}{T}\frac{dT}{dx} \qquad (12.10)$$

where T = temperature
dT/dx = temperature gradient
k_1 = constant, can be determined experimentally

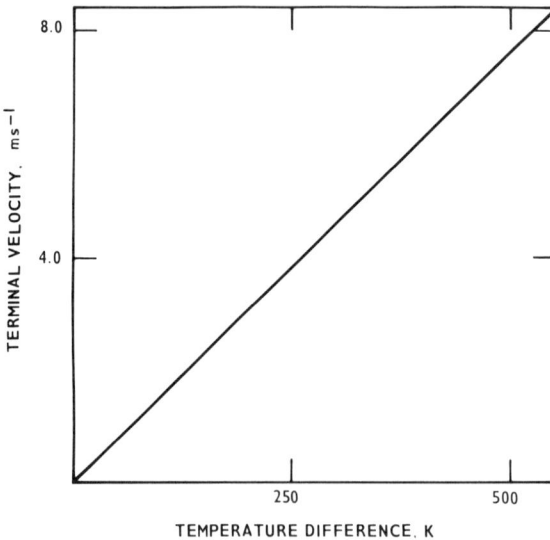

Figure 12.1 The effect of boundary-layer temperature gradient on thermophoresic deposition; particle diameter 1 μm.

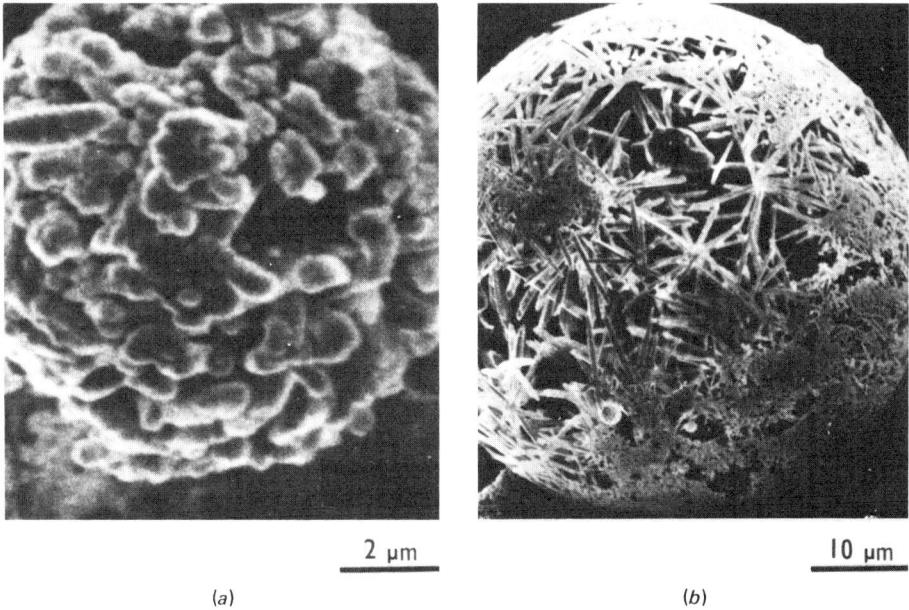

2 μm   10 μm

(a)   (b)

Figure 12.2 Crystalloid skeletons of ash particles revealed on dissolution of glassy matrix. (a) Quartz microcrystals. (b) Mullite needles.

Exact calculations of the rates of thermal deposition of the flue-gas-borne ash on cooled boiler tubes cannot be carried out due to uncertainty of the temperature gradient in the boundary layer of gas surrounding the tubes. However, the plot in Fig. 12.1 suggests that the thermophoresic deposition is likely to play a significant role in transporting small (below 5 μm diameter) particles from the flame to the furnace-wall tubes of coal-fired boilers. This is particularly relevant in the initial stages of deposit formation, when the surface temperature of clean boiler tubes is around 700 K and that of the flame exceeds 1700 K. As a layer of deposit of low thermal conductance builds up on the furnace-wall tubes, its surface temperature increases sharply, and as a consequence the thermophoresic deposition of flame-borne ash particles becomes less significant.

The moving particles in the flue gas can acquire an electrostatic charge as a result of flame ionization, or the charge may arise from frictional forces on particle collision. Deposition of the charged particles as a result of electrophoresic force takes place on a target, situated in the electric field, that has a potential gradient. Penney and Klinger (1962) have suggested the particles on nonhomogeneous composition can acquire a dipole moment. This is likely to occur with silicate ash particles, as these consist chiefly of microcrystals and needles dispersed in a silicate-glass matrix, as discussed by Raask and Goetz (1981) and shown in Fig. 12.2, a and b. Dalmon and Raask (1972) have shown that silicate mineral particles have a high electrical resistivity, above 10^{12} ohm m^{-1}, after being heated to 1700 K. The high resistivity and the nonuniform composition of partially devitrified surface layer of ash result in particles acquiring and retaining a dipole moment.

Small particles with strong dipole moments are aligned under the influence of electrostatic potential and thus deposit with a negative charge on one particle adjacent to the positive charge area on another particle. This orientation results in a coulombic force that can be responsible for an adhesive bond holding the deposited particles onto the earthed conductor, e.g., boiler tubes. The electrophoresic force F_e acting on the particles is given by

$$F_e = 0.5 \; \epsilon_o \left[E^2 - \left(\frac{J_{\rho\epsilon_1}}{\epsilon_o} \right)^2 \right] \tag{12.11}$$

where F_e = force per unit area (a positive force tends to pull the ash particles off
       the collector)
   ϵ_o = permittivity of free space
   $J_{\rho\epsilon_1}$ = permittivity coefficient of ash
   E = potential gradient adjacent to boiler-tube surface
When $E = J_{\rho\epsilon_1}/\epsilon_o$, the particles are no longer attracted to the collector surface. It is therefore evident that ash particles of certain optimum resistivity and permittivity characteristics are attracted and deposited on earthed boiler tubes. The potential gradient E in the flue gas around the boiler tubes is much less than that induced in an electrical precipitator; it is therefore likely that ash particles of comparatively high conductivity are preferentially deposited on boiler tubes as a result of the electrophoresic effect. The silicate ash surface, enriched by alkalis and iron-oxide particles, has a higher conductivity when compared with that of the noncontaminated aluminosilicate and quartz particles (Dalmon and Raask, 1972). It is therefore possible that the electrophoresic deposition is partly responsible for the enrichment of alkali metals and iron-rich constituents of boiler deposits.

12.3 DEPOSITION BY INERTIAL IMPACTION AND SUMMARY OF DIFFERENT MECHANISMS

The mechanism of inertial impaction is the principal mode of transport of the flue-gas-borne ash to boiler tubes, but most particles escape from the "noncaptive" surface. A target will cause the gas stream to diverge (Fig. 12.3a), and the change in direction of the gas flow will exert a drag influence on ash particles. On small particles the drag effect is sufficient to make them follow the streamlines, while the particles just smaller than a critical diameter will crowd towards the outer edges of the target. Thus, the chance of a particle touching the target depends on its inertial momentum, the drag force on the particle, and its position relative to the central line in the flow pattern, as depicted in Fig. 12.3a.

If particles of a given diameter initially at a distance of $0.5D'$ from the central streamline just graze the surface of a cylinder of diameter D, then D' is said to be the characteristic target diameter, and the value of D'/D is referred to as the target or collection efficiency e:

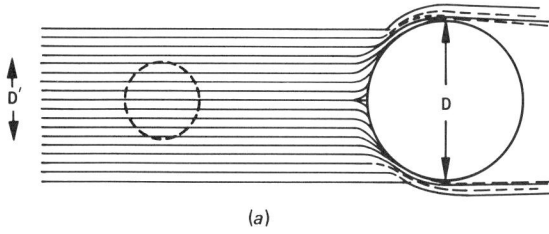

(a)

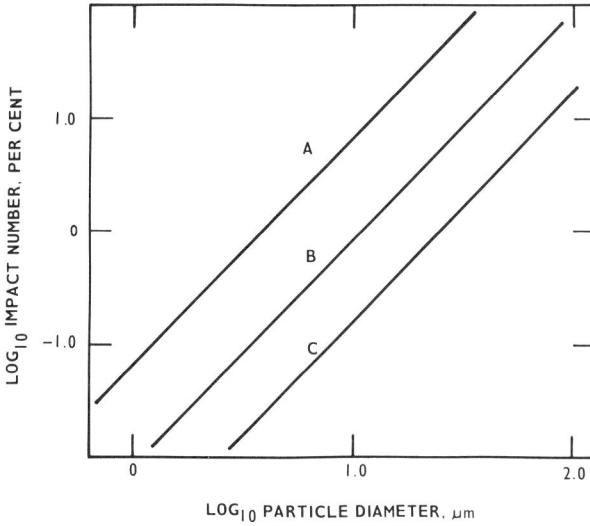

(b)

Figure 12.3 Inertial impaction, with flue gas velocity: A, 300 m s^{-1}; B, 50 m s^{-1}; and C, 10 m s^{-1}. (a) Impaction and deviation of gas-borne particles around tube target of diameter D with impaction zone D'. (b) Dependence of impact number on particle size.

$$e = \frac{D'}{D} \tag{12.12}$$

Based on his experimental results, Sell (1931) has assessed the collection efficiency e of spherical particles in terms of a nondimensional group:

$$e = \frac{mv}{k_2 D} \tag{12.13}$$

where m = mass of particles

v = velocity of gas stream

D = diameter of collecting target

k_2 = resistance coefficient of Stokes' flow law

Later Stairmand (1950) obtained a different expression for the collection efficiency e':

$$e' = \frac{Dg}{vf} \tag{12.14}$$

where g is the gravitational force constant and f is the free-falling velocity of particles. These two expressions are in fact reciprocal:

$$\frac{mv}{k_2 D} = \frac{v_f}{Dg} \tag{12.15}$$

The dimensionless group is referred to as the impact number or particle parameter and is related to the collection efficiency, as shown in Fig. 12.3b. A comprehensive review of impaction efficiency for single collectors is given by Golovin and Putman (1962). Taylor (1940) in his assessment of impact number I arrived at the following relationship:

$$I = \frac{\rho_p d^2 \mu \text{Re}}{9\rho D^2} \tag{12.16}$$

where ρ_p, ρ = density of particles and gas stream, respectively
      μ = viscosity (absolute) of the gas
     d, D = diameter of particles and collecting target, respectively
Re is the Reynolds number and is defined for a gas flow:

$$\text{Re} = \frac{\text{tube diameter} \times \text{velocity} \times \text{density}}{\text{viscosity}} \tag{12.16a}$$

The flue-gas velocity in pulverized coal fired boilers is usually between 10 and 25 m s^{-1}, and the plots in Fig. 12.3b show that ash particles below 5 μm in diameter do not possess sufficiently high inertial momentum to penetrate the boundary layer and will not reach a tubular target. The velocity of gas-borne particles in gas turbines is above 100 m s^{-1}, and thus the threshold particle diameter for inertial deposition is reduced to about 1 μm.

A high rate of impaction of large particles on gas-turbine blades and boiler tubes does not necessarily lead to a rapid build-up of ash deposit. Large particles have inertial momentum in excess of what can be dissipated on impaction and thus they rebound. As a result of the selective retention, the deposit on a gas turbine has a size distribution curve, as shown by curve A in Fig. 12.4, where the particles of 1 to 5 μm in diameter contribute the bulk of captive ash. The corresponding size distribution curve for the particles in initial deposit on boiler tubes (curve B) shows that lower impact velocity results in a capture of particles up to 10 μm in diameter.

The review of possible mechanisms of transporting different ash constituents from the flue gas to cooled boiler tubes can be summarized as follows:

1. Flame-volatilized species, chiefly sulfates, are deposited on cooled boiler tubes via the mechanism of vapor diffusion. This mechanism is more efficient than the particulate ash deposition, and as a result there is usually an enrichment of condensible salts, chiefly sulfates, in boiler deposits.
2. Particle diffusion (Brownian movement) may account for deposition of some

fume particles below 0.1 μm in diameter. In comparison with the mechanism of vapor diffusion and particle deposition, however, the amount of material transported to boiler tubes by this route is small.

3. The thermophoresic and electrophoresic deposition mechanisms are likely to have a marked influence in transporting 0.1- to 5-μm particles from the flue gas to cooled boiler tubes.

4. Inertial impaction is the dominant mechanism in transporting particles above 1 μm in diameter to gas turbine blades, where the flue-gas velocity is above 100 m s^{-1}. The stream velocity of the flue gas in a pulverized-coal-fired boiler is between 10 to 25 m s^{-1}, and correspondingly the gas-borne ash particles have a threshold size of about 5 μm for deposition on water and steam tubes by the mechanism of inertial impaction. Particles above about 5 μm in gas turbines and 10 μm in diameter in the conventional boilers have kinetic energy in excess of what can be dissipated at impaction, resulting in their reentrainment in the flue gas.

5. Combined effects of deposition mechanisms 1-4 result in a preferential deposition of 0.1- to 5-μm diameter particles on gas turbine blades in the absence of molten sulfate or viscous slag deposit. The corresponding range for the ash particles preferentially deposited on boiler tubes is between 0.1 to 10 μm.

The review of different deposition mechanisms gives a useful guidance for assessing the composition of the initial deposit landing on cooled (clean) boiler tubes. However, the rate of initial deposition does not necessarily relate quantitatively to the build-up of boiler deposits. A thin layer of loose particles of ash would

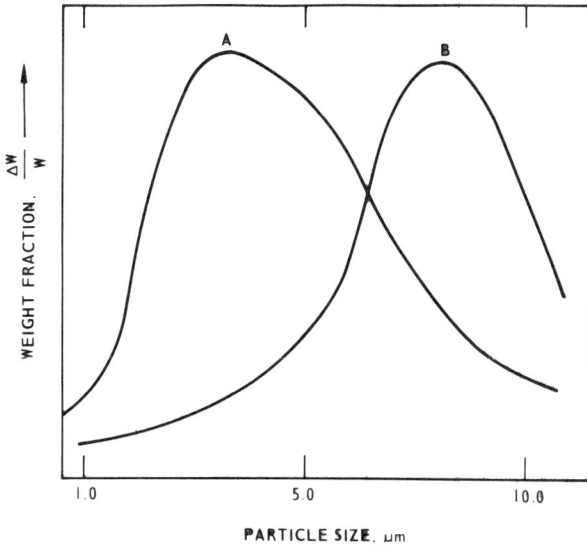

Figure 12.4 Selective deposition by inertial impaction on "noncaptive" surfaces: A, in gas turbines; B, in coal-fired boilers.

largely terminate the thermophoresic and electrophoresic deposition of small particles, and the large, inertially impacted particles would escape from a "dry" surface. It is therefore evident that a captive surface is an essential requirement for rapid build-up of boiler deposits.

12.4 RETENTION OF IMPACTED PARTICLES ON CAPTIVE SURFACES

Ash particles impacting on the surface of fused silicate slag or molten sulfate will lose their kinetic energy by resistance of viscous drag as the particles penetrate the liquid. The approximate depth of penetration s is given by the equation (Raask, 1966a)

$$s = \frac{2 v r^2 \, \rho_p \, \sin \theta}{9 \eta} \qquad (12.17)$$

where v, r, ρ_p = velocity, radius, and density of the impacting particles, respectively
θ = angle of impaction
η = viscosity of the surface layer

The surface tension force F acting on the particle that has partially penetrated in liquid to a depth s is given by

$$F = 2\pi(2sr)^{1/2} \, \gamma \qquad (12.18)$$

where r is the radius of particle and γ is the surface tension of the liquid. When this force is equal to or greater than the gravitational force, the particle is retained by liquid:

$$2\pi(2sr)^{1/2}\gamma = \frac{4\pi r^3 \rho_p g}{3} \qquad (12.19)$$

where g is the gravitational force constant and other notations are as already defined. Thus the penetration s of a particle when the surface-tension and gravitational forces balance is given by

$$s = \frac{2r^5 \rho_p^2 g^2}{9\gamma^2} \qquad (12.20)$$

Impacting particles have a threshold velocity to produce the depth of penetration [Eq. (12.17)] that is equal to that required for the surface tension to balance the gravitational pull on the particles [Eq. (12.20)]. Thus, the threshold velocity v_c is given by

$$v_c = \frac{\rho_p g^2 r^3 \eta}{\gamma^2} \qquad (12.21)$$

where the notations are as before. The density of ash particles and the surface tension of slag can be taken as 2500 kg m^{-3} and 0.32 N m^{-1}, respectively (Chap. 6), and the gravitational force g is 9.81 m s^{-2}. Thus, the threshold velocity v_c in m s^{-1} units is given in a logarithmic form by

$$\log v_c = 6.39 + 3 \log r + \log \eta \qquad (12.22)$$

where r is the radius of particles in meters, η is the viscosity of the captive surface layer in N s m^{-2} units, and θ is $\pi/2$ rad (normal impaction).

Plots in Fig. 12.5 show that the surface of viscous slag is an effective medium for capture of impacting ash particles when the slag viscosity is well above 100 N s m^{-2}, which is a limiting value for flow under the action of gravity. This is in accordance with the results of deposition measurements discussed in Sect. 12.6 and with observations in boiler operation, that a rapid build-up of ash deposit occurs when there is first a layer of semimolten material firmly anchored to boiler tubes.

Molten sodium and potassium sulfate have viscosity values around 2×10^{-3} N s m^{-2} (Dantuma, 1928; Glover, 1969), and the surface tension is between 0.1 and 0.2 N m^{-1} (Bertozzi and Soldani, 1967). It is therefore evident that a layer of molten sulfate on boiler tubes can capture and engulf quantitatively all impacting ash particles, irrespective of their size or the velocity and angle of impaction. Of course, such mobile liquids flow readily on boiler tubes under the action of gravity, together with captured particles. This can be observed occasionally in practice, where molten sulfate deposit flows on the high-temperature superheater tubes and is unable to support large masses of sintered ash or fused slag.

A more usual occurrence is that as a result of dissolution of oxides of boiler-tube metals and ash constituents, chiefly iron, aluminum, and calcium (Adams and Raask, 1963), viscosity of the sulfate melt will increase as depicted in Fig. 9.8. Also, when a silicate ash of low thermal conductivity (Chap. 13) deposits on the sulfate layer, it provides a thermal shield; thus the molten phase will be cooled below its freezing temperature. It is therefore evident that an initial layer of molten sulfate will freeze or disappear as the silicate ash deposit grows on boiler tubes.

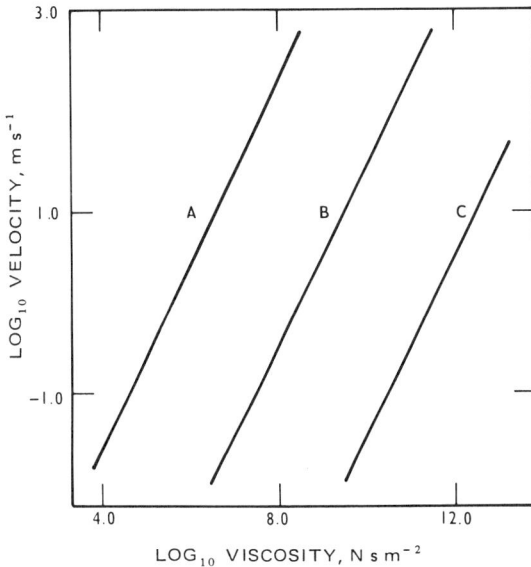

Figure 12.5 Impact velocity/ viscosity relationship for particle capture at the surface of viscous slag: Particle radius A, 100 μm; B, 10 μm; C, 1.0 μm.

12.5 THE RATE MEASUREMENTS AND MODE OF DEPOSITION OF SULFATES ON COOLED TARGETS IN BOILER PLANT

Measurements of the rate of deposition of sulfates have been made in pulverized coal fired boilers (Raask, 1962, 1963a, b), in a cyclone-fired boiler (Raask, 1961a, 1963b) and in a residual oil-fired boiler (Raask, 1961b). A high-temperature deposition probe described by Jackson and Raask (1961) was used for the measurements. Briefly, it consisted of an air-cooled probe made of an austenitic, corrosion-resistant steel, 25 mm outside diameter with 3 mm wall thickness and 3.5 m long. Figure 12.6a shows the method of mounting the sheathed, chromel/alumel thermocouples in the probe wall.

The probe was inserted in the flue gas of the superheater duct (Fig. 12.6b) and kept at the required temperature by controlling the flow of cooling air. Duration of sampling runs was varied from 0.05 to 24 h, and after allowing the probe to cool in air the deposited material was removed from different sections of the probe by brushing and by washing with distilled water. The samples were filtered and the filtrate and solid residue were analyzed for sulfate, chloride, alkali metals, and other major constituents.

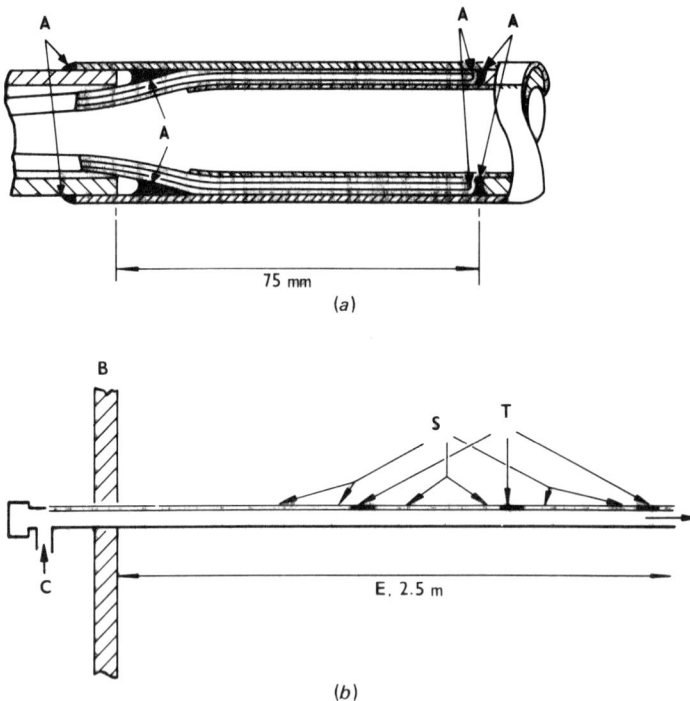

(a)

(b)

Figure 12.6 Air-cooled probe for deposition measurements. (a) Method of mounting thermocouple, where A marks welds. (b) Exposure (E) of probe in flue gas: C, cooling air; B, boiler duct wall; S, sampling sections; T, thermocouples.

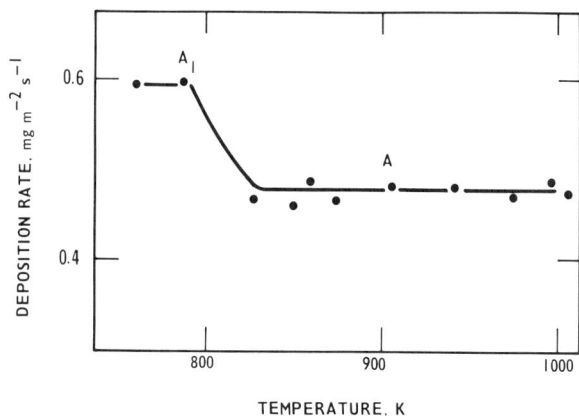

Figure 12.7 Deposition of sodium sulfate in oil-fired boiler (A) and of $Na_2SO_4 + SO_3$ (pyrosulfate) (A_1), with sodium content of oil 100 ppm.

Figure 12.7 shows that the rate of deposition of sodium sulfate was virtually independent of the probe temperature in the range of 825 to 1000 K. Below 825 K the sulfate deposit had absorbed an additional amount of SO_3 to form pyrosulfate, $Na_2S_2O_7$. The deposition measurements were made in an oil-fired boiler (Raask, 1961b) where the flue-gas temperature at the sampling point was 1340 K and the probe exposure time was varied from 0.5 to 24 h. During that time interval the rate of Na_2SO_4 deposition decreased from 0.5 mg m^{-2} s^{-1} to 0.35 mg m^{-3} s^{-1}. This may have been due to an increase in the surface temperature of deposit material on the probe, or it may have been caused by reactions between sulfate and vanadium oxides in the deposit material.

The residual fuel oil contained 100 ppm sodium, chiefly in the form of chloride, and 4.2 percent (42,000 ppm) sulfur. There was thus a large excess of sulfur gases in the flue gas when compared with concentration of NaCl vapor, and as a result the latter was quantitatively converted to sulfate; no chloride was found in the probe deposits. Further, there was no significant capture of sodium by silicates; it is therefore possible to calculate the concentration of sodium sulfate in the flue gas and also the rate of deposition of Na_2SO_4 on cooled boiler tubes by using Eq. (12.1). The calculated rate of Na_2SO_4 deposition, assuming a streamlined flow of the flue gas, was 0.13 mg m^{-2} s^{-1}, and the measured rates were 0.35 mg m^{-2} s^{-1}. The disparity between the predicted and measured rates could have been due to a high degree of turbulence in the flue gas at the sampling point. That is, constant A in Eq. (12.1) had a value of 1.3.

The rate of deposition of alkali-metal sulfates in coal-fired boilers is dependent on the amount of volatile sodium chloride in the coal, but the sulfate deposition/ chloride concentration curves are nonlinear, as shown in Fig. 12.8. This is a result of capture of volatilized alkali-metal species by fused silicate ash, as discussed in Chaps. 6, 7, and 8. When the chlorine content of coal is below 0.4 percent by weight, equivalent to about 0.2 percent of the volatilizable alkalis, the deposition rates of Na_2SO_4 and K_2SO_4 are below 0.1 mg m^{-2} s^{-1}. Once the surface of silicate particles becomes saturated by absorbed alkalis, the excess sodium and potassium is converted to sulfates. Thus the rate of deposition of sulfates is greatly enhanced when the chlorine content of coal is above 0.5 percent, as shown in Fig. 12.8.

Curve C_1 in Fig. 12.9 shows that the amount of chloride deposited on the probe was insignificant in comparison with that of sulfate when a high-chlorine coal was burned in a 200-MW boiler. In contrast, when the same coal was burned in a 15-MW boiler, a significant quantity of chloride was deposited (curve C_2 in Fig. 12.9) because of the short residence time for completion of the sulfation reaction. A similar coprecipitation of chloride with sulfate on cooled targets was found by Bishop (1968) in his combustion rig fired with a high-chlorine coal.

It appears that the degree of conversion of sodium chloride to sulfate is critically dependent on the length of flame path. When the residence time of volatilized alkali-metal species is not less than 2 s, the sulfation is virtually complete. It is therefore unlikely that significant quantities of chloride are deposited on high-temperature superheater tubes in large (above 100 MW) capacity boilers fired with pulverized coal. However, substantial amounts of chloride may be deposited on combustion-chamber walls, in particular on the tubes near the burners. This may have a significant bearing on the rate of corrosion of furnace-wall tubes, as discussed in Chap. 17.

As in oil-fired boilers, the rate of sulfate deposition on a cooled target in coal-fired boilers was not markedly influenced by the probe temperature below 1100 K (Figs. 12.9 and 12.10). However, the deposition rate decreased rapidly as

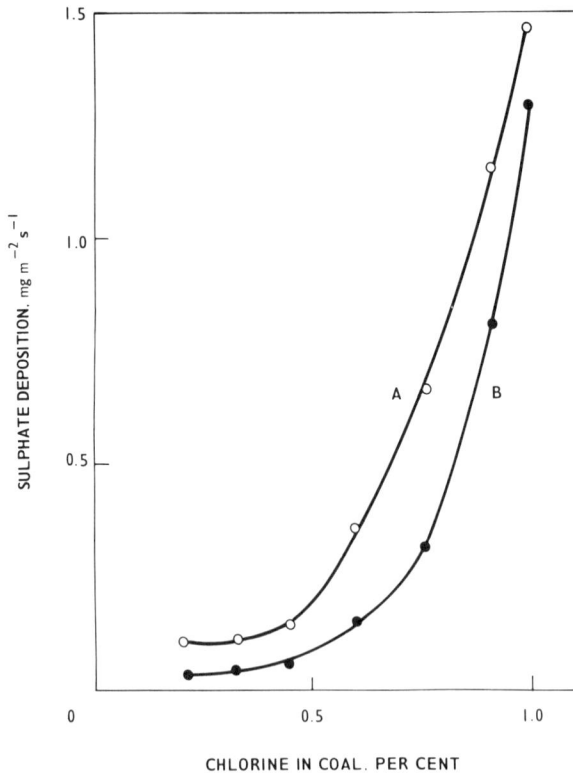

Figure 12.8 Deposition of alkali-metal sulfates with different chlorine-content coals; A, Na_2SO_4; B, K_2SO_4.

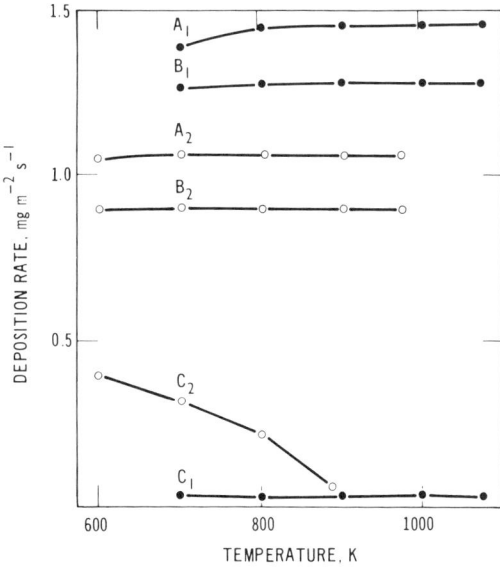

Figure 12.9 Deposition of sulfates and chlorides in pulverized-coal-fired boilers, with 1.0 wt. % chlorine and 1.1 wt. % sulfur in coal. In 200-MW boiler, A_1, Na_2SO_4; B_1, K_2SO_4; C_1, NaCl. In 15-MW boiler, A_2, Na_2SO_4; B_2, K_2SO_4; C_2, NaCl.

the temperature increased above 1100 K, and curves A and B give a dewpoint temperature of around 1160 K for both Na_2SO_4 and K_2SO_4. This dewpoint temperature is consistent with that calculated from concentration of the volatilized alkali-metal species in the flue gas and the saturation vapor pressure of Na_2SO_4 and K_2SO_4, as shown in Fig. 7.5.

The results of sulfate deposition measurements depicted in Fig. 12.10 were obtained in a cyclone boiler fired with a coal of moderate chlorine content, 0.28 percent, which is equivalent to about 0.15 percent of volatile sodium. The rate of

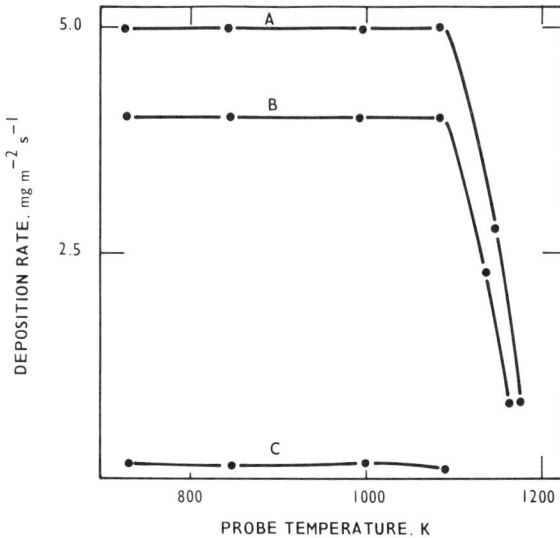

Figure 12.10 Deposition of sulfate and chloride in cyclone-fired boiler, 0.28% chlorine in coal: A, Na_2SO_4; B, K_2SO_4; C, NaCl.

sulfate deposition was about three times higher than that measured in pulverized-coal-fired boilers, burning coals of the same chlorine content as shown in Fig. 12.8. A possible reason for the high rate of sulfate deposition in the cyclone-fired boiler was that, as a result of high flame temperature, large fractions of the total sodium and potassium were volatilized and were available for the formation of sulfates.

An interesting feature of the results presented in Figs. 12.8, 12.9, and 12.10 is that the molar ratio of Na_2SO_4 to K_2SO_4 is approximately 2 to 1. This is consistent with the findings on relative distribution of the alkali metals in silicates and sulfates, as discussed in Chaps. 6 and 7. That is, potassium introduced as silicates into a coal flame, where some of it is released, always gives a molar concentration of volatilized species less than that for sodium. Nevertheless, sufficient amounts of potassium are released in the flame that, together with sodium sulfate, they can form a highly corrosive boiler deposit as discussed in Chap. 17.

Raask and Bhaskar (1981, unpublished results) have studied the mode of deposition of sulfates on a cooled target. A sample of Na_2SO_4 and K_2SO_4 was evaporated in a small furnace at 1500 K and then deposited from nitrogen carrier gas on the glass disc target at 500 K. The gas stream cooled at a rate about 500 $K s^{-1}$ before the deposition took place. The initial Na_2SO_4 deposit on the target is shown in Fig. 12.11a, where the smallest discernable particles were 0.05 μm in diameter, and these, together with 0.1 and 0.2 μm particles, are neatly arranged around larger particles 0.5 to 2 μm in diameter. The larger particles appear to be agglomerates of 0.1- to 0.2-μm-diameter particles. It is probable that as a result of the rapid rate of cooling the agglomerates were frozen before the constituent particles were able to coalesce to a single sphere. That is, the time interval from when the particles first formed as an agglomerate to its freezing was less than 1 ms, which can be calculated from the sintering equation, Eq. (10.2).

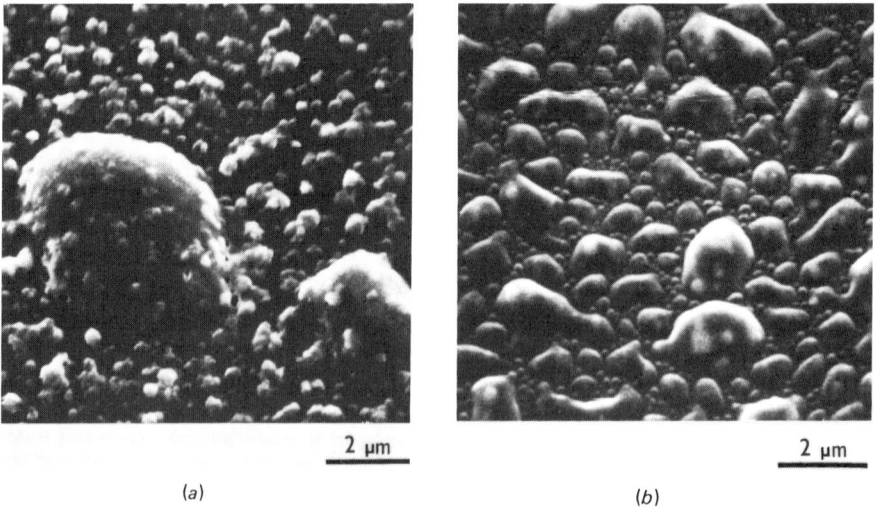

(a)   (b)

Figure 12.11 The mode of deposition of alkali-metal sulfate particles and agglomerates. (a) Sodium sulfate. (b) Potassium sulfate.

Deposition of volatilized potassium sulfate, K_2SO_4, showed a pattern similar to that produced by Na_2SO_4. Figure 12.11b shows that the majority of particles were in the size range of 0.05–0.2 μm, and there were a few agglomerates up to 5 μm in diameter. Because of the high melting temperature of K_2SO_4, the degree of agglomeration was less extensive than that of Na_2SO_4.

The mode of sulfate deposition in experimental combustion rigs has been studied by Bishop (1968) and by Brown and Ritchie (1968). They have found that Na_2SO_4 deposit on a target cooled below 775 K consisted of submicron-size particles with some agglomerates. The degree of agglomeration was more extensive at higher target temperatures, and on cooling, Na_2SO_4 crystals 0.5–5.0 μm in size were found in the deposit. The observations on the mode of deposition of sulfate in the flue gas were consistent with the findings of Buckle (1978). He has observed that when fume was formed from the metal and salt vapors in an argon-carried gas at temperatures below the melting point, the particles were submicron in size and spherical in shape. However, when formed above their melting temperature, the particles grew to a size of 1–10 μm and on cooling they took their crystal habit shape.

12.6 RATE MEASUREMENTS OF SILICATE–ASH DEPOSITION AND FORMATION OF MONOLITHIC AND LAYER-STRUCTURED DEPOSITS

The mode of silicate ash capture has been investigated and the deposition rate measurements have been carried out by inserting a cooled probe in pulverized-coal- and cyclone-fired boilers, as described by Raask (1961a, 1962, 1963a, b). Figure 12.12, a and b, depicts schematically the pattern of ash deposit on the probe as the exposure time was increased from a few seconds to 30 min. Figure 12c depicts the formation of a wedge-shaped deposit of sintered ash frequently found on super-heater tubes.

The mode of deposition with respect to particle size and the amount of ash captured were in accordance with the mechanisms of transporting particulate ash from the moving flue gas to a cooled target, as discussed in Sects. 12.1 to 12.4. The initial deposit on both the exposed and lee sides of the probe consisted chiefly of submicron-size particles, as shown in Fig. 12.13a. A possible explanation for formation of the thin layer of dust around the circumference of the probe was that the clean metal target attracted the small particles of ash as a result of electro-phoresic deposition, as discussed in Sect. 12.2. Subsequent deposition ceased on the sheltered side of the probe after an insulating layer of ash had formed and the electrostatic attraction force ceased to be effective. On further exposure, deposition of ash by inertial impaction continued on the side of probe facing the flue-gas flow. Figure 12.13b shows a typical sample of the inertially impacted ash where the particle diameter is between 0.1 and 10 μm.

Figure 12.14 shows the average rate of ash deposition in the superheaters of pulverized-coal-fired boilers burning a typical selection of bituminous coals. The

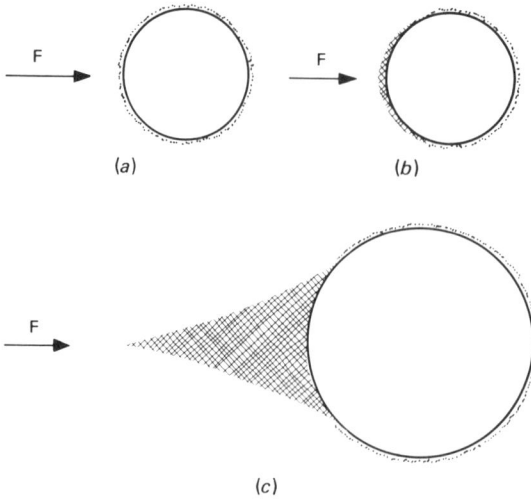

Figure 12.12 The mode of deposit build-up on cooled probe and boiler tubes, where F indicates flows of flue gas, dotted lines are unsintered ash (<1.0 μm) particles, and cross-hatching shows sintered ash (0.1 to 10 μm) particles. (a) Initial deposit of submicron-size ash particles. (b) Probe deposit after 15 min. (c) Aerodynamic wedge-shaped deposit on superheater tube.

composition of laboratory-prepared ashes and the coals burned during the deposition tests is given in Table 12.1.

The velocity (15 m s^{-1}) and temperature (1250 K) of the flue gas at the sampling point were the same in the three boilers. It could be therefore expected that the rate of deposition should have been proportional to the ash burden. However, this was not the case, as shown by the calculated values of the ash burden, the rate of inertial impaction, and the capture efficiency given in Table 12.2.

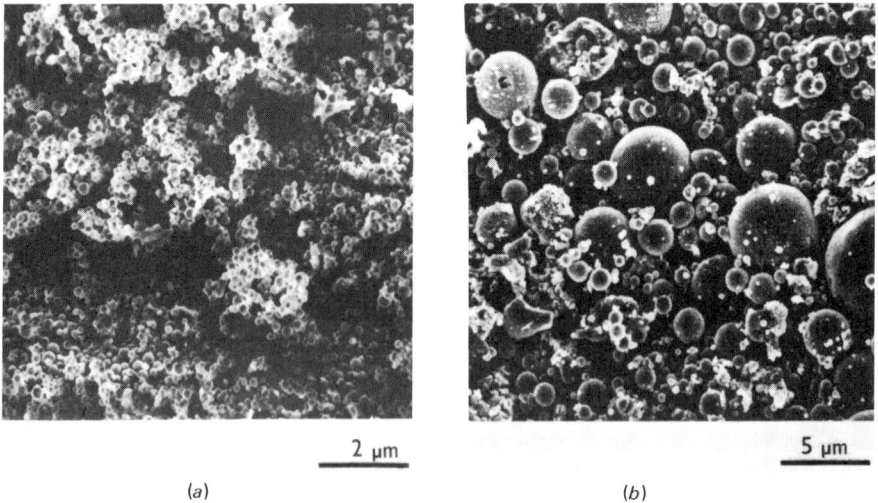

Figure 12.13 Ash deposit on cooled probe in coal-fired boiler. (a) Initial deposit. (b) Inertial impaction of ash.

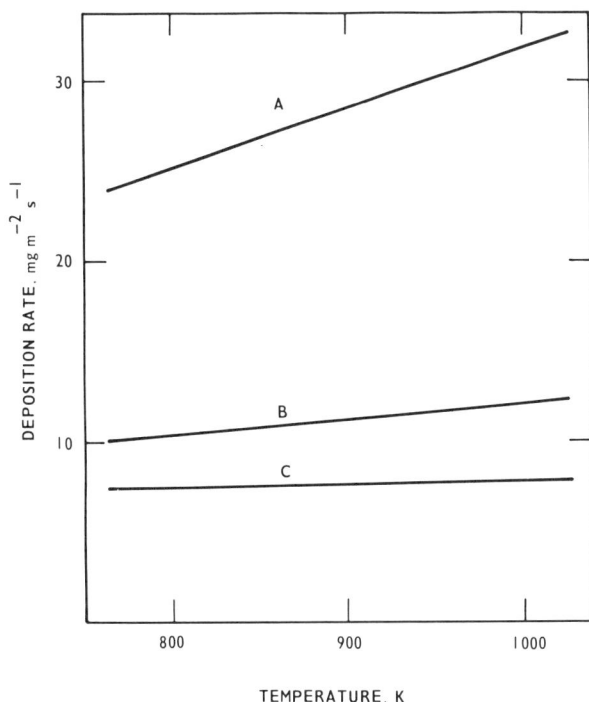

Figure 12.14 Rate of ash deposition on cooled superheater probe in pulverized-coal-fired boilers: percent CaO + F_2O_3 in ash; A, 21.3; B, 13.4; C, 12.8.

It was assumed that the impact number was 1.5% (Fig. 12.3*b*) and that the deposition rates given in Table 12.2 are twice as high as those given in Fig. 12.14; the latter were calculated from deposit weight measurements by assuming that the deposited ash is evenly distributed around the circumference of the probe. In fact, the deposits by inertial impaction form only on the exposed side of a probe target or boiler tube, as depicted in Fig. 12.12*b*.

The rates of captive deposition given in Table 12.2 and illustrated in Fig. 12.14 were consistent with the deposit-forming propensity of the three ashes. The analysis

Table 12.1 Ash composition (weight percent) of deposition test goals

Ash constituent	Coal A (20.2% ash)	Coal B (17.8% ash)	Coal C (21.6% ash)
SiO_2	40.1	46.7	51.2
Al_2O_3	22.8	24.2	27.3
Fe_2O_3	11.9	6.6	5.3
CaO	9.4	6.8	7.5
MgO	3.1	2.7	2.7
Na_2O	2.0	3.0	2.0
K_2O	1.0	3.0	1.0
SO_3	8.5	5.7	3.9

Table 12.2 Ash burden in flue gas, and inertial and captive impaction

Ash code	Ash burden in flue gas ($g\,m^{-3}$)	Inertial impaction ($mg\,m^{-2}\,s^{-1}$)	Captive deposition ($mg\,m^{-2}\,s^{-1}$)	Capture efficiency (%)
A	3.9	560	48 to 65	8.6 to 11.6
B	3.2	460	20 to 25	4.3 to 5.4
C	4.2	600	15	2.5

results in Table 12.1 shows that the $(CaO + Fe_2O_3)$ content of ash A was above 20% by weight, and thus it has a relatively high fouling index when assessed by the formulas discussed in Chap. 10. High rates of deposition on steel tubes and ceramic probes with high-sodium and high-calcium coal ashes have been reported by Tufte et al. (1976), Sondreal et al. (1978), and Hein (1976).

It is therefore evident that the sintering and fusion characteristics of different ashes determine the rate of build-up of probe deposit and also the boiler-tube deposit. That is, when the deposited ash particles form a sinter matrix within a few minutes, as was the case with ash A in Fig. 12.14, the result is a 10% retention of total ash deposited on the probe. Further, the rate of captive deposition increased with temperature, as shown by the plot. One of the most rapid build-ups of superheater deposit has been discussed by Jackson and Ward (1956). They found that a deposit about 100 mm thick and of 1000 kg m^{-3} density was formed in 76 h operation as a result of the high flame-volatile sodium content of coal.

The rate of slag build-up in the combustion chamber can be much more rapid. This was demonstrated by Jackson and Raask (1957) by inserting a high-temperature sampling probe in the combustion chamber of a pulverized-coal-fired boiler. The surface temperature of the probe was 1400 K and that of the flame at the sampling point was 1710 K. A layer of slag deposit, 5 to 10 mm thick and of a density of about 1000 kg m^{-3}, formed on the probe during 10-min exposure, as shown in Fig. 12.15, a and b. The rate of captive ash deposition was between 5.4 and 10.7 g m^{-2} s^{-1}, which was some 200 to 300 times higher than the rate of "dry" ash deposition given in Table 12.2. This clearly demonstrates the efficacy of a layer of molten slag for capture of impacting ash particles, as discussed in Sect. 12.4. It is also evident that the low-viscosity slag on cyclone boiler tubes captures quantitatively the ash particles thrown to the walls by the centrifugal force, and thus up to 85% of total ash is retained as slag.

The slag deposit formed on high-temperature probes (Fig. 12.15) and that taken from the furnace wall of pulverized-coal-fired boilers (Fig. 12.16a) may vary in composition but there is no layered structure in the appearance. In fact, the only distinguishable feature in appearance of slag deposit formed under different conditions is the number, size, and distribution of gas holes in the fused ash material. In the initial stage of slag formation, coal particles are encapsulated in the deposit matrix, which generates CO and CO_2 as discussed in Chap. 6, and frozen slag is highly porous, as shown in Fig. 12.15. Once all of the encapsulated coal is consumed and the gas bubbles escape, the density of slag will increase with time, and Fig. 12.16a shows a typical slag sample from a pulverized-coal-fired boiler.

The sintered ash deposits formed on boiler tubes, and in particular those formed on the high-temperature superheater tubes (Fig. 12.17) usually have a layered structure, as discussed by Gumz et al. (1958) and Reid (1971). One of the principal reasons for formation of layer-structured deposit in coal-fired boilers is that the chief constituents of ash—silicates and sulfates—are immiscible. That is, the SiO_2-rich silicates and the alkali-metal and calcium sulfates do not form a single phase, as discussed in Chap. 7. The formation of a layer-structured deposit is enhanced as a result of high porosity of the sintered matrix and the temperature gradient across the ash material on cooled boiler tubes, as depicted in Fig. 12.17.

The transport of alkali-metal species through a porous deposit matrix toward the surface-cooled boiler tube has been discussed by Gumz et al. (1958), Jackson (1963), and Reid (1971). The vapor-phase diffusion routes are considered to constitute the main pathway for the enrichment of alkali metals in the deposit layer next to tube surfaces. The diffusable species may be sulfates, the chlorides, oxides, and hydroxides, but the thermodynamic data as discussed by Halstead and Raask (1969) and the results of deposition measurements discussed in Sect. 12.5 suggest that sodium and potassium sulfates are the principal vapor species that diffuse through a porous matrix of silicate ash deposit and condense on cooled boiler tubes.

Figure 12.17 depicts in sequential stages, first, the formation of sulfate-rich

4 mm

(a)   (b)

Figure 12.15 Slag build-up on combustion chamber probe. (a) With steam injection. (b) Normal operation (Chap. 14).

(a)

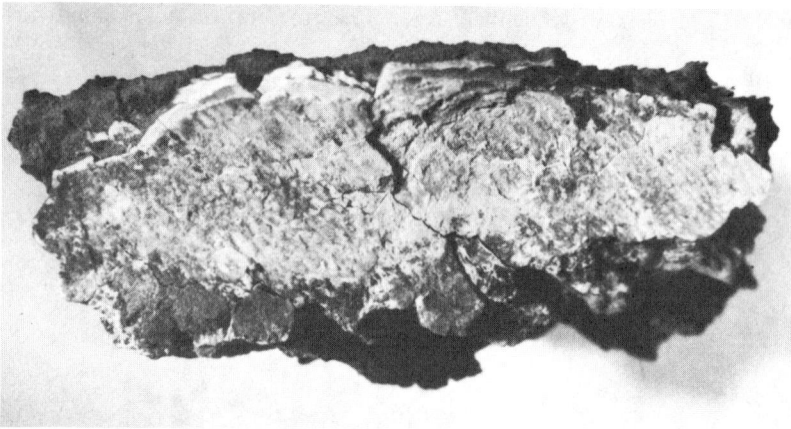

(b)

Figure 12.16 Typical samples of slag and sintered ash in pulverized-coal-fired boiler. (a) Furnace-wall slag. (b) Superheater deposits.

deposit and, subsequently, strongly adhering deposit on a ferritic-steel superheater tube with the metal surface temperature of 900 K. The relative amounts of different sulfates, chiefly Na_2SO_4 and K_2SO_4, that diffuse through the sintered matrix of silicate ash depend on the temperature gradient across the deposit layer, vapor pressure of the species, and thermodynamic stability of the sulfates in the presence of silicates. Potassium sulfate has a higher temperature stability limit when compared with that of sodium and calcium sulfates (Chap. 7), and as a result

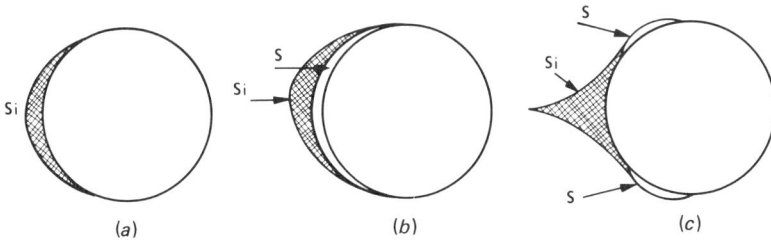

Figure 12.17 Sequential stages in the formation of layer-structured and firmly adhering superheater deposit. (*a*) Initial unsegregated ash deposit. (*b*) Separated sulfate (S) and silicate (Si) phases. (*c*) Displaced sulfate phase and silicate–metal oxide bond.

K_2SO_4 can be preferentially transported to the surface of cooled boiler tubes when there is a steep temperature gradient across the ash deposit. The K_2SO_4-rich phase, when molten, can cause severe corrosion of tube metal, as discussed in Chap. 17.

In contrast, Na_2SO_4 and $CaSO_4$ are decomposed by silicates when the deposit temperature exceeds 1100 K, as summarized by the expression

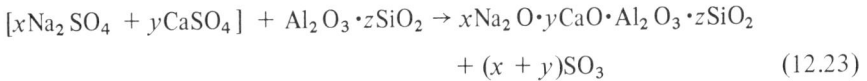

$$[xNa_2SO_4 + yCaSO_4] + Al_2O_3 \cdot zSiO_2 \rightarrow xNa_2O \cdot yCaO \cdot Al_2O_3 \cdot zSiO_2$$
$$+ (x + y)SO_3 \qquad (12.23)$$

The above reaction takes place at the surface of silicate ash particles and results in a reduction of viscosity of the glassy material, thus enhancing particle-to-particle sintering (Chap. 10). The sintering of surface layers of silicate ash particles, rich in sodium and calcium, is therefore largely responsible for a rapid build-up of deposits frequently found in the top half of combustion chamber and in superheaters. Figure 12.18*a* shows a typical segment of sintered superheater deposit where the large

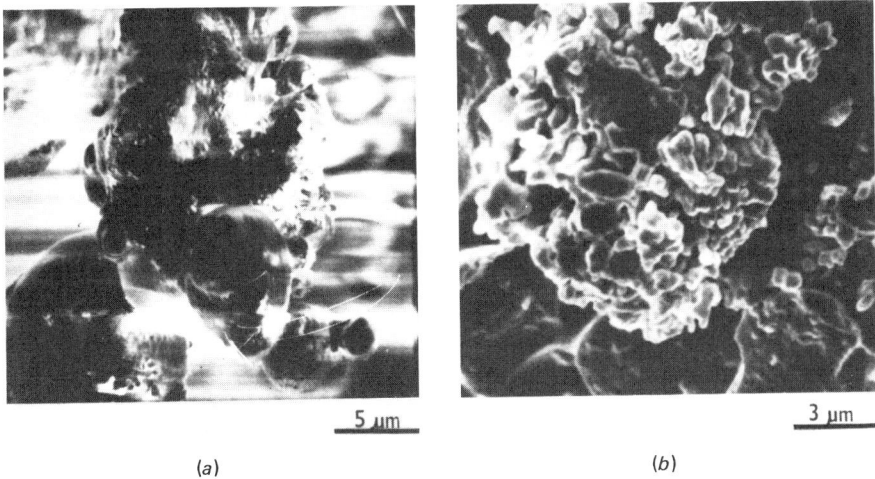

Figure 12.18 Sintered ash matrices formed at 1300 K. (*a*) Superheater deposit. (*b*) Laboratory ash sinter.

spherical particles have been sinter-bridged by small particles. Figure 12.18*b* shows the matrix of ash sinter formed at the same temperature, 1300 K, in a laboratory furnace. The sintered clusters are made up from nonspherical particles, which illustrates the difference in the formation of this matrix from that of the boiler deposit formed from the flame preheated ash. As the thickness of silicate ash deposit on boilers tube increases, the sulfate phase in the middle section will be displaced toward edges of the deposit (Fig. 12.17*c*) as a result of the temperature gradient. This allows the formation of a strong bond between the silicate ash deposit and the oxide surface of boiler tubes, as discussed in Chap. 11. It is therefore evident that formation of the layer-structured and strongly adhering deposits is intimately related to the behavior, separation or decomposition, of sulfates when in contact with silicates in the sintered ash matrix.

Figure 12.19 depicts the composition of mineral matter in a bituminous coal (A), the flue-gas-borne ash captured in the electrical precipitator (B), superheater probe deposit (C), and the superheater tube deposit. The last was separated into the silicate-rich outer layer (D_1) and the sulfate layer (D_2) next to tube metal before analysis. A solution of perchloric acid, 0.04 M $HClO_4$, was chosen to dissolve the soluble constituents in ash and deposit by boiling a dispersion of ground sample for 10 min. A solution of $HClO_4$ is a useful extraction media for coals, ashes, and deposits. The weakly acidic medium dissolves the chlorides and carbonates in coal, and the sulfates in ash and deposits.

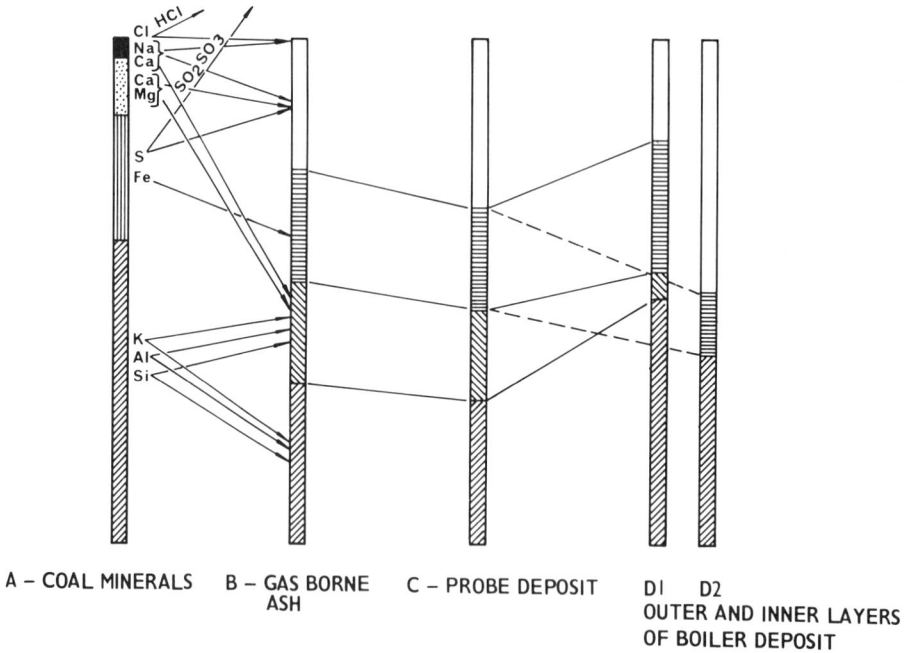

A – COAL MINERALS   B – GAS BORNE ASH   C – PROBE DEPOSIT   D1   D2 OUTER AND INNER LAYERS OF BOILER DEPOSIT

Figure 12.19 Ash composition changes on route from mineral matter to boiler-tube deposits: insoluble silicates; soluble silicates; pyrites; iron oxides; carbonates; chlorides; and sulfates.

As expected, the chief soluble constituents in coal mineral matter were chlorides and carbonates, and sulfates in ash and deposits. The sulfate content of the outer and inner fraction of superheater tube deposit (columns D_1 and D_2) clearly demonstrates the degree of separation of the sulfate and silicate phases as a result of temperature gradient across the deposit layers. The degree of solubility of silicates in a weakly acidic solution is also of interest when assessing deposit-forming propensity of the ash deposited on a cooled probe. The silicate minerals in coal when unheated are practically insoluble in dilute acids, but as a result of flame vitrication and reactions with alkalis, a significant fraction of the glassy material is changed to a soluble form, as depicted in Fig. 12.19. A useful criterion here is that the soluble fraction of silicate ash is likely to be reactive at boiler deposit temperatures. The soluble glassy material, rich in alkali metals and calcium, is likely to have sufficiently low viscosity in the range of 10^6-10^{10} N s m^{-2} at deposit-forming temperatures to result in a rapid rate of sintering. It appears therefore that the soluble silicate fraction in the short-exposure (1-10 min) probe ash is a useful criterion of its deposit-forming propensity. As yet, there are insufficient data for establishing the relationship between the degree of boiler fouling and the amount of soluble silicates in the deposited ash.

Figure 12.19 shows the relative proportion of different mineral species in coal, and the ratio of silicate, iron oxide, and sulfate constituents of the flue gas borne ash and deposit samples taken from a cyclone-fired boiler, as discussed by Raask (1963a). The high flame temperature in the cyclone boiler and a high sodium content, 3.4 percent Na_2O in laboratory-prepared ash and 5.6 percent Na_2O in the superheater probe deposit, result in extensive vitrification and the formation of sodium-rich glassy material. Consequently, nearly a 40 percent fraction of the total silicates in the probe deposit was soluble when extracted with a 0.04 M acid solution. It is to be expected that in pulverized-coal-fired boilers where the peak flame temperature is lower, the soluble silicate fraction will be between 1 and 10 percent of total silicates initially deposited on water and steam tubes with different coal ashes, and their fouling "index" would vary accordingly.

The original bonding matrix of silicate glassy material, rich in sodium and calcium, may devitrify in boiler deposits, and the crystalline species have been identified by Mazza and Wilson (1977). Tufte and Beckering (1975) have identified melilite, which is a solid solution of sodium silicate in gehlenite, $2CaO \cdot Al_2O_3 \cdot SiO_2$, as discussed by Rigby (1953). Gehlenite can take up to 4% of Na_2O into solid solution and the melilite species thus formed have a chemical composition $xNa_2O \cdot 2CaO \cdot Al_2O_3 \cdot SiO_2$. An important point to note here is that the crystalline material and the annealed glass, rich in sodium, are changed into a form which is insoluble in water or in a dilute acid solution. It is therefore no longer possible to determine the amount of the original bonding phase of low-melting glass in an aged boiler deposit by extraction technique. Thus, analysis of the initial deposit material collected with a cooled probe as described previously is likely to give more useful information than that of boiler deposits on the sintering and slagging characteristics of ash. The thermodynamic and experimental aspects of the formation of sodium silicates in boiler deposits have been further discussed by Wibberley and Wall

(1982). Borio and Narciso (1978) and Hazard et al. (1979) have discussed the enrichment of iron in furnace-wall deposits, which can lead to severe slagging. Wynnyckyj and Rhodes (1981) have suggested that deposition of high-density particles and the formation of "pyroclastic lumps" are symptomatic features in the mechanism of deposit build-up. The enrichment of iron in furnace slag with U.S. high-sulfur coal is discussed in Chap. 15.

12.7 THE ROLE OF MINOR COAL IMPURITY CONSTITUENTS IN FORMATION OF BOILER DEPOSITS

In addition to the nine major elements that, together with oxygen, usually account for over 98 percent of the bulk of ash (Tables 3.2 and 3.3), there is a large number of other constituents present in concentrations from 0.5 percent down to less than 1 ppm. Recently the behavior of the trace-element constituents in boiler flame and their distribution in the chimney emission solids have been investigated by many researchers, as exemplified by publications of Abel and Rancitelli (1975), Davison et al. (1974), Gladney et al. (1976), Kaakinen et al. (1975), Klein et al. (1975), Linton et al. (1976), Natusch et al. (1974), Ondov et al. (1979), and Raask and Goetz (1981).

In contrast, the possible role of minor constituent elements in promoting the formation of bonded deposits has been much less extensively studied. An exception has been phosphorus, and its role in formation of bonded deposits in stoker-fired boilers has been discussed by Crossley (1952). He found that a coal containing only 0.04 percent phosphorus resulted in a formation of economizer deposit with a phosphorus content of 20 percent. Thus there was a 50 times enrichment of phosphorus from about 0.4 percent in ash to 20 percent in deposits. Such high enrichment of phosphorus in boiler deposit is likely to take place via a volatilization/deposition route similar to that of alkali-metal salts discussed in Chap. 7. Much of phosphorus occurs in coal as apatite, $CaF_2 \cdot 3Ca(PO_3)_2$, and on heating the mineral with carbon in the hot fuel bed of a stoker-fired boiler, the fluoride fraction as well as a part of the phosphorus is volatilized:

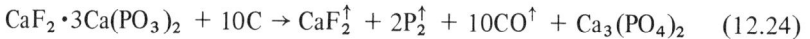

$$CaF_2 \cdot 3Ca(PO_3)_2 + 10C \rightarrow CaF_2^\uparrow + 2P_2^\uparrow + 10CO^\uparrow + Ca_3(PO_4)_2 \quad (12.24)$$

According to studies of Hoffmann and Hoffmann (1948), two-thirds of phosphorus in the apatite mineral can be transferred into the vapor phase by the reduction route. Subsequently, the metal vapor readily oxidizes in the flue gas where oxygen is present in excess:

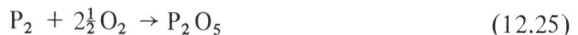

$$P_2 + 2\tfrac{1}{2}O_2 \rightarrow P_2O_5 \quad (12.25)$$

and the oxide depositing together with ash can function as a high-temperature cement in the formation of a strongly bonded deposit matrix.

Volatilization of phosphorus is likely to be insignificant in an oxidizing flame of a pulverized-coal-fired boiler, since calcium phosphate, $Ca(PO_3)_2$, is said to be thermodynamically stable in an oxidizing atmosphere at temperatures up to 2100 K, as discussed by Reid (1971). In cyclone-fired boilers, the coal particles are larger and the flame temperature is higher than in the pulverized-coal-fired boiler, and as a result some of the phosphate mineral may be reduced by carbon to volatile phosphorus metal.

Arsenic and lead are other coal trace elements that have been found occasionally to be present in high concentrations in boiler deposits, as reported by Kirsch et al. (1968) and Raask et al. (1970). The arsenic and lead content in ash is usually between 10 and 100 ppm, whereas in boiler deposits the elements may be present in concentrations of 1–5%. So far there is no experimental evidence or plant observations to show that arsenic and lead compounds could play a significant role in enhancing the formation of bonded deposits. In fact, some flame-volatilized compounds when added to coal in trace quantities may produce the opposite effect: that is, they may function as antislagging additives, as discussed in Chap. 16. The possible significance of high concentrations of arsenic and lead in boiler deposits to fireside corrosion of superheater and furnace-wall tubes is discussed in Chap. 17.

NOMENCLATURE

A	constant
D	diameter of collection target
D'	characteristic diameter of collecting target
D_v	diffusion coefficient
d	diameter of particle
E	potential gradient of electric field
e	collection efficiency
e'	modified collection efficiency
e_d	deposition efficiency
F	surface tension force
F_e	electrophoresic force
F_t	thermal force
f	free-falling particle velocity
g	gravitational force constant
I	impact number
$J_{\rho\epsilon_1}$	permittivity coefficient of ash
k	Boltzmann constant
k_1	constant
k_2	resistance coefficient
M_g	concentration of sodium sulfate vapor in flue gas
M_s	concentration of sodium sulfate vapor in target surface
m	particle mass
Nu	Nusselt number
Pr	Prandtl number
R	thermodynamic gas constant
R_D	rate of deposition

Re	Reynolds number
r	particle radius
Sc	Schmidt number
s	depth of penetration
s_g	thickness of gas layer
T	temperature
T_g	gas temperature
T_t	target temperature
t	time
v	flue gas velocity
v_c	threshold velocity
v_τ	terminal velocity
x	molar fraction of sodium sulfate
y	molar fraction of calcium sulfate
z	molar fraction of silica
γ	surface tension coefficient
ϵ_o	permittivity of free space
η	deposit surface layer viscosity
θ	angle of impaction
κ_g	thermal conductivity of gas
κ_p	thermal conductivity of particle
λ	molecular mean free path
λ_o	molecular mean free path at NTP
μ	gas viscosity
ρ	gas density
ρ_o	gas density at NTP
ρ_p	particle density

THERMAL RADIATION AND HEAT TRANSFER PROPERTIES OF BOILER DEPOSITS

13.1 EMITTANCE, ABSORPTANCE, AND REFLECTANCE MEASUREMENTS

Heat transfer from the flame to furnace-wall tubes in a coal-fired boiler is effected chiefly by radiation. In previous discussions it was established that the heat-exchange surfaces in coal-fired boilers will have an ash deposit in the form of loose dust, sintered matrix, or fused slag. It is therefore imperative that the relevant thermal properties—i.e., radiative absorptance and thermal conductance—of ash deposit should be known. Kirschoff's third law on radiation states that the absorptance of a body is equal to its emittance. The latter is easier to measure, and thus the emittance of boiler ashes and deposits is usually determined, rather than absorptance.

In literature the specific terms conductivity, absorptivity, emissivity, and reflectivity are often synonymously used for the generic terms conductance, absorptance, emittance, and reflectance. There should not be any confusion in the use of heat flow terms: thermal conductance has units of $W\,K^{-1}$, whereas thermal conductivity has units of $W\,m^{-1}\,K^{-1}$. The heat radiation terms, however, are expressed in dimensionless ratios, and the choice of the generic and specific terms by different researchers has been rather arbitrary. In this chapter the generic terms absorptance, emittance, and reflectance are used for the total radiation (i.e., radiation at all wavelengths), and the specific terms absorptivity, emissivity, and reflectivity are used for radiation at a given wavelength.

The methods of measuring high-temperature emittance of porous bodies have been described by Eckert et al. (1956). The emittance property of refractory materials has been determined by Pattison (1955), and by Robijn and Angenot (1963). Agababov (1957) in his measurements assumed that ash deposit on furnace walls is approximate to gray-body radiators, but Mulcahy et al. (1966) have suggested that this assumption is better avoided since the temperature of ash surface influences the intensity of radiation at different wavelengths. They have used a pyrometric technique described by Drury et al. (1951) for measurements of the surface emittance of ashes and deposits.

The measuring assembly consisted of an open hollow hemisphere, 50 mm in diameter, gold-plated on its inner surface and fitted with a thermopile detector behind a fluorite window at the pole of the hemisphere. When the instrument is held close to the surface of the test sample, the enclosed space approximates to a black-body radiation cavity and the output voltage V_G from the detector is given by

$$V_G = \kappa \epsilon_e (T_s^4 - T_o^4) \qquad (13.1)$$

where κ = calibration factor
T_s, T_o = temperatures of the test surface and detector, respectively
ϵ_e = effective emittance

$$\epsilon_e = \frac{\epsilon_r}{(1 - P_S)(1 - \epsilon_r)} \qquad (13.2)$$

where ϵ_r is the emittance of the test surface and P_S is the reflectance of the gold surface, which is close to unity. The instrument is provided with a black, nonreflecting hemisphere that can be inserted to cover the gold surface, and the output voltage V_B then becomes

$$V_B = \kappa \epsilon_r (T_s^4 - T_o^4) \qquad (13.3)$$

Equations (13.1) and (13.2) enable ϵ_r and T_s values to be calculated from V_G and V_B readings after making the necessary corrections as described by Goard (1966).

Mulcahy et al. (1966) have also measured the monochromatic emissivity of ash and deposits at 0.9 μm wavelength. They claim that the data are necessary to correct the results of surface temperature measurements of ash deposits on boiler walls. The pyrometer instrument was the same as that used for total emittance measurements except that it was fitted with a glass window. A silicon detector was employed, sensitive to radiation in the wavelength range of 0.5 to 1.1 μm. The spectral emissivity at 0.9 μm (ϵ_Λ) is given by

$$\epsilon_\Lambda = \exp \frac{C_2}{\Lambda} \left(\frac{1}{T_G} - \frac{1}{T_B} \right) \qquad (13.4)$$

where C_2 = second radiation constant
Λ = wavelength
T_G, T_B = sample and black body temperatures, respectively

Godridge and Morgan (1971) have used similar pyrometric techniques for the emittance measurements of coal ashes and deposits from oil-fired boilers. Becker et al. (1979) have pointed out that coal ash has a low thermal conductivity, and there is a drawback that the surface temperature of the sample varies continuously during the measurement. It is therefore unavoidable that the output variation merges with the relatively slow response of the thermopile detector. The authors describe the arrangement where they have used a narrow-angle radiometer with a faster response.

A different arrangement for measurements of coal ash emittance has been used by Mitor and Konopel'ko (1970), Abryutin and Karasina (1970), and Batch et al. (1970). The apparatus consisted of a heated surface onto which the sample is placed and the emitted flux ϵ_r is compared with that of a black body ϵ_b at the same surface temperature:

$$\epsilon_r = \frac{\epsilon_b}{\phi T_s^4} \tag{13.5}$$

where ϕ is the radiation constant.

A shortcoming of the method is that a temperature gradient is established across the sample and thus the surface temperature cannot be measured accurately with a thermocouple. Optical methods of surface temperature measurements necessitate prior knowledge of emissivity properties of the sample.

Abryutin and Karasina (1970) have measured absorptance α_s of coal ash in an arrangement where the heat transmitted from a black-body radiation source to the sample was measured by a calorimeter. For gray bodies—i.e., when the absorptance equals emittance—the heat transfer q is given by

$$q = \alpha_s \epsilon_b \tag{13.6}$$

where ϵ_b is the black-body emittance. The shortcomings of the method are, first, the assumption that coal ash is a gray-body radiator, which is best avoided as suggested by Mulcahy et al. (1966); and second, there are substantial errors in the heat-flow measurements by a calorimetric method.

Mitor and Konopel'ko (1966, 1969, 1970), and Abryutin and Karasina (1970) have used a reflectance technique for determining the absorptance of coal ashes. The spectral properties of a body are given by the relationship

$$\epsilon_\Lambda = \alpha_\Lambda = 1 - \rho_\Lambda \tag{13.7}$$

where ϵ_Λ, α_Λ, and ρ_Λ are the emissivity, absorptivity, and reflectivity, respectively, at a given radiation wavelength. The arrangement for reflectance measurements consists of a black-body source that beams radiation to the sample surface. A hemispherical collector focuses the reflected energy onto a detector, and the reflectance (ρ_s) is given by

$$\rho_s = \frac{V_d}{k \epsilon_b} \tag{13.8}$$

where V_d = detector output voltage
ϵ_b = black-body emittance
k = calibration constant

The shortcoming of the method is that for a non-gray surface, reflectance will depend on the temperature of the black-body source as well as the surface temperature of ash.

Wall et al. (1979b) have carried out a critical review of measuring the emittance, absorptance, and reflectance properties of coal ashes. They concluded that of the methods for determining the thermal radiation properties of coal ashes and deposits, the reflectance technique should be the least prone to experimental error. The reported emittance values obtained by Mitor and Konopel'ko (1966, 1969, 1970), and by Abryutin and Karasina (1970) using the reflectance technique are significantly higher than other reported data. Their emittance values measured by the reflectance method are given as 0.85 to 0.90, whereas the widely quoted emittance data published by Mulcahy et al. (1966), Boow and Goard (1969), and Godridge and Morgan (1971) are in the range of 0.55–0.70. Wall et al. (1979b) consider that these values may be low and that the data should be reexamined and compared with the results of reflectance measurements.

13.2 EMITTANCE CHARACTERISTICS OF PARTICULATE ASH AND BOILER DEPOSITS

Pulverized-coal ash consists largely of 0.5- to 50-μm-diameter particles of silicate, iron oxide, and unburnt carbon residue, together with 0.5–2.5 percent sulfate of 0.1 to 0.3 μm in diameter, as discussed by Raask (1980, 1981a) and Raask and Goetz (1981). The sulfate content, chiefly $CaSO_4$, of sub-bituminous coal ashes can be much higher, as shown by the analysis results published by Sondreal et al. (1968). The silicate particles either are colorless or are colored by iron varying from pale red to black when viewed under the optical microscope. Magnetic iron oxide and coke residue particles in ash are black, whereas the submicron-size sulfate particles are colorless.

Raask and Street (1978) have shown that at room temperature the appearance and light-reflecting property of ash depends chiefly on the amount and particle size of the coke residue. The results of measurements with mixtures of white magnesia and black coke are plotted in Fig. 13.1 and show that the mix became reflectance-saturated when the 2- to 5-μm coke particles were present in a quantity between 6 and 8 percent by weight. The influence of particle size of coke on the reflectance of magnesia/coke mixes can be seen from Fig. 13.2, and it is evident that it is inversely related to particle size.

In the absence of coke residue the appearance of ash depends largely on the amount and the mode of distribution of iron in ash. Iron in the ferrous valency state (Fe^{2+}) gives a black color to wustite (FeO), magnetite (Fe_3O_4), and to silicate particles when dissolved in concentrations above about 3% by weight. The ratio of ferrous to ferric oxides in silicate-dissolved species is governed by the oxygen partial pressure and temperature, as discussed in Chap. 9 [Eq. (9.9)].

Johnson (1952, 1964) has examined the optical absorption and diffuse reflectance from the surface of a powder of monochromatic particles in terms of

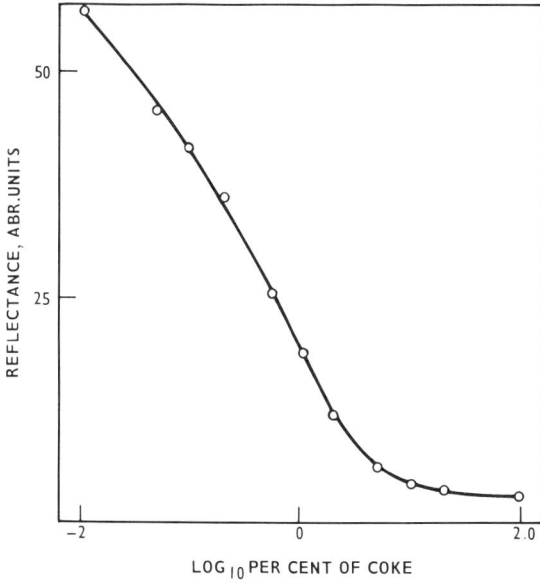

Figure 13.1 Reflectance of white magnesia and black coke powder mixtures.

refractive index η, particle diameter d, and absorption coefficient α. The reflectance increases, and hence the emittance ϵ_r decreases, as the product αd decreases. When the particle diameter is less than half the radiation wavelength the effect of Rayleigh scattering must also be considered. As the product αd approaches unity, the reflectance depends only on the refractive index. The change in silicate ash composition does not have a marked influence on the refractive index and,

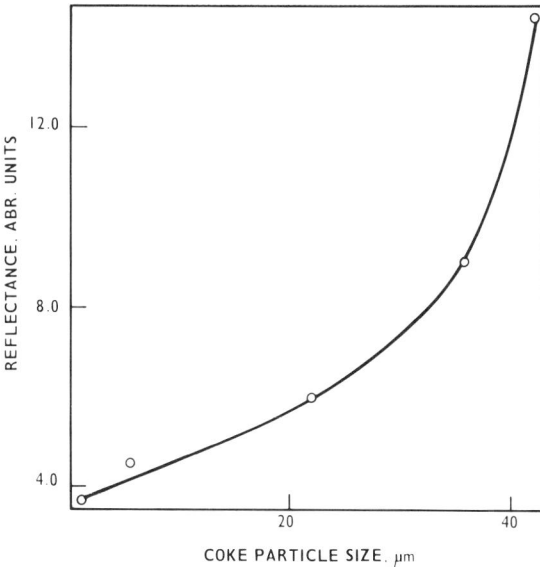

Figure 13.2 Effect of coke particle size on reflectance of 6.5 percent coke/magnesia mixture.

according to data published by Morey (1954), the values for silicate glasses of coal ash and deposits should be between 1.5 and 1.6. Boow and Goard (1969) have determined the emittance of glass powders, and of laboratory-prepared and boiler ashes, of different sizes at 775 K. From the results they obtained two expressions:

$$\epsilon_r = 0.25 \log d + 0.13 \qquad (13.9)$$

$$\epsilon_r = 0.30 \log d + 0.16 \qquad (13.10)$$

where ϵ_r is the total emittance expressed in dimensionless ratio and d is the particle diameter in μm.

Equations (13.9) and (13.10) are applicable when the particle size is in the range of 7 to 420 μm. Equation (13.9) applies for colorless glass powders, and Eq. (13.10) can be used to calculate the emittance of typical silicate ash deposit on boiler tubes before a sinter matrix will develop. These equations are not applicable for calculating the initial deposit on a clean metal target. The ash consists largely of submicron-size particles, chiefly sulfates, as discussed in Chap. 12, and the colorless small particles constitute a highly reflecting surface; thus the emittance of the initial deposit material on boiler tubes is likely to be low, between 0.1 and 0.2 units.

Mulcahy et al. (1966) have measured the emittance of coal ash and boiler deposit at different temperatures, and a typical emittance/temperature relationship is shown in Fig. 13.3. Curve A shows that the emittance of particulate ash decreased with the increase in temperature up to the sinterpoint, and the sintered ash (curve B) and fused slag (curve C) had significantly higher emittance values. Similar temperature–emittance and sintering–emittance relationships have been found by Boow and Goard (1969), Godridge and Morgan (1971), and Wall et al. (1979b). The marked increase in emittance of sintered and slagged ash is caused partly by reduction in the reflective surface of particulate ash and partly by the change from ferric to ferrous state of iron at high temperatures, as depicted in Fig. 13.4.

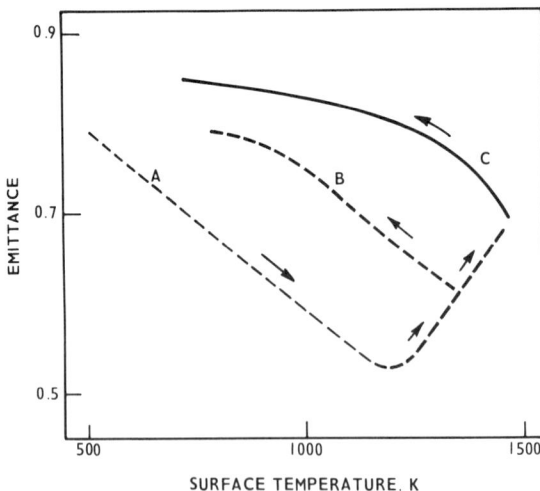

Figure 13.3 Emittance characteristics of boiler deposit: A, particulate ash; B, sintered deposit; and C, slag. Arrows indicate direction of heat flow.

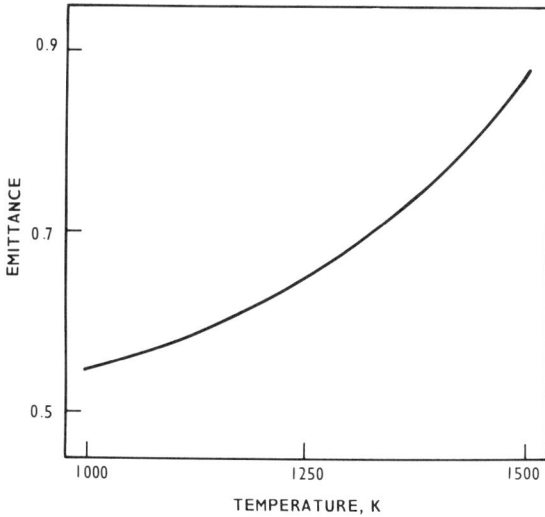

Figure 13.4 The effect on emittance of iron-rich deposit caused by increased ferrous/ferric ratio with temperature.

The emittance characteristics of ash deposit as it builds up on boiler tubes are depicted in Fig. 13.5. The initial deposit of submicron-size particles of silicates and sulfates (Chap. 12) constitutes a low-emittance surface, and thus a thin layer of particulate ash can significantly reduce the radiative heat transfer to boiler tubes, as discussed in the following section. In subsequent deposition, large ash particles, 1 to 50 μm in diameter, are captured and a sintered or slagged matrix is formed, resulting in a marked increase in the surface emittance.

The deposit on furnace-wall and superheater tubes in coal-fired boilers is periodically removed by sootblowing with steam, air, or water jets, usually once in 8 h. It is therefore likely that the surface emittance property of boiler tubes shows cyclic variation, as depicted in Fig. 13.5. The surface of oxide layer on boiler steels

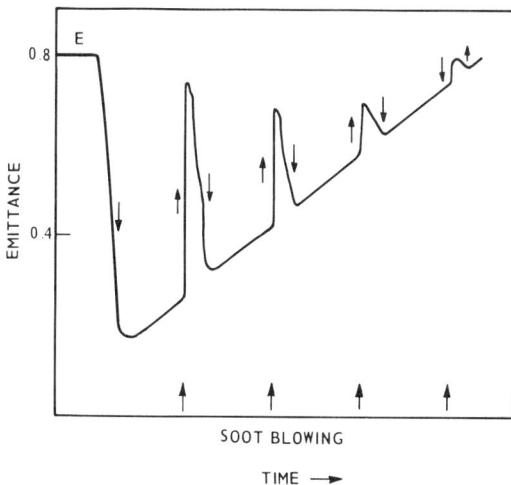

Figure 13.5 Schematic representation of the effect of deposition and sootblowing on emittance of boiler-tube surface, where E is emittance of deposit-free surface, arrows down show deposition of fine ash, and arrows up show removal by sootblowing.

free from ash deposit, has an emittance value of around 0.8 at 675 K, and the surface of fused ash deposit at 1375 K has approximately the same emittance value, as reported by Mulcahy et al. (1966). During the initial stages of deposit formation, sootblower jets remove most of the slightly sintered ash, causing a marked fluctuation in the emittance of tube surface. There may remain a residue of sintered ash that resists the action of sootblowers, and consequently the effect on the emittance is less marked with time, as depicted in Fig. 13.5.

13.3 THERMAL CONDUCTIVITY OF PARTICULATE ASH AND BOILER DEPOSITS

The ash deposit on boiler tubes can be considered as an insulating material that reduces the heat transfer from the flame and hot gases to the Rankine fluid. Clements (1966) has shown that there is an approximately linear relationship between the thermal conductivity and bulk density of sintered products, as shown in Fig. 13.6. The scatter in the results was considered to be caused largely by the difference in size of pores in the sintered matrix. Figure 13.7 shows the relationship between the pore size and thermal conductivity in the temperature range of 475–1275 K. The plots show that at temperatures above 1000 K an increase in pore size will raise the effective thermal conductance, whereas at 500 K large pores would decrease the heat flow through porous bricks. It is therefore evident that the mode of heat transfer through a porous body changes with temperature.

The heat passes through an insulating material by solid- and gas-phase conductance, and by radiation across the pores. A fourth possible mode of heat transfer, convection within the pores below about 5 mm in size, is likely to be negligible. The heat transfer q by radiation through a porous body is influenced by the mean pore diameter d_p, the operating temperature T, and the emittance of the

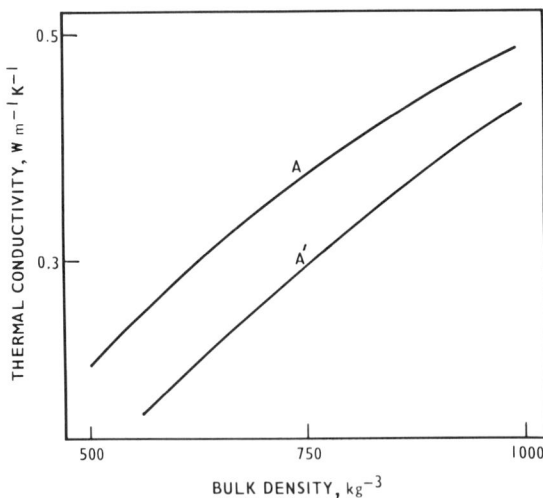

Figure 13.6 Thermal conductivity of heat-insulating bricks at 775 K; region between A and A' is scatter band.

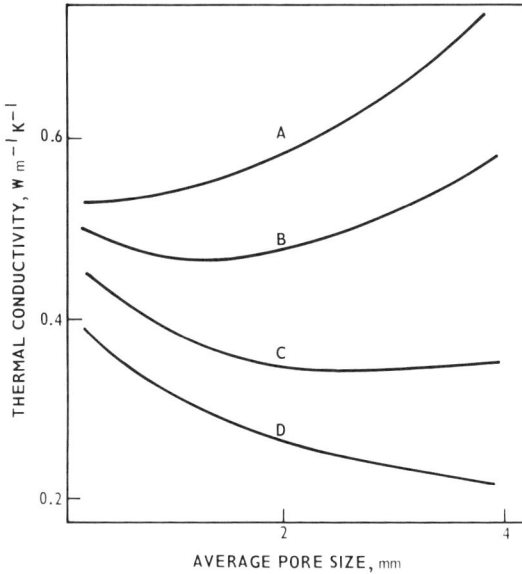

Figure 13.7 The effect of pore size on thermal conductivity of insulating bricks with nominal bulk density 640 kg m^{-3}: A, at 1275 K; B, at 1075 K; C, at 775 K; D, at 475 K.

matrix material ϵ_r; an approximate relationship is

$$q = \frac{Cd_pT^3}{(2/\epsilon_r) - 1} \qquad (13.11)$$

where C is a constant. It is evident that a large size of pores and a high emittance value would enhance the heat transfer by radiation at high temperatures. Table 13.1 shows the relative contributions in heat transfer through a porous refractory material by different modes of energy transportation when the surface temperature increases from 775 to 1775 K.

The above values were obtained from the data published by Barrett (1949) and show that conduction through the solid phase was the predominant mode of heat transfer; at high temperatures conduction by radiation became significant, and accounts for about 20% of the total heat conducted through a porous refractory material. Young et al. (1964) have suggested that the radiative heat transfer makes a more significant contribution, and up to 40 percent of the total heat passing through insulating bricks is conducted by this mechanism.

Table 13.1 Heat transfer through a body of 70% volume porosity

Temperature (K)	Thermal conductivity (W m^{-1} K^{-1})			
	Total measured	Conduction through solid	Conduction through gas	Conduction through radiation
775	0.36	0.29	0.04	0.03
1275	0.49	0.36	0.06	0.07
1775	0.62	0.43	0.07	0.12

Mulcahy et al. (1966) have determined the effective conductance across the layers of unsintered ash, sintered deposit, and fused slag. The temperature gradient across the sample, 3.5 to 4 mm thick, was established by thermocouple measurements, and the surface emittance of ash was determined by the pyrometric method described in Sect. 13.1. Heat flow q_r through a layer of ash or deposit is given by

$$q_r = \epsilon_r \phi (T_e^4 - T_s^4) = \frac{k}{s} (T_s - T_c) \qquad (13.12)$$

where ϵ_r = emittance of ash surface at temperature T_s
$\quad T_e$ = temperature of radiation source
$\quad T_c$ = temperature of cold face of the ash layer
$\quad k$ = thermal conductivity
$\quad s$ = thickness of ash material
$\quad \phi$ = Stefan-Boltzmann constant

Their results gave the thermal conductivity of unsintered ash between 0.08 and 0.3 W m^{-1} K^{-1} with the temperature increase from 475 to 1275 K. They compared the results with thermal conductivity data of some other materials—e.g., slag wool, 0.04 W m^{-1} K^{-1}; building brick, 0.6 W m^{-1} K^{-1}; and mild steel, 40 W m^{-1} K^{-1}.

Boow and Goard (1969) have shown that the effective thermal conductivity of unsintered ash depends on the particle size, and from the experimental results they deduced two equations:

$$\log k = (0.48 \log d) - 1.75 \qquad (13.13)$$

for colorless glass particles, and

$$\log k = (0.56 \log d) - 1.63 \qquad (13.14)$$

for iron-containing ash, where k is the thermal conductivity (W m^{-1} K^{-1}) and d is the particle diameter (μm). These equations are applicable for ashes of the particle size of 20-300 μm at 975 K. With laboratory-prepared ashes Boow and Goard obtained exceptionally low values of thermal conductance, 0.02 to 0.06 W m^{-1} K^{-1}, in the temperature range of 500 to 1000 K. Golovin (1964) has made measurements on samples of anthracite ash and reported values between 0.012 and 0.023 W m^{-1} K^{-1}, which are considerably less than that of air. The thermal conductivity of air in the temperature range of 500 to 1000 K at atmospheric pressure is between 0.03 and 0.07 W m^{-1} K^{-1}, according to the data published by Hilsenrath (1960). Prasolov and Vainshenker (1960) have measured the thermal conductivity of a variety of ashes, and their results were in the range of 0.04-0.26 W m^{-1} K^{-1}.

It is remarkable that the thermal conductivity of a powder material can be as low as or lower than that of air. Further examples of such low values have been observed by Kistler (1931) and White (1939) in porous silica gels that had a bulk density of 100 kg m^{-3}. Gurvich and Prasolov (1960) have observed that the initial ash deposit of submicron-size particles on boiler tubes had a thermal conductivity below 0.1 W m^{-1} K^{-1}. It appears that the small particles of unsintered ash deposited on boiler tubes have a dominant role in determining the rate of heat transfer from the flame and hot flue gases to the Rankine fluid.

Raask (1981a) has shown that the particles formed in the coal flame, chiefly sulfates of diameter 0.1–0.3 μm, are either captured on 0.5–50-μm-diameter silicate particles, or dispersed as individual particles or as agglomerates in the flue gas, as discussed in Chap. 8. The submicron-size particles can form voids in an ash deposit where the cavity dimensions are comparable with that of the mean free path of the gas molecules. This allows the molecular or Knudsen mode of heat conductance to dominate in the initial deposit of ash, which consists chiefly of submicron-size particles. The gas molecules in small pores create a thermal barrier known as temperature "jump" or "slip," which is known as the Knudsen effect in the heat transfer. The retarding effect occurs at all gas–solid interface boundaries, and the presence of small pores in a low-density material can be regarded as having created a large number of boundary barriers to heat flow. Boow and Goard (1969) have suggested that in order to reduce the thermal conductivity of a porous body below that of air or boiler-flue gas, the pore sizes created by small particles in an ash deposit must be below 0.5 μm.

Mulcahy et al. (1966) have shown that there was a marked increase in the effective thermal conductance through a layer of ash above its sinterpoint temperature, as depicted in Fig. 13.8. In analogy to the electrical conductance through a layer of unsintered and sintered ash as discussed in Chap. 10, an equation for the heat conductance can be written in terms of density and temperature:

$$q = B_1 \left(\frac{D}{D_o} - 1 \right) \exp \frac{-E}{RT} \tag{13.15}$$

where q = conductance
D_o, D = ash density before and during sintering
T = temperature
E = exponential coefficient of conductance
R = thermodynamic (gas) constant
B_1 = constant related to thermal characteristics of the ash

When the degree of sintering does not change—e.g., on cooling as depicted by curve B in Fig. 13.8, where the density of sintered ash remains constant—Eq. (13.15) reduces to

$$q = B_2 \exp \left(\frac{-E}{RT} \right) \tag{13.16}$$

where B_2 is a constant and other terms are as above.

The change in bulk density of boiler deposits with temperature is depicted in Fig. 13.9. The density of initial, unsintered ash deposit is likely to be between 500 and 600 kg m^{-3}, which corresponds to a volumetric porosity of 70–80%. Curve A shows that iron-rich deposits have a lower sinterpoint temperature above which the density increases rapidly, and the nonporous slag has a slightly higher density than that of low-iron slag (curve B). Completely pore-free slag is rarely formed in pulverized-coal-fired boilers, and the highest bulk density is usually around 2200 kg m^{-3}, representing 10 to 15 volume porosity. The reason is that the slag is highly viscous and the gas bubbles trapped in the matrix cannot escape. In contrast, molten slag flowing on the furnace walls of cyclone-fired boilers has a viscosity

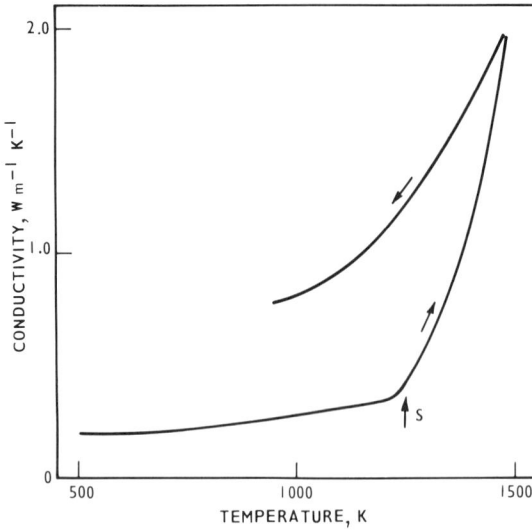

Figure 13.8 Changes of the effective thermal conductivity of coal ash on heating and on cooling, where S is sinterpoint temperature.

below 100 N s m^{-1}, allowing the gas bubbles to escape, and usually a pore-free slag is discharged from the furnace.

Boow and Goard (1969) have found that the effective thermal conductivity k of unsintered ash depended on the emittance ϵ_r, and the relationship at 775 K is

$$\epsilon_r \log k = 0.75\epsilon_r - 1.55 \tag{13.17}$$

where k is given in $\text{W m}^{-1} \text{ K}^{-1}$.

A significant correlation coefficient (0.94) was found to hold for a variety of ashes and powdered glasses, despite wide variation in composition and thermal history. Similar relations were found to hold for other surface temperatures, but scatter of points increased as the ash particles started to sinter. The authors have added a cautionary note that care should be exercised in extrapolating the experimental data outside the range of particle size 7–420 μm. For larger particle sizes the emittance will depend on the refractive index of the material. With smaller particles, Rayleigh scattering would have a marked effect in changing the relationship between the conductivity and the emittance.

A marked dependence of the effective thermal conductivity on the ash surface emittance characteristics is shown in Fig. 13.10; the data were taken from the results published by Boow and Goard (1969). The plots show that there was an order of magnitude difference in the conductivity of laboratory-prepared ash and that of furnace-wall deposits in the temperature range of 500 and 1000 K. At higher temperatures this difference decreased rapidly, and it disappeared at 1450 K, as shown in Fig. 13.10.

The thermal conductivity measurements are usually carried out in air, using laboratory heat sources as discussed in the previous section. The boiler-flue gas has relatively high concentration of CO_2 and water vapor, 12 to 14 percent and around

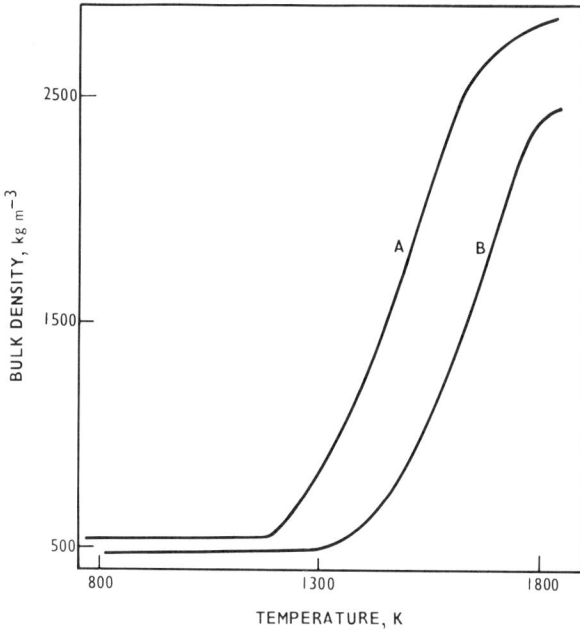

Figure 13.9 Change in bulk density of ash deposits with temperature: A, iron-rich ash; B, low-iron ash.

8 percent by volume, respectively. Young et al. (1964) have suggested that these gases may adsorb at the surface of silicate materials, thus changing the emittance and hence the radiative heat transfer properties. It is, however, unlikely that errors introduced by the change of atmosphere is significant when compared with those resulting from possible differences in the density of boiler deposits.

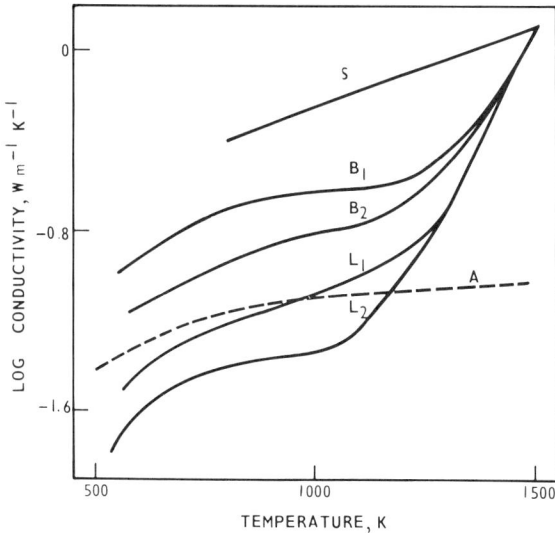

Figure 13.10 Thermal conductivity of laboratory ashes (L_1, L_2), boiler ashes (B_1, B_2), sintered ash (S), and air (A).

13.4 THERMAL BARRIER EFFECT OF ASH DEPOSITS ON HEAT TRANSFER

The heat transfer from the flame to the surface of water and steam tubes in the combustion chamber takes place chiefly by radiation, with less than 5 percent of the heat transferred by convection, as discussed by Godridge and Read (1976). The equation of energy balance for a steady operation of boiler plant is

$$Q - q_r - q_s = 0 \qquad (13.18)$$

where Q, q_r, and q_s are, respectively, the rates at which heat energy is released in the combustion chamber, absorbed by furnace-wall tubes, and passed out of the furnace as the sensible heat of the flue gas.

The heat transfer through the porous deposit of ash on boiler tubes takes place by conduction, convection, and radiation, then by conduction through the metal oxide and metal, and finally by convection to the fluid (water or steam) flowing in the tubes. The tube fluid temperature T_w is related to the outer surface temperature of deposit T_s by the relationship

$$q_r = U(T_s - T_w) \qquad (13.19)$$

where U is the overall heat transfer coefficient and can be expressed in the form

$$U = \left(\frac{1}{h} + \frac{s_m}{k_m} + \frac{s_{ox}}{k_{ox}} + \frac{s_1}{k_1} + \frac{s_2}{k_2} + \cdots \right)^{-1} \qquad (13.20)$$

where h = coefficient of convective heat transfer from tube to fluid
s_m, s_k = thickness and thermal conductivity of the boiler steel, respectively
s_{ox}, k_{ox} = thickness and thermal conductivity for its oxide
The terms $s_1, s_2, \ldots$ and $k_1, k_2, \ldots$ denote the thickness and thermal conductivity of different layers of the ash deposit as summarized in Table 13.2.

Table 13.2 Thermal data for calculating the heat-flow barrier of ash deposit

Convective heat coefficient[a] (W m^{-2} K^{-1})	Thermal conductivity (W m^{-2} K^{-1})			Thickness (mm)		
	Metal[b]	Oxide[b]	Deposit[c]	Metal	Oxide	Deposit
1.1×10^4–1.7×10^4	40	10	0.03–3.0	6–8	0.01–0.1	0–50

[a]Data were taken from publication of Wall et al. (1979b).
[b]The thermal conductivity values of steels and oxides at 800 K were taken from the data published by Powell et al. (1966) and Kaye and Laby (1978).
[c]The conductivity values of boiler deposit were taken from the references given earlier in Sect. 13.3 and are as follows: initial deposit of submicron sized particles, 0.03 W m^{-1} K^{-1}; inertially impacted particles, 0.1 W m^{-1} K^{-1}; sintered deposit with a fused surface layer, 1.1 W m^{-1} K^{-1} [this value was taken by Neal and Northover (1980) in their measurements of radiant heat flux in large boilers]; and dense iron-rich slag, 3 W m^{-1} K^{-1}.

It is evident that ash deposits on boiler tubes have a dominant role in determining the U value, i.e., the overall heat transfer coefficient in Eq. (13.19). The thickness of initial ash deposit that has the same heat barrier effect as that of the metal-to-fluid convective coefficient, and conductivity of tube metal and its oxide, can be obtained from

$$\frac{1}{h} + \frac{s_m}{k_m} + \frac{s_{ox}}{k_{ox}} = \frac{s_1}{k_1} \tag{13.21}$$

where the terms were defined for Eq. (13.20).

The typical thickness and thermal properties of tube metal and oxide given in Table 13.2 show that the initial ash deposit of submicron-size particles, of about 10 μm thick, is sufficient to double the resistance to heat flow. This is in accordance with observations by Gurvich and Prasolov (1960) and Becker (1971) that the initial deposit on cooled metal surfaces had a marked effect in reducing the heat transfer. This significant reduction in heat flow by the thin layer of deposit is caused by the low emittance characteristics and the Knudsen heat barrier effect of particles below 0.5 μm in diameter, as discussed in Sects. 13.2 and 13.3.

Mulcahy et al. (1966) have estimated the possible effects of ash deposits in reducing the heat transfer from the flame to boiler tubes in the manner depicted in Fig. 13.11. Curves A_1 and B_1 represent approximately the decrease in the rate of heat flow as deposits of unsintered ash or fused matrix builds up on boiler tubes. The corresponding increase by as much as 300 K in the temperature of the flue gas leaving the combustion chamber is depicted by curves A_2 and B_2. In boiler design practice, an allowance is made for the heat barrier effect of ash deposits and thus the surface area of the combustion chamber of a pulverized-fuel-fired boiler is approximately 40% larger than that of an oil-fired boiler of the same capacity (Godridge and Read, 1976). Curves A_1 and B_1 show that a 3-mm-thick layer of loose ash or a 10-mm-thick layer of fused deposit covering the boiler tubes would reduce the heat transfer of about 40%. In some pulverized-coal-fired boilers this degree (average) of slagging is not exceeded and little on-load cleaning is required. However, in most boilers extensive sootblowing of the combustion chamber is required in order to maintain the appropriate rate of heat transfer from the flame to the steam-generating tubes.

Wall et al. (1979b) have pointed out that it is difficult to apply laboratory data of the emittance and thermal conductivity of ash for estimating the heat resistance of boiler deposits. They suggest that the effective emittance of *in situ* deposits may be significantly higher than that indicated by the laboratory measurements. For example, a thin layer of deposit on boiler walls will follow the curvature of the tubes, so the effective emittance ϵ_e is related to the flat surface emittance ϵ_r by the expression

$$\epsilon_e = \frac{\pi\epsilon_r}{1 + (\pi - 1)\epsilon_r} \tag{13.22}$$

Further, an ash deposit on the walls of the combustion chamber of a pulverized-coal-fired boiler rarely has a smooth surface, and the contoured surface will increase

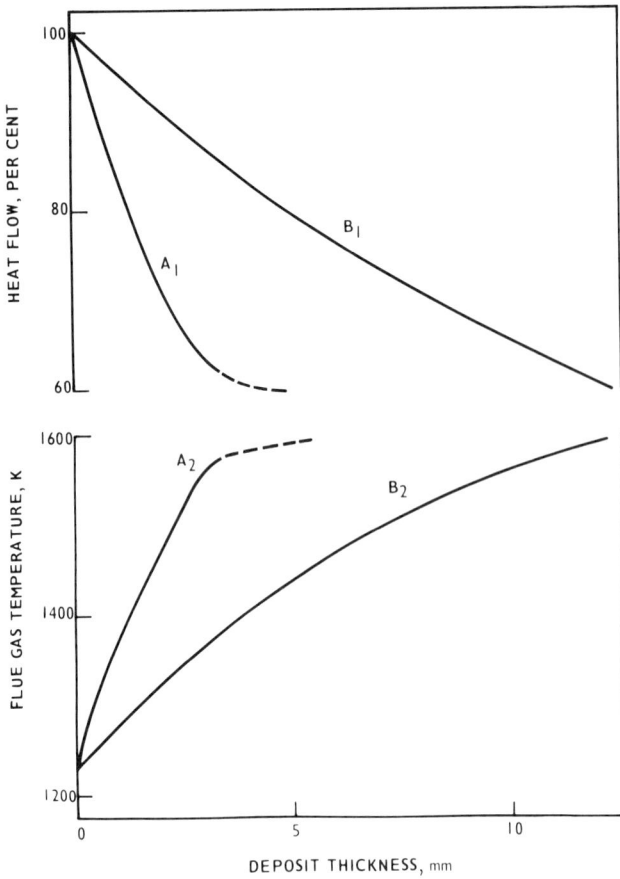

Figure 13.11 The effect of furnace-wall deposit on heat transfer and flue-gas exit temperature: A_1, A_2, particulate ash; B_1, B_2, fused deposit.

the effective emittance well above that obtained in laboratory measurements. They advocate that both the emittance and thermal conductivity measurements should be made in operation boilers. However, the measuring sensors in the combustion chamber will be covered with a deposit that may not necessarily have the same heat conductance property as that of the ash deposit on boiler tubes. The heat-flux meters used in boiler plant are described in the next section.

13.5 HEAT–FLUX PROBES AND MEASUREMENTS IN COAL–FIRED BOILERS

There are several essential needs for measuring the flux on the furnace-wall tubes of coal-fired boilers. For example, heat-flow mapping during the commissioning stages of a new plant will indicate problem areas that have to be corrected in the existing plant and should be avoided in subsequent designs. Excessive heat fluxes in the primary combustion zone (Chap. 14) may lead to formation of massive deposits or may enhance the rate of tube-metal corrosion. Further, computer programs that have been developed for optimizing the boiler operation require reliable heat flux data.

The method of construction of heat-flux meters usually makes use of the temperature gradient generated along a cylindrical piece of metal alloy of known thermal conductivity. The temperature gradient is measured by thermocouples at known spacing along the cylinder, and for a measuring device in large furnaces the heat sink is usually the boiler tubes. To ensure that the heat flow in the measuring head is essentially axial, it must be surrounded by a thermal guard that exhibits a temperature gradient similar to that along the measuring cylinder. Figure 13.12, *a* and *b*, show a surface-mounted device and a thickened tube attachment. The latter concept has the advantage of having a minimum of change in the tube profile. It is therefore to be expected that the amount and the nature of ash deposit at the surface of the device is similar to that formed on boiler tubes.

A variety of heat-flux measuring devices have been described by Lucas (1963), Northover and Hitchcock (1967), and Anson and Godridge (1967, 1975). More recently, Neal et al. (1980) have described the development of three different arrangements of measuring heat fluxes in boiler furnaces, each to fulfill a specific role.

The Fluxtube

This is a comprehensive instrument and it measures, in addition to heat flux, the tube-metal and water–steam temperatures. The Fluxtube can be used as a "standard" instrument against which other heat-flux measuring devices may be compared. The instrument has six thermocouples, where the paired thermocouples are connected differentially to give two signals related to heat flux, and the individually measured thermocouples give the metal and tube-fluid temperatures. Northover (1977a) has given a detailed description of the construction and mounting of the instrument.

The Dometer

The Dometer is a boiler tube surface-mounted instrument, and when installed has a dome-shaped profile. It gives also the surface temperature of tube metal in addition

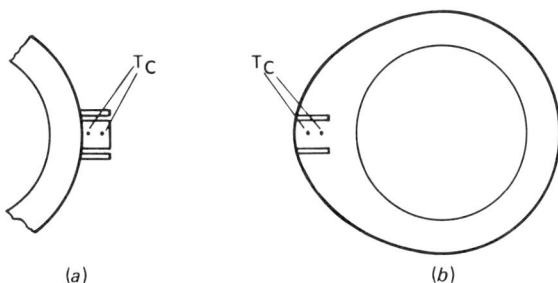

Figure 13.12 Heat-measuring devices on boiler tubes. (*a*) Surface-mounted and (*b*) mounted in thickened tube, where T_C show thermocouples.

to the heat-flux measurements. The device, as described in detail by Northover (1978), is suitable for installation in large numbers for flux mapping in power station boiler furnaces. In this comparatively simple arrangement, perfect matching between the thermal gradient in the measuring core and that in thermal guard is not possible. However, calibration studies by Neal et al. (1980) have shown that the performance of the device was acceptable when the welding instructions for mounting of measuring head and thermal guard were closely followed. There may be errors introduced at high heat fluxes as a result of high surface temperature of the measuring section. For example, with heat flux of 600 kW m^{-2} there could be a difference of 150 K between the probe surface temperature and that of boiler tubes. This may significantly increase the rate of build-up of ash deposit on the flux metering device. It is therefore essential to carefully observe the mode of formation and characteristics—e.g., thickness, density, chemical composition, and appearance of the deposit—on boiler tubes fitted with heat-flux measuring devices.

The Fluxprobe

The Fluxprobe, as described by Northover (1977b), is a portable, direct-reading instrument for measuring the incident heat flux from the flame to boiler furnace walls. It has been designed primarily for use in membrane-construction boilers where adjacent tubes are joined along their length by continuous joint strips, some 13 mm wide and 6 mm thick. The sensing head of the Fluxprobe is small enough to insert through 7-mm holes in the membrane joint. Alternatively, the Fluxprobe may be used through any available boiler port.

The sensor head of the Fluxprobe is made of steel type EN 56A, which has a near-constant thermal conductivity over the range of probe working temperature. This, together with the use of nickel–chromium/nickel–aluminum thermocouples having a near-linear temperature/EMF characteristic, ensures a virtually constant calibration over a wide range of heat fluxes. The sensor head has a double heat guard arrangement and this has reduced the errors in calibration to a maximum of 4 percent.

In order to obtain accurate measurements of the incident radiation heat flux, it it essential to have low sensor-surface temperature so that back radiation is negligible. For example, at an incident heat flux of 500 kW m^{-2}, a sensor surface temperature of 875 K produces a back radiation of 6.6 percent, rising to 15 percent at 1075 K. A low surface temperature is achieved by using a heat conductance pipe to provide a short thermal path between the sensing head and a thinned natural-convection heat dissipator. The sensor is coated with a layer of nickel oxide, using a flame spraying technique in order to give a surface of high emittance, about 0.95. Calibration of the Fluxprobe is performed in a black-body furnace at 1825 K. In general any flux meter can be calibrated to read correctly under clean conditions, but in the deposit-forming and corrosive environments of coal-fired boilers, large and often unpredictable errors can occur, as discussed by Morgan (1974). Neal and Northover (1980) have used a computer program based on the finite-element analysis model as described by Fullard (1971) to evaluate the effect of ash deposits

Table 13.3 Changes in heat transfer rates and steam conditions in five power-station boilers

Condition	Tidd, U.S.A. (commissioned 1948)	Paddy's Run, U.S.A. (commissioned 1950)	Brunswick Wharf, U.K. (commissioned 1960)	Ferrybridge C, U.K. (commissioned 1970)	Fawley, U.K. (commissioned 1970)
Heat flux (kW m^{-2})	160–270	140–260	150–270	250–350	300–550
Furnace projected area (m^2)	710	990	570	3100	2000
Steam output (kg s^{-1})	51	72	40	430	430
Steam pressure (MN m^{-2})	9.6	6.4	6.3	16.5	16.5
Steam temperature (K)	769	755	769	841	811
Reference	Reid et al. (1948)	Corey and Cohen (1950)	CEGB (1959)	Godridge and Read (1976)	

on the rate of heat flow, as measured by the flux metering devices mounted on boiler tubes. They have found that an increase in the thickness of ash deposit from 1 to 2 mm resulted in a 28 percent reduction of heat transfer from flame to boiler tubes. This is in accordance with the assessment data published by Boow and Goard (1969) and depicted in Fig. 13.11. Of course, the results of any computer evaluation depend on reliability of the primary data, in this case the emittance and effective thermal conductivity of different layers of a boiler deposit. In practical terms, Neal et al. (1980) suggest that combined use of the *in situ* heat flux devices and the portable Fluxprobes can give a useful assessment of the effect of boiler deposits on heat transfer.

Godridge and Read (1976) have reviewed the changes in the unit size and heat transfer rates in power-station boilers commissioned in Britain during the 10-yr period 1960–1970. The unit output increased from 60 to 660 MW, and there are at present 900- and 1200-MW units in operation in the United States. In contrast, the increase of average heat transfer rate in the combustion chamber has been less spectacular, as shown in Table 13.3. For comparison, the transfer rates and the steam conditions in a residual-oil-fired boiler are also given.

Fawley is a residual-oil-fired station and the 500-MW output boilers have average heat transfer rates 50 percent higher than those in Ferrybridge coal-fired boilers of the same capacity. That is, the combustion chamber of a pulverized-coal-fired boiler is substantially larger than an oil-fired boiler with comparable output. This is one of the many difficulties encountered when it becomes necessary to convert oil-fired boilers to coal firing as a result of the high price of oil, as discussed by Laire (1980).

NOMENCLATURE

B_1	constant
B_2	constant
C	constant
C_2	second radiation constant
D	density of sintered ash
D_o	original density of ash
d	particle diameter
d_p	pore diameter
E	exponential coefficient of thermal conductance
h	coefficient of convective heat transfer
k	thermal conductivity
k_m	thermal conductivity of tube metal
k_{ox}	thermal conductivity of oxide layer
k_1, k_2	thermal conductivity of deposit layers
n	refractive index
Q	rate of heat release
q	heat transfer

q_r	heat transferred through boiler wall deposits
q_s	heat content of flue gas
R	thermodynamic gas constant
s	thickness of ash material
s_m	thickness of boiler tube wall
s_{ox}	thickness of oxide layer
s_1, s_2	thickness of deposit layers
T_B	black body temperature
T_c	cold face temperature
T_e	radiation source temperature
T_G	sample temperature
T_o	detector temperature
T_s	hot face temperature
T_w	fluid temperature
U	heat transfer coefficient
V_B	calibration voltage
V_d	detector voltage
V_G	test voltage
α	absorption coefficient
α_s	absorptance
α_Λ	monochromatic absorptivity
ϵ_b	black body emittance
ϵ_e	effective emittance
ϵ_r	surface emittance
ϵ_Λ	monochromatic emissivity
η	refractive index
κ	calibration factor
Λ	wavelength
ρ_s	surface reflectance
ρ_Λ	monochromatic reflectivity
ϕ	Stefan–Boltzman constant

FOURTEEN

MEASURES TO COMBAT BOILER FOULING AND SLAGGING

14.1 CHOICE OF BOILER SYSTEMS: PULVERIZED FUEL OR CYCLONE FIRING

The brief resumé on the development of electricity utility boiler designs in Chap. 2 shows that there are two different combustion systems for large generating units in current use. The most widely used system is that of pulverized-coal firing, where the aim is to keep the flame temperature and that of deposit-collecting surfaces sufficiently low to avoid the formation of massive build-up of sintered ash and fused slag. In contrast, the principal feature of cyclone-fired boilers is the high flame temperature, which causes ash to melt, and up to 85% of the total coal mineral impurities is removed from the system in the form of molten slag.

The first task facing the design engineers when considering the layout of a new power plant for a given fuel supply is to decide whether pulverized-coal firing or the high-temperature cyclone combustion system is appropriate with a minimum of slagging problems. Evaluations of the slagging propensities of different coal ashes are set out in Chaps. 9 and 10, and these should give some useful guidelines to the design engineers. Pulverized-coal-fired boilers can be designed to utilize almost any coal of an ash content up to 50% by weight, and the amount and composition of ash largely dictates the size and the temperature regime of boiler plant. Coals with a slagging ash require a large combustion chamber when burned in pulverized-coal-fired systems in order to prevent build-up of massive deposits of sintered ash and fused slag on the furnace-wall and superheater tubes. Exceptionally slagging coals where the ash has a high "flux" content, chiefly $CaO + FeO + Fe_2O_3$, between 30 and 40%, are difficult to burn in a pulverized-coal-fired boiler and are more suitable for cyclone or slag tap firing. There are, however, a number of other considerations that must be evaluated in the planning and design stage.

Consistency of Coal Ash Composition

Boiler units of 500 to 1200 MW output capacity of 2000- to 4000-MW power stations consume large quantities of coal during their operational life, which may extend over 30 yr. It is therefore inevitable that there will be some changes in the quality of coal delivered to large power stations. In these circumstances a pulverized-coal-fired system is a better choice, as it is more tolerant to changes in the composition of coal mineral matter compared with the slag discharge boilers. For example, it would be difficult to operate a cyclone boiler where there is a marked change from a slagging to an infusible ash in coal. It would therefore be unwise to plan large coal-fired boilers based on a slag discharge design for operation with imported fuels. Also, introduction of high-alkali coals to cyclone-fired boilers may cause severe superheater fouling and corrosion as a result of the high ratio of sodium to ash in the deposited material (Chap. 12). The high flame temperature and the mode of combustion in cyclone fired boilers would enhance the formation of submicron-size particles, which are difficult to capture in the electrical precipitators (Chap. 16).

Boiler Construction and Capital Expenditure

Design and construction of cyclone boilers require specialist skills over and above those required to build pulverized-coal-fired boilers. In particular, the choice and mode of attachment of suitable refractory lining materials in the cyclone boiler furnace need a great deal of attention. Silicon carbide materials with antislagging additives, usually vanadium oxide or zirconia, are most frequently used in this application. Keying a refractory lining material to boiler tubes requires extensive stud welding, which is an expensive and time-consuming process. This, together with the costs of high-grade refractory materials and the complicated furnace construction, increases the capital layout of the cyclone boiler by a significant margin.

The high costs of cyclone boiler construction are partly balanced by the smaller size of the electrical precipitator plant because of the lower ash burden of the flue gas, and by some savings in the maintenance of the coal milling plant: cyclone boilers burn a coarser grade fuel than is required for pulverized-coal-fired boilers.

Combustion Efficiency

One of the justifications usually put forward in favor of cyclone boiler design is the high combustion efficiency. The highly turbulent flow of the combustion air and the high flame temperature ensure that near-complete combustion of coal can be achieved with about 10 percent excess of combustion air, whereas the amount of excess combustion air in pulverized-coal-fired boilers is usually around 20 percent. This difference represents about a 0.5 percent loss in combustion efficiency. Further, with precipitated ash refiring, there is practically no loss of unburnt carbon in cyclone-fired boilers, whereas in pulverized-coal-fired boilers usually between 0.5 and 1.5 percent of coal energy is lost in the form of unburnt carbon. Against that, there is the heat loss as a result of the hot slag discharge from cyclone boilers, which may amount to about 0.5 percent of coal energy. Thus the net advantage in combustion efficiency of cyclone boilers over that of pulverized-coal-fired boilers amounts to about 1 percent.

14.2 BOILER TEMPERATURE REGIME CONSISTENT WITH ASH SLAGGING CHARACTERISTICS

Discussions in Chaps. 9–12 have established that the rate of ash sintering (Fig. 10.3) and the formation of adhesive bond between the ash deposit and tube metal (Fig. 11.7) increase exponentially with temperature. Formation of slag deposit in the combustion chamber could therefore be expressed by the general rate equation

$$R_s = A e^{ET_a} \tag{14.1}$$

where R_s = rate of slagging
T_a = temperature of ash deposit
A = preexponential constant
E = rate coefficient
A and E values characterize the slagging propensity of ash.

The temperature T_a of ash deposit at a given depth in the matrix can be represented by the expression

$$T_a = T_s + x(T_f - T_s) \tag{14.2}$$

where T_s = surface temperature of boiler tubes
T_f = temperature of deposit surface exposed to the flame
x = variable parameter, depending on the thickness and on the effective thermal conductivity of the deposit layer

Equations (14.1) and (14.2) show that slag formation is a self-accelerating process. That is, the increase in deposit thickness results in a higher temperature, so the rate of coalescence of ash particles at the exposed surface increases exponentially.

The surface temperature of steam generating tubes in a combustion chamber depends on the Rankine cycle fluid temperature and on the rate of heat transfer through the tube wall. Sur (1979) has discussed the development of coal-fired plant from low-pressure and low-temperature boilers to the supercritical units with a corresponding rise in the superheated and reheated steam temperatures, as shown in Fig. 14.1. For boilers with natural water circulation, economic pressures of the Rankine cycle are between 16 and 18.5 MN m^{-2}, but at higher pressures,

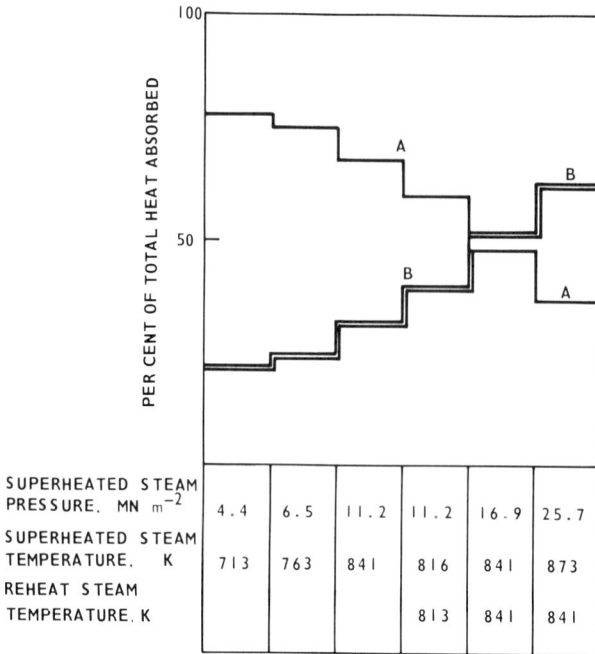

Figure 14.1 Heat requirements for steam generation (A) and for superheat + reheat (B) at different Rankine cycle pressures.

once-through boiler systems can have a supercritical pressure over 25 MN m^{-2}. Eydam (1978) has discussed the merits of double reheater systems of a high steam heating requirement.

The temperature gradient across the wall of boiler tubes, free from ash deposit, depends on the rate of heat transfer and also on the thickness of metal oxide layer on the inside and outside surfaces. Thermal conductivity data for steel and oxides are given in Table 13.2, and from these values the flame-exposed surface temperatures of boiler tubes for different heat transfer rates were calculated as shown in Fig. 14.2. The temperature increase across the wall of most furnace tubes is likely to be around 50 K, but it may rise to 100 K across the tubes situated in the burner zone of coal-fired boilers.

High heat flux will thus increase both the temperature at the deposit–tube-surface interface and the temperature gradient with the ash material on the tubes, as discussed in Chap. 13. High tube-surface and deposit temperatures will increase the rate of ash adhesion, sintering, and slagging reaction exponentially, and it is therefore imperative that maximum heat flux rates in the combustion chamber of pulverized-fuel-fired boilers are consistent with the deposit-forming propensity of the ash of the given coal supply. This and other features in the combustion chamber design that influence furnace-wall slagging and superheater fouling are discussed next.

Limitation of Heat Flux in the Burner Zone

Burners in a pulverized-coal-fired boiler are usually grouped in the lower section of the combustion chamber (Fig. 14.3), and this section can be described as a burner zone as shown in Fig. 14.4. From the point of view of boiler slagging, both the volumetric heat input—i.e., the ratio of coal energy input rate to volume of the burner zone—and the heat flux per unit area of the zone need to be considered in the design stage. Table 14.1 gives some guidance values regarding the maximum energy input and heat flux rates in the burner zone of pulverized-coal-fired boilers that are consistent with the slagging propensity of coal ashes.

It should be noted that the energy input rates and heat fluxes in the burner zone given in Table 14.1 should be taken only as a general guidance. The burner zone of a pulverized-coal-fired boiler is an arbitrary concept (Fig. 14.4), and its size depends on the inclusion of a space immediately below and above the burner levels. Also, the heat flux in the combustion zone depends on the fineness of pulverized coal, the degree of mixing of fuel with combustion air, and combustion characteristics of the fuel.

Combustion Efficiency and Size of Combustion Chamber

Coal combustion in pulverized-fuel-fired boilers should be completed by the time the flue gas enters the convection heat transfer sections of superheaters and reheaters. Incomplete combustion leads to a loss of coal energy in the form of unburnt carbon in ash and CO in the flue gas. Further, local and intermittently reducing atmosphere in the superheater chamber will enhance the deposit forming

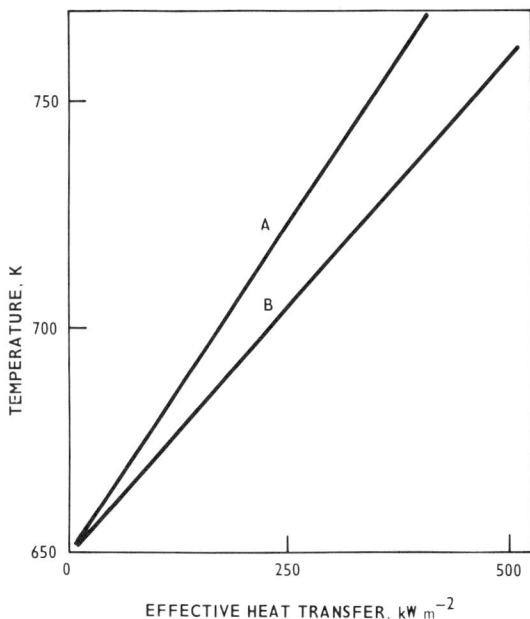

Figure 14.2 The relationship between exposed surface temperature and heat transfer through furnace tube walls: A, 6 mm tube wall + 0.5 mm oxide; B, 6 mm tube wall + 0.025 mm oxide.

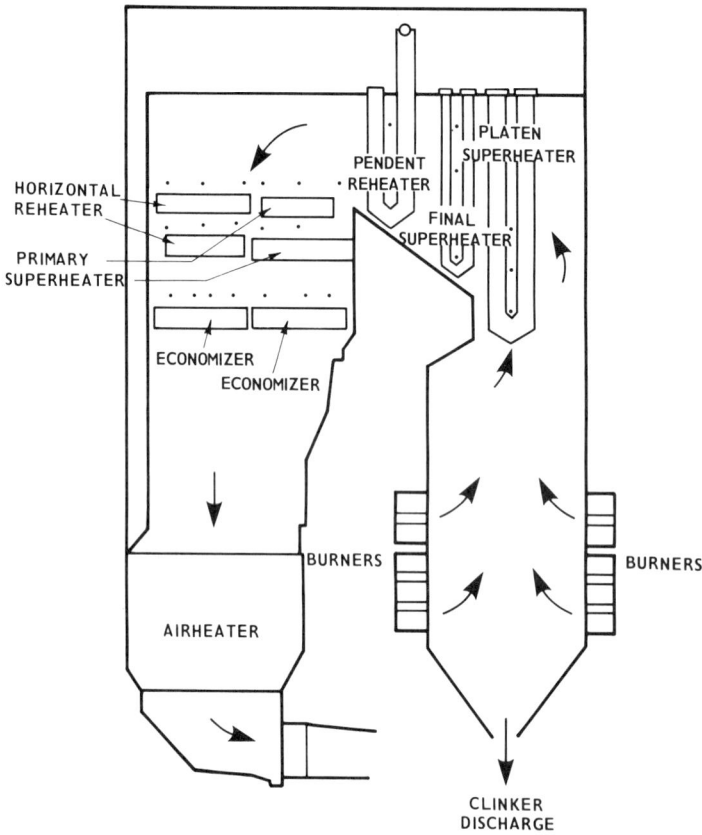

Figure 14.3 Outline of 660-MW pulverized-coal-fired boiler with opposite-wall firing; dots indicate sootblowers.

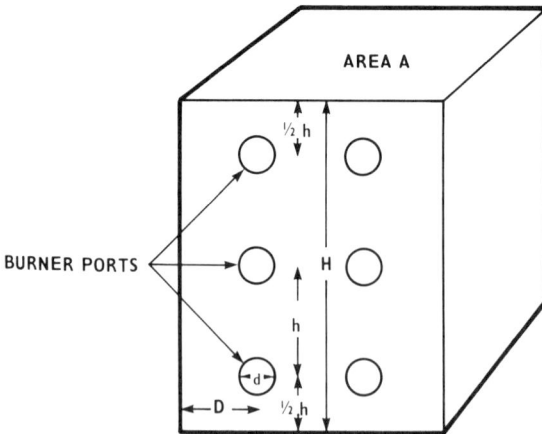

Figure 14.4 Burner zone of front-wall-fired boiler: D, distance between burner and side wall; H, height of burner zone; A, cross-sectional area; $A \times H$, volume of burner zone; d, diameter of burner port; h, distance between burner rows.

Table 14.1 Energy input and heat flux rates in the burner zone for coal ashes of differing slagging propensity

Slagging propensity of coal ash[a]	Typical energy input $(kW\ m^{-3})$	Heat flux $(kW\ m^{-2})$
Slight	550 to 600	410 to 440
Moderate	500 to 550	380 to 410
Severe	450 to 500	350 to 380

[a]As rated in Chap. 10.

propensity of ash (Chaps. 9 to 12) and high-temperature corrosion, as discussed in Chap. 17.

Minimum path length of the flame above the top burner level depends on the combustibility of coal (Chap. 5) and the slagging characteristics of ash. For example, the combustion-chamber layout of a 660-MW unit shown in Fig. 14.3 allows for a distance of 13 m between the top burner level and the superheater entry. The boiler burns low-rank bituminous coal with a comparatively high volatile (combustible) matter content between 26 and 30% by weight, and the coal ash has a slight to moderate slagging propensity. The flame path above the top burner level is barely adequate for this type of fuel, and for more slagging coals it would be necessary to increase the distance between the burner and superheater chamber by 2 to 5 m.

Coals low in combustible volatiles—i.e., below 20%—require a special design where the flame path in the temperature range of 1500–1750 K is increased in order to achieve the required degree of combustion. This has led to development of downshot combustion chambers as discussed by Sur (1979), but a problem could arise with this type of furnace when low-volatile coals have a slagging ash resulting in a formation of large masses of semifused ash deposits in the lower half of the combustion chamber where the flame path changes its direction. The high-rank coals of a low volatile-matter content and highly slagging ash are difficult to burn in a pulverized-fuel-fired boiler, and for these exceptional coals it would be expedient to consider a cyclone boiler design.

Temperature Regime of Radiant and Convection Superheaters

In current designs of large coal-fired boilers the high temperature superheaters and reheaters require more heat energy than that available in the convection passages of the flue gas, as shown in Fig. 14.1. It has therefore been necessary to introduce superheater sections into the top half of the combustion chamber where they are exposed to flame radiation. The superheater tubes are accommodated partly in the combustion chamber walls above the burners and partly as platen banks as shown in Fig. 14.3.

In order to avoid severe slagging and fireside corrosion of the radiant superheater tubes, it is necessary to consider carefully the maximum tube-metal temperature compatible with the flame temperature. For this reason the steam outlet (secondary) section of the superheater with a metal temperature in the range of 850–925 K cannot be exposed to the flame radiation. Table 11.1 showed the metal temperature of water and superheater tubes, and the corresponding flame and flue-gas temperatures of a typical pulverized-coal-fired boiler. In each design the layout of coal burners and positioning of the radiant wall, platen, and secondary superheaters and reheaters need to be carefully considered according to the slagging characteristics of coal ashes. The guiding principle is that the rate of slagging depends on the tube surface temperature and also on the temperature gradient in the deposited material [(Eqs. 14.1 and 14.2)]. The latter depends on the exposure temperature, and it is therefore essential to ensure that the high-temperature steam tubes are not exposed to flame radiation.

14.3 PREVENTION OF DEPOSIT BUILDUP IN BURNER QUARLS

The use of refractory materials in pulverized-coal-fired combustion chambers is limited to quarls of the burners of premixed fuel/air flow type, which are usually mounted in one of the four walls, as discussed by Godridge and Read (1976); large (above 500 MW) output boiler units may have the burners placed in two opposite walls, as shown in Fig. 14.3. Refractory materials are used to give burner quarls the required aerodynamic shape and to protect adjacent boiler tubes against fuel-particle impact erosion. The inner face of the burner quarl is cooled by the flow of combustion air, but the outer edge is exposed to flame radiation, and the surface temperature can exceed 1200 K.

Neal et al. (1980) have measured the heat flux at the top burner level of the combustion chamber of a 500-MW pulverized-coal-fired boiler and found that the heat flux in the middle of the wall was around 500 kW m^{-2}; near the corners it was much lower, about 200 kW m^{-2}. It is therefore obvious that sintered and semifused slag deposits usually form first on the burner quarls in the top middle section of the wall. The initial deposit may take a form of "eyebrows" shaped by the turbulent stream of the flue gas, and subsequently the growth of deposit may extend to the inside face of the quarl, thus partially or completely blocking the burner port. The blockage by ash deposit is most likely to occur when a burner is taken out of service and there is not a sufficient flow of air to cool the quarl refractory material.

In order to minimize adhesion of ash deposit on the flame-exposed side of the burner quarl, the refractory lining material should have a noncompatible surface for bonding ash deposits, should be nonporous, and should have comparatively high thermal conductivity. The latter is a rather unusual requirement for refractory materials, as these are used in most applications as a heat barrier. A dense graphite material would meet the above requirements, but it oxidizes readily above 1000 K and has a low resistance to erosion wear caused by fuel-particle impaction. A compromise choice of material for use in lining the burner quarls in pulverized-coal-fired boilers is silicon carbide, originally developed for application in cyclone boilers.

Silicon carbide refractory materials are prepared from granulated SiC mixed with a silicate binder, which gives early strength when heat-cured. Above 1400 K, grain-to-grain sinter bond will develop, giving the ultimate strength to the material. The strength of sinter-bonded silicon carbide material depends on the purity and size of SiC-grains, the cement binders used and the firing temperature. The aim should be to have a product of high sinter strength and a low porosity. A dense silicon carbide material has a high wear resistance to particle impact erosion as shown by the results given in Table 5.3. The weight loss of silicon carbide was 15 times less than that of mild steel when erosion-tested with 75- to 150-μm quartz grains impacting at a velocity of 27.5 m s^{-1}. Allowing for the difference in density of the two materials, the results show that the wear rate of silicon carbide on a thickness basis was over 6 times less than that of mild steel. However, the wear resistance decreases rapidly with an increase in porosity, as discussed by Vaux et al. (1981), and thus the porous silicon carbide refractory materials are unsuitable for the use in burner quarls because of the erosion loss and poor mechanical strength. Failure of the quarl lining material will lead to rapid erosion of the exposed water tubes by impaction of the fuel stream, and the flame "back-suction" can result in gross overheating and failure of the unprotected tubes, as shown in Fig. 14.5.

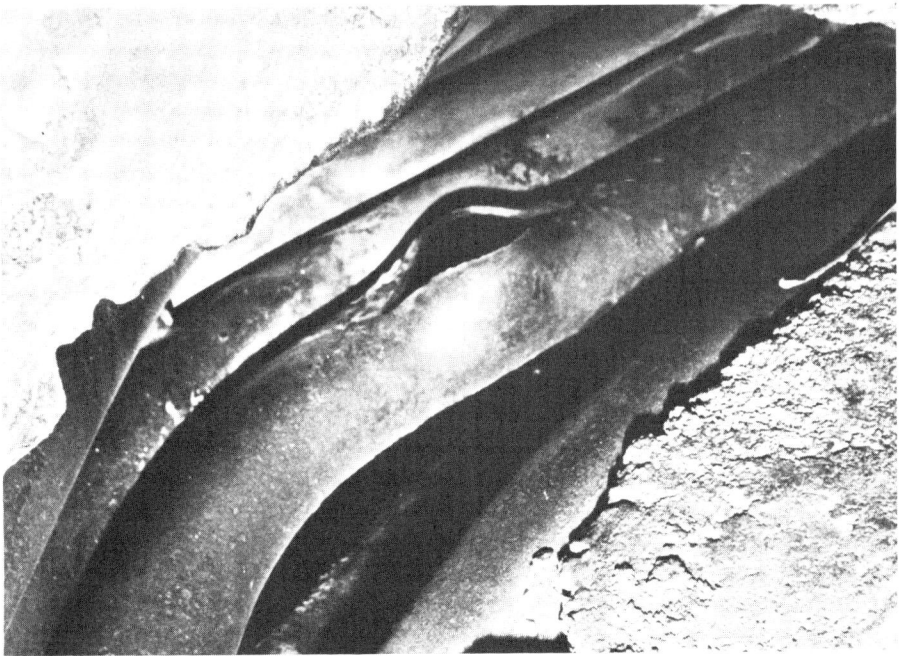

Figure 14.5 Ruptured water tube in burner quarl.

Ash adhesion on silicon carbide refractory material can be reduced by incorporating some additive materials, usually vanadium and zirconium oxides. The mechanism by which the additives can reduce the strength of the adhesive bond between the surface of silicon carbide and ash deposit is not fully explained, but it may be that the additives will enhance crystallization in the glassy material of silicate ash deposits, as discussed in Chap. 16. As a result, there is a weak adhesive bond between the surface refractory and deposits, and the latter become detached on thermal cycling.

The data published by Grayson and Eckroth (1978) and Stein et al. (1979) show that sintered silicon carbide has a thermal conductivity value around 25 $W m^{-1} K^{-1}$, which is considerably higher than that of ash deposit material; the latter has the conductivity between 0.03 to 3.0 $W m^{-1} K^{-1}$ as given in Table 13.2. The disparity in thermal conductivity of the refractory material ensures that the surface temperature of dense silicon carbide refractory material remains low and that the sharp increase in temperature takes place within the ash deposit. This reduces thermal stresses in the refractory lining material and thus increases its working life.

14.4 POSSIBLE APPLICATION OF AUSTENITIC-STEEL TUBES AND SURFACE TREATMENT TO REDUCE DEPOSIT ADHESION

Low-carbon steel alloys (mild steels) have traditionally been used to fabricate steam-generating tubes located in combustion chamber walls. For this application mild steel is heat-treated (killed) to remove slag inclusions that would otherwise prevent the formation of a protective layer of magnetite oxide, which slows down the rate of further oxidation of tube metal according to a parabolic expression as discussed in Chap. 17. At the same time the strongly adhering layer of magnetite is compatible with silicate ash for formation of a keying bond at the tube deposit interface, as discussed in Chap. 11. In contrast, the surface of an austenitic steel is less likely to provide a compatible surface for strongly adhering deposits as a result of the thermal expansion and chemical incompatibility of an austenitic alloy and ash deposit. Austenitic alloy and coextruded tubes are frequently used to prevent excessive fireside corrosion in high-temperature superheater and reheaters, and more recently also in the combustion chamber, as discussed in Chap. 17. In contrast, application of tube metal alloys with a noncompatible surface for prevention of adhesion of ash deposit has been rarely considered.

Massive quantities of sintered ash and fused slag deposit usually accumulate on water tubes in the burner zone facing the high flame temperature, and on the radiant superheaters in the combustion chamber. It is therefore expedient to consider in the boiler design stage the possible application of coextruded tubes with an austenitic outer layer to prevent the build-up of large masses of strongly adhesive deposits—in particular, on the coextruded tubes immediately above the top burners in the combustion chamber and in bottom bends of the radiant superheater. In subsequent boiler operation when a more definite pattern in deposit build-up is established, it may be possible to extend the "antislagging" areas by the application of flame- or plasma-sprayed coatings. An austenitic alloy could readily be sprayed on the furnace tubes, but it has not been established that austenitic alloys in the form of a thin surface layer on mild steel will have the same low adhesive characteristics for bonding of ash deposits as that of austenitic steel tubes. It would not be difficult to test the efficacy of different metal-sprayed coatings for the effectiveness of reducing the strength of adhesive bond at the tube surface–ash deposit interface.

Build-up deposits on the pendant superheater loops usually show a consistent pattern. The leading tube in each loop receives unhindered impaction of the gas-borne ash and is exposed to flame radiation. Combined effects of the high rate of deposition and high temperature frequently result in either a rapid build-up of ash deposit or severe corrosion of the leading tubes. The deposit build-up and corrosion become progressively less on the sheltered tubes. It is therefore practical to consider the use of austenitic steel tubes in the leading position to prevent build-up of massive deposits on pendant superheater elements, which have austenitic tube sections where the deposit build-up and ash deposit and high-temperature corrosion are expected to be most severe, as discussed in Chap. 17.

14.5 THE SCOPE FOR REDUCING BOILER SLAGGING BY COAL CLEANING AND BLENDING

The quality of coals utilized for electricity generation at large power stations has been discussed in Chap. 4. The relationship between boiler slagging and the composition of mineral impurities in coal is extensively discussed in Chaps. 9–12. Practical experience has shown that the amount of ash in coal also has an important bearing on the degree of interference caused by ash deposits in boiler operation. It has been observed that pulverized-fuel-fired boilers designed to burn British coals of 16–20 percent ash had more pronounced slagging problems when the ash content increased to a value between 22 and 25 percent. In particular, difficulties occur in removing the sintered ash clinker and fused (furnace-bottom ash) slag accumulating on the bottom slopes of the combustion chamber. The results of a survey on slagging of electricity utility boilers in the United States reported by Dimmer (1980), showed that a number of slagging boilers had been fired with coals having an ash content up to 30 percent above the original design specifications. It was claimed that this had markedly accentuated the slagging problems.

An increase in the ash content of coal results in a lower calorific value, as was shown in Fig. 4.1, and thus it is necessary to increase the fuel input to maintain the boiler load. The increased ash input to the boiler and the enhanced rate of slagging can result in doubling the amount of furnace-bottom ash to be removed from the system. This will grossly overload the clinker handling plant, and in extreme cases the furnace bottom exit can be bridged by the semifused slag.

The rate of overall boiler slagging R_s can be expressed in terms of the slagging propensity of ash s_p, as discussed in Chap. 10, and the ash content a_c of coal

$$R_s = f(s_p, a_c) \tag{14.3}$$

Modern methods of machine mining can result in a high ash content of raw coal. For example, British coals as mined for power-station utilization have an overall average ash content of about 30% as reported by Chironis (1981). By coal washing and by selective blending, the average ash content of coals delivered to power stations is reduced to around 17 percent. That is, about 40 percent of the mineral matter mined with coal is removed at collieries, and there is a continued debate on whether a further reduction in the ash content of power station coals would significantly reduce boiler slagging.

Phillips and Cole (1979) have calculated the empirical benefits as a result of cleaning coals mined in the United States for use in electricity utility boilers. The ash plus sulfur content was taken as a criterion, and it was related to the costs of coal transport, ash disposal, and coal plant and boiler maintenance, and to the availability at peak-load and reduced-load requirements. The authors concluded that boiler availability and maintenance costs were not significantly influenced by changes in the amount or nature of coal mineral matter when the ash plus sulfur content was below 12.5 percent. When the ash plus sulfur content exceeded 12.5 percent, plant maintenance costs increased linearly with the mineral content. When the ash plus sulfur content exceeded 17.5 percent, boiler capacity started to become more markedly affected. For each percentage point increase above 17.5 percent, the rated capacity was reduced by 3 percent, the peak load capacity was reduced by 0.3 percent, and boiler availability was reduced by 1 percent.

The results of a survey published by Vivenzio (1980) showed that with bituminous coals, a reduction in ash content below 17 percent resulted in a decrease of boiler slagging. In particular, removal of large particles of pyrites in pulverized coal should markedly reduce furnace-wall slagging, as discussed by Bryers (1976). In contrast, Suydam and Duzy (1977) have reported that a reduction of the ash content of sub-bituminous coals in Arizona from 23.6 to 13.7 percent was accompanied by an increase in the deposit-forming propensity of the residual ash in cleaned coal. As a result of coal washing, the sodium oxide content of the ash increased from 1.8 to 2.2 percent, and the test firings confirmed that the slagging propensity of the ash of washed coal increased in relation to the sodium oxide content. This was a result of nonsodium mineral matter being preferentially removed during the coal wash process. In order to remove sodium selectively from the nonbituminous coals, washing should be carried out with dilute acid or solutions containing divalent ions (Ca^{2+} or Mg^{2+}) or trivalent ions (Fe^{3+} or Al^{3+}). The ion exchange reactions have been studied by Paulson and Fowkes (1968), Crystal (1970), and Sondreal et al. (1978).

Blending of high-ash and low-ash coals is widely practiced by the fuel suppliers, such as the National Coal Board in the U.K., as discussed by Chironis (1981). The principal aim is to ensure that the ash content of coal supplies does not exceed the stipulated limit and that the calorific value of the fuel does not fall below the value set in the boiler design stage. An effective blending of high- and low-ash bituminous coals should markedly reduce the overall rate of boiler slagging. That is, continuous operation of a pulverized-fuel-fired boiler with a coal mix of 15–20 percent ash is significantly preferable to firing of 10–15 percent ash coals, followed by firing of 20–25 percent ash coals. Blending coals can be undertaken on the basis of the slagging propensity of different coal ashes, as defined in Chap. 10.

The concept of blending the low- and high-sulfur coals is attractive for the purpose of reducing the SO_2 emission, as discussed by Loeffler and Vaka (1979) and Morgan and Mahr (1981). The calcium and sodium in low-sulfur coals are utilized to increase the sulfur (sulfate) capture in ash when burnt with high-sulfur coals. Sub-bituminous coal ash may have a high fouling index because of its high sodium and calcium contents, and the ash is highly basic as discussed in Chap. 10. Addition of silicious ash from a bituminous coal may increase the fouling propensity of the non-bituminous coal ash, as evident according to Eq. (10.14). The deposit-forming characteristics of blended coals are further discussed by Markley (1968), Stewart and Shou (1975), and Mozes (1982).

Blas (1970) has discussed the problems on burning low-grade coals mined in Spain that have an ash content, moisture-free basis, between 16 and 55 percent with an average value around 45 percent. His conclusions were that boiler slagging and coal mill wear were the chief causes for the loss of availability of the generating plant. There is an additional problem that some of the high-ash coals have a low content of combustible volatile matter and hence are difficult to burn. Boiler trials were made with blended fuels of coals of low and medium volatile matter, aiming for 12–16 percent volatiles and limiting the maximum ash content of the blended fuel to 45 percent. The results showed that there were significant benefits because of improved combustion efficiency and increased boiler availability when a blended fuel was used.

14.6 COMBUSTION CONTROL FOR COMBATING BOILER SLAGGING

The first aim of combustion control in coal-fired boilers should be to avoid high peak temperatures of the flame and of the flue gas in the convection superheater passage. This is clearly evident from Eq. (14.1), which states that the deposit-forming propensity of coal ashes is exponentially related to temperature. Thus, a reduction of 100 K in the peak flame temperature and of 50 K in the superheater chamber gas temperature may significantly reduce the rate of boiler slagging. Further, elimination of a reducing atmosphere in the deposit-forming zones can also markedly reduce the rate of boiler slagging. This can be achieved by increasing the amount of combustion air, and improving coal milling and distribution. The change from a reducing to oxidizing atmosphere increases the viscosity of iron-rich slags; hence it reduces the rate of slagging. With calcium-rich ashes, an oxidizing atmosphere prevents the formation of low-melting calcium sulfide (CaS), which can function as a binding constituent in deposit build-up. Possible measures in boiler operation to combat excessive slagging are summarized in Table 14.2.

Table 14.2 Combustion control measures to combat boiler slagging

Measure	Effect
Consistently fine coal milling	Minimizes the risk of a reducing flame entering the superheater chamber. It may, however, increase the peak flame temperature in the burner zone.
Improved mixing of pulverized coal and combustion air by changes in burner and secondary air-vane setting	As above
Changes in flame pattern by reshaping the burner quarls and modifications in burner settings	Reduces the risk of flame impingement on boiler-wall and superheater tubes.
Choice of burners when available	Peak thermal loading of burner zone can be markedly reduced, as shown in Fig. 14.6.
Increase of combustion air	Reduces the flame temperature and the flue gas temperature entering the superheater. Penalties are increased loss of sensible heat of the flue gas and increased air fan requirement, and higher velocity of the flue gas may cause boiler tube erosion by ash impaction (Chap. 18).
Decreasing the temperature of preheated combustion air	As above, but the penalties are less severe.
Flue gas recirculation into the bottom section of combustion chamber or into the superheater duct	Reduces the flame and flue-gas temperatures. Penalties are that the expenditure of fans and the increased velocity of the flue may cause boiler tube erosion. Also, there are difficulties of mixing large volumes of flue gas, and could lead to an increase of carbon in ash.
Reduction in boiler load	Effective in decreasing the flame temperature, and reducing zones in the flue gas can be more easily avoided. There is a penalty due to the loss of plant output.

Effectiveness of the measures listed in Table 14.2 to combat boiler slagging can be assessed by monitoring the flame and superheater flue-gas temperatures, and by the heat-flux measurements in the combustion chamber. The latter in particular gives a useful guide for a possible decrease in the rate of boiler slagging, since the temperature of ash deposit on a cooled surface is closely linked to the radiative heat flux, as discussed in Chap. 13. Chambers et al. (1981) have described a differential arrangement of two heat-flux meters: one of these was kept clean by air purge, and thus the output signal remained constant at a given boiler load. The rate of build-up of ash deposit on the other probe was approximately the same as that of the ash deposit on the furnace walls, and thus the differential signal from the two meters was proportional to the extent of boiler slagging at any given time. The authors claim that the system will be valuable in improving efficiency of boiler operations and in minimizing slagging and fouling problems when firing difficult coals.

Most large pulverized-coal-fired boilers have an excess of burner capacity so that routine maintenance on the burners and coal mills can be carried out without the necessity of reducing the boiler load. Thermal loading in the combustion chamber depends markedly on the combination of burners in service, as shown in Fig. 14.6. Subject to availability, the most favorable (i.e., with lowest heat load combination) should be used for reducing the rate of slagging in the furnace-wall tubes in the burner zone. Should severe slagging occur in the platen superheater section (Fig. 14.3), a different burner combination, such as leaving out the top burners, should reduce the rate of deposit build-up. Similarly the slag formation on the bottom slopes of the combustion chamber should decrease by leaving out the lower-level burners.

The thermal flux on furnace walls is reduced correspondingly by decreasing the boiler load. In practice it has been found that many 500-MW pulverized-coal-fired boilers perform without significant slagging difficulties on loads up to 420 or 450 MW. The problem of slagging, however, may become acute when the last 50 to 80 MW output is required. This is a result of a more intense combustion, hence the higher flame and deposited ash temperatures, and is partly due to increased ash burden in the combustion, as discussed in Sect. 14.5 [Eq. (14.3)].

The efficacy of finer milling of coal, increasing the excess of combustion air, and better mixing of pulverized coal and combustion air can be best monitored by measurements of carbon monoxide in the flue gas, as discussed by Godridge and Read (1976) and Ormerod and Read (1979). Pulverized-coal-fired boilers are usually

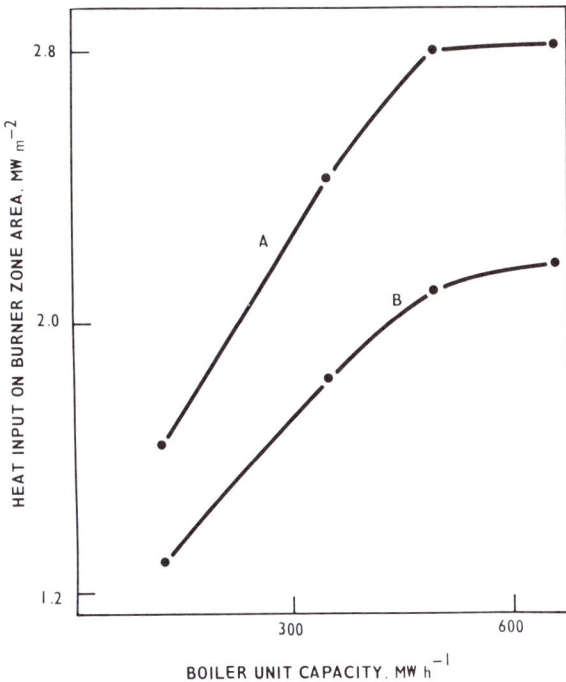

Figure 14.6 Heat energy input in burner zone of 120- to 660-MW boilers on full load: A, least favorable burner combination; B, most favorable burner combination.

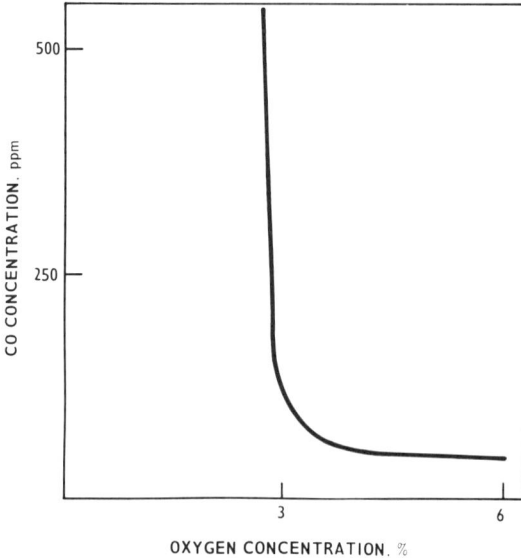

Figure 14.7 The relationship between oxygen and carbon monoxide in pulverized-coal-fired boilers.

operated with an oxygen excess between 3 and 5 percent by volume, which is equivalent to 15-20 percent of combustion air. Figure 14.7 shows that the CO content of the flue gas measured in the superheater increases rapidly when the oxygen excess falls below 3 percent. This is a characteristic feature of pulverized-coal combustion, that a near stoichiometric burn-out cannot be achieved. In contrast, in residual-oil-fired boilers the excess oxygen level can be reduced to 0.5 percent before CO concentration of the flue gas rises above the 50 ppm level as discussed by Anson et al. (1971).

Carbon monoxide concentration of pulverized coal flame can be above 1 percent, and a high CO content of the flue gas near furnace wall tubes is a sensitive indicator for enhanced slagging and high temperature corrosion (Chap. 17). Routine gas analysis can show up possible malfunction of a mill or burner when there is a marked increase in CO content of the flue gas, as discussed by Hadrill (1979). Ormerod (1981) has described an automatic trimming system for adjusting the flow of combustion air using the CO concentration of the flue gas as an activating signal.

14.7 STEAM AND AIR SOOTBLOWERS

All coal-fired boilers require a reliable service of sootblowers to prevent the build-up of excessive deposits of sintered ash and slag in the high-temperature section and accumulation of ash in the lower-temperature passages. Traditionally, sootblowers were used in railway locomotive engines and other small boilers to blow off the deposits of soot that accumulated on water boxes and tubes as a result of rapid flame quenching. In contrast, very little soot is deposited on water and steam tubes in large coal-fired boilers of current design, and thus a more correct use would be "ash blower" or "deposit blower." The concept of sootblower, however, is firmly established and to change the term now would lead to confusion.

When a high-pressure gas jet from the steam or compressed-air sootblower nozzle expands, it mixes with an approximately equal volume of the ambient gas for a travelling distance of the nozzle diameter. It follows therefore that by the time the expanded jet from the sootblower reaches the boiler tubes at a distance between 40 and 100 nozzle diameters, the impacting fluid consists largely of the flue gas. The expanded jet pressure p at the point of impact is given by

$$p = \tfrac{1}{2}\, \rho v^2 \qquad (14.4)$$

where ρ and v are the density and velocity, respectively, of the impacting gas.

Characteristics of diverging jets have been investigated by Anderson and Johns (1955), Vulis and Terekhina (1955), Warren (1955), and Sarjeant (1973). Figure 14.8 shows the logarithmic decay of the dynamic pressure with distance, and it is evident that the location of sootblowers with respect to tube surfaces to be cleaned is a critical parameter. Different sootblowers and their deployment in the boiler plant have been fully discussed in the CEGB (1971) Handbook, and further by Brook (1962), Genck (1964), and Koonce (1968). Table 14.3 gives the number of sootblowers, steam requirement, and operating and impact pressures, in a typical 500-MW pulverized-coal-fired boiler.

Sootblowers are usually operated once in every 8 h, and it is therefore appropriate to consider the strength of bonding of ash to boiler tubes that develops during this time interval. Raask (1973) has measured the rate of bond development between the ash deposit material and tube metal surface, as discussed in Chap. 11. Figure 11.6 showed that at a given temperature the bond strength increased

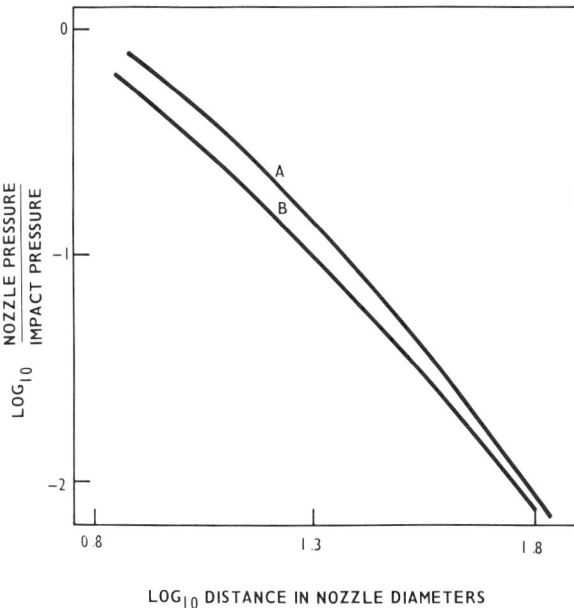

Figure 14.8 Decay of dynamic pressure at sootblower jet axis: A, air; B, steam.

Table 14.3 Sootblowers in 500-MW boiler

Location	Number	Duration of operation of each blower (min)	Steam consumption (kg min^{-1})	Steam pressure at nozzle (MN m^{-2})	Impact pressure (kN m^{-2})^a
Furnace	46	0.25	68	1.0	22
Platen superheater	6	17	132	0.7	16
Secondary superheater	8	17	85	0.8	18
Primary superheater	8	11	73	0.8	18
Secondary reheater	8	17	85	0.8	18
Primary reheater	8	11	73	0.8	18
Economizer	4	6	73	0.8	18
Air heater	8	20	53	0.7	16

^{a}For 25-mm nozzle at 2 m.

approximately linearly with time, and it increased exponentially with temperature as shown in Fig. 11.7. The results suggest that usually the strength bond between ash deposit and tube surface is likely to remain below 5 kN m^{-2} during a time interval of 8 h. Under exceptional circumstances—e.g., an ash of high slagging propensity (Chap. 10) or high deposit temperature—the rate of formation of bond between ash deposit and tube surface can be much faster. Honea and Persinger (1981) in their study of the ash-related outages and load curtailments of boilers burning low-rank coals have emphasized the need for adequate sootblowing. Frequently the problem is not the rate of formation of boiler deposit, but the strength with which it adheres to the water and steam tubes. The authors have suggested that further investigations are required on the topic of deposit adhesion related to the performance of sootblowers.

Hadrill (1979) has described a case of severe slagging of the bottom bend of the platen superheater where the sootblowers at a distance of 2.5 m were ineffective. A marked improvement in preventing the slag formation in this sensitive area was achieved by installing additional sootblowers nearer the superheater tube bends and thus reducing the distance of sootblower jet travel by half. This has made a significant contribution in preventing excessive slagging and thus increasing load availability of 500-MW boilers.

A frequent shortcoming in the design of large coal-fired boilers is that insufficient provisions are made for the installation of sootblowers in the combustion chamber. The sootblower requirements of new boilers should be based on a systematic study of existing boilers of a similar design and burning coals of the comparable ash slagging characteristics. The design of sootblowers should be flexible so that the duration of blowing cycle could be adjusted by changing the speed of traverse and rotation. Further, the impact force of steam or air jets may need to be adjusted according to the rate of build-up of ash deposits in different localities. This would require provisions for changeable nozzles of different diameter and means of adjusting the nozzle outlet pressure.

Traditionally, the sootblowing medium has been steam bled from the super-heater or reheater circuit and reduced to required pressure. More recently claims have been made that compressed-air sootblowers require less energy to operate, and a number of power station boilers have been provided with the air lancers. Further, it has been claimed that the compressed-air sootblowers need less maintenance than the steam blowers, although the capital layout for air compressors and reservoirs is comparatively high. However, Sarjeant (1974), after his study of economics of sootblowing in a 500-MW boiler, concluded that the energy requirement for the steam and compressed-air sootblowers was the same, amounting to about 0.1% of boiler heat energy.

The long, retractable sootblowers required to clean platen superheaters (Fig. 14.3) are exposed to flame radiation, and adequate cooling by the steam or air flow is a problem. Shenker (1980) has suggested that water tempering of the compressed air sootblowers is one solution to this problem. He describes a system where the rate of coolant air to the sootblowers working at 1700 K could be cut by introduction of water at 0.04 m^{-3} min^{-1}. Extensive tests in a power plant have shown that the possible adverse effect of thermal shock on tube life was minimal, and that the reduction in steam or air flow provides greater flexibility in blower operation, decreases energy consumption, and extends sootblower life. There is a risk of boiler-tube erosion caused by impaction of the water droplets accelerated by the air jet, and long-term assessments are required to establish the possible rate of erosion. Further discussions on water-tempered sootblowers have been published by Shenker et al. (1981).

14.8 WATER–JET SOOTBLOWERS

Water lancing has been extensively used to remove large masses of sintered-ash clinker and slag accumulating on the bottom slopes of the combustion chamber above the ash pit (Fig. 14.3). The concept of water-jet cleaning has now been extended to clean the boiler tubes higher up on the combustion chamber walls where massive quantities of deposit can form, in particular in boilers fired with sodium-rich nonbituminous coals (Chap. 10). The merits and possible risks of water-jet sootblowers have been discussed by Bieber (1968), Ellery et al. (1973), Bieber and Herrmann (1974), Wilhelm et al. (1975), and Vasil'ev and Guzenko (1977). The principal features of water-jet sootblowers are high efficacy and low operation costs. According to Eq. (14.4), the impact pressure is directly propor-tional to the density of impacting fluid. The density of water is at least 1000 times higher than that of air, flue gas, or steam, and it is therefore evident that a water jet has a high impact pressure. Further, a water jet has a marked quenching effect on hot ash deposits, resulting in fracture of the deposit mass or rupture of the adhesive bond between the slag and boiler tubes.

The water-jet sootblowers have not yet been universally accepted in boiler operation practice because of a risk of tube failures as a result of thermal shock. Ellery et al. (1973) have carried out the water-jet impact experiments, and typical

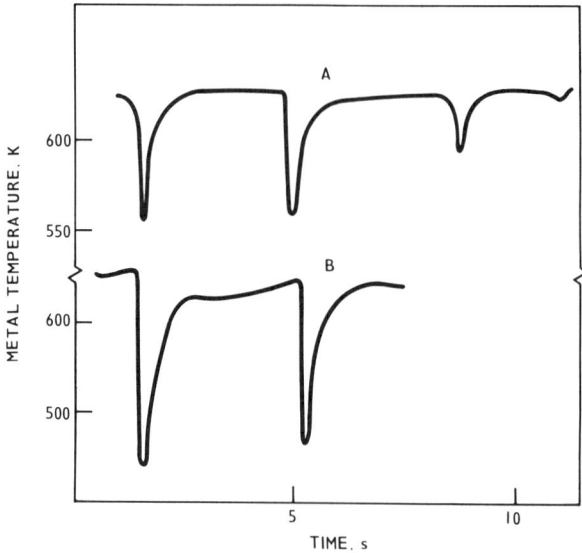

Figure 14.9 The cooling effect of twin-nozzle water-jet furnace sootblower: A, advancing; B, retracting.

cooling effects caused by a two-nozzle lancer are shown in Fig. 14.9. The water jets from a rotating sootblower cool boiler tubes from about 625 to 525 K at a rate of 200 K s^{-1}, followed by an almost equally rapid rise in temperature. Ellery et al. (1973) have found that 0.1-mm-deep cracks had developed in the mild steel tube after 16,000 thermal shocks. Subsequently the crack propagation slowed down, and after 64,000 impactions the cracks were not significantly deeper. Bieber and Herrmann (1974) carried out tests where a boiler-tube test piece was cooled from 650 to 400 K by a water jet. There was no marked deterioration after 93,000 thermal shocks, but 2.5-mm-deep cracks developed after 121,000 chilling treatments, and the cracks had reached to a depth of 5 mm after 140,000 spray tests. In practice the thermal shock damage may 'be less since the rig tests were carried out on clean tubes, whereas the ash deposit on boiler tubes offers some cushioning against thermal shock. Nevertheless, it is currently recommended that water-jet sootblowers should be used only in areas where there is severe slagging, and that the operation of the sootblowers should be carefully controlled. It is imperative that a fail-safe device cuts off the water supply when faults in the drive stop the blower rotating or prevent its withdrawal.

Nelson and Koester (1980) have discussed the sootblowing experience with water lancers in the combustion chamber and the retractable blowers in convection passages. They describe also the use of half-cycle type blowers to clean the heavily slagged pendant superheaters. The cycle mechanism directs the water jets in a predetermined arc so that the tubes at the lee side, free from deposit, are not subjected to thermal shock.

14.9 NOVEL SUGGESTIONS FOR COMBATING BOILER SLAGGING

There have been many ingenious suggestions made for combating coal-fired boiler slagging that have not yet been extensively proven in practice. Among these are the proposals to use chemical additives, as discussed in Chap. 16; other suggestions are outlined briefly below.

Thorough Off-Load Cleaning

Claims have been frequently made that new coal-fired boilers have an "immunity" period lasting a year or longer, when the degree of boiler fouling and slagging is much less than that on subsequent operation. One reason for this may be that it takes several months to form an interface of metal oxide and ash constituents that can bond massive deposits to boiler tubes. The argument is therefore that when this interface of metal oxide and the residue of ash deposit is removed by sandblasting, the immunity conditions of a new boiler are restored.

Webner (1969) has reported the results of a trial where the entire combustion chamber of a 150-MW boiler was sandblasted to clean the tubes to bare metal. It was observed in subsequent operation that the deposits on furnace-wall tubes caused much less interference to boiler operation. The superheated and reheated temperatures could be more easily maintained, 20–30 percent less sootblowing was required, and there were no massive slag falls. The beneficial effects of sandblast cleaning were observed for a period of 5 yr, after which the thorough cleaning of the combustion chamber was repeated.

The boiler was tangentially fired from burners situated in the four corners. The sandblasting experiment was carried out on two other boilers of front-wall firing design. The effects of thorough cleaning of these two boilers were much less noticeable. As an explanation it was suggested that slagging in the front-wall-fired boilers was limited to a small area and sandblast-cleaning the rest of the furnace areas was unnecessary. Further trials of thorough off-load cleaning of furnace-wall and superheater tubes are required to assess the possible beneficial effects of removing the ash and oxide residue on tubes in reducing the rate of deposit formation in subsequent operation.

Pulsating Acoustic Sootblowers

Bargstedt (1979) and Werthaum and Sogndal (1982) have described a method of boiler cleaning by pulsating air jets causing the break-up of deposit as a result of the oscillation effect on the layer of ash. The frequency generators of 250 and 360 Hz were used; these frequencies were sufficiently higher than the natural oscillation of boiler plant, 25–30 Hz, in order not to cause any damage by resonance. The waveguide of a frequency generator working at 250 Hz is 0.5 m long; it can be made of a corrosion-resistant alloy and cooled by compressed air for application up to 1250 K. The energy requirement is 0.5–1 kW, and the noise level at the source is 130 dB(A).

A system of seven frequency generators has been installed on a waste-heat boiler of a sulfate manufacturing plant. The generators work in sequence, each generating two 15-s emissions with a 15-s interval in every 6 min. The acoustic cleaning system has proved to be effective in removing the ash deposits on boiler walls, and it is claimed that the capital and running costs are significantly lower than those of the conventional sootblowers.

Repulsion Electrophoresis

The possible effects of electrophoresic influence in transporting and adhesion of small ash particles to boiler tubes have been discussed in Chaps. 11 and 12. As a result of the electrostatic force between boiler tubes and the friction-charged particles of ash, the latter are attracted to and held at the surface for subsequent formation of more strong chemical and mechanical bonds at the metal oxide–ash deposit interface. It is argued that if the deposit collecting target could be surface charged to above 1 kV, having the same polarity as that of ash particles, the latter would not be attracted to the target. Thus, the rate of formation of ash deposit could be significantly reduced.

Steel and Brandes (1963) have made some experiments with an electrically isolated steel probe charged to 3 kV. When the probe surface charge was of the same polarity as that of the gas-borne particles, the rate of ash deposit was markedly reduced. There has not been sufficient experimental work carried out with charged probes to assess the attraction and repulsion effects of an electric field in high-temperature deposition. This is presumably because it is considered that the concept of repulsion electrophoresis has no practical application in large coal-fired boilers. It would be impractical to isolate and charge the surface of water and steam tubes to a potential over 1000 V.

NOMENCLATURE

A preexponential constant
a_c ash content of coal
E exponential coefficient of deposit formation
p jet pressure
R_s rate of boiler slagging
s_p slagging propensity of coal ash
T_a deposit temperature
T_f temperature of exposed surface of deposits
T_s surface temperature of boiler tubes
v jet impact velocity
x boiler deposit parameter
ρ density of impacting gas jet

FIFTEEN

SOME SPECIFIC ASH-RELATED PROBLEMS WITH U.S. COALS

15.1 COMPOSITION CHARACTERISTICS OF THE MINERAL MATTER IN LOW– AND HIGH–RANK COALS

A variety of solid fuels, lignites, sub-bituminous and bituminous coals, and anthracites is utilized in the United States for power generation. Each type of fuel can have specific problems related to the mineral matter in boiler operation. The analytical data of U.S. coals are available, and it is therefore expedient to discuss the specific features of the mineral matter in low- and high-rank fuels, followed by an assessment of the deposit-forming propensity of lignite and of sub-bituminous and bituminous coal ashes.

Tables 15.1 and 15.2 show examples of the ash composition and mineral species analysis of lignites and sub-bituminous coals; Tables 15.3 and 15.4 give the data for bituminous fuels. The analytical data were selected from the publication by O'Gorman and Walker (1972) in order to illustrate the wide variety in composition of the mineral matter in U.S. coals.

The analysis results in Table 15.2 show that the sulfur species gypsum (calcium sulfate), pyrite (iron sulfide), and thenardite (sodium sulfate) were the principal constituents of the low-temperature ash. An exception was Texas lignite ash, sample 141, which had kaolinite and quartz as the chief mineral species. It is interesting to note that some thenardite (sodium sulfate) was found in the mineral-matter residue

Table 15.1 Composition of laboratory combustion (875 K) ashes of lignite and sub-bituminous coals

Sample number[a]	Constituent weight percent of ash								
	SiO_2	Al_2O_3	Fe_2O_3	TiO_2	CaO	MgO	Na_2O	K_2O	SO_3
87	20.0	9.3	7.5	0.5	20.8	6.0	11.5	0.4	21.9
88	23.5	13.6	5.7	0.5	18.3	5.4	8.3	0.3	21.7
91	17.0	14.6	1.2	0.4	34.6	13.3	0.3	0.1	14.3
92	12.5	13.3	6.8	0.3	26.1	10.4	0.3	0.1	26.17
99	39.8	19.4	3.9	0.9	15.0	2.4	0.2	1.1	16.2
100	9.8	12.9	6.7	1.0	33.4	7.9	1.6	0.2	21.7
140	38.4	21.0	11.2	1.1	10.0	1.7	0.7	0.5	14.6
141	62.1	16.8	3.9	0.9	5.2	1.1	0.5	0.6	8.9

	Origin	Comments
87 } 88 }	North Dakota, Zap lignite	Exceptionally high in sodium, fairly high in calcium, and low in silica and alumina
91 } 92 }	Montana, Savage lignite	Low in sodium, exceptionally high in calcium, and low in silica and alumina
99 } 100 }	Wyoming, Glenrock and Gilette sub-bituminous	Glenrock fuel-ash sample moderately high in calcium; Gilette sample exceptionally high in calcium, and exceptionally low in silica and alumina
140 } 141 }	Texas, Darco lignite	Sample 140 moderately high in calcium and iron; sample 141 exceptionally high in silica

[a]Sample number of ashes are the same as those in the publication of O'Gorman and Walker (1972).

Table 15.2 Mineral species in 470 K combustion residues of lignites and sub-bituminous coals (weight percent of mineral matter)

Species	Sample number							
	87	88	91	92	99	100	140	141
Kaolinite	1–10	10–20	10–20	1–10	20–30	10–20	20–30	30–40
Illite	1–10	1–10	10–20	1–10	1–10	1–10	–	1–10
Chlorite	Trace	Trace	–	–	1–10	–	–	–
Quartz	10–20	1–10	1–10	1–10	10–20	1–10	10–20	30–40
Aragonite	1–10	1–10	1–10	1–10	–	–	–	–
Ankerite	–	–	1–10	1–10	1–10	–	–	–
Gypsum	30–40	30–40	30–40	50–60	30–40	40–50	10–20	10–20
Pyrite	1–10	1–10	1–10	1–10	1–10	1–10	1–10	1–10
Thenardite	1–10	1–10	–	–	–	1–10	–	–
Rutile	–[a]	–	–	–	Trace	Trace	1–10	Trace

[a]Denotes species not detected.

Table 15.3 Composition of laboratory combustion (875 K) ashes of bituminous coals and anthracites

Sample number[a]	Constituent weight percent of ash								
	SiO_2	Al_2O_2	Fe_2O_3	TiO_2	CaO	MgO	Na_2O	K_2O	SO_3
2	59.6	26.7	4.2	3.4	2.3	0.7	0.8	0.1	1.8
3	68.5	20.8	2.6	3.6	1.4	0.4	0.6	<0.1	1.7
6	17.5	9.2	64.1	0.4	4.1	0.4	0.3	1.0	1.8
67	25.6	21.0	20.5	0.6	15.8	0.7	2.0	0.2	10.9
124	42.0	26.5	7.5	1.2	7.9	1.6	1.2	1.6	9.5
127	46.1	34.5	7.5	1.5	2.5	0.7	0.3	2.0	3.0
142	19.4	10.5	36.6	0.4	8.9	5.8	0.4	1.1	15.9
143	29.2	14.2	27.8	0.5	7.1	5.1	0.6	1.5	14.0

	Origin	Comments
2 } 3	Kentucky, Deane	High in silica
26 } 67	Illinois, Carrier Mills, Utah, Horse Canyon	Low in silica, sample 26, exceptionally high in iron; sample 67 moderately high in iron and calcium
124 } 127	West Virginia, Bickmore, Pennsylvania, Ebensburg	Moderately high in silica
142 } 143	Oklahoma, Redstone	Low in silica, exceptionally high in iron

[a]Sample numbers of ashes are the same as those in the publication of O'Gorman and Walker (1972).

Table 15.4 Mineral species of 470 K combustion residues of bituminous coals and anthracites (weight percent of mineral matter)

Species	Sample number							
	2	3	26	67	124	127	142	143
Kaolinite	40–50	30–40	Trace	10–20	40–50	60–70	1–10	10–20
Illite	Trace	Trace	1–10	1–10	20–30	10–20	1–10	1–10
Chlorite	_[a]	Trace	1–10	–	1–10	–	1–10	1–10
Montmorillonite	–	–	Trace	–	–	–	–	–
Illite/ montmorillonite	1–10	Trace	1–10	1–10	1–10	1–10	–	–
Quartz	30–40	40–50	1–10	1–10	1–10	1–10	1–10	1–10
Calcite	–	–	–	–	1–10	1–10	–	–
Dolomite	–	–	–	–	1–10	–	10–20	1–10
Siderite	–	–	1–10	1–10	–	–	30–40	30–40
Pyrite	1–10	1–10	60–70	20–30	1–10	1–10	1–10	1–10
Gypsum	1–10	1–10	1–10	20–30	10–20	1–10	1–10	1–10
Jarosite	–	–	1–10	–	–	–	–	–
Rutile	1–10	1–10	–	–	1–10	1–10	–	–

[a]Denotes species not detected.

of high-sodium North Dakota lignite. It is not certain whether the sodium sulfate species was originally present in the mineral matter or formed during the low-temperature combustion of lignite in the atmosphere of activated oxygen before the mineral species analysis was carried out. Sodium sulfate can form at low temperatures when coal together with its mineral matter is heated in an oxidizing atmosphere as discussed by Daybell and Pringle (1958).

The mineral matter in most bituminous coals consists largely of the alumino-silicate clay species and quartz, where the silicious material constitutes between 60 and 90 percent of the ash residue left on combustion (Chap. 4). There are, however, a number of major bituminous coal fields in the United States where the fuel mineral matter has a large fraction of iron compounds, chiefly in the form of pyrite and siderite. The amount of silicious species is correspondingly less than "normal." The data published by O'Gorman and Walker (1972) given in Tables 15.3 and 15.4 are used to show the composition of low-silica (<35% SiO_2), medium-silica (35–50% SiO_2) and high-silica (>50%) ashes.

It is evident that the kaolinite and quartz species together constitute between 70 and 80 percent of the mineral matter when the silica content is over 55 percent of ash (samples 2 and 3 in Tables 15.3 and 15.4). The bituminous coal mineral matter low in silica has a high iron content, chiefly in the form of pyrite (samples 26 and 67) or siderite (samples 142 and 143).

The analysis results in Tables 15.1–15.4 show that the composition of mineral matter in U.S. coals depends on the locality of coalfields and also on fuel rank. The deposit-forming propensity of coal ashes depends largely on the presence of sodium, calcium, and iron species, and the mode of occurrence of these key constituents in the mineral matter of U.S. low- and high-rank fuels will be examined in some detail.

15.2 THE MODE OF OCCURRENCE OF SODIUM AND CALCIUM IN U.S. LOW–RANK FUELS

The alkali metals can be present in coals in the form of organometal salts, chlorides, sulfates, carbonates, and silicates. The distribution of sodium and potassium between the coal substance and different inorganic species changes with coal rank as summarized in Table 15.5.

The fuel matrix of lignites and sub-bituminous coals is highly porous and can hold a large quantity of sodium in the form of exchangeable ions. As a result, sodium in flood and ground waters is readily transferred to the coal substance. It therefore follows that the amount of sodium in lignite and sub-bituminous fuels largely reflects the availability of metal in the geological environment. Table 15.6 gives some data on sodium and calcium in low-rank U.S. fuels.

It is evident from Table 15.6 that lignites and sub-bituminous coals in the Northern Plain Areas, Wyoming, and North Dakota have ashes rich in sodium, and the Na_2O content can exceed 10 percent (Table 15.1). In contrast the low-rank fuels in Montana, New Mexico, and Texas have an ash of moderate or low sodium content. Table 15.6 shows that most of the low-rank fuels in the United States have

Table 15.5 Alkali-metal species in low- and high-rank coals

Fuel deposits	Sodium and potassium species	Comments
Vegetable matter and flood and ground waters	Organometal salts, chlorides, sulfates, and carbonates	The vegetable matter is rich in potassium, but flood and ground waters can be rich in sodium
Lignites and sub-bituminous coals	Chiefly organometal salts and soluble inorganic salts	Frequently rich in sodium, potassium content is usually low
Low-rank bituminous coals	Chiefly chlorides and silicates	Sodium is present partly as chloride and partly in silicates, potassium has a high affinity to silicates
High-rank bituminous coals and anthracites	Chiefly silicates with some chloride	High-rank, low-porosity coals have a limited capacity to hold chlorides

an ash rich in calcium, which is present largely in the form of organometal salts. This is evident from the results of extraction experiments with the solution of ammonium acetate, which showed that calcium was largely soluble in this medium (Sondreal et al., 1968; Miller and Given, 1978). The circulating flood and ground waters in low-grade fuel deposits had facilitated interchange reactions between alkali

Table 15.6 Sodium and calcium oxide contents of U.S. lignite and sub-bituminous coal ashes (arithmetric mean values)

		Weight percent		
Fuel[a]	Locality	Ash of coal	Na_2O of ash	CaO of ash
1. Lignite and sub-bituminous	Northern Great Plains	10.1	3.2	15.0
2. Sub-bituminous	Wyoming	6.0	4.8	16.7
3. Lignite	North Dakota	6.4	5.5	24.3
4. Lignite	Montana	–	1.8	21.3
5. Sub-bituminous	Montana	9.3	1.7	15.6
6. Sub-bituminous	New Mexico	25.7	1.4	3.6
7. Lignite	Texas	11.4	1.1	12.6
8. Lignite	Gulf Province	27.6	0.3	7.1

[a]Sources:

1. 93 samples–Swanson et al. (1976).
2. Typical sample–Honea et al. (1979).
3. 432 samples–Cooley and Ellman (1979).
4. 93 samples–Cooley and Ellman (1979).
5. 125 samples–Cooley and Ellman (1979).
6. Typical sample–Honea et al. (1979).
7. 10 samples–Russel (1979).
8. 34 samples–Swanson et al. (1976).

metals and alkaline earth metals as depicted in Eq. (15.1).

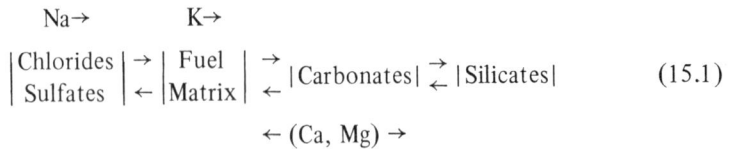

$$\begin{array}{c} \text{Na} \rightarrow \qquad\qquad \text{K} \rightarrow \\ \left| \begin{array}{c} \text{Chlorides} \\ \text{Sulfates} \end{array} \right| \begin{array}{c} \rightarrow \\ \leftarrow \end{array} \left| \begin{array}{c} \text{Fuel} \\ \text{Matrix} \end{array} \right| \begin{array}{c} \rightarrow \\ \leftarrow \end{array} |\text{Carbonates}| \begin{array}{c} \rightarrow \\ \leftarrow \end{array} |\text{Silicates}| \\ \leftarrow (\text{Ca, Mg}) \rightarrow \end{array} \qquad (15.1)$$

This scheme suggests that the original source of calcium found in lignites and sub-bituminous coals was chiefly the carbonates deposited together with the organic matter.

15.3 THE MECHANISM OF FORMATION OF SINTERED ASH DEPOSITS WITH LIGNITE AND SUB–BITUMINOUS COALS

The mineral-matter analysis data discussed in Sects. 15.1 and 15.2 show that the amount of the nonsilicate minerals and organometal species in U.S. low-rank coals is frequently high. The characteristic feature of the nonsilicate species, carbonates, chlorides, sulfides and sulfates, and the organometal salts and complex compounds is that they produce a large number of fume particles in the pulverized coal flame, as discussed in Chap. 8. In the case of sodium, volatilization takes place before the formation of fume particles, whereas calcium remains largely involatile and the small particles are formed on the thermal dissociation of carbonate and organometal species (Chap. 7).

Sodium present in lignite and sub-bituminous coal is largely volatilized in the flame and the initial species are those of hydroxide, oxides, and metal, the last only in a strongly reducing atmosphere. The volatile species are partly captured by the flame-heated silicate particles and partly sulfated, as depicted schematically by Eq. (15.2).

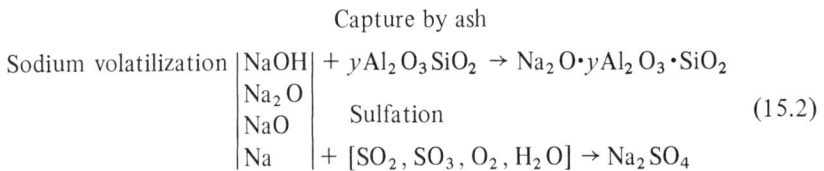

$$\text{Sodium volatilization} \left| \begin{array}{l} \text{NaOH} \\ \text{Na}_2\text{O} \\ \text{NaO} \\ \text{Na} \end{array} \right| \begin{array}{l} \overset{\text{Capture by ash}}{+ y\text{Al}_2\text{O}_3\text{SiO}_2 \rightarrow \text{Na}_2\text{O} \cdot y\text{Al}_2\text{O}_3 \cdot \text{SiO}_2} \\ \\ \quad\text{Sulfation} \\ + [\text{SO}_2, \text{SO}_3, \text{O}_2, \text{H}_2\text{O}] \rightarrow \text{Na}_2\text{SO}_4 \end{array} \qquad (15.2)$$

The behavior of sodium in the bituminous coal flame has been discussed in Chaps. 6 and 7. The only significant difference is that sodium is present in British bituminous coals in association with chlorine and sodium chloride is a volatile species. However, the end products of the flame reactions are the same, sodium-enriched silicate particles and sulfate fume particles.

The microscopic examination of fly-ash samples from lignite and sub-bituminous coal-fired-boilers showed coal particle characterization similarities and differences compared with those of bituminous coal ashes. Figure 15.1a shows a typical sample of lignite fly ash collected from the electrical precipitator, and it is

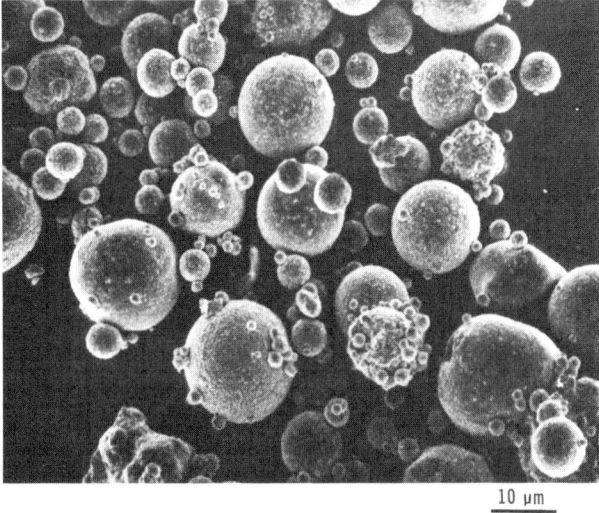

(a)

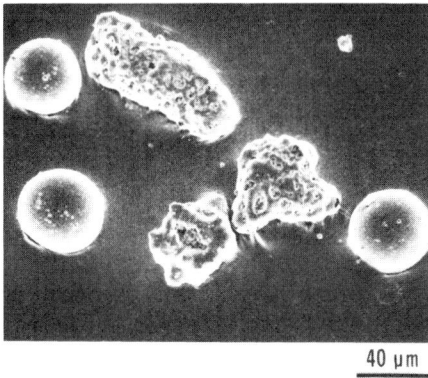

(b)

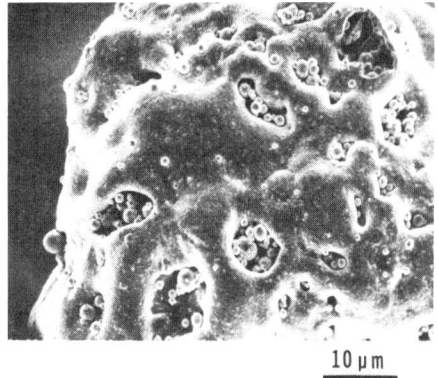

(c)

Figure 15.1 North Dakota lignite fly ash. (a) Typical sample. (b) Spherical and fusion enveloped particles. (c) Close-up ash agglomerate.

evident that with the exception of some above 10 μm diameter, particles all others are spherical in shape. Figure 15.1, b and c, shows that the nonspherical particles are agglomerates of small particles enclosed in an envelope of fused silicate. It is probable that the fused agglomerates are formed in a manner similar to that described for the creation of ash plerospheres in a bituminous coal flame, described in Chap. 6. That is, an envelope of fused silicate forms as an ash-rich particle burns, encapsulating the small particles of clay dispersed in the fuel substance. The agglomerated particles are not found in the initial deposit on boiler tubes because of their large size (Chap. 12).

The silicate particles of lignite fly ash rich in sodium and calcium have a large number of small particles (0.1 to 0.3 μm in diameter) on the surface, as shown in Fig. 15.2a. This is probably because of the high ratio of sulfate to silicate species in the ash. That is, a comparatively small number of silicate particles provide a platform for a large number of sulfate fume particles. The EDAX analyses showed that the small captive particles and the ash particles below 2 μm in diameter (Fig. 15.2b) were enriched in sodium and calcium sulfate. The lignite fly ash, rich in calcium when dispersed in water, gave a highly alkaline solution, indicating that the ash contained a significant quantity of unreacted calcium oxide (CaO).

The mechanism of formation of boiler-ash deposits when burning low-rank fuels has been extensively studied at the Grand Forks Energy Center, North Dakota. Notable papers on this topic have been published by Duzy (1965), Gronhovd et al. (1967, 1968, 1970, 1973), Sondreal and Ellman (1975), and Sondreal et al. (1977, 1978).

The results of studies of boiler fouling with sub-bituminous coals in Victoria, Australia, have been reported by Brown et al. (1962), Grant and Weymouth (1962, 1964), Wolfe (1962), Dunderdale et al. (1963), Procter and Taylor (1966), Bonafede (1975), and Wall et al. (1979a). The deposit-forming propensity of low-rank fuels in Germany has been discussed by Hein (1976, 1979), and those in Canada by Sekhar (1978, 1980), Jackson (1980), Chawla (1982), and Mozes (1982).

The key role of sodium in the formation of sintered ash deposits in lignite and sub-bituminous coal-fired boilers has been frequently emphasized. It is evident that the amount of the flame-volatile sodium transferred to silicate ash particles before and also after deposition largely determines the deposit-forming propensity of lignite and sub-bituminous coal ashes.

Tufte and Beckering (1975) have found that the initial deposit material had a relatively high sodium content and the average size of deposited ash particles was

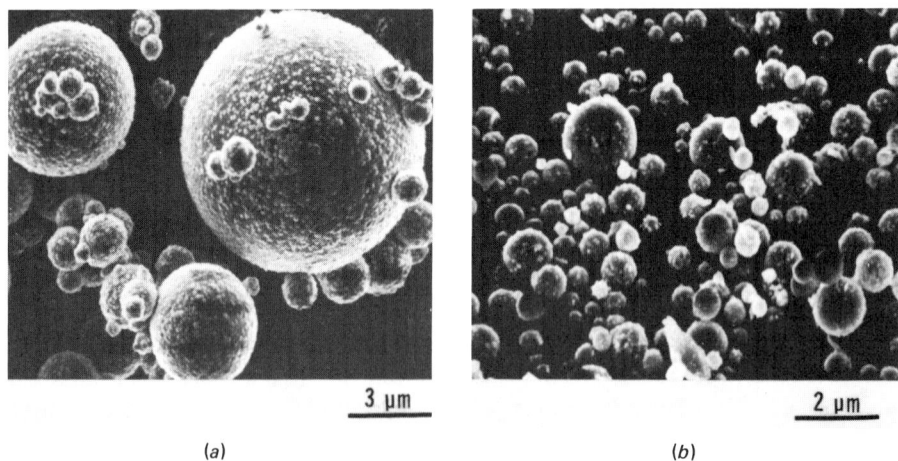

| 3 μm | 2 μm |
| (a) | (b) |

Figure 15.2 Small particles in sub-bituminous coal fly ash. (a) Surface captive particles. (b) Ash particles <2 μm.

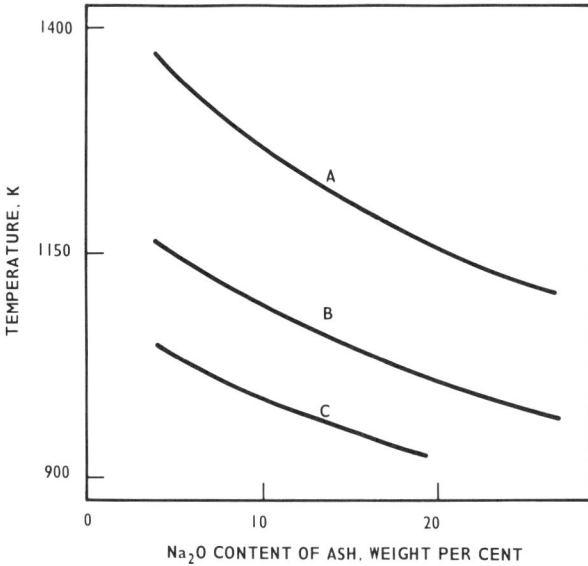

Figure 15.3 The effect of added sodium on ash viscosity: original Na_2O content, 4.4%; viscosity, A, 10^4 N s m^{-2}, B, 10^6 N s m^{-2}, C, 10^8 N s m^{-2}.

smaller than that of fly ash. Their observations are in accordance with enhanced deposition of the volatile and small-particle species on a cooled target, as discussed in Chap. 12. Sodium sulfate can form low-temperature melting phases in the presence of potassium sulfate (Chap. 17), but these have not been found in the calcium-rich deposits of lignite and sub-bituminous coal ashes. It is therefore likely that the initial deposit matrix forms as a result of bridging of sodium-rich silicate particles, as discussed by Brown et al. (1962) and Tufte and Beckering (1975).

The effect of enrichment of sodium in silicate ash on the viscosity, hence the rate of sintering (Chap. 10) has been demonstrated experimentally by Boow (1972). He carried out the viscosity measurements with added amounts of sodium in a coal-ash slag by the technique of needle penetrometer (Chap. 9) in the temperature range of 900–1375 K. The coal ash contained originally 4.4 percent sodium oxide (Na_2O), and Fig. 15.3 shows the effect of lowering the viscosity with added amounts of sodium. The results show that there was an order of magnitude reduction in viscosity at a given temperature when the sodium oxide content of ash was increased by about 10 percent from 4.4 to 15 percent. There was a corresponding increase in the ash sintering propensity, since the growth of contact area between the particles is inversely proportional to the viscosity [Eq. (10.2)]. The viscosity data were used to calculate the rate of sintering—i.e., the increase in contact area of 5-μm particles with added amounts of sodium. The results are shown in Fig. 15.4, and it is evident that an enrichment of sodium in the silicate ash has a marked effect in increasing the rate of formation of sintering deposits.

The low-rank fuels rich in sodium usually have also high amounts of calcium,

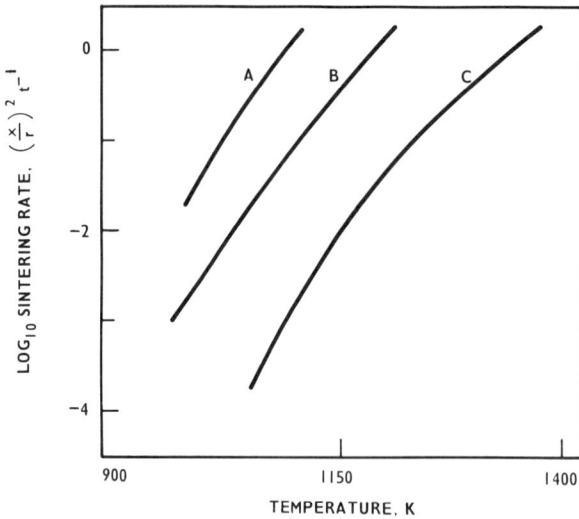

Figure 15.4 The effect of added sodium on ash sintering [Eq. (10.2)], with surface tension 0.32 N m^{-1} and particle radius 2.5 μm: A, 20% Na$_2$O added; B, 10% Na$_2$O added; C, original 4.4% Na$_2$O in ash.

chiefly in the form of organometal complexes, sulfates, and carbonates (Sects. 15.1 and 15.2). The presence of calcium in sodium-rich ashes may enhance the rate of sintering, but in large concentration calcium may retard the process of deposit formation. Initially calcium reacting together with sodium will reduce the viscosity of flame-heated silicate particles and increase the rate of sintering of deposited ash. Calcium sulfate is frequently found in boiler deposits (Procter and Taylor, 1966), but it is considered that sulfate does not take part in the initial sintering process. However, the growth of crystalline material inside the porous matrix of sintered ash may increase the strength of deposits.

The crystalline species of calcium silicates found in ash deposits include plagioclase (Rost and Ney, 1956; Kiss et al., 1972; Gibb, 1982) and melilite (Tufte and Beckering, 1975). The crystalline silicate species also incorporate sodium in the lattice structure, thus reducing the alkali-metal content of glassy material. As a result the viscosity of residual glass will increase with the corresponding decrease in the rate of sintering. This may be a possible reason for the reduced rate of formation of boiler deposits when the ratio of calcium to sodium is high, as discussed by Sondreal et al. (1978). Figure 15.5 shows that the amount of deposit formed with lignite ash containing 20 percent CaO (curve A) was 2 to 3 times higher than that formed when the CaO content was over 30 percent.

It is interesting to note that the deposits with both bituminous coals and sub-bituminous fuels rich in sodium and calcium occur chiefly in the top half of the combustion chamber and on the superheater tubes. A rapid build-up of superheater deposits in boilers fired with high-sodium (chlorine) bituminous coals has been discussed by Jackson and Ward (1956). The conditions for formation of superheated deposits are characterized by a comparatively high temperature of the tube surface–deposit interface (Table 11.1), and a relatively long residence time of ash particles in the flame before deposition. Thus, the flame-borne silicate ash captures volatile sodium, resulting in reduced viscosity of the surface layer of the particles and rapid sintering on deposition.

An ash deposit rich in sodium and calcium, but low in iron, requires a comparatively high temperature of the tube surface–deposit interface for adhesion to take place. This is because of the composition incompatibility of silicate ash deposits with the oxidized surface of boiler tube steels, as discussed in Chap. 11. It is therefore a usual occurrence to find a build-up of sodium- and calcium-rich deposits on the superheater tubes having the surface temperature above 800 K in lignite and sub-bituminous coal-fired boilers.

The other characteristics of sintered deposits formed in the superheater of a boiler burning lignite and sub-bituminous coals are low density and a high degree of resilience to thermal and mechanical shocks. The flue-gas temperature in the superheater duct is usually below 1300 K; thus the maximum deposit temperature is not sufficiently high to densify the deposit by slagging. The low-density deposit does not require a firm intrinsic bonding to boiler tubes in order to support "nesting" type growths in sizes sufficiently large to block the flue-gas passages.

The deposit-forming propensity of lignite and bituminous coal ashes has been discussed in Chap. 10, chiefly Eqs. (10.13) and (10.14). The principal conclusions are that build-up of ash deposits in boilers fired with low-rank U.S. coals increases with the sodium content. Burning the high-sodium coals, the rate of boiler fouling increases also with the ash content of coals. Sondreal et al. (1978) have stated that North Dakota lignites high in sodium are likely to cause troublesome boiler deposits. Texas lignites of moderate amounts of sodium and high in ash can also form troublesome boiler deposits, but the rate of build-up can be controlled by effective on-line boiler cleaning. Low-sodium, high-ash Montana lignites should not cause any deposit problems.

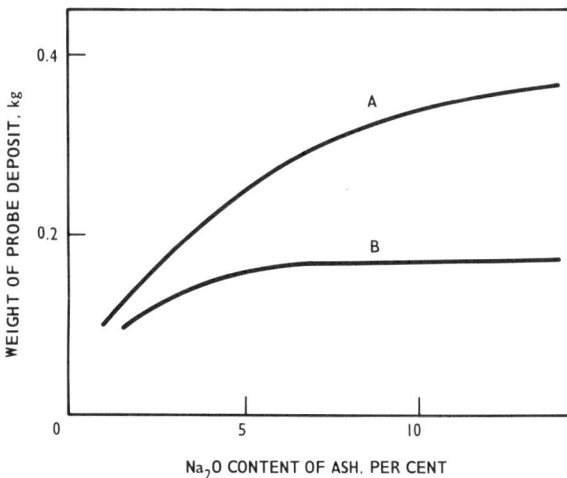

Figure 15.5 The effect of calcium on formation of sintered deposits with sodium-rich ashes: A, Beulah lignite ash, 19–20% CaO; B, Velva ash, 32–39% CaO.

Table 15.7 Deposit-forming propensity of lignite and sub-bituminous coals

Analysis data	References
Lignites and sub-bituminous coals in United States	
Sodium in ash	Duzy and Walker (1965)
	Gronhovd et al. (1970)
	Gray and Moore (1974)
The ratio of basic to acidic oxides in ash multiplied	
by sodium in ash	Attig and Duzy (1969)
Sodium, calcium, and silicate in ash; ash in coal	Sondreal et al. (1978)
Lignites in Canada	
Basic oxides in ash	Mozes (1982)
Sub-bituminous coals in West Germany	
Sodium, potassium, calcium, and sulfur in coal	Hein (1979)
Sub-bituminous coals in Australia	
Sodium in ash and ash in coal	Durie and Swaine (1971)
Major constituents in ash and deposits (rational analysis)	Drummond et al. (1978)
Sub-bituminous coals in New Zealand	
Calcium, magnesium, and boron in ash	Bibby (1977)
Calcium and iron in ash	Bibby et al. (1978)

The analytical data used to assess the ash deposit propensity of lignite and sub-bituminous coals in the United States and other countries are summarized in Table 15.7.

The remedial measures to combat excessive ash deposits in lignite and sub-bituminous coal-fired boiler are mostly the same as those discussed in Chap. 14. That is, the design of the boiler and the high-temperature regime should be compatible with the deposit-forming propensity of fuel ash. The requirements of adequate coal milling and fuel–air mixing have been discussed by Gray and Moore (1974). These measures would reduce the possibility of reducing atmosphere in the superheater zone, which may enhance the rate of build-up of ash deposits. Hein (1976) has suggested that low-temperature-melting sulfides may be deposited or formed on boiler tubes in a reducing atmosphere.

The possibility of reducing boiler fouling by coal cleaning and blending has been discussed in Chap. 14. The main point to note with high-sodium coals is that the conventional methods of coal washing do not reduce the alkali-metal content of fuel, and the result may be enhanced boiler fouling, as discussed by Suydam and Duzy (1977). Removal of sodium from lignite with a solution of calcium chloride in an ion exchange assembly has been discussed by Sondreal et al. (1978) and Paulson and Futch (1979). The possible methods of up-grading low-rank coals for power generation by physical cleaning, drying, and sodium and fine clay removal have been evaluated in the EPRI Report compiled by Ruby and Huettenhain (1981).

15.4 THE MODE OF OCCURRENCE OF IRON IN U.S. COALS

Iron occurs chiefly in the mineral-matter fraction of bituminous coals (Chap. 3), but in the low-rank fuels, lignites and sub-bituminous coals, some iron can be chemically associated with coal substance in the form of organometal complexes (Francis, 1961). The principal iron mineral species in U.S. coals are sulfides (pyrite) and carbonates (siderite), as was shown in Table 15.4, but some iron is also present in silicates, chiefly in chlorite and illite minerals.

The amount of pyritic iron in coal can be determined by chemical analysis (ASTM, 1969) or it can be estimated from the sulfur content. The analysis data on U.S. coals published by Swanson et al. (1976) showed that with the exception of lignites there was a linear relationship between the total sulfur (average values) and pyritic sulfur, as shown in Fig. 15.6. The graph gives the pyritic sulfur content S_p of coal in terms of total sulfur S_t,

$$S_p = 0.7(S_t - 0.3) \qquad (15.3)$$

where the sulfur contents are expressed in weight percent of coal.

The pyrite sulfur is combined chiefly with iron in sulfide (FeS_2) in the weight ratio of 1.9 to 1. Thus, from Eq. (15.3), the pyrite content of mineral matter in coal is given by

$$FeS_2 = \frac{1.9 \times 0.7(S_t - 0.3) \times 100}{M} = \frac{130(S_t - 0.3)}{M} \qquad (15.4)$$

where the pyrite content FeS_2 of mineral matter, the total sulfur content S_t of coal, and the mineral matter (ash) content M of coal are given in weight percent.

The estimated pyrite content (average values) of the mineral matter in U.S. low- and high-rank coals shown in Fig. 15.7 are based on the sulfur and ash analysis data published by Swanson et al. (1976). The plot shows that the average pyrite content of the mineral matter in low-sulfur sub-bituminous coals was only 5 percent, whereas the bituminous coals with the average sulfur content of 2.8 percent contained a mineral matter that had over 25 percent by weight of pyrite.

The pyrite iron in coal is related to pyritic sulfur (Fe to S_2) in the weight ratio

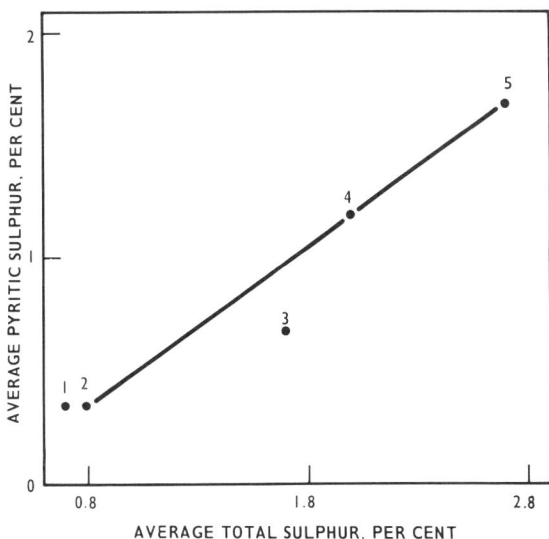

Figure 15.6 The relationship between pyritic sulfur and total sulfur in U.S. coals: 1, sub-bituminous; 2, anthracite; 3, lignite; 4, all coals; 5, bituminous.

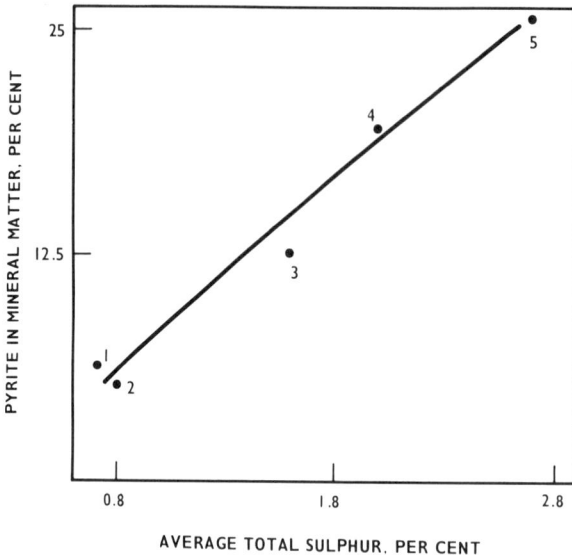

Figure 15.7 Estimated pyrite content (average) of mineral matter in U.S. coals: 1, sub-bituminous; 2, anthracite; 3, lignite; 4, all coals; 5, bituminous.

of 0.87. Thus, the pyritic iron content (Fe_p) of coal from Eq. (15.3) is

$$Fe_p = 0.87 \times 0.7(S_t - 0.3) = 0.61(S_t - 0.3) \qquad (15.5)$$

where the iron and sulfur contents are in weight percent of coal.

The analysis data of total iron in U.S. coals (Swanson et al., 1976) and Eq. 15.5 were used to calculate the ratio of pyritic iron to total iron in different-rank fuels, as depicted in Fig. 15.8. The plots show that in lignites only about 30 percent of iron was present as the pyrite mineral increasing to a value between 70 and 100

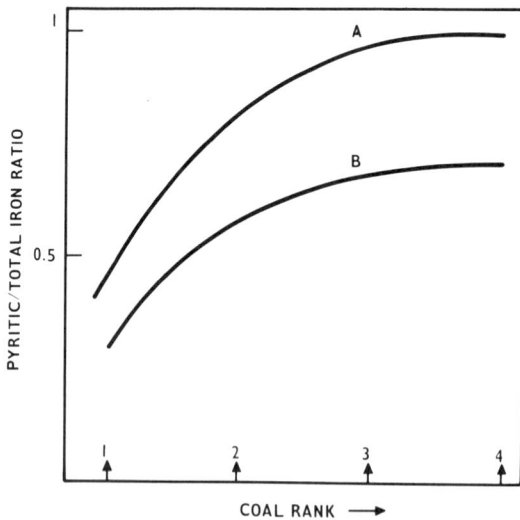

Figure 15.8 The increase in pyritic iron fraction with coal rank: 1, lignite; 2, sub-bituminous coal; 3, bituminous coal; 4, anthracite; A, high-sulfur coals; B, low-sulfur coals.

percent in bituminous coals, depending on the sulfur content. There are some notable exceptions to this pattern of distribution of iron species in U.S. coals. For example, Oklahoma low-sulfur bituminous coals can have a mineral matter that consists largely of iron and calcium carbonates (samples 142 and 143 in Table 15.4). However, the majority of bituminous coals in the United States have pyrite mineral as the principal species of iron.

15.5 THE MECHANISMS OF FORMATION OF IRON–RICH SLAG DEPOSITS

The flame reactions of iron mineral species in pulverized-coal-fired boilers are summarized in Table 15.8 with references to the relevant discussions in previous chapters.

Table 15.8 shows that the initial products of the flame-heated iron-containing species are the fume and large particles of oxide, molten particles of pyrite residue, and fused particles of silicates with dissolved iron. The organo–iron complex compounds and also carbonates do not procreate any molten species, whereas the pyrite mineral goes through a molten phase before oxidation to wustite and magnetite (solid) particles. The iron-rich silicate minerals, chiefly chlorite and also some species of illite, change in the flame to glassy particles of comparatively low viscosity (Chap. 6). All the principal species of iron formed in the flame have their own role to play in the formation of boiler deposits. The mechanism of deposition and the relative catch efficiency of different species on cooled boiler tubes are given in Table 15.9.

It would be difficult to determine experimentally the relative rates of deposition of different iron-containing species on boiler tubes. However, the boiler slagging experience with U.S. high-sulfur (pyrite) coals (Bryers, 1976) suggests that deposition of molten or semifused particles of pyrite decomposition residue may have a significant role to perform in the initial stage of slagging. The ash deposits in the form of semifused or fused slag with high-pyrite coals form chiefly on the combustion chamber walls, in contrast to high-sodium low-rank coals, which can cause severe superheater fouling. That is, the reactions between the flame-borne

Table 15.8 Reactions of iron species in coal flame

Species	Flame changes	Products	Reference
Organometal compounds	Dissociation and oxidation	Oxide fume	
Carbonates	Dissociation	Fume and large oxide particles	Chap. 8
Sulfides, chiefly pyrite	Dissociation and oxidation	Molten (transient) sulfide and oxide particles	Chap. 7
Silicates	Vitrification	Fused silicate particles with dissolved iron	Chap. 6

Table 15.9 Deposition of iron-rich species

Species	Mechanism of deposition	Catch efficiency
Oxide fume, <5 μm in diameter	Thermophoresic and electrophoresic deposition; inertial impaction	Fairly high
Oxide particles, >5 μm in diameter	Inertial impaction	Low
Molten sulfide residue particles	Inertial impaction	High
Fused silicate particles rich in iron	Inertial impaction	Moderate

silicate particles and volatile alkali-metal species before deposition enhance the formation of sintered deposits in the superheater zone, as discussed in Sect. 15.4. The formation of iron-rich deposits in the combustion chamber is favored by the short reaction time before deposition, a reducing atmosphere, and a coarse milling. These conditions retard the oxidation of pyrite mineral, resulting in deposition of the reactive FeS residue.

The adhesion of iron-rich deposit on the furnace-wall tubes is enhanced by the composition compatibility of the deposit and the oxidized surface layer of boiler steels (Chap. 11). Also, the high radiative heat flux in the combustion chamber results in a rapid increase in the deposit temperature with thickness. The high deposit temperature promotes the subsequent slagging reactions between iron oxide fume particles, iron-rich silicate particles, and other silicate species. In contrast, large particles of iron oxide formed on dissocation of carbonates and oxidized pyrite species in the superheater flue gas are comparatively inert in an oxidizing atmosphere. Bibby (1977) has shown that removal of the magnetic iron fraction from the fly ash reduced both the softening and hemisphere temperatures by 10–15 K in an oxidizing atmosphere. In a reducing atmosphere the magnetite particle removal increased the softening temperature by about 10 K and the hemisphere temperature by 30 K. This suggests that the magnetite particles behave as an inert dilutant material in an oxidizing atmosphere, whereas in a reducing atmosphere they enhance slagging.

The degree of enrichment of iron in boiler deposits has been investigated by Bryers (1976). Table 15.10 shows the composition of laboratory-prepared ash, fly ash, and furnace-wall slag for a pulverized coal-fired boiler burning high-sulfur coals.

The average enrichment factor of iron in boiler deposits can be estimated from the Fe_2O_3 content of laboratory-prepared ash and that of fly ash. The total amount of deposits in a pulverized-coal-fired boiler is usually between 15 and 30 percent of the ash, the remainder being discharged as fly ash. The balance of iron in the laboratory-prepared ash, fly ash, and deposit is given by

$$x = wx_1 + (1 - w)x_2 \tag{15.6}$$

where the weight of coal ash entering the boiler is taken to be unity; and w and $1 - w$ are the weight fraction of fly ash and deposit, respectively; and x, x_1, and x_2

Table 15.10 Enrichment of iron in boiler slag

Constituent	Weight percent		
	Laboratory ash	Fly ash	Furnace slag
SiO_2	36.5	54.4	14.5
Al_2O_3	12.5	26.1	6.0
Fe_2O_3	24.0	13.9	71.2
CaO	8.4	1.7	<0.1
MgO	<0.1	0.5	0.2
Na_2O	<0.1	<0.1	<0.1
K_2O	1.8	3.0	2.9
SO_3	12.1	0.9	6.5

Iron enrichment in slag and deposit

$$\frac{Fe_2O_3 \text{ in furnace slag}}{Fe_2O_3 \text{ in laboratory ash}} = 3.0$$

$$\frac{Fe_2O_3 \text{ in total deposit}}{Fe_2O_3 \text{ in laboratory ash}} \text{ (calculated)} = 2.3$$

are the Fe_2O_3 content of laboratory ash, fly ash, and deposit, respectively. From Eq. (15.6), the Fe_2O_3 content of deposit is

$$x_2 = \frac{x - wx_1}{1 - w} \tag{15.7}$$

and the enrichment factor is

$$\frac{F_2O_3 \text{ in deposit}}{Fe_2O_3 \text{ in laboratory ash}} = \frac{x_2}{x} = \frac{x - wx_1}{x(1 - w)} \tag{15.8}$$

A highly slagging coal ash gives about 25 percent deposit and 75 percent fly ash, i.e., $w = 0.75$. The Fe_2O_3 analysis data in Table 15.10 gives an iron enrichment factor of 2.3.

The observation reported by Bryers (1976) that the high degree of enrichment of iron in boiler slag occurs with pyrite-rich coals supports the premise that significant quantities of molten or semifused residue are deposited on the furnace-wall tubes (Table 15.9). The molten particles deposited on boiler tubes do not escape from the surface, whereas much of the "dry" ash material rebounds and is reentrained in the flue gas, as discussed in Chap. 12.

15.6 ASH SLAGGING PROPENSITY BASED ON THE PYRITIC IRON CONTENT

The formulas for assessing the slagging propensity of coal ashes have been discussed in Chap. 10, chiefly Eqs. (10.12), (10.13), and (10.18). Here an assessment is made of the slagging characteristics of U.S. bituminous coals based on the pyritic iron content of ash, that is, the fraction of iron oxide in ash that originates from pyrite (FeS_2) mineral in coal is taken as a criterion for the ash slagging propensity. The amount of iron oxide $(Fe_2O_3)_p$ of pyritic origin can be calculated from the pyritic sulfur S_p and ash contents A of coal:

$$(Fe_2O_3)_p = \frac{1.25 \, S_p \times 100}{A} = \frac{125 \, S_p}{A} \qquad (15.9)$$

where 1.25 is the factor for converting iron in pyrite (FeS_2) to Fe_2O_3, $(Fe_2O_3)_p$ is expressed in weight percent of ash, and S_p and A are given in weight percent of coal.

Table 15.11 gives the ash and sulfur contents of U.S. bituminous coals together with the ash analysis and the calculated values of iron oxide derived from pyrite [Eq. (15.9)] and the ratio of pyritic iron to total iron in the oxide. The data—the arithmetic mean values and typical samples—were taken from the publication of Swanson et al. (1976) and refer to the bituminous coals in the eastern and interior states of the United States.

The data in Table 15.11 show that the ash from the bituminous coal in the eastern states have usually low calcium and sodium contents. It therefore follows that the slagging characteristics of the coal ashes depend largely on the iron oxide content, more specifically on the iron oxide fraction derived from the pyrite mineral in coal. The coals in the interior states have an ash moderately high in calcium, but the iron oxide content is significantly higher and the slagging propensity is largely governed by the pyrite mineral in coal. Thus, the ash and

Table 15.11 Analysis of U.S. bituminous coals and ash by region

Constituent	Eastern states		Interior states	
	Mean	Typical	Mean	Typical
Weight percent of coal				
Ash	13.3	11.1	15.7	11.7
Total sulfur	2.3	2.1	3.9	4.2
Pyritic sulfur	1.6	1.2	2.4	2.1
Ash constituents, weight percent of ash				
SiO_2	44	36	27	32
Al_2O_3	23	23	13	12
Fe_2O_3 (total)	21	19	30	27
Fe_2O_3 (pyritic)	15	13.5	19	22.5
Pyritic Fe_2O_3/total Fe_2O_3	0.72	0.71	0.64	0.83
CaO	1.9	2.4	10	9.7
MgO	0.9	0.4	1.3	0.5
Na_2O	0.4	0.2	0.4	0.2
K_2O	1.8	0.9	1.3	1.3
SO_3	2.3	0.8	7.0	3.5

Table 15.12 Ash sulfur and iron oxide data of U.S. bituminous coals[a]

Coal, locality, and sulfur level	Percent of coal			Percent of ash		$\dfrac{Fe_2O_3 \text{ pyritic}}{Fe_2O_3 \text{ total}}$
	Ash	Total sulfur	Pyritic surfur	Total Fe_2O_3	Pyritic Fe_2O_3	
Eastern states						
Pennsylvania						
Sulfur, high	13.1	4.7	3.30	37	31.4	0.85
medium	16.5	2.0	1.42	16	10.8	0.67
low	10.1	1.2	0.39	9.0	4.8	0.54
Ohio						
Sulfur, high	13.1	4.0	2.96	29	28.2	0.97
medium	7.1	2.6	1.67	33	29.4	0.29
West Virginia						
Sulfur, high	17.7	3.2	2.0	14	13.6	0.91
low	5.1	0.5	0.07	6.0	1.7	0.29
Virginia						
Sulfur, high	9.0	2.6	1.37	21	19.0	0.91
medium	5.1	1.4	0.52	18	12.7	0.71
low	8.1	0.5	0.09	9.3	1.4	0.15
Kentucky						
Sulfur, high	18.2	4.5	3.66	38	25.1	0.66
medium	13.6	2.0	1.29	12	11.9	0.99
low	6.9	0.8	0.19	9.9	3.4	0.35
Alabama[b]	5.0	0.1	0.36	14.4	9.0	0.63
Interior states						
Michigan						
Sulfur, medium	5.8	1.4	0.99	31	21.3	0.69
Indiana						
Sulfur, high	11.7	4.2	2.06	27	25.8	0.95
medium	10.8	2.5	1.44	17	16.6	0.98
Illinois[b]						
Sulfur, high	11.2	4.0	2.6	26.5	26.5	1.00
medium	14.4	2.6	1.5	10.8	10.8	1.00
Iowa						
Sulfur, very high	10.6	6.3	3.49	52	41.2	0.79
high	18.7	3.4	1.38	14	9.0	0.64
Nebraska						
Sulfur, high	7.8	4.3	0.71	24	11.3	0.47
low	8.5	1.4	0.07	19	1.0	0.05
Missouri						
Sulfur, high	11.8	4.5	2.73	32	28.9	0.90
medium	11.2	2.8	1.06	12	11.8	0.99
Kansas						
Sulfur, very high	17.2	8.3	6.64	52	48.3	0.93
high	12.8	3.4	2.52	24	24.0	1.00
low	6.7	1.6	0.57	11	11.0	1.00
Oklahoma						
Sulfur, high	16.9	4.7	3.87	29	28.6	0.99
medium	10.2	1.6	0.82	50	10.0	0.20
low	5.6	0.8	0.27	25	6.0	0.24
Arkansas						
Sulfur, medium	7.0	1.3	0.75	23	13.4	0.58
low	6.4	0.7	0.16	25	3.1	0.13

[a]Swanson et al. (1976).
[b]Data from the publication of O'Gorman and Walker (1972).

sulfur (pyrite) contents of coal and the iron oxide content of ash can be used for an assessment of the slagging propensity of the majority of U.S. bituminous coals. The data for high-, medium-, and low-sulfur coals in different localities are given in Table 15.12.

Previously it was shown that the amount of pyritic iron increased with coal rank (Fig. 15.8), but in bituminous coals it increases also with sulfur content, as shown in Fig. 15.7. That is, a high-sulfur bituminous coal is likely to contain most of the iron in the form of pyrite, a highly slagging ingredient, whereas the same amount of iron in low-sulfur coals when present largely as carbonates and silicates would cause less slagging. Table 15.13 gives an assessment of slagging propensity of U.S. bituminous coals based on the sulfur (total) content of coal and the iron oxide (originated from pyrite) content of ash.

According to numerous literature data, an iron-rich ash is the characteristic feature of a large number of U.S. bituminous coals, and thus moderate or severe slagging can occur in the majority of pulverized-fuel-fired boilers. The rate of build-up of iron-rich slag deposits depends not only on the amount but also on the particle size of pyrite in pulverized coal, as discussed by Bryers (1976). He recommended that the fineness of pulverized fuel of high-sulfur coals should be controlled to ensure the absence of large pyrite particles. Other measures to promote rapid oxidation (combustion) of pyrite particles in the flame include adequate supply of combustion air, and thorough mixing of fuel and air, as discussed in Chap. 14.

The possibility of removing pyrite mineral from high-sulfur coals and hence reducing boiler slagging has been frequently discussed. Saltsman (1968) has investigated the degree of liberation of pyrite in crushed coals, and Deurbrouck (1972) has reported the results of washing experiments with 322 coals. The average total sulfur content of raw coals was 3.2 percent, of which 67 percent was pyritic, the remainder was organically combined in the coal substance. On average, 50 percent of the pyritic sulfur and also the pyrite iron was removed from coal by washing, and as a result the slagging propensity of the cleaned coal ashes was reduced from the severe to moderate category as given in Table 15.13. Further reduction in boiler slagging with iron-rich coals may be achieved by the use of chemical additives, as discussed in Chap. 16.

Table 15.13 Slagging propensity of U.S. bituminous coal ash

Total sulfur in coal (weight percent)	Pyrite originated Fe_2O_3 in ash (weight percent)	Slagging propensity
<1.5	<10	Marginal
1.5–2	10–15	Moderate
2–2.5	15–20	High
>2.5	>20	Very high

NOMENCLATURE

A	ash content of coal
Fe_p	pyritic iron content of coal
FeS_2	pyrite content of coal
$(Fe_2O_3)_p$	iron oxide content of ash of pyritic origin
M	mineral matter content of coal
S_p	pyritic sulfur content of coal
S_t	total sulfur content of coal
w	weight fraction of fly ash
$(1-w)$	weight fraction of boiler deposits
x	iron oxide content of laboratory prepared ash
x_1	iron oxide content of fly ash
x_2	iron oxide content of boiler deposits

SIXTEEN
USE OF ADDITIVES IN COAL-FIRED BOILERS

16.1 THE AIMS AND PREMISE

The use of high-temperature additives in coal-fired boilers has been less extensive than in oil-fired boilers. There are, however, several acidic and basic chemicals that have been thoroughly investigated with the aim of improving the performance of the electrical precipitators in removing particulate ash from the flue gas before it is discharged into the atmosphere. The premise is that for an efficient performance of the electrical precipitators the ash must have minimum conductivity of about 10^{-8} S m^{-1}. This corresponds to a resistivity value of 10^8 Ω m when measured across a loosely compacted column of ash at precipitator working temperatures around 400 K.

The surface resistivity of some ashes, in particular those from low-sulfur coals, can be 2 to 3 orders of magnitude above 10^8 Ω m, and it may be necessary to use an additive in order to increase the particle-to-particle charge leakage when a layer of ash builds up on the precipitator collector plate. For an additive to be effective, a sufficient quantity must be absorbed at the surface of ash particles to increase the surface conductivity by 2 to 3 orders of magnitude.

The premise for using high-temperature additives to combat slagging and fireside corrosion in coal-fired boilers is less definable than it is in oil-fired boilers, as discussed by Locklin et al. (1980). The amount of ash in coal is over 100 times higher than in oil, and it is not realistic to consider using additives of the same proportion as for the ash in coal. It follows therefore that the amount of an additive used is not likely to be sufficient to markedly change the overall composition of ash. However, it may be possible by the aid of additives to decrease the deposit-forming propensity of ash as a result of surface catalysis and selective chemical reactions.

With the declining use of residual oil for electricity generation in large power stations, the specialist additive suppliers have focused their attention on coal-fired utilities. There have always been conflicting reports on the efficacy of additives to control boiler fouling and corrosion, particularly in coal-fired boilers. Comparative boiler trials with and without additives are notoriously difficult to carry out in coal-fired units. First the quality of coal—i.e., the amount and composition of ash—is much more variable than that of the residual oil. Further, comparatively minor changes in boiler operation, such as the performance of coal milling plant, boiler load, and the operating effectiveness of sootblowers, can negate or magnify the possible beneficial effects of additives in preventing excessive boiler slagging. It is therefore imperative to consider carefully the scientific premise for use of high-temperature additives in coal-fired boilers.

The ash resistivity characteristics and application of the precipitation aiding additives are discussed in the next two sections, followed by an assessment of high-temperature additives.

16.2 ELECTRICAL RESISTIVITY OF PARTICULATE COAL MINERALS AND PRECIPITATOR ASH

The removal of gas-borne ash particles in the electrical precipitator is accomplished by charging the particles, then passing them through an electric field where they are accelerated toward an oppositely charged collecting surface by coulombic forces. The captured particles form a layer of ash, which periodically is removed by rapping and transferred into a hopper. Electrical resistivity is an important property of ash and influences the precipitator performance, as discussed by Tassicker (1975).

The resistivity of the gas-borne ash in coal-fired boilers is in the range of 10^5 to 10^{12} Ω m as discussed by White (1963). Pulverized-coal ash after passing through the boiler flame consists largely of silicate, iron oxide, sulfate, and coke residue particles (Raask, 1980), and the particles may have some adsorbed gases at the surface (Raask, 1981a). Dalmon and Raask (1972) have measured the electrical resistivity of ground silicate minerals before and after heating during the passage through a laboratory furnace. The results of chemical analysis of the silicate species were given in Table 6.2.

The minerals, with the exception of kaolin, which had a particle size already below 105 μm, were ground and the size cut between 45 and 105 μm was used in the experiments. The particles were entrained in air and passed through a vertical furnace, the residence time in the hot zone being 0.5 ($\pm$0.2) s. The furnace temperature was increased until the particles started to fuse, fusion being defined as the point where rounding of an irregular particle occurs, as discussed in Chap. 6. For quartz and kyanite the temperature was raised to 1925 K, the limit of the furnace, but the majority of the particles remained unfused.

For the resistivity measurements it was necessary to construct a miniature cell as the quantity of heated material was small (2 to 3 g). It consisted of a cylindrical Teflon container, with 25 mm diameter and 3 mm high, having brass ends which formed the electrodes. The cell was mounted on a Teflon block to obviate stray currents when the measuring potential of 85 V dc was applied. The samples were compacted by an electric vibrator to obtain a packing density of about 900 kg m^{-3}. The cell was placed in an oven at 417 (± 2 K) and left for 1 h before readings were taken.

There was a marked difference in the resistivity of the crystalline silicate minerals before heat treatment. The resistivity values were between 2.6×10^{11} and 2.5×10^{12} Ω m for muscovite, albite, quartz, and kyanite, whereas the values were 2 or 3 orders of magnitude less for kaolin, illite, and chlorite. The latter minerals have a high degree of hydration: that is, water molecules are incorporated in the crystalline structure. When a dc potential is applied, hydroxyl (OH$^-$) radicals are likely to contribute towards enhanced surface conductivity. On heating to 1000 K, water in the hydrated silicates was lost, as shown for kaolin in Fig. 16.1, resulting in an increase of the resistivity by 3 orders of magnitude with a slight decrease after being heated to 1800 K, as in Fig. 16.2. Above 1225 K, kaolin is converted into mullite and silica, and neither product has a tendency to rehydrate on cooling. The same applies to all other hydrated silicates in coal, such as illite and chlorite, and after being heated at high temperatures the products do not readily reabsorb water vapor. The behavior of both quartz and kyanite is that of a good insulating material: their resistivity remains around 10^{12} Ω m after the trace amount of adsorbed water is removed, as shown in Figs. 16.3 and 16.4. On the other hand, the two micas, illite and muscovite, display a marked difference: after heating to 1000 K, both had approximately the same resistivity (Fig. 16.5), but illite had a significantly lower resistivity before heating because of water in the crystalline

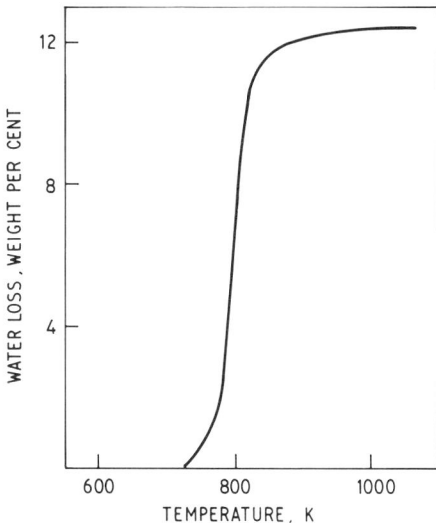

Figure 16.1 Loss of water from kaolin on heating.

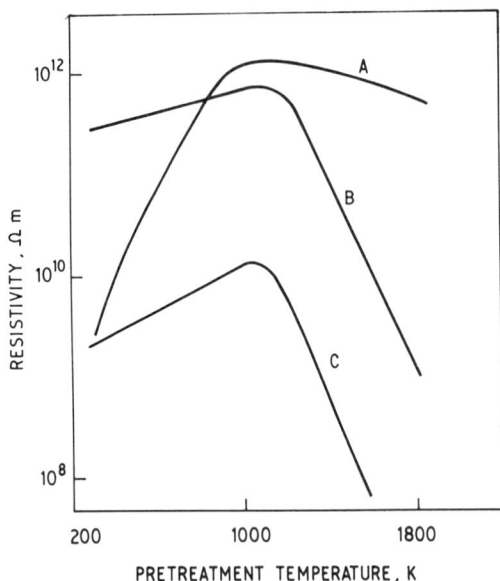

Figure 16.2 Resistivity of heated kaolin (A), albite (B), and chlorite (C) measured at 417 K.

structure, and also after heating to 1715 K. At this temperature both illite and muscovite had been fused, but in the case of muscovite, fusion did not result in a reduced resistivity value.

The low resistivity of illite after heating was likely to be due to iron present in the mineral, as shown in the analysis given in Table 6.2. A high concentration of iron in fused silicate or at the surface of particles should markedly reduce the

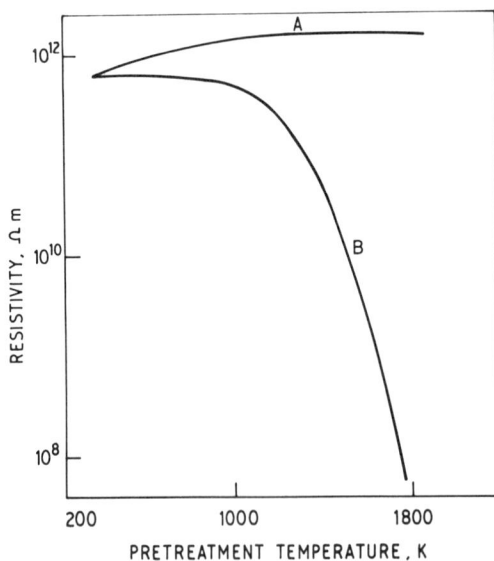

Figure 16.3 Resistivity of quartz at 417 K: A, no additive; B, FeO additive.

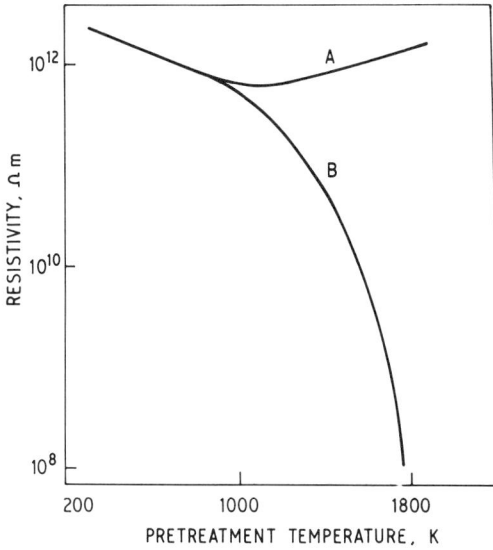

Figure 16.4 Resistivity of kyanite at 417 K: A, no additive; B, FeO additive.

resistivity. This was borne out by the resistivity values obtained on chlorite, and by the experiments where the particles of quartz and kyanite had been coated with a film of iron oxide shown in Figs. 16.3 and 16.4. A coherent layer of iron oxide, about 1 μm thick, on the silicate particles with high insulation property was sufficient to reduce their resistivity from 10^{12} to 10^6 Ω m; trace amounts equivalent to a 0.01-μm-thick layer had no effect. Conductance in iron oxide depends on its oxidation state, as shown by the results of Dalmon and Raask (1972). The resistivity of hematite (Fe_2O_3) was 6×10^{10} Ω m; for magnetite (Fe_3O_4) it was 2.5×10^3 Ω m, and the very low value of 4.5 Ω m was measured

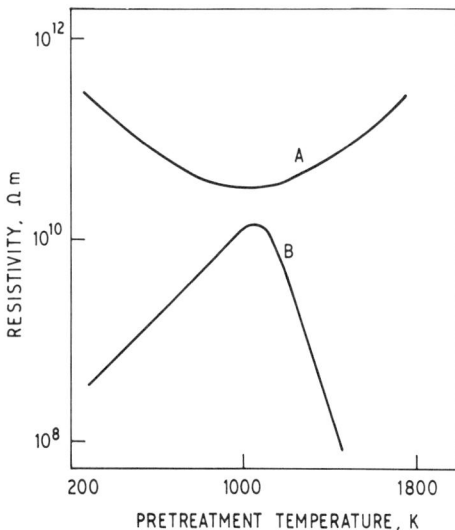

Figure 16.5 Resistivity of (A) muscovite at 417 K and (B) illite.

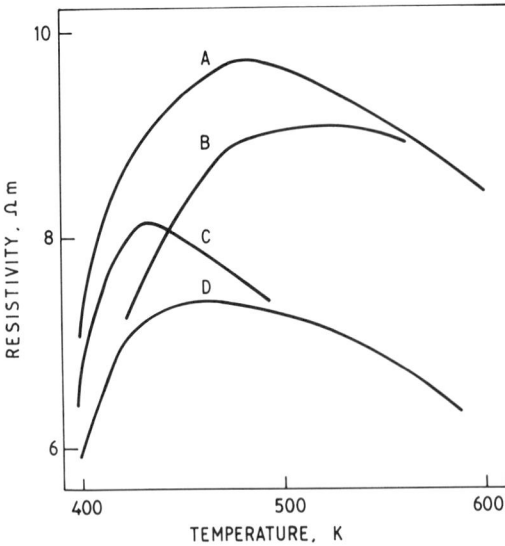

Figure 16.6 Electrical resistivity of coal ashes: A, low-sodium, low-sulfur coal; B, low-sodium, high-sulfur coal; C, high-sodium, low-sulfur coal; D, high-sodium, high-sulfur coal.

for the iron oxide residue obtained on an incomplete oxidation of pyrite (FeS_2). In contrast to iron oxide, calcium oxide was ineffective in modifying the electrical properties of quartz or kyanite (Figs. 16.3 and 16.4). Neither the silicates nor the oxide layer had been fused, and diffusion of calcium into the unfused silicate was insufficient for the formation of a glassy layer of enhanced conductance.

Oglesby and Nichols (1977) have shown that sodium and sulfur contents of coal had a marked effect on the conductivity of ash, as depicted in Fig. 16.6. Also, Selle et al. (1972) have established that for sub-bituminous coals mined in the western states of the United States the ratio of ($Na_2O + SO_3$) to ($CaO + MgO$) in ash determines the resistivity, as shown by the data given in Table 16.1.

The above data show that sodium rather than calcium or magnesium additives are likely to reduce the resistivity of coal ash. Additional data on the relationship between the resistivity and ash composition have been published by Bickelhaupt (1975, 1979). He has found that the bulk resistivity of ash depends on sodium,

Table 16.1 Electrical resistivity of non-bituminous coal ashes

Mole ratio, $\dfrac{Na_2O + SO_3}{CaO + MgO}$	Resistivity at 420 K (Ω m)
0.20	$>10^8$
0.23	5×10^7 to 10^8
0.38	10^7 to 5×10^7
0.71	3×10^6 to 10^7
1.43	$<3 \times 10^6$

lithium, calcium, magnesium, and iron contents, whereas the surface resistivity is largely influenced by the composition and specific surface (ratio of surface area to mass) of ash and by the concentration of water vapor in the atmosphere. He has introduced a concept of "acid resistivity" that depends on the SO_3 concentration and the amount of alkalis available for sulfation reaction, as discussed in Chap. 7. The use of SO_3 as an additive in coal-fired boilers is discussed in the next section.

16.3 ADDITIVES FOR IMPROVING THE PERFORMANCE OF ELECTRICAL PRECIPITATORS

It has been shown by Deutsch (1922) that the collection efficiency E of charged particles in an electrical precipitator can be related to the particle effective migration velocity v:

$$E = 100 \left[1 - \exp\left(-\frac{Av}{V_g}\right) \right] \qquad (16.1)$$

where A is the collector surface area and V_g is the volume of gas passing through the system. The migration velocity v can be taken to represent a precipitation rate parameter, and it is related to the resistivity of collected dust as discussed by White (1963) and shown in Fig. 16.7.

The decrease in the migration velocity, and hence in the precipitator performance, with dust of high resistivity (above 10^8 Ω m) is a result of insulating effect on the collecting surface, as discussed by Lowe et al. (1965). A layer of insulating dust on the collecting plates will prevent the ionic current leaking away to earth, and a high voltage gradient builds up across the ash layer. This layer will eventually break down, causing the emission of positive ions, a phenomenon known as back discharge or back ionization. The positive ions act to reduce the effective

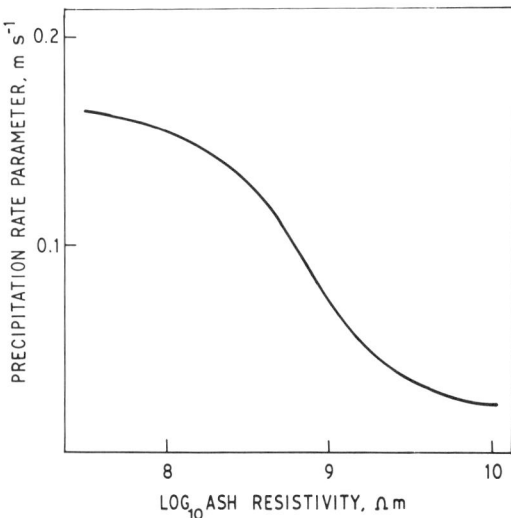

Figure 16.7 The effect of resistivity on precipitation rate parameter.

particle charge, and the breakdown in the dust layer also triggers flashover across the interelectrode gap. A reduction in flashover voltage causes a reduction in the electric field and consequent lowering of effective migration velocity of the particles, hence reducing the catch efficiency. Dalmon (1980) has discussed the relationship between the migration velocity and flashover voltage, as shown in Fig. 16.8, and McCain et al. (1975) have shown that the electrical precipitators collect preferentially the ash particles above 1 μm in diameter.

Brown et al. (1978a) have studied the efficiency of different additives in lowering the resistivity of an ash from bituminous (Luscar) coal mined in western Canada. The results in Fig. 16.9 show that SO_3 and sulfuric acid (H_2SO_4) were more effective than ammonia. Their experimental data indicated that the resistivity decreased to a limiting value as the gas-phase concentration of additive was increased. The equation that describes the effect of additives is

$$\log R - \log R_s = ae^{-kc} \tag{16.2}$$

where R and R_s are the resistivity of ash at a given additive concentration c and at saturation respectively, and a and k are constants. When c is zero,

$$a = \log R_o - \log R_s \tag{16.3}$$

where R_o is the resistivity of ash in the absence of conditioning agent.

Brown et al. (1978b) have shown that with each additive—i.e., SO_3, sulfuric acid, sulfamic acid, and ammonia—there was an optimum concentration for producing the maximum effect in increasing the performance of the experimental precipitator. Their data show that the acidic conditioning additives were significantly more effective than ammonia, as depicted in Fig. 16.10.

Castle (1980) has carried out a review of different mechanisms of ash

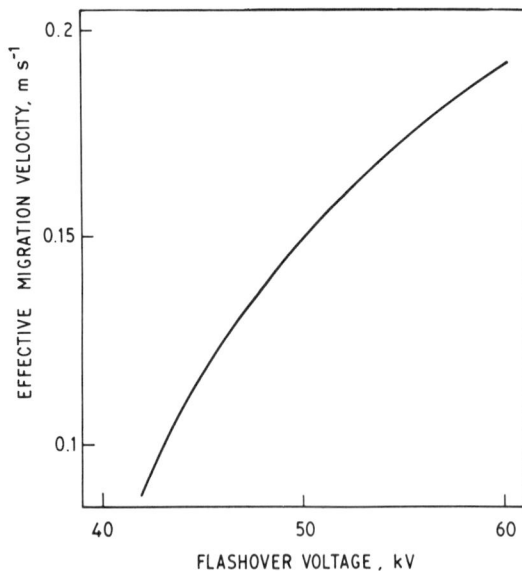

Figure 16.8 The relationship between migration velocity and flashover voltage of treated ash.

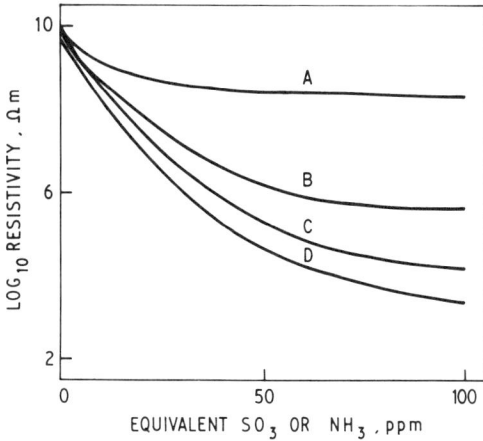

Figure 16.9 The effect of additives on ash resistivity: A, ammonia; B, sulfur trioxide; C, sulfuric acid; D, sulfamic acid.

precipitation in the presence of additives, as summarized in Table 16.2. Castle concluded that the principal effect of precipitation-aiding additives was to reduce the surface resistivity of ash, but some chemicals may cause additional changes in the electrical field and in ash adhesion. These include the change in the electrical breakdown strength of the flue gas, space charge enhancement of the collecting field, increase in the particle size by agglomeration, increase of the adhesive and cohesive properties of the ash, and decrease or increase in the acid dewpoint temperature of the flue gas.

The decision when to use an additive for improving the performance of boiler plant precipitators is usually based on the sulfur content of coal. Dalmon (1980) has suggested that with British coals a precipitation-aiding additive may be necessary when the sulfur content of the fuel is below 1.3 percent. Schrader (1970) has

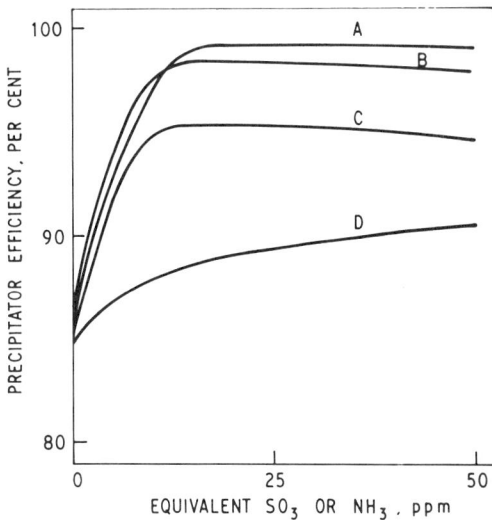

Figure 16.10 The effect of additives on ash precipitation efficiency: A, sulfur trioxide; B, sulfuric acid; C, sulfamic acid; D, ammonia.

Table 16.2 Possible mechanistic effects of additives on ash precipitation

Additive activity	Electrical effect	Adhesive effect	Effect on efficiency[a]	Comments
Adsorbs on surface of ash and reduces surface resistivity	Reduces the voltage drop in the dust layer Delays the onset of back corona Increases the sparkover voltage		++	Useful for high resistivity dusts: Increases charging and precipitation field strength, reduces the electrical adhesion on the wall and thus improves the effectiveness of rapping
		Reduces the electrical adhesion effect	+ or −	Beneficial for high resistivity dusts: if used with low- or medium-resistivity dusts, further lowering of adhesion forces could lead to reentrainment losses
Adsorbs on ash and changes cohesiveness or "stickiness"		Aids agglomeration and increases mean particle size	+	Agglomeration may occur independently of resistivity change and thus improve migration velocity Larger size fraction also aids removal by rapping
		Dust layer on plates becomes more cohesive	+	Cohesive dust layer tends to shear off collecting plate with less reentrainment losses
		Dust layer has stronger adhesion to plates	+ or −	Stronger adhesion is an advantage for low-resistivity dusts Could be a disadvantage for high-resistivity dusts

Increases fume (<0.5 μm) particle concentration, e.g., $(NH_4)_2SO_4$	Reduces ion density and thus current due to space charge suppression	+ or −	The current reduction could decrease charging effectiveness On the other hand, the lower current density will alleviate field reduction problems caused by the voltage drop through a high-resistance dust layer
	Increases collection field strength due to space charge enhancement	+	Space charge increases the field strength near the collecting electrode
	Increases sparkover voltage	+	A slight increase in sparkover voltage usually results from increased space charge
Increases electrical breakdown strength of flue gas	Increases sparkover voltage Delays onset of back corona	++	The breakdown characteristics of the gases are very sensitive to minor concentrations of electronegative species and to surface conditions of the dust layer; this can be independent of fly-ash resistivity
Neutralization of acid in flue gas	Decreases acid dewpoint: this reduces surface "tracking" on high-voltage insulators, allowing higher voltages to be applied	++	With some high-sulfur coals, the sulfuric acid concentration in the flue gases is so high that the acid dewpoint may be above the precipitator temperature; this may result in acid condensation on support insulators

[a]Key: ++, indicates strong tendency to increase efficiency; +, indicates moderate tendency to increase efficiency; −, indicates tendency to decrease efficiency.

claimed, however, that SO_3 as an additive produced a significant improvement in the performance of the electrical precipitators with 1.5–1.7 percent in coal. It appears that the necessity of precipitation-aiding additives depends more on the SO_3 content of the flue gas rather than the sulfur content of coal. The extent of SO_2 oxidation, hence the SO_3 content, of the flue gas in coal-fired boilers is more difficult to predict than that in oil-fired boilers. That is, the relationship between SO_3 concentration and the oxygen content of the flue gas is less predictable than that in oil-fired boilers. The excess oxygen in coal-fired flue gas is higher, between 3 and 5 percent, compared with that in oil-fired boilers. The high oxygen content and a large amount of gas-borne ash particles should enhance the rate of oxidation of SO_2 to SO_3. The SO_3 thus formed, however, is largely captured by the coal alkalis, chiefly sodium and calcium, resulting in the formation of sulfates in the flue gas and in boiler deposits. Raask (1982a) has estimated that with some bituminous coals over 10 percent of total coal sulfur is retained as sulfates. With some sub-bituminous fuels of exceptionally high calcium content (Gronhovd et al., 1973), the sulfur retention in ash and boiler deposit in the form of sulfates is even higher. The formation of SO_3 and sulfates in pulverized-coal-fired boilers can be summarized as follows:

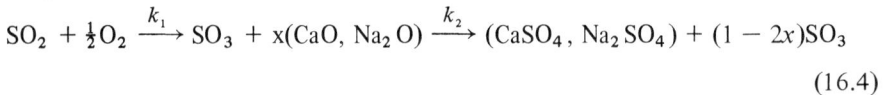

$$SO_2 + \tfrac{1}{2}O_2 \xrightarrow{k_1} SO_3 + x(CaO, Na_2O) \xrightarrow{k_2} (CaSO_4, Na_2SO_4) + (1 - 2x)SO_3$$

$$(16.4)$$

where k_1 and k_2 are the reaction rate constants of SO_2 oxidation and alkali sulfation respectively. The measured concentration values of SO_3 in the flue gas of the coal-fired boiler show a wide variation range, from less than 1 ppm in some sub-bituminous coal-fired boilers reported by Gronhovd et al. (1973) to 80 ppm found by Fielder (1958) in a stoker-fired boiler.

Another useful method of assessing whether precipitation-aiding additives are required is to determine the pH–time relationship of freshly captured ash dispersion in CO_2-free water. Pulverized-coal ash usually gives an alkaline solution as a result of dissolution of the leachable alkalis and CaO in silicate glass particles, as discussed by Raask (1981a). There is, however, initially an acidic reaction as a result of dissolution of the adsorbed SO_3 and HCl at the surface of ash particles. Figure 16.11 shows the typical acid dip and the alkaline reaction of size-graded precipitator ash. The sample was taken from the electrical precipitator of a 500-MW pulverized-coal-fired boiler, burning 1.6 percent sulfur in coal. The performance of the precipitators was satisfactory without an additive. It appears that an ash will have a sufficiently low resistivity for a high precipitator efficiency when it has surface adsorbed SO_3 and other acidic gases to give a pH value below 6 when dispersed in CO_2-free distilled water when the ash to water ratio is 1 to 10 by weight.

Neat SO_3 and SO_3-producing additives are widely used to improve the performance of the electrical precipitators in coal-fired boilers, as discussed by Morris and Schumann (1974), Archer (1977), and Locklin et al. (1980). The additives are usually injected at a rate that gives SO_3 concentration in the flue gas between 25 and 30 ppm. This concentration is equivalent to adsorbed acid on the

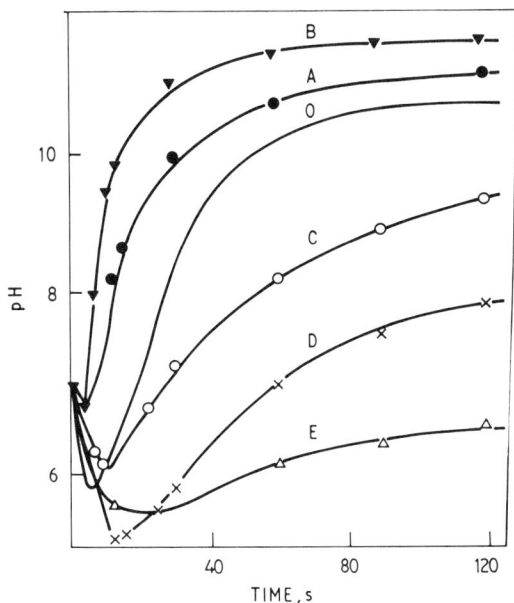

Figure 16.11 Acidic dip in pH–time plots of precipitator hopper ash/water slurry. Particle sizes: A, >45 μm; B, 10–45 μm; C, 5–10 μm; D, 2.4–5 μm; E, <2.4 μm; 0, ungraded.

surface of ash particles in a quantity less than 10 mg m^{-2} SO$_3$, as shown in Table 16.3. Two typical values were taken for dust burden in the flue gas entering the precipitators at 400 K. Further, it was assumed that the specific surface (ratio of surface area to mass) of ashes was 2500 m^2 kg^{-1} (Raask and Bhaskar, 1975; Raask, 1981a). The rate of SO$_3$ injection was taken to be 25 ppm by volume.

Table 16.3 shows that the ratio of the injected quantity of SO$_3$ to the specific surface of ash is comparatively low, i.e., a few milligrams SO$_3$ per square meter of ash. It is thus found in practice that the ash will adsorb quantitatively the injected SO$_3$. That is, there is no significant increase in the SO$_3$ content of the flue gas leaving the precipitators when the additive is well mixed upstream.

The SO$_3$ additive can be injected directly upstream of the precipitator, or it can be formed by catalytic oxidation of SO$_2$. The latter is purchased as liquid SO$_2$, or it may be generated by burning sulfur. Kanowski and Coughlin (1977) have suggested that SO$_3$ can be produced by a catalytic oxidation of the required amount of SO$_2$ in the boiler flue gas. The SO$_3$ can be also produced by thermal

Table 16.3 Adsorbed layer of SO$_3$ on ash particles

Ash in coal (%)	Ash retained as slag (%)	Dust burden of flue gas (g m^{-3})	Specific surface of ash per m^3 flue gas (m^2 kg^{-1} m^{-3})	SO$_3$ absorbed on ash (mg m^{-2})
12.5	20	6.7	16.8	5.3
25	25	11.9	30.0	3.0

Table 16.4 Relative costs of SO_3-producing additives

Additive	Relative material cost/mol SO_3 ($SO_3 = 1$)	Relative cost of material and application ($SO_3 = 1$)
Sulfur trioxide, SO_3	1	1
Sulfuric acid, H_2SO_4	0.6	1.1
Ammonium sulfate, $(NH_4)_2SO_4$	1.1	2.4
Ammonium bisulfate, NH_4HSO_4	5.5	11.1
Sulfamic acid	5.7	11.4

decomposition of sulfuric acid. Dalmon and Tidy (1972) have compared the efficacy and cost of five additives: SO_3, sulfuric acid, ammonium sulfate and bisulfate, and sulfamic acid. They have found all the additives to be equally effective, but the relative cost of SO_3 and sulfuric acid was significantly less than that of the other additives, as shown in Table 16.4. Further investigations of the efficacy and cost of SO_3 additives have been reported by Cook (1975), Brennan and Reveley (1977), Breisch (1978), and Patterson et al. (1978).

In the application of acidic additives, care should be exercised that only enough SO_3 is provided to control the electrical resistivity of ash; any excess is objectionable. The condensed sulfuric acid leads to excessive cohesiveness of the collected ash and cannot be removed from the collector plates by normal rapping. Reese and Greco (1968) have reported that as little as 5 ppm of the condensable acid in the flue gas lowered the efficiency of ash precipitation from 90 to 50 percent. Further condensed acid may lead to severe corrosion of the precipitator elements and casing, and the excess SO_3 increases the acidity of chimney-discharged flue gas.

Bennett and Kober (1978) have pointed out that the performance of ash precipitators in boilers burning high (above 3 percent) sulfur coals can be poor. They investigated a case where the ash was overconditioned with SO_3 formed in the flame, to the degree that there was a rapid loss of charge from the particles on collector plate resulting in the material being reentrained in the flue gas. Ammonia or magnesium compounds [MgO or $Mg(OH)_2$] can be used as precipitation aids in these cases, but an injection of alkali compounds upstream of the air heater may lead to fouling of the heat-exchange plates by bonded deposits, as discussed in Chap. 19.

White (1963) and Baxter (1968) have found that the resistivity of a catalyst material and coal ashes were significantly reduced by the ammonia treatment. In other reported cases, little or no change in resistivity was observed, as discussed by Watson and Blecher (1966) and Dismukes (1975). It appears therefore that ammonia as an additive does not always result in lowering the resistivity of ash. Reese and Greco (1968) and Baxter (1968) have suggested that ammonia and its reaction products, chiefly $(NH)_4SO_4$, on the surface of ash particles causes "stickiness" under certain temperature conditions. This could lead to enhanced particle agglomeration, resulting in an improved performance of ash precipitation.

Dismukes (1975) has reported that the ammonia additive was successful in increasing the efficiency of precipitators in boilers burning low- as well as high-sulfur coals. For example, 10 ppm added ammonia increased the precipitator efficiency from 90 to 98 percent in a boiler burning 0.9 percent sulfur fuel. The corresponding increase in efficiency with the additive was from 87 to 99 percent with high-sulfur (3.5 percent) coal.

Triethylamine, $(C_2H_4)_3N$, is another basic organic compound that has been frequently included among the additives tested in experimental rigs and boiler plants. Potter and Paulson (1974) have reported that triethylamine as an additive was more effective than either SO_3 or ammonia in improving the precipitator performance. They suggest that the additive greatly enhances agglomeration of the gas-borne ash particles, thus increasing the precipitator performance.

Triethylamine as an additive has been also investigated in an EPRI study reported by Bickelhaupt et al. (1978). The results showed that the additive was more effective than ammonia in lowering ash resistivity. Addition of 25 ppm of the additive decreased resisitivity by 2 orders of magnitude, depending on temperature and ash composition. If added at a temperature above about 600 K, triethylamine breaks down and is ineffective. The authors state that the exact way in which triethylamine functions is not known, but it may serve as a charge carrier. If this is so, its effectiveness depends upon the absorption on the surface of ash particles, but the temperature must be limited to minimize dissociation of the additive. In contrast, Brown et al. (1978b) have found that the triethylamine additive was less effective than either SO_3 or ammonia in improving the performance of the electrical precipitators.

Solid additives, such as alkali-metal carbonates and sulfates, and calcium and magnesium compounds have been investigated for use as ash precipitation aids. Among sodium-rich silicates, albite has a comparatively low resistivity of 10^9 Ω m after being heated at 1800 K, as shown in Fig. 16.2 (curve B). In contrast, the resistivity of heated kaolin (curve A), quartz (Fig. 16.3), and kyanite (Fig. 16.4) was around 10^{12} Ω m when an additive was not used. It appears that sodium dissolved in silicate particles results in lowering the surface resistivity by a significant margin. Further, Selle et al. (1972) and Bickelhaupt (1975) have shown that the resistivity of coal ashes decreases as the sodium content increases. Schliesser (1978) has reported that sodium carbonate as an additive reduced the resistivity of ash from coals low in sulfur and alkalis. The injection of anhydrous carbonate into the flue gas below 1000 K at a rate of 1–1.5 percent of the ash, expressed as Na_2O, significantly increased the efficiency of the electrical precipitators. The collection efficiency for 3-μm particles increased from 98.9 to 99.8 percent. Similarly beneficial results with sodium carbonate powder injection have been reported by Lederman et al. (1978).

Gooch et al. (1981) have investigated the performance of a hot-stage precipitator when sodium sulfate was added to coal low in sodium and sulfur. The rate of sulfate addition was 0.2 and 0.5 percent of coal, which increased the sodium (Na_2O) content of ash from 0.3 to 1 percent and from 0.3 to 2 percent, respectively. The results showed that the sulfate additive was effective both in reducing the resistivity of ash and in improving the performance efficiency of the electrical precipitators. Petersen (1978) has shown that injection of potassium sulfate to the off-take gas from a cement kiln at a rate that raised the K_2O content of the gas-borne ash by 0.4 percent, decreased the ash resistivity from 10^{11} to 10^9 Ω m. There was also a corresponding improvement in the performance of precipitators.

It should be emphasized here that addition of alkali-metal compounds to coal for the purpose of enhancing ash precipitation can be considered only in coals of exceptionally low-alkali fuels—that is, when the sum of $Na_2O + K_2O$ in ash is below 2 percent. There is always a risk, even with low-alkali coals, that the flame–volatile sodium and potassium compounds added to coal may cause some boiler fouling (Chaps. 10 and 15) and high-temperature corrosion (Chap. 17). The addition of carbonates is preferable to that of sulfate or chlorides. There is a risk of fireside corrosion when alkali-metal sulfate or chloride is injected into the flame.

Iron compounds also may be considered as ash precipitation aids. The results in Fig. 16.2 (curve C) and Fig. 16.5 (curve B) show that iron-containing silicates (chlorite and illite) had comparatively low resistivity after being heated at 1800 K. Further, Figs. 16.3 and 16.4 show that 1 percent iron oxide distributed on the surface of quartz and kyanite decreased the resistivity from 10^{12} to below 10^8 Ω m. There is more than 1 percent iron dioxide in all coal ashes, but much of it is present as discrete particles of magnetite and is not distributed on the surface of high-resistivity silicate particles. A sufficient quantity of iron together with alkalis could be recovered from ash in an acid leach process, but there would be a risk of enhanced boiler fouling and slagging. Thus, from ash-recovered iron and alkali additives can be considered only with coals low in iron and alkalis.

Injection of a calcium compound into the flame or adding it to coal is unlikely to be effective in enhancing ash precipitation. The results, plotted in Figs. 16.3 and 16.4, show that in contrast to iron oxide, calcium oxide distributed on the surface of high resistivity silicate particles had no appreciable effect in reducing the resistivity. This is in accord with the results of dolomite injection tests carried out in Australia (CSIRO, 1969), where the results showed no improvement with the additive. During the flue-gas desulfurization trials with dolomite reported by Walker (1971) it was observed that the resistivity of ash increased from 10^6 to 10^{10} Ω m, and the collection efficiency of the precipitators decreased from 78.3 percent without additive to 55 percent with the additive. This experience showed that large quantities of alkaline-earth compounds, which capture the SO_3 in the flue gas, can strikingly reduce the efficiency of the electrical precipitators.

Following the successful applications of magnesium additives, chiefly oxide and hydroxide, in oil-fired boilers it has been suggested that the same compounds can cure many ash-related problems in coal-fired boilers, e.g., improving the performance of ash precipitators. There are no unequivocal experimental data to show that a magnesium additive will improve the performance of the precipitators in coal-fired boilers burning low-sulfur coals. Magnesium as an additive could further reduce the low amount of SO_3 in the flue gas, thus impairing the precipitator performance. A magnesium additive may be beneficial with high-sulfur coals to prevent the condensation of sulfuric acid in the precipitators, as discussed by Bennet and Kober (1978).

The results reported by Radway and Rohrbach (1976) seem to contradict the premise that magnesium oxide or hydroxide is unlikely to reduce the resistivity of ash. They found that MgO additive in a cyclone-fired boiler decreased the ash resistivity from 4×10^9 to 7×10^8 Ω m. The decrease in resistivity is, however, comparatively small, and it may have been a result of magnesium oxide reacting with silicate ash at the high cyclone boiler temperatures, thus releasing alkali metals, chiefly potassium from the slag, which were captured by the flue-gas-borne ash on cooling.

The experience with additives to improve the performance of electrical precipitators in coal fired boilers can be summarized as follows:

1. The precipitator performance can be severely impaired when the surface resistivity of particulate ash at precipitation temperatures is above 10^8 Ω m. Trace amounts of acidic and alkaline additives when adsorbed or deposited on the surface of ash can reduce the resistivity by 2 to 3 orders of magnitude.
2. The most commonly used additive for this purpose is sulfur trioxide (SO_3). The additive may be injected as neat SO_3, or generated from sulfuric acid or by catalytic oxidation of SO_2.
3. Solutions of ammonium sulfate and sulfamic acid are frequently used, which decompose to SO_3 and ammonia in the flue gas. Despite the higher cost, the solution additives are preferred because they are more easily handled, but they may cause fouling by sticky deposits.
4. The effective dose rate of SO_3 in the flue gas depends on the ash composition and the sulfur content of coal, but is usually between 5 and 30 ppm. At these levels, SO_3 is quantitatively adsorbed by the flue-gas-borne ash and there should be no significant increase in the SO_3 content of chimney-emitted flue gas.
5. Ammonia may be injected directly into the flue gas or it can be generated from thermal decomposition of sulfamic acid and triethylamine. The effective dose rates were found to be between 10 and 50 ppm NH_3.
6. Solid additives are less extensively used as ash precipitation aids. The experimental work has shown that alkali-metal compounds can be used with some low-sodium coals. Iron compounds may also have some use, but calcium compounds do not enhance ash precipitation. There are doubts that magnesium compounds can increase the precipitator performance with high-resistivity ashes.

Locklin et al. (1980), in their survey on additives, have commented that surprisingly only about one-third of the successful trials resulted in continued use of additives. In some cases different coals were used and additives were not needed to reduce the resistivity of ash. In other cases the precipitator plant was improved in order to meet the required ash catch efficiency. This illustrates clearly the reluctance by power-station engineers to use any kind of additives; they place their trust in improving the engineering hardware rather than on "magic" chemicals. In any case, whatever additive they may choose to use, someone is bound to object on environmental grounds or for other reasons.

16.4 DEVITRIFICATION–ENHANCING ADDITIVES TO COMBAT BOILER SLAGGING

The flame-fused particulate ash is partially devitrified on cooling and the deposit-forming propensity is markedly influenced by the ratio of glass to crystalline material of the silicate ash (Fig. 12.2). The glassy fraction may have a viscosity in the range of 10^6 to 10^{10} N s m^{-2} at deposit-forming temperatures, and the rate of ash sintering and slagging is governed by the viscosity parameter, as discussed in Chaps. 9 and 10. In contrast, viscosity of the crystalline material is above 10^{13} N s m^{-2}; thus it does not take part in the coalescence of deposited particles by viscous flow, which is the dominant mechanism in the formation of boiler deposits. The deposit-forming characteristics of silicate ash could be summarized as follows:

$$R_d = a \left(\frac{x_g}{1 - x_g} \right)^b \tag{16.5}$$

where R_d = rate of deposit formation
x_g, $1 - x_g$ = the glassy and crystalline fractions of the ash, respectively
a = constant
b = rate index

Watt and Thorne (1965) have shown that the glassy fraction of typical coal (bituminous) ashes from pulverized-fuel-fired boilers was between 71 and 88 percent by weight. They estimated the glass content by a difference method:

$$G = 100 - (X + C + S) \tag{16.6}$$

where G, X, C, and S denote, respectively, the glassy material, crystalline species, and carbon and sulfate contents of the ash. The crystalline species, chiefly mullite, quartz, and spinels, are usually determined by X-ray diffraction. It is probable that the amount of the crystalline fraction thus determined is low because much of the quartz and mullite species is embedded in a glassy matrix of silicate ash particles and may not be detected by the X-ray diffraction technique. As a result, the glass content of ash as given by Eq. 16.6 is high. Figure 12.2 shows the quartz microcrystals and mullite needle skeletons in typical silicate ash particles after the glassy material had been dissolved in a dilute hydrofluoric acid solution, as discussed by Hulett and Weinberger (1980) and Raask (1981a).

From the point of view of sintering and slagging it is important to discern whether the surface layer of deposited ash particles on boiler tubes consists largely of glassy material of comparatively low viscosity or of cluster and needles of nonsintering microcrystals. The appearance of the crystalline skeleton shown in Fig. 12.2a suggests that a glassy material constituted the surface layer, which had dissolved in hydrofluoric acid solution, revealing the center packed with microcrystals. In contrast, the surface of the particle shown in Fig. 12.2b consisted largely of crystalline needles, and a glassy material constituted the interior. The premise that ash sintering is greatly enhanced by a glassy material but is reduced by crystalline species at the surface particle offers a basis for the possible use of minute quantities of antislagging additives in the coal-fired boiler. The rate of ash sintering could be significantly reduced when an additive captured on flame-borne ash acts as a nucleating agent, resulting in the transformation of glass to crystalline species on the surface layer of silicate ash particles.

The advent of glass ceramics—e.g., microcrystalline material bonded in matrix of residual glass—has led to considerable interest in the nucleation and phase-separation phenomena in many glassy systems, as discussed by McMillan (1979). Transition metals and other metals of multivalency characteristics are added to melts to perform the function of nucleating the transition from the glass to crystalline phase. MacKenzie and Brown (1975) have shown that the additives can both reduce the temperature of glass devitrification and increase the rate of crystallization.

Copper is considered to be a nucleation-enhancing metal and its compounds have been used as coal additives to combat boiler slagging, as discussed by James and Fisher (1967), Muyl (1971), and Kiss et al. (1972). Originally a proprietary additive, a mixture of magnesite and copper oxychloride, was used, but subsequently the copper compound only was added to coal as it was considered to be the reactive compound in the proprietary mix. O'Connor (1970) has reported the results of using copper additive in peat-fired boilers. He has claimed that both copper oxychloride and sulfate were equally effective in preventing boiler slagging.

The amount of nucleating additives used in the manufacture of glass ceramics is around 1 or 2 percent by weight. This amount is far too high for use in coal-fired boilers. It would mean that 2–3 kg of an additive would be necessary to dose 1 t coal. The aim is to induce crystallization in the surface layer of silicate particles, and thus large quantities of the additive may not be necessary. Accordingly, the dosage rate of copper additive is usually selected to give the metal (Cu) concentration of ash a range of 5 to 45 ppm. In the late 1960s and the early 1970s several copper additive trials were carried out in small pulverized-coal- and stoker-fired boilers. Neat copper oxychloride was used, added to coal in the form of dry powder or as a water slurry. The results are summarized in Table 16.5, taken from the work published by Kiss et al. (1972).

It can be seen from Table 16.5 that the beneficial effects of copper oxychloride additive vary from a marginal effect to marked improvement. There was a general trend that the additive was more effective with high-iron ashes, 10–20 percent Fe_2O_3, as exemplified by the results of additive trials with two coals of different ash compositions given in Table 16.6.

Muyl (1971) has given a detailed account of the improvements in boiler operation and cost benefits that result from the use of copper oxychloride as an antislagging additive in a 250-MW pulverized-coal-fired boiler. The additive was injected in the form of a water slurry at rates of 2 and 4 ppm copper to coal over a period of 18 mo. It was observed that the strongly fused deposits on combustion chamber walls changed to friable sintered material, which could be much more easily removed by sootblowers. As a result the radiative heat transfer in the combustion chamber was markedly improved and the amount of water required to desuperheat the steam for controlling the outlet temperature was correspondingly reduced. The cost benefits with the additive accrued from less frequent reductions in boiler load, less frequent use of sootblowers, and reduced maintenance on clinker (furnace-bottom ash) crusher plant.

Table 16.5 The effect of copper oxychloride as an additive on boiler slagging

Type of boiler	Capacity (t steam/h)	Position of slagging difficulties	Additive dosage		Effect of additive
			ppm per coal	ppm Cu per ash	
Pulverized fuel	342	Furnace and screen tubes	1.1	4	*Marked*: ashing of furnace bottom, easier and less maintenance on ash crushers.
	113	Furnace tubes	1.2–2.3	4–7	*Moderate*: improved availability furnace ashing, easier and less maintenance on ash plant.
	136	Furnace and superheater tubes	1.1	4	*Marked*: improved availability furnace ashing, easier and less maintenance on ash plant.
	136	Burner ports, screen, and superheater tubes	4.6	20	*Marked*: less draught loss in superheater, less maintenance on ash plant and off-load cleaning of superheater easier.
Chain grate	83–165	Furnace tubes, bottom slopes	1.9	4	*Marked*: slagging of furnace bottom prevented.
	82	Steam-generating and super-heater tubes	1.1	7	*Moderate*: off-load cleaning easier.
	45	Lower side walls and rear tubes	2.8	15	*Marginal*: less deposit on furnace tubes.
	82	Side walls above grate	13.9	15	*Moderate*: less deposit on furnace tubes, less attack on refractory.

Firing		Component			Effect
	18–91	Furnace bottom	4.6	35	*Marked*: amount of dense slag on brickwork much less, prevents boiler shutdown.
	45	Side walls	5.5	25	*Moderate*: furnace slag and economizer deposit more friable.
	32–113	Side walls, generating, and superheater tubes	4.6	20	*Marked*: enables running of boilers on full load, off-load cleaning easier.
Pulverized fuel	136	Screen and superheater tubes	9.3–13.9	30–45	*Marginal*: not sufficient to warrant further use.
	250	Screen tubes	1.7–3.4	6–11	*Moderate*: soot-blowing frequency reduced and less maintenance on ash crushers.
	73	Screen and superheater tubes	5.5	15	*Moderate*: off-load cleaning easier and less maintenance on ash crushers.
Chain grate	159	Rear furnace wall	1.1	4	*Marginal*: not sufficient to warrant further use.
		Furnace superheater and airheater tubes	5.9	27	*None*.
Pulverized fuel	14–145	Furnace walls above burners	1.1–2.2	4–8	*Moderate*: off-load cleaning easier.
	386	Furnace walls and primary superheater	2.2	10	*Marked*: fouling of furnace and superheater was reduced, but economizer became blocked. Use of additive discontinued.
Chain grate	55	Furnace walls	1.1	8	*Marginal*.
	68–82	Furnace walls	5.5	40	*Marked*: boiler cleaning much easier.
	36–82	Furnace walls and superheater	8.0	27	*Moderate*: fused slag on sintered deposit; boiler cleaning easier.

Table 16.6 Composition of ashes, constituent oxides by weight percent: copper additive effective in A, not effective in B

	Ash	
Ash	A	B
SiO_2	52.6	49.3
Al_2O_3	22.7	25.5
Fe_2O_3	13.6	5.6
CaO	5.0	8.1
MgO	1.1	2.7
K_2O	1.4	3.4
Na_2O	0.8	2.0

Kiss et al. (1972) have carried out diagnostic examinations and tests on boiler slag and sintered deposit samples taken before and during the use of copper oxychloride as an additive. Microscopic examinations revealed that the crystalline species found in slag without the additive appeared to have been precipitated from a solution (Fig. 16.12a). This suggests that formation of the crystalline species had taken place after the deposited ash particles had formed a slag matrix. In contrast, it appeared that the additive had induced some crystallization in ash particles before the formation of slag deposit. That is, spherical "ghost" structures of the devitrified

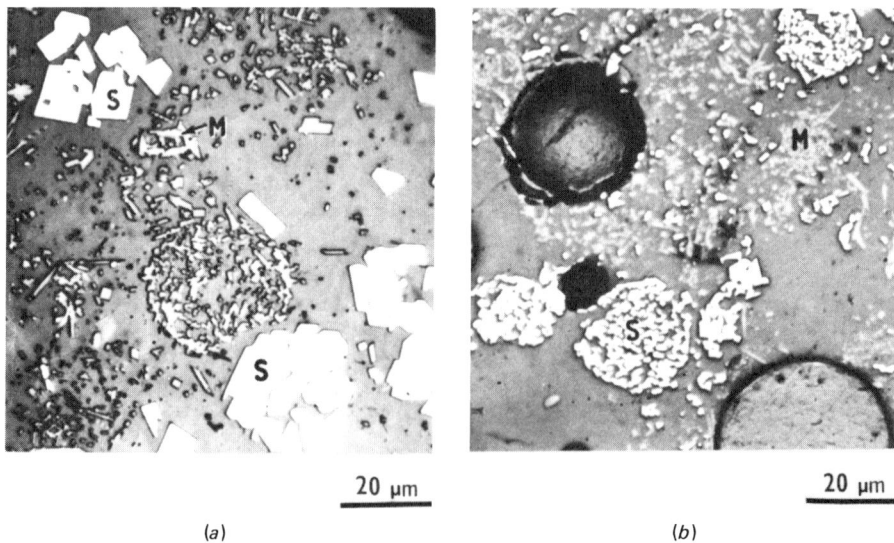

20 μm

(a)

20 μm

(b)

Figure 16.12 Crystalline species in furnace-wall deposits, M = mullite, S = spinel. (a) Nucleation from solution without additive. (b) Nucleation at the surface of ash particles with copper.

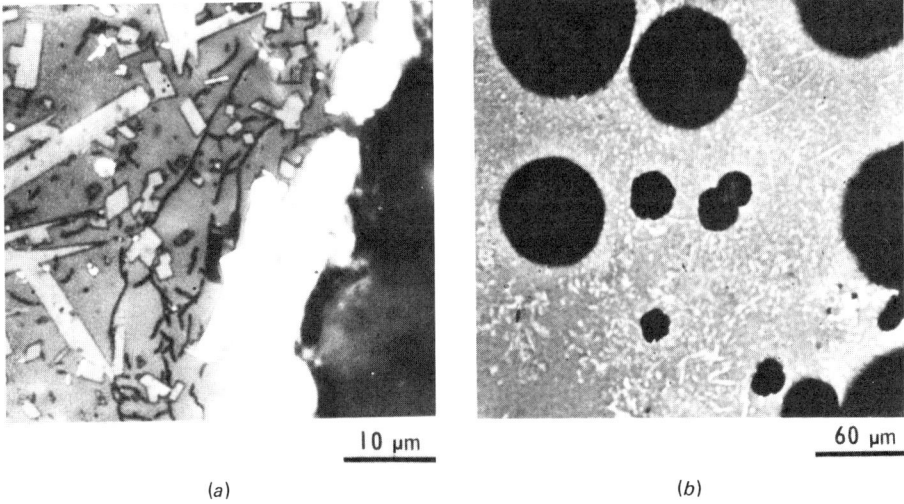

10 μm

(a)

60 μm

(b)

Figure 16.13 Furnace-wall deposits doped with copper additive. (*a*) Intergranular cracks. (*b*) Gas bubbles in glassy crystalline matrix.

ash particles were discernable, as evident in Fig. 16.12*b*. Also, there were some intergranular cracks in the residual glass of doped slag, as shown in Fig. 16.13*a*. It was further observed that the copper-doped slag had a large number of voids, 20–30 μm in diameter, as shown in Fig. 16.13*b*. The microprobe analysis showed that there was a high concentration of sulfur in the slag around the cavities.

The microscopic observations that the copper-doped deposits had a high degree of crystallization were confirmed with the results of viscosity measurements by the needle penetration method (Chap. 9) and sintering-rate measurements by the neck growth technique (Chap. 10). Figure 16.14 shows that viscosity of the additive-doped slag was about an order of magnitude higher than that of undoped slag in the

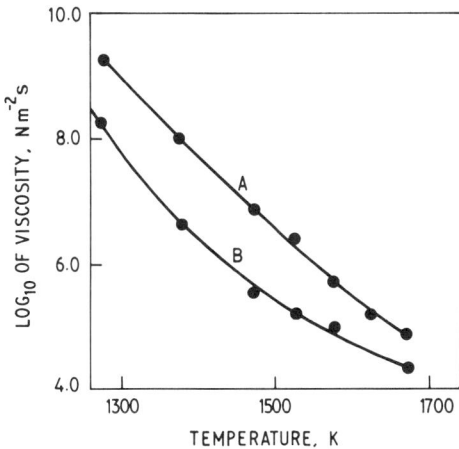

Figure 16.14 The effect of copper additive on slag viscosity measured by the needle penetration method: A, doped slag; B, undoped slag.

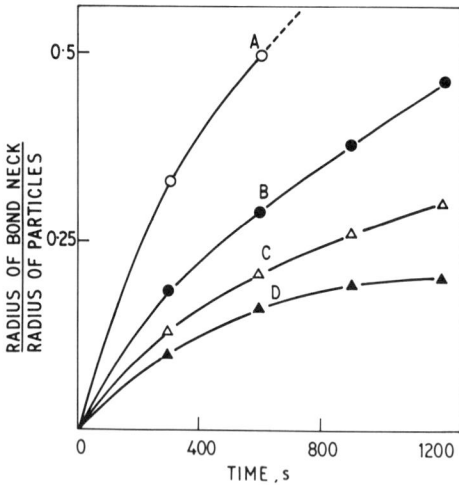

Figure 16.15 The effect of copper additive on sintering of spherical particles: A, undoped slag at 1425 K; B, doped slag at 1425 K; C, undoped slag at 1375 K; D, doped slag at 1375 K.

temperature range of 1300–1700 K. There was a corresponding difference in the rate of sintering of the two slag deposit samples, as shown in Fig. 16.15.

Based on these experimental results, Kiss et al. (1972) have discussed the possible mechanisms where copper additives could be effective as nucleating agents for crystallization. Copper oxychloride is partially volatilized in coal flame in the form of chloride, as discussed by Kiss and Lloyd (1975); subsequently, copper is captured and dissolved in the surface layer of silicate ash particles. Cuprous oxide (Cu_2O) dissolved in a silicate melt can be reduced to metallic copper by ferrous oxide:

$$2FeO + Cu_2O \rightarrow Fe_2O_3 + 2Cu \qquad (16.7)$$

The free energy change (ΔG_o) of the reaction is −131 kJ at 1240 K and −184 kJ at 1650 K, calculated from the thermodynamic data published by Elliott and Gleiser (1960). Richardson and Billington (1956) and Johnston and Chelko (1966) have shown that copper has a low solubility in silicate metals, so the precipitated metal particles could initiate crystallization.

Alternatively, cuprous oxide could be reduced to metal by sulfide dissolved in silicate:

$$\bar{S} + 2Cu_2O \rightarrow 4Cu + SO_2 \qquad (16.8)$$

where $\bar{S}$ denotes the sulfide dissolved in the slag. The solubility of SO_2 (sulfate) in a silicate slag is low (Fincham and Richardson, 1954), resulting in the formation of gas bubbles 20–80 μm in diameter, as shown in Fig. 16.13*b*.

The possible effects initiated by the copper additive that may lead to lowering the cohesive strength of boiler deposits can be summarized as follows:

1. The additive acts as a nucleating agent for the formation of crystalline species, so the rate of ash sintering and slagging is reduced.
2. It initiates the growth of gas bubbles in the slag.

3. It may cause crack propagation from the voids in the deposit matrix, as shown in Fig. 16.13a. Rieke and Steinbock (1933) have shown that Cu_2O in silicate glasses lowers the coefficient of thermal expansion. Thus an uneven distribution of copper on a microscale may induce crack propagation and produce a friable slag deposit.

Kiss et al. (1972) were unable to confirm that copper induced a nucleating mechanism by electron microprobe analysis for the additive metal. The detection limit for copper was around 500 ppm, which was far too insensitive to show up any enrichment of the metal or the surface layer of ash particles or in the walls of cavities in the slags, nor were Kiss et al. able to define closely the conditions—e.g., ash composition, temperature, and the oxygen content of flue gas—for the additive to be most effective. The difficulty was that most coal ashes, without the additive, contain over 100 ppm copper (O'Gorman and Walker, 1972; Brown et al., 1976). That is, the amount of copper in oxychloride additive is small compared with the quantity already present in ash.

Bethell (1963), in his survey on coal trace elements, has shown that copper is frequently associated with iron pyrites (FeS_2) in the form of chalcopyrite, copper pyrites. About 1 percent of iron in coal FeS_2 is substituted by copper in the form of double sulfide, $CuFeS_2$. It can be argued that copper associated with iron pyrites becomes trapped in iron oxide particles that are formed on oxidation of coal pyrites in boiler flame. In contrast, copper in the oxychloride additive is largely distributed on the surface of silicate ash particles, as suggested by Kiss et al. (1972).

The use of copper oxychloride as an antislagging additive has declined in Britain since early 1970s with closure of small coal-fired power stations where it was applied. Since then some trials have been made with the additive in large, 500-MW pulverized-coal-fired boilers with inconclusive results. There is one large station where the antislagging additive has been used with some success when the problem of boiler slagging could not be controlled by other measures discussed in Chap. 14.

16.5 USE OF ALKALINE-EARTH ADDITIVES IN CYCLONE-FIRED AND PULVERIZED-COAL-FIRED BOILERS

The slagging characteristics of coal ashes (Chaps. 9 and 10) can be summarized by an expression related to the ratio of fluxing constituents, chiefly Na_2O, CaO, and FeO, to the infusible fraction, chiefly SiO_2 and Al_2O_3:

$$R_d = a \left(\frac{y_f}{1 - y_f} \right)^b \tag{16.9}$$

where R_d = rate of slagging
$y_f, 1 - y_f$ = fluxing and infusible fractions of ash, respectively
a = constant
b = rate index

The above relationship should hold for the majority of bituminous coal ashes, but it does not apply for the sub-bituminous coal ashes high in calcium and low in silica.

Discussions in Chaps. 9 and 10 have shown that the viscosity of coal-ash slag is the rate-controlling parameter both in the formation of sintered ash and slag deposits and in the flow of molten slag. It is evident that for an additive to be effective in decreasing the viscosity and thus increasing the rate of slag flow it must increase the ratio of flux to infusible constituents, as defined by Eq. (16.9).

The fluxing, slag viscosity-reducing additives are occasionally used in cyclone and slag-tap boilers in order to maintain the flow of molten slag, i.e., to keep the slag viscosity below 25 N m^2 s^{-1}. Limestone is the usual fluxing material, and it is added to coal and mixing takes place on milling. Locklin et al. (1980) cited one case where the limestone additive was necessary to reduce the viscosity of slag when low-sulfur coals were burned in cyclone-fired boilers. The required dosage rate of additive was high, between 6 and 10 percent of the coal.

Iron ore, fluorspar (CaF$_2$), and boron-containing compounds would be more effective fluxing agents than limestone; fluorspar has been widely used in the metallurgical industry to reduce slag viscosity. Bennett and Kukin (1977) have described a proprietary additive based on a borate mineral containing sodium and calcium. They had claimed that the borate additive is highly efficient in reducing the coal ash viscosity in cyclone boilers, but the material is expensive.

The use of fluorine and boron fluxing additives cannot be recommended on grounds of the possible environmental impact. Fluospar, CaF$_2$, will react with silica to form volatile silicon fluoride, SiF$_4$, which may subsequently be emitted from power-station chimneys. Any increase in the emission of potentially toxic fluorine compounds is highly undesirable, and this rules out the use of fluorspar as a fluxing agent in cyclone-fired boilers. Boron is less volatile, but high concentrations of boron in ash may cause toxic symptoms to plant life when boiler slag and ash are used for land reclamation. The boron content of coal ash without the additives is already comparatively high, above 100 ppm, as discussed by Brown et al. (1976) and Furr et al. (1977), and the use of boron-containing additives would increase the risk of damage to plant life.

The use of fluxing additives in cyclone-fired boilers can be regarded as an emergency measure, applied only when the quality of coal—i.e., the ash fusion characteristics—show a marked change. Also, when a fluxing additive is required, a calcium compound, limestone or chalk, will be the first choice. Iron ore is comparatively cheap and would be highly effective, but iron-rich coal-ash slag is notoriously reactive. It can attack refractory lining materials in cyclone-fired boilers, and it can be reduced to liquid iron under unstable firing conditions.

Calcium and magnesium additives have been used spasmodically in pulverized-coal-fired boilers with the object of reducing fouling of the furnace-wall and superheater tubes. Michel and Wilcoxson (1955) and Barnhart and Williams (1956) claimed some success with magnesium oxide as an additive in reducing boiler fouling. The quantity of additive was rather large, between 0.4 and 3 percent by weight of coal, and the method of combating boiler fouling was considered to be uneconomical.

Doxey et al. (1967) have investigated the effect of adding calcium chloride on boiler fouling. The chloride was added to wet coal to prevent freezing during transport of fuel to power stations in frosty weather. Doxey et al. observed that with the additive, boiler deposits were more friable and easily removed by sootblowing, and this effect was noticeable with as little as 100 ppm $CaCl_2$ added to coal. There was no significant enrichment of calcium in the boiler deposit, but there was some evidence that the additive reduced the SO_3 content in the flue gas. A similar "scavenging" effect in lowering the amount of SO_3 in the flue gas has been claimed for magnesium additives by Radway and Boyce (1978). They emphasize that the additive must have a small particle size, 0.5–5 μm, since its effectiveness is proportional to the specific surface. However, the reactivity of MgO decreases above 1200 K, as found by Goldberg and Kittle (1979), who have reported that magnesium oxide as an additive undergoes a sharp reduction in the "reactive" surface at boiler-flame temperatures due to the formation of nonporous crystalloids. The implication is that resultant "dead-burned" particles have a much diminished reactivity for combining with SO_3 in the flue gas.

Dixit and Cuisia (1977) have suggested that magnesium additives containing alumina are effective in reducing boiler fouling. It was claimed that these two compounds form a spinel, $MgAl_2O_4$, which makes the deposit friable and easily removed by sootblowing. There is no unequivocal experimental evidence for this claim. Dixit (1978) has proposed a "friability" index of ash deposits of boiler tubes to explain the effect of magnesium additive:

$$I_f = \left(\frac{B_a}{A_a}\right) S_{sul} \tag{16.10}$$

where I_f = friability index

(B_a/A_a) = ratio of basic oxide to acidic oxide constituents of deposits (Chap. 10)

S_{sul} = sulfate content of deposit

It has been suggested that magnesium as an additive decreases boiler fouling by reducing the sulfate content of deposit, but this view has not been universally accepted.

There have been several specific studies made to investigate the effect of calcium and magneisum compounds when added to sub-bituminous coals of high sodium content. These coal ashes are highly basic, and according to the fusibility/temperature curve in Fig. 10.18, further addition of calcium or magnesium oxide should increase the fusion temperature of ash deposit. Anderson et al. (1977) have found that boiler fouling with high-sodium, low-silica lignite fuel was significantly reduced by lime sludge and also by dolomite addition. The additives, however, increased the rate of boiler fouling when the lignite fuel contained clay partings as a result of reaction with the silicious material in fuel. This is in agreement with the results published by Szulakowski (1968), who found that addition of finely ground silica to low-silica coal ashes markedly reduced the fusibility temperature. Sondreal et al. (1978) have found that calcium and magnesium additives had little effect on the rate of ash deposition, but the strength of adhesion of deposit matrix on collecting target was significantly reduced. The authors have also tested aluminum compounds to establish whether sodium in ash would react with added alumina to tie up sodium in a harmless form. They found that the additive reduced neither the amount nor the adhesive strength of the deposits.

The use of calcium and magnesium additives to prevent excessive boiler fouling in bituminous-coal-fired boilers has been viewed with much skepticism by many power-station engineers. This skepticism is justified in the absence of conclusive results and a scientific premise that would explain how a fluxing additive can increase the ash fusion temperature. An exception is the basic ash of some sub-bituminous fuels, where the roles of calcium or magnesium oxides and silica are reversed. That is, calcium and magnesium oxides constitute the infusible fraction and silica can be regarded as a fluxing agent, as discussed in Chap. 10. With silica-rich ashes, it is to be expected that introduction of a basic oxide additive (calcium and magnesium compounds) would lower the ash fusion temperature and hence increase the rate of boiler fouling. The only slender evidence is that the basic additives can reduce the SO_3 content of the flue gas, so that the formation of "sticky" sulfate constituents in deposited ash is prevented. The same premise applies for the use of the basic oxide additives to prevent high-temperature corrosion, which is discussed in Chap. 17.

16.6 INTERMITTENT APPLICATION OF ADDITIVES

In most applications it is essential to achieve a uniform distribution of additive in the flue gas. For example, the efficacy of additives applied in oil-fired boilers to reduce the SO_3 content of chimney emission depends largely on the degree of mixing of these chemicals with the flue gas. Similarly, SO_3 injected in coal-fired boilers to improve the performance of the electrical precipitators has to be well dispersed in the flue gas.

In contrast, there are two possible methods of applying the slagging enhancing additives in the cyclone-fired boiler and the slagging-preventing additives in pulverized-coal-fired boilers. An additive can be mixed with coal or may be injected into the flue gas continuously, or it may be added intermittently. The attraction of the latter method is that for short periods a high ratio of additive to coal ash will be achieved that would markedly change the composition of ash deposited on boiler tubes. This is a practice in many power stations operating cyclone-fired and slag-tap boilers. They keep a stock of limestone additive for use in an emergency when a large amount of slag builds up in the cyclone or slag-tap furnace.

The concept of intermittent application of additives may also be appropriate for pulverized-coal-fired boilers, in this case to prevent boiler fouling and slagging. In pulverized-coal-fired boilers, 75–85 percent of ash is carried through the system by flue gas and is captured in the electrical precipitators. It is thus evident that only 15–25 percent of the ash takes part in the formation of boiler deposit. As a result most of the additive which is mixed with coal or injected in the flame continuously is carried through the system and is thus "wasted." It would therefore be logical to consider the possibility of applying additives directly to the areas of furnace walls and superheater section where the build-up of large deposit masses occur as indicated in Fig. 14.3.

Clark (1966) has reported on the successful results of intermittent application of magnesium oxide additive on furnace-wall tubes in recovery boilers burning black liquor. This combustible material contains up to 40 percent ash with low melting point, around 1000 K. Without the additive, hand lancing was required and it was difficult to control the temperature of superheated steam. A slurry of magnesium oxide was sprayed on boiler tubes using retractable sootblowers. The operation problems were thus cured and the availability was much improved. It was suggested that a layer of magneisum oxide at the tube surface–ash deposit interface constituted a weak nonbonding zone that prevented adherence of the low-melting ash to boiler tubes.

Of course, the firing conditions and ash deposit are entirely different in large pulverized-coal-fired boilers, and it would be an extensive task to cover the boiler tubes with a layer of additive. This approach is, however, more attractive than the attempts to modify the deposit-forming propensity of entire ash. The idea is to introduce material at the tube surface–ash deposit interface that constitutes a nonadhesive layer. Materials that may be considered for this purpose include magnesium, aluminum and titanium oxides, and silica. Obviously the nonadhesive characteristics of an additive would be maintained longest at low temperatures, and thus it should be applied on clean boiler tubes.

Alternatively, more reactive chemicals—e.g., nucleating agents for crystallization of ash silicates—could be introduced. The results of some preliminary experiments on bond strength measurements (Raask, 1973) showed that nucleating additives such as copper compounds can reduce the adhesive bond between the oxidized steel specimen and the low-melting fraction of boiler deposit.

Application of antislagging additives directly to furnace-wall and superheater tubes in large coal-fired boilers would require a new technology where the injection of chemicals in the form of solution, slurry, or powder needs to be carefully controlled. Water and air sootblowers are, when appropriately modified, a possible way of injecting the additives to boiler tubes. Of course, it would be necessary to show that the additive will not enhance high-temperature corrosion and that there will be no tube erosion as a result of particulate or slurry additive impaction.

NOMENCLATURE

A	collector surface area
A_a	acidic oxide content of deposits
a	constant
B_a	basic oxide content of deposits
b	slagging rate index
C	carbon content of ash
c	additive concentration
E	collection efficiency

G	glass content of ash
I_f	deposit friability index
k_1, k_2	reaction rate constants
n	power index
R	ash resistivity with additive
R_o	ash resistivity without additive
R_s	saturation resistivity of ash with additive
R_d	rate of deposit formation (slagging)
S	sulfate content of fly ash
S_{sul}	sulfate content of deposit
V_g	gas volume
v	particle migration velocity
X	crystalline materials content of ash
x_g	glass fraction of silicate ash
$(1 - x_g)$	crystalline fraction of silicate ash
y_f	flux fraction of ash
$(1 - y_f)$	infusible fraction of ash

HIGH-TEMPERATURE CORROSION
IN COAL-FIRED PLANTS

17.1 THE FUEL IMPURITY ENVIRONMENT
AND TEMPERATURE REGIME

There exists extensive literature on the subject of high-temperature corrosion in fossil-fuel-fired systems. It was therefore inevitable that in condensing the subject matter in a single chapter much of the valuable material had to be left out. The topic is known also as external corrosion, fireside corrosion, or gas-side corrosion. Formerly some boiler operators referred to furnace-wall corrosion as flame erosion, but the term is no longer used. The high-temperature corrosion discussed here can be defined as a process of tube-metal wastage enhanced by nonsilicate impurity species that volatilize and deposit in the coal-fired plant. It is distinct from low-temperature corrosion caused by condensation of sulfuric acid (discussed in Chap. 19) and from the internal, or water and steam-side, corrosion, which is outside the scope of this book.

In the presence of gaseous oxidants, oxygen, carbon dioxide, and water vapor, the rate of oxidation of boiler-tube metals,

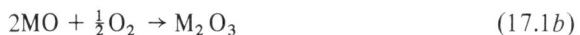

$$M + \tfrac{1}{2}O_2 \rightarrow MO \tag{17.1a}$$

$$2MO + \tfrac{1}{2}O_2 \rightarrow M_2O_3 \tag{17.1b}$$

313

depends largely on the protective characteristics of the oxide layer thus formed. In the presence of corrosive constituents of boiler deposits and the flue gas, chiefly sulfur and chlorine compounds, the rate of metal wastage can be significantly higher than that of oxidation in the atmosphere of gaseous oxidants. The corrosion-enhancing reactions include:

Enhanced oxidation	$M + [SO_3] \rightarrow MO + SO_2$	$(17.2a)$
Destruction of the	$M_2O_3 + 3\ [SO_3] \rightarrow M_2(SO_4)_3$	$(17.2b)$
protective oxide layer by molten sulfate and		
by reducing gas	$M_2O_3 + [CO] \rightarrow 2MO + CO_2$	$(17.2c)$
Sulfidation	$M + [S] \rightarrow MS$	$(17.2d)$
Chloridation	$M + 2[Cl] \rightarrow MCl_2$	$(17.2e)$

where $[SO_3]$, $[S]$, $[Cl]$, and $[CO]$ denote, respectively, the concentration of sulfate, sulfide, chloride, and reducing species in boiler deposits and in the flue gas.

The temperature regime of boiler tubes is governed by the Rankine cycle requirements for maximum possible efficiency of the power generation plant. Figure 17.1 shows the relationship between the plant efficiency and the superheater steam temperature and pressure. The layout of steam generating, superheater, and reheater tubes in a coal-fired boiler is dictated by the high-temperature corrosion and by boiler slagging: the latter is discussed in Chaps. 9–14. Table 11.1 showed the metal surface temperature of steam-generating tubes in the combustion chamber, and of the superheater and reheater tubes with the corresponding environment temperature, i.e., that of the flame or flue gas.

In Chap. 12 it was shown that the rate of boiler fouling and slagging depends on the temperature of the deposit-collecting surface and on the exposure tempera-

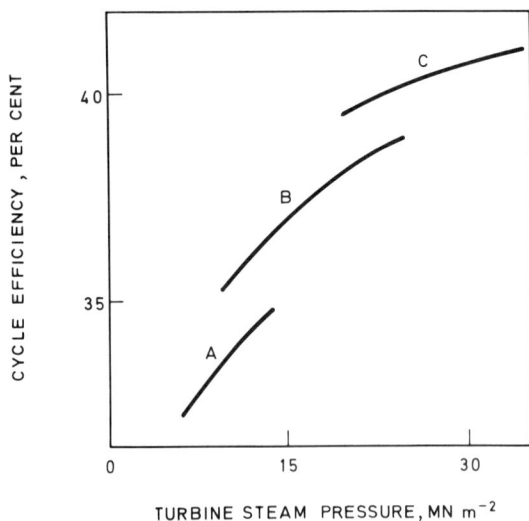

Figure 17.1 Rankine cycle efficiency at different steam conditions: A, without reheat at 803 K; B, single reheat at 803/803 K; C, double reheat at 873/833/833 K.

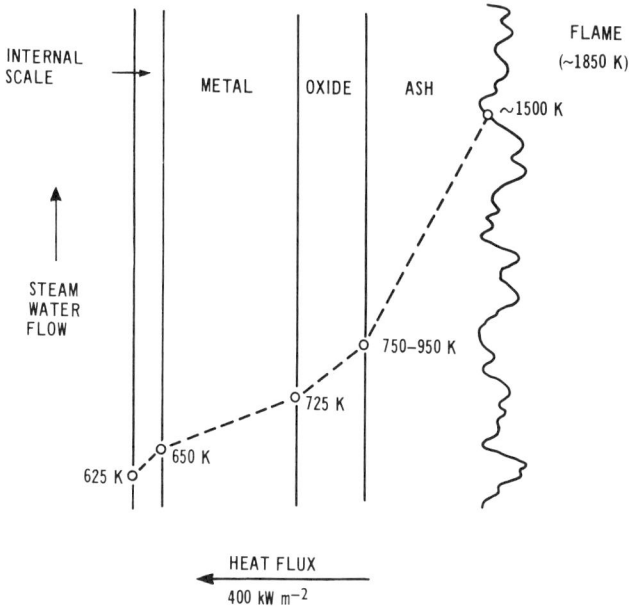

Figure 17.2 Thermal gradient across furnace-wall tube and deposit.

ture, i.e., on the temperature gradient across the deposit layer. Similarly, it is to be expected that the rate of corrosion of boiler tubes is strongly influenced by the temperature gradient across the metal oxide and deposit layers, as depicted in Figs. 17.2 and 17.3; thus the corrosion-defining expression is:

$$C_r = f(T_m, \Delta T) \tag{17.3}$$

where C_r = rate of corrosion
T_m = metal temperature
ΔT = the temperature difference $T_e - T_m$
T_e = exposure (i.e., flame or flue gas) temperature

A high-temperature gradient across the protective oxide layer may increase its porosity and could reduce the adhesive bond between the oxide scale and parent metal. As a result of the temperature gradient across the deposit there is enhanced diffusion of the condensible species of flame-volatilized impurities toward the tube surface. There are thus four parameters that largely determine the rate of high-temperature corrosion of boiler tubes:

1. Composition of the protective oxide on metal, M_{ox}
2. Metal temperature, T_m
3. Exposure (i.e., flame or flue gas) temperature, T_e
4. Composition of the flame-volatile ash deposits, D_c

These can be summarized in an expression defining the corrosion rate R_c:

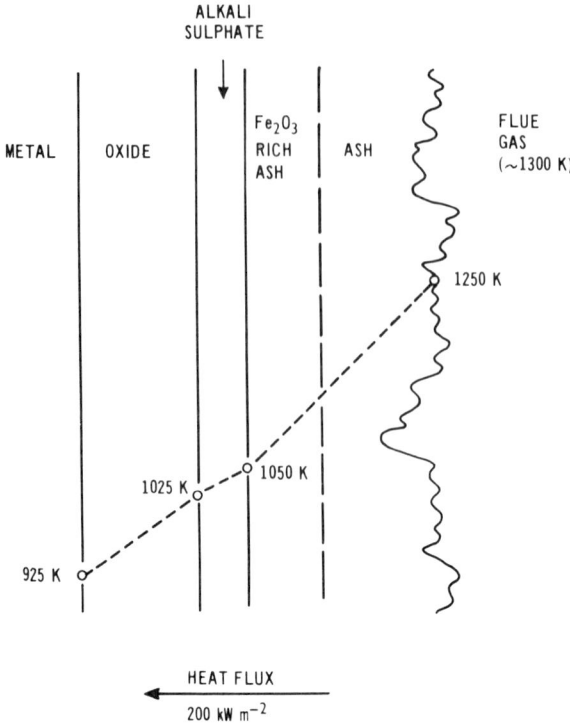

Figure 17.3 Thermal gradient across superheater tubes and ash deposit.

$$R_c = f(M_{\mathrm{ox}}, T_m, T_e, D_c) \qquad (17.4)$$

This approach will be used in the following sections to discuss the formation of aggressive deposits and the mechanism of corrosion of boiler tubes.

17.2 FORMATION OF CORROSIVE DEPOSITS

The formation of corrosive deposits depends largely on the reactions of nonsilicate impurity species in the flame and their subsequent deposition on cooled boiler tubes. The volatilization and flame reactions of the chlorine and sulfur compounds have been discussed in Chap. 7, and the formation of layer-structured boiler deposits have been elaborated on in Chap. 12. The principal findings relevant to the formation of potentially corrosive deposits are:

1. Volatilization of sodium and release of potassium from silicates. The sodium fraction associated with coal chlorine (water-soluble sodium) is rapidly volatilized in the flame, where it is in part captured by the fused silicate particles; the remainder is converted to sulfate. Potassium in coal is present as nonvolatile aluminosilicates (muscovites and illites) but as a result of an exchange reaction with volatilized sodium up to 40 percent of the silicate potassium is released, as

discussed in Chap. 6. Potassium thus released and the residual volatilized sodium are sulfated (Chap. 7), and the initial deposit on boiler tubes contains sodium and potassium sulfates in an approximate molar ratio of 2 to 1.

2. Thermal stability of sulfate and silicate phases. Molten alkali-metal sulfates and silicates separate to two immiscible phases, as discussed in Chap. 7. The sulfate phase is thermodynamically stable at temperatures below 1300 K in the presence of SO_2 and SO_3 in the flue gas, as illustrated in Fig. 7.7 for potassium sulfate. In boiler deposits the alkali metals, in particular potassium, can be transferred from the silicate to sulfate phase, leading to the formation of a complex system of sulfates as discussed subsequently (chiefly Eq. 17.10).

3. Sulfate "dewpoint" and the deposition rates. The dewpoint temperature for deposition of sodium and potassium sulfates is around 1150 K, as shown in Fig. 12.10, and the rate of sulfate deposition depends on the amount of volatile (water-soluble) sodium associated with coal chlorine, as evident in Fig. 12.8. The sulfates are transported from the flue gas to cooled boiler tubes by the mechanism of vapor-phase diffusion, which is more efficient than the inertial impaction of silicate ash particles. As a result the initial deposit material on boiler tubes is enriched in sulfate, as shown by the analysis of probe deposit depicted in Fig. 12.19.

Measured rates of sulfate deposited on a cooled metal probe in coal-fired boilers were between 0.15 and 3.5 mg m^{-2} s^{-1}. This means that the formation of a layer of sulfate deposit 1 mm thick, frequently found in boiler deposit next to the tube surface, would require a minimum of 2000 to 4600 h of continuous operation. In practice, the rate of sulfate and silicate ash capture can be significantly higher as a result of formation of a molten or fused layer of deposit of a high catch efficiency on boiler tubes, as discussed in Chap. 12. Silicate ash deposit falls off or is periodically removed by sootblowing, but some sulfate diffuses through the ash deposit (Figs. 17.2 and 17.3) and remains attached to the surface of cooled boiler tubes.

The sulfate mix in the initial deposit on boiler tubes consists of neutral salts of sodium (Na_2SO_4), potassium (K_2SO_4), and calcium ($CaSO_4$). That is, the complex sulfates [e.g., $K_3Fe(SO_4)_3$] and pyrosulfates (e.g., $Na_2S_2O_7$) are thermodynamically unstable above 1000 K and thus cannot form in the flue gas. The molten sulfate phases frequently found in boiler deposits must therefore form *in situ* as a result of absorption of SO_3 from the flue gas and dissolution of transit element oxide, chiefly those of iron and aluminum. There has been much discussion as to whether significant quantities of sodium and potassium pyrosulfates form in boiler deposits:

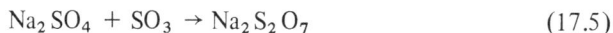

$$Na_2SO_4 + SO_3 \rightarrow Na_2S_2O_7 \tag{17.5}$$

$$K_2SO_4 + SO_3 \rightarrow K_2S_2O_7 \tag{17.6}$$

Kirsch (1963) and Crossley (1967) have suggested that pyrosulfates, in particular potassium pyrosulfate ($K_2S_2O_7$), may play a significant role in corrosion of combustion-chamber walls where the metal temperature does not exceed 770 K. Coats et al. (1968) have studied the thermal stability of the solid and liquid phases of pyrosulfates in an SO_3-containing atmosphere. The results showed that a minimum concentration of 150 ppm gaseous SO_3 was necessary for the formation of molten $K_2 S_2 O_7$ at 680 K, whereas sodium pyrosulfate ($Na_2 S_2 O_7$) required a much higher concentration of SO_3 (about 2000 ppm) for the molten phase to form.

Reid (1971) has reviewed the arguments concerning the possible role of pyrosulfates in high-temperature corrosion of furnace-wall and superheater tubes and concluded that there is no direct evidence for the formation of $K_2 S_2 O_7$ or $Na_2 S_2 O_7$ in boiler deposits. Thus more emphasis has been placed on the formation of molten sulfates with absorption of SO_3 and dissolution of transition metal oxides:

$$3Na_2 SO_4 + Fe_2 O_3 + 3SO_3 \rightarrow 2Na_3 Fe(SO_4)_3 \qquad (17.7)$$

$$3K_2 SO_4 + Fe_2 O_3 + 3SO_3 \rightarrow 2K_3 Fe(SO_4)_3 \qquad (17.8)$$

$$K_2 SO_4 + Al_2 O_3 + 3SO_3 \rightarrow 2KAl(SO_4)_2 \qquad (17.9)$$

Of course, the $(Na,K)(Fe,Al)(SO_4)_3$ system can be regarded as pyrosulfates stabilized by the dissolved transition metal oxides, thus increasing the temperature range of the molten phase. The argument for or against the formation of "pure" pyrosulfates in boiler deposits does not have any significant relevance, as it would be unrealistic to assume that pyrosulfates could be formed on the surface of oxidized boiler tubes without dissolution of some iron oxides.

Since the pioneering work by Corey et al. (1945) on the sulfate systems of type $(Na,K)Fe(SO_4)_3$, these have been studied by numerous researchers. Cain and Nelson (1961) have demonstrated that a minimum melting point of 825 K occurred in a $K_3 Fe(SO_4)_3 / Na_3 Fe(SO_4)_3$ mixture when the molar ratio of potassium to sodium was between 1 to 1 and 4 to 1, as shown in Fig. 17.4. Hendry and Lees (1980) have reviewed the work on complex sulfates in boiler deposits and have

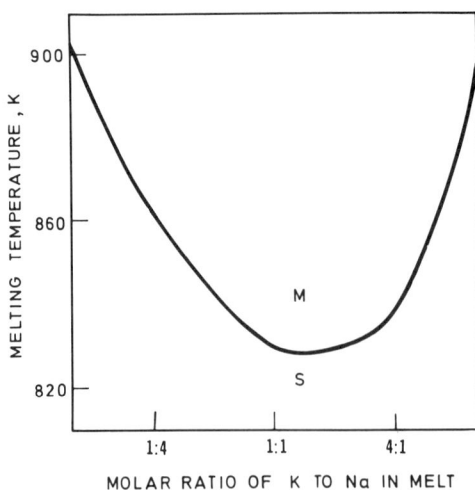

Figure 17.4 Solid/liquid equilibrium of $K_3 Fe(SO_4)_3 / Na_2 Fe(SO_4)$ system. M, molten phase; S, solid phase.

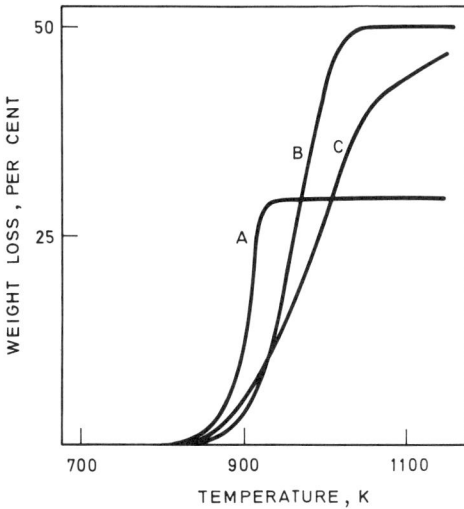

Figure 17.5 Thermogravimetric analysis of complex sulfates and corrosive superheater deposit: A, $K_3Fe(SO_4)_3$; B, $KAl(SO_4)_3$; C, sulfate in deposit.

carried out additional work on the $Na_3Fe(SO_4)_3/K_3Fe(SO_4)_3$ system. Their results on the melting characteristics of the sulfate system were close to those published by Cain and Nelson (1961).

The analytical data published by Adams and Raask (1963) show that the main constituent of highly corrosive deposit was a potassium aluminum sulfate, $KAl(SO_4)_2$, rather than the much discussed iron-containing sulfates. Figure 17.5 shows that the thermal stability of $KAl(SO_4)_2$ is higher by some 50 K than that of $K_3Fe(SO_4)_3$, so that the potassium/aluminum complex can store up to 50 percent by weight of labile sulfate (SO_3), as shown by curve B. Thermal decomposition of the corrosive, sulfate-rich deposit (curve C) was closer to that of synthetic $KAl(SO_4)_2$ (curve B) rather than to that of $K_3Fe(SO_4)_3$ (curve A). The results are in accordance with subsequent findings published by Anderson and Goddard (1968), who have measured the dissociation partial pressure of SO_3 when sulfate-rich deposits were heated in a Knudsen cell fitted to a time-of-flight mass spectrometer.

The iron oxide dissolved in molten sulfates [Eqs. 17.7 and 17.8)] could have originated from the oxide layer of boiler tubes, or it may have been extracted from the deposited ash. Some iron oxide fume of submicron-size particles is formed on dissociation of carbonates and pyrites in the flame, as discussed in Chaps. 7 and 8, and the oxide material thus formed dissolves readily in molten silicates and sulfates. The aluminum constituent of molten sulfate originates from the aluminosilicate ash, probably partly via a flame-volatile species, e.g., chloride. This is consistent with the findings of Allen et al. (1982) that the surface layer of alumino-silicate particles showed a depletion in aluminum. Subsequently more aluminum can be extracted, together with potassium from the aluminosilicate ash particles trapped in molten sulfate:

$$K_2O\cdot 3Al_2O_3\cdot 6SiO_2 + 4SO_3 \rightarrow 2KAl(SO_4)_2 + 2Al_2O_3\cdot 6SiO_2 \quad (17.10)$$

Hosegood and Raask (1964) have shown that potassium and aluminum are extracted from the surface layer, about 1 μm deep, of the flame-treated ash particles when dispersed in water. It is therefore reasonable to assume that some potassium and aluminum are extracted from the aluminosilicate ash particles when engulfed in the molten sulfate layer of boiler deposit.

The enrichment of potassium in a layer next to tube metal is enhanced by a high-temperature gradient across the silicate ash deposit on boiler tubes. Figure 7.7 showed that potassium sulfate on contact with silicates is thermodynamically stable at temperatures up to 1450 K, whereas sodium, calcium, and magnesium sulfates can be decomposed by the silicate ash deposit in the temperature range of 100 to 1300 K. It is thus evident that potassium sulfate can diffuse preferentially through a high-temperature silicate ash deposit without decomposition and will condense on the cooled surface of boiler tubes.

Thermal stability of complex sulfates—i.e., the solubility-limiting temperature of dissolved Fe_2O_3 and Al_2O_3 in the sulfate phase—depends on the gaseous SO_3 concentration within the deposit matrix:

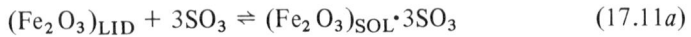

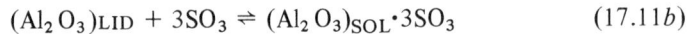

$$(Fe_2O_3)_{LID} + 3SO_3 \rightleftharpoons (Fe_2O_3)_{SOL} \cdot 3SO_3 \qquad (17.11a)$$

$$(Al_2O_3)_{LID} + 3SO_3 \rightleftharpoons (Al_2O_3)_{SOL} \cdot 3SO_3 \qquad (17.11b)$$

where SOL and LID denote, respectively, the dissolved and solid state of Fe_2O_3 and Al_2O_3. The average concentration of SO_3 in the flue gas in coal-fired boilers is around 10 ppm, but it can be much higher within the sintered ash deposit matrix that usually covers the sulfate-rich layer depicted in Figs. 17.2 and 17.3. First, as a result of catalytic oxidation of SO_2 to SO_3 by iron oxide in the deposit, the SO_3 concentration could approach the equilibrium value between 50 and 100 ppm in the temperature range of 650–1000 K, as discussed by Hedley (1967). Further, some SO_3 could be formed on decomposition of sulfates by silicates in the flame-exposed layer of deposits:

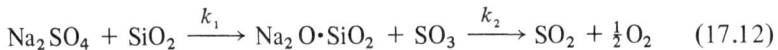

$$Na_2SO_4 + SiO_2 \xrightarrow{k_1} Na_2O \cdot SiO_2 + SO_3 \xrightarrow{k_2} SO_2 + \tfrac{1}{2}O_2 \qquad (17.12)$$

The rate of decomposition of SO_3 can be comparatively low, as discussed by Hedley (1967)—i.e., the $k_1 > k_2$ relationship may apply, and it is therefore possible that the SO_3 concentration at the surface of sulfate-rich layer can be well over 100 ppm.

The temperature range of molten sulfates with dissolved iron and aluminum extends from 750 to 950 K, and the thermal stability of potassium sulfates is about 100 K higher than that of the corresponding sodium sulfate systems. This is because the solvated potassium ions have a high ratio of size to charge—i.e., a high coordination number—and are thus capable of stabilizing the potassium–aluminum and potassium–iron (ferric) sulfate systems at temperatures up to 950 K.

17.3 DIAGNOSTIC ANALYSIS OF CORROSIVE DEPOSITS

An assessment of the corrosion propensity of boiler deposits is usually based on the results of chemical, X-ray diffraction, and thermal analysis. A general premise is that the sulfate fraction is potentially corrosive whereas the silicate fraction is practically noncorrosive, although the latter may be a source for some corrosive ingredient, e.g., potassium found in the sulfate phase. The corrosive sulfate mix in coal-fired boiler deposit consists largely of complex alkali-metal sulfates, which are readily soluble in a dilute acid solution. A sufficient quantity of acid must be added to water to prevent hydrolysis of iron and aluminum sulfates:

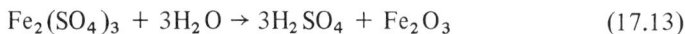

$$Fe_2(SO_4)_3 + 3H_2O \rightarrow 3H_2SO_4 + Fe_2O_3 \qquad (17.13)$$

Table 17.1 shows the results of analysis of dilute acid-soluble sulfates in highly corrosive and marginally corrosive deposits published by Adams and Raask (1963). Ground sample (100 mg) was dispersed in 100 ml of 0.02 M HClO$_4$, boiled for 10 min, cooled, and filtered. The filtrate was analyzed for the constituents shown in Table 17.1. The results of X-ray diffraction analysis of the same deposit samples is given in Table 17.2.

Based on the results of the chemical and X-ray diffraction analysis, the relative concentration of different sulfate species in the highly corrosive and marginally corrosive deposit can be estimated as shown in Table 17.3.

The results of chemical analysis given in Table 17.1 and summarized in Table 17.3, and those published by Reese et al. (1961), Anderson and Goddard (1968), and Blazewicz and Gold (1979), show that the ratio of potassium to sodium in the sulfate layer characterizes the corrosive propensity of boiler deposit. The weight ratio of K to Na in the highly corrosive deposit was between 5 to 1 and 7 to 1 (Table 17.1), whereas the ratio of the alkali metals in the marginally corrosive deposit was between 0.6 and 1.5. The corrosive deposits of high potassium content contained also comparatively large amounts of aluminum and iron in the form of complex sulfates. Thus the characteristics of corrosive sulfate deposits formed in coal-fired boilers can be summarized as follows:

$$(K_2O)_{SOL}/(Na_2O)_{SOL} > 1.5 \qquad (17.14)$$

Table 17.1 Sulfate (dilute-acid soluble) constituents of boiler deposits

Deposit sample	Steel type	Weight percent of sample								
		SO$_3$	K$_2$O	Na$_2$O	CaO	MgO	Al$_2$O$_3$	Fe$_2$O$_3$	FeO	Total
				Highly corrosive deposit						
A	2¼Cr–Mo	37.0	11.0	1.9	1.4	0.5	6.5	3.9	1.0	63.3
B	Steel A,									
	chromized	37.1	10.9	1.7	1.3	0.4	6.2	3.5	1.0	62.1
C	5Cr–1Mo	30.2	9.5	1.7	1.3	0.5	5.4	2.9	1.1	52.7
D	18Cr–12Ni–Nb	37.2	10.5	2.0	1.7	0.6	6.2	3.8	0.9	63.0
				Marginally corrosive deposit						
E	2¼Cr–Mo	11.8	2.5	4.0	2.5	0.3	0.3	0.5	1.2	23.8
F	7Cr–2¼Mo–Ti	17.2	1.4	0.9	7.9	2.4	0.2	<0.1	0.1	30.4
G	18Cr–11Ni–Ti	5.6	1.0	1.0	1.3	0.7	0.7	0.1	1.6	13.5

Table 17.2 X-ray diffraction analysis of sulfate species in boiler deposits

Deposit sample	Steel type	Relative intensities of patterns[a]							
		Na_2SO_4	K_2SO_4	$K_3Na(SO_4)_2$	$K_2Mg_2(SO_4)_3$	$CaSO_4$	$KAl(SO_4)_2$	$K_3Fe(SO_4)_3$	$Na_3Fe(SO_4)_2$
					Highly corrosive deposit				
A	$2\frac{1}{4}$Cr–1Mo As (A),	ND	ND	ND	ND	ND	S	VW	W
B	chromized	ND	ND	ND	ND	ND	S	ND	VW
C	5Cr–$\frac{1}{2}$Mo	ND	ND	ND	ND	ND	M	VW	W
D	18Cr–12Ni–Nb	ND	ND	ND	ND	ND	M	VW	MW
					Marginally corrosive deposit				
E	$2\frac{1}{4}$Cr–1Mo	W	W	M	ND	ND	VW	VW	W
F	7Cr–$2\frac{1}{4}$Mo–Ti	M	ND	W	W	M	ND	ND	ND
G	18Cr–11Ni–Ti	W	ND	W	ND	ND	VW	ND	ND

[a]Notation: S, strong; M, medium; W, weak; VW, very weak; ND, not detected.

Table 17.3 Relative concentration of sulfate species in boiler deposits

Sulfate species	Weight percent of total sulfate	
	Highly corrosive deposit	Marginally corrosive deposit
$KAl(SO_4)_2$	50	2
$(Na,K)Fe(SO_4)_2$	30	23
$Na_2SO_4 + K_2SO_4$	5	25
$CaSO_4 + MgSO_4$	10	50

$$[(Al_2O_3)_{SOL} + (Fe_2O_3)_{SOL}] > 0.1\ SO_3 \qquad (17.15)$$

where SOL denotes the dilute-acid-soluble alkali and transit metal oxides in the sulfate deposit, and SO_3 denotes the sulfate content.

An assessment of the corrosive characteristics of sulfate deposits can be substantiated by thermogravimetric analysis (TGA). The corrosive deposit is characterized by a high weight (SO_3) loss in the temperature range of 750–950 K. The weight-loss plot should be based on the amount of sulfate in the sample as in curve C of Fig. 17.5, rather than on the total weight of sample. This obviates chance variations introduced by inclusion of the thermally inert silicate and metal scale material in the sample. A weight loss over 10 percent in the temperature range of 750–1000 K will indicate a low-temperature melting and corrosive phase.

There is another thermal test that gives some useful information on the potential corrosiveness of sulfate-rich deposits. The corrosion phenomenon in molten salts is associated with the transport of ionic species in the melt, as discussed in Sect. 17.4. The ionic activity can be determined by measuring the electrical conductance of sulfate deposit. This can be done by the crucible method as described in Chap. 10 or by a sandwich test where the deposit material is wedged between two steel discs, as discussed in Chap. 11.

Curve A in Fig. 17.6 shows the conductance–temperature plot of a sulfate-rich deposit, and curve B shows the weight loss as a result of SO_3 evolution when heated in a thermobalance furnace purged with air. The thermal decomposition of the sulfate deposit commenced at 800 K (curve B) but there was no significant ionic activity below 850 K, as shown by curve A. The electrical conductance of the sulfate-rich deposit is a measure of ionic activity of the sulfate phase. It is therefore to be expected that the rate of transport of metal ions (Fe^{3+} and Cr^{3+}) and oxidants (SO^{2-} and O^{2-}), and hence the rate of corrosion will show a similar exponential increase with temperature. That is, the corrosion propensity C_s of sulfate-rich boiler deposits can be assessed by the electrical conductance measurements,

$$C_s = f(\Lambda) \qquad (17.16)$$

where Λ is the conductance at a given temperature. The mechanism of corrosion in molten sulfates is discussed in the next section.

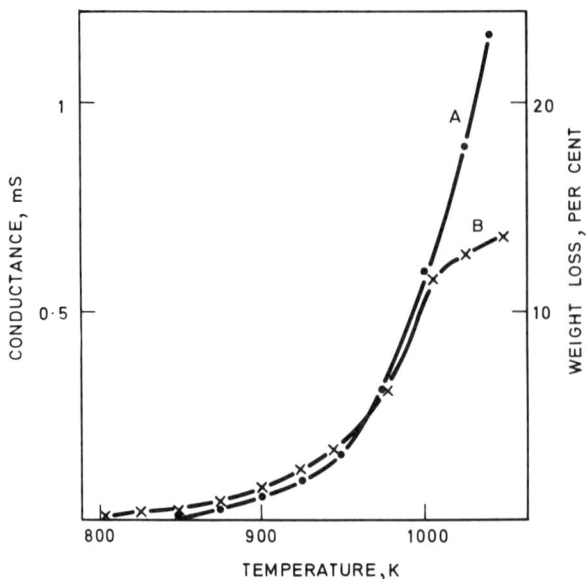

Figure 17.6 Thermal analysis of sulfate-rich superheater deposit: A, conductance; B, weight loss.

17.4 THE MECHANISM OF TUBE–METAL CORROSION IN MOLTEN SULFATE

The loss of boiler tube metal as a result of oxidation or high temperature corrosion is usually expressed by a power-law equation:

$$W = At^n \tag{17.17}$$

or

$$d = A_1 \rho_m^{-1} t^n \tag{17.18}$$

where W = decrease in mass of metal per unit area in time t
d = depth of corrosion in time t
A, A_1 = temperature-dependent coefficients
ρ_m = density of metal
n = rate coefficient

The parameter n characterizes an oxidation or corrosion process: for a diffusion-controlled reaction, chiefly oxidation, the value of n is 0.5, but for severe corrosion–i.e., a kinetically controlled reaction–the value of n approaches unity. The temperature-dependent parameter, coefficient A_1 in Eq. (17.18), usually shows an exponential increase with temperature:

$$A_1 = A_o \exp\left(-\frac{E}{RT}\right) \tag{17.19}$$

where T = temperature
E = coefficient of corrosion (energy of activation)
R = gas constant
A_o = preexponential parameter

The diffusion-resistant property, i.e., the protectiveness of the oxide layer on metal, changes with temperature, and thus "constants" n and A_o vary accordingly:

$$\ln d = \ln (A_o T \rho_m^{-1}) - E(RT)^{-1} + nT \ln t \qquad (17.20)$$

For a first approximation it can be assumed that $A_o T$ is constant and

$$nT = a + bT \qquad (17.21)$$

Thus Eq. (17.20) can be changed to

$$\ln d = \ln (A_o T \rho_m^{-1}) - E(RT)^{-1} + (a + bT) \ln t \qquad (17.22)$$

The kinetic coefficients—i.e., A_o, E, and n (a and b)—in Eqs. (17.20) and (17.22) defining the process of metal wastage depend largely on the type of steel and on the composition of boiler deposit and that of the flue gas.

The corrosion of boiler-tube steels can be considered as a series of electrochemical reactions where the anodic charge (e) transfer processes are summarized by

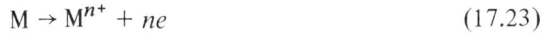

$$M \rightarrow M^{n+} + ne \qquad (17.23)$$

and the counterpart cathodic changes are:

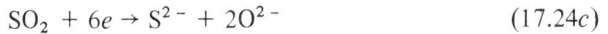

$$O_2 + 4e \rightarrow 2O^{2-} \qquad (17.24a)$$

$$SO_3 + 2e \rightarrow SO_2 + O^{2-} \qquad (17.24b)$$

$$SO_2 + 6e \rightarrow S^{2-} + 2O^{2-} \qquad (17.24c)$$

Hills (1963) has pointed out the complexity of the electrochemical reactions but was able to show that the relative rate of corrosion can be related to the electrode potential series in a solvent of alkali-metal sulfates. Burrows and Hills (1966) showed that the rate of corrosion of metals such as iron in molten sulfate was dependent on the concentration of SO_3 dissolved in the melt. SO_3 is more soluble than oxygen in the molten sulfate as a result of formation of the pyrosulfate ion ($S_2O_7^{2-}$), as discussed in Sect. 17.2. It is therefore evident that SO_3 plays an important role as the oxidizing species in molten salt corrosion.

As a result of reactions at the metal oxide-molten sulfate interface, SO_3 (i.e., pyrosulfate ion) can be reduced to sulfide, which allows sulfidation of metal to take place, as depicted in Fig. 17.7. Cutler and Grant (1975) have made an assessment of the relative rates of oxidation and sulfidation of metals in terms of transport of the oxygen-sulfur species in molten sulfates. Severe corrosion of boiler tubes is frequently associated with the formation of sulfide phases at the metal-oxide interface, even though thermodynamically the partial pressure of oxygen in the flue gas is sufficiently high to prevent the formation of sulfides. The morphology of the sulfides and consequential effects in enhancing the rate of corrosion have been discussed by Erdos (1973).

A number of investigators—Bombara et al. (1968), Rahmel (1968), Cutler (1971), Cutler and Grant (1973), and Mansfield (1973)—have employed electrochemical measuring techniques to elucidate the mechanism and measure the relative rates of corrosion of different steel alloys in molten sulfates. Measurements of the

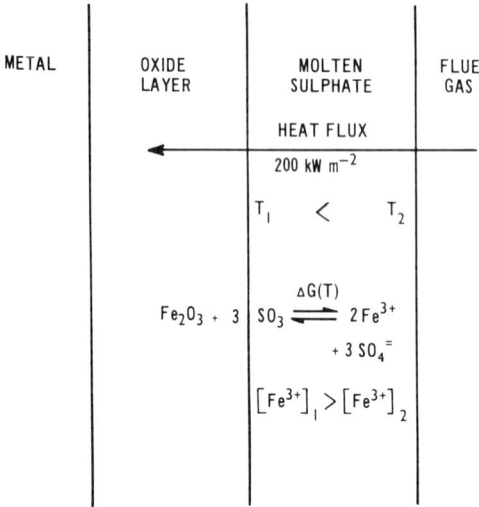

Figure 17.7 The mechanism of corrosion in molten sulfate.

corrosion potentials for iron and nickel alloys in the molten sulfate electrolyte give initial values for the relatively clean metal surface in the region of 1.2 V negative with respect to an inert platinum or gold electrode responding to the redox potential of the melt. Over a period of time the corrosion potential gradually approaches the redox potential to a greater or lesser degree, depending on the protectiveness of the oxide layer that is progressively formed on the surface of the metal. A comparison between the values for the corrosion potential and the polarization studies for the inert electrodes suggests that the reduction of SO_3 described by Eq. (17.24b) is the major cathodic counterpart in the corrosion process to the oxidation of the metal described by the anodic charge transfer process, Eq. (17.23). The corrosion mechanism can therefore be represented in its simplest terms by the polarization diagram shown in Fig. 17.8. The corrosion current I_c, which corresponds to the rate of the corrosion process, is determined by the balance between the anodic and the cathodic charge transfer processes at the corrosion potential E_c.

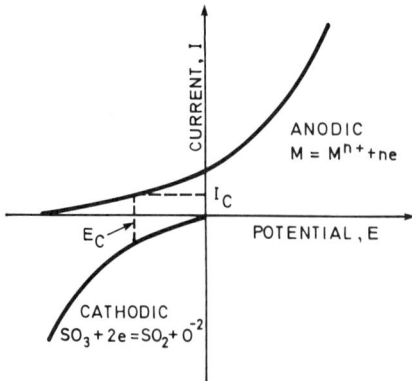

Figure 17.8 Polarization curves for the anodic and cathodic charge transfer in molten sulfate.

The corrosion rate measured in terms of the corrosion current can be derived from an analysis of the current–potential relationship for the polarized specimen electrode. If the rates of the anodic and cathodic processes are controlled by the activation energies for the respective charge transfer steps, the corrosion current at the electrode I_c is described by the equation

$$I = I_c \left[\exp b_a(E_a - E_c) - \exp b_c(E_a - E_c) \right] \qquad (17.25)$$

where E_a and I are the applied voltage and measured current, and b_a and b_c are related to the slopes for the anodic and the cathodic charge transfer processes, respectively. The equation for the current can be written in the same form if the rate of the charge transfer processes is controlled by diffusion of the reaction products rather than the activation energy, and the equation can be extended to describe charge transfer rates controlled by reactants diffusion for zero b_a and b_c. Cutler and Grant (1973) have suggested that diffusion through the oxide layer and the molten sulfate deposit controls the rate of the corrosion process, and the nonzero values of b_a and b_c that are observed experimentally suggest that the rates of the charge transfer processes are in fact controlled by diffusion of the reaction products away from the charge transfer interface.

Figure 17.9 shows typical values of the corrosion current measured over a period of several hundred hours for the austenitic steel (18Cr-8Ni) electrode in molten sulfate, as published by Cutler and Raask (1981). The system was kept at 873 K under an atmosphere containing 0.01 percent O_2 and 260 and 240 ppm of SO_2 and SO_3, respectively. The corrosion rate first decreased as the "protective" layer of oxide was formed according to the parabolic decay of a diffusion-controlled process [Eq. (17.17)]. Subsequently the corrosion rate increased, followed by a period of reduced attack by the molten sulfate medium. The cyclic variations in the measured corrosion rates are a typical feature: there are the long-term variations as indicated in Fig. 17.9, and short-term rate changes that are more difficult to discern

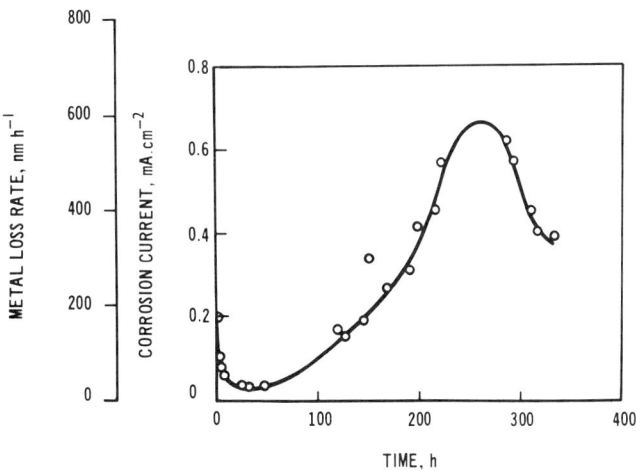

Figure 17.9 Corrosion current and metal loss of 18Cr/8Ni steel in molten sulfate at 873 K.

from the experimental scatter. These variations are considered to be associated with changes in the degree of protectiveness of the oxide layer, which controls the rate of diffusion of the oxidant and sulfide ions. The enhanced corrosion corresponds to periods when the oxide layer is porous, followed by periods when the "self-healing" effect reduces the porosity of oxide scale and hence the rate of corrosion.

The dissolution reactions of the protective oxide layer on boiler-tube metals in molten sulfates given in Sect. 17.3 and reiterated here,

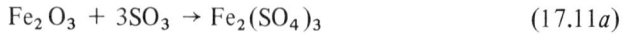

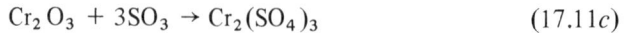

$$Fe_2O_3 + 3SO_3 \rightarrow Fe_2(SO_4)_3 \qquad (17.11a)$$

$$Cr_2O_3 + 3SO_3 \rightarrow Cr_2(SO_4)_3 \qquad (17.11c)$$

have been repeatedly discussed since being initially proposed by Corey et al. (1945), and Reid (1971) has summarized the views of different researchers. The thermal stability of iron and chromium sulfates, hence the corrosion rate, is increased by the formation of complex sulfates, as discussed by Rahmel and Jaeger (1960). However, Cutler and Raask (1981) have pointed out that the observed rates of corrosion of boiler tubes can be significantly higher than those measured in the laboratory under isothermal conditions. The temperature gradient across the molten sulfate layer on superheater tubes as depicted in Fig. 17.3 establishes a concentration gradient of the iron and chromium sulfates as shown in Fig. 17.10. This chemical gradient results in a continued dissolution of the protective oxide scale on the metal surface and the deposition of nonprotective oxide in the interfacial region between the molten sulfate layer and the porous ash deposit.

The dissolution of the protective oxide layer under these conditions will increase the corrosion rate and a steady state must then be established when the corrosion rate just balances the dissolution of the protective oxide. The rate can be

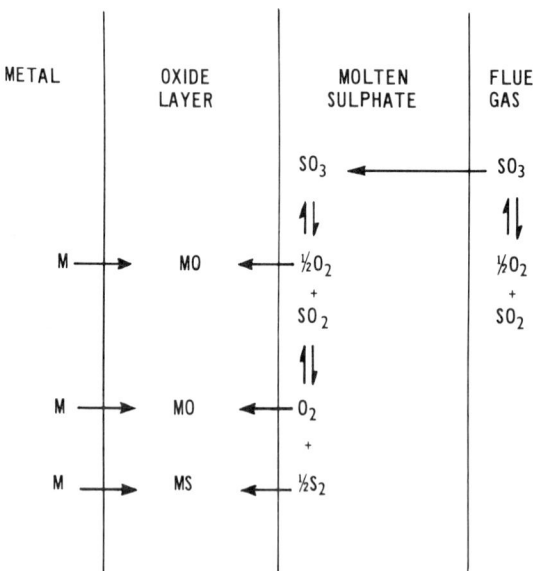

Figure 17.10 Enhanced corrosion in molten sulfate as a result of temperature gradient.

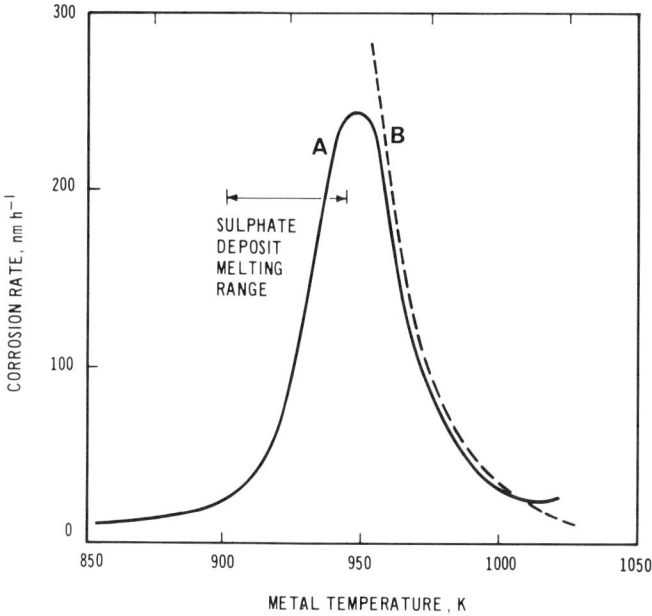

Figure 17.11 The bell-shaped curve of superheater steel corrosion (A) and the curve from protective oxide dissolution model (B).

calculated as a function of the heat flux and the thickness of the molten sulfate layer from data for the thermal stability and diffusivity of the ionic species in the melt. Curve A in Fig. 17.11 depicts the typical "bell-shaped" curve of the corrosion rate/temperature plot and curve B shows the corrosion rates calculated from the upper temperature limit of molten sulfates.

The bell-shaped curve showing a decrease in the rate of corrosion of oxidation resistant boiler steels above 900 K was observed by Koopman et al. (1959), Jonakin et al. (1959), Sedor et al. (1960), and Cain and Nelson (1961). The explanation offered at that time for the rather unexpected phenomenon was that the corrosive medium of molten sulfates was unstable above 900 K. The explanation has proved to be essentially correct, but the bell-shaped corrosive rate–temperature curve may not be universally applicable. Effertz and Wiume (1979) have suggested that the gaseous SO_3, together with SO_2 and oxygen, rather than molten sulfates, are the principal corroding agents. The presence of complex sulfates in boiler deposits is merely an indication for a locally high concentration of SO_3, which can occur also in the absence of sulfates as a result of oxidation catalysis SO_2 by iron oxide in the oxide scale on boiler tubes. Rehn (1980) has pointed out the importance of pitting corrosion of superheater alloys when the metal temperature changes periodically in service. A severe pitting corrosion is likely to occur when the complex sulfate formed below 900 K is heated above 975 K. That is, decomposition of the sulfate and the release of SO_3 in locally high concentrations results in a severe attack on the tube metal.

The rate of tube-metal corrosion is unlikely to show a reduction above 900 K in an atmosphere of changeable oxygen concentration and in the presence of chlorides, as discussed in the next section.

17.5 CHLORIDE–ENHANCED CORROSION

Shirley (1956, 1963) and Alexander (1963) have shown that in their laboratory experiments the rate of oxidation of boiler steels and gas turbine alloys was markedly increased by the presence of sodium chloride. It was claimed that the chloride prevented the formation of a protective layer of oxides on metal, and there was an exponential increase in the rate of metal wastage with temperature. Further results on chloride-induced corrosion have been published by Cutler et al. (1970), Fruehan and Martonik (1973), Hurst et al. (1973), and Mansfield (1973).

There has been much discussion on the mechanism of chloride-enhanced corrosion in coal-fired boilers and in gas turbines. The analysis of boiler deposits usually shows negligible amounts of chloride, but occasionally small quantities of chloride have been detected at the metal–scale interface of corroded furnace-wall tubes, as discussed by Mortimer and Raask (1968). The absence of chloride in boiler deposit is consistent with the high-temperature behavior of coal chlorine compounds, as discussed in Chaps. 7 and 12. The chlorides are vaporized in pulverized-coal flame partly as HCl (hydrogen chloride) and partly as sodium and calcium chlorides. The latter are rapidly sulfated and thus no significant quality of chloride was found in the material deposited on a cooled probe inserted in the superheater flue gas of coal-fired boilers, as discussed in Chap. 12. An exceptional case was when a high (1 percent) chlorine coal was burned in a 15-MW boiler when a significant quantity of chloride was deposited on a cooled probe, as shown in Fig. 12.9. It is therefore likely that some chloride is deposited on the superheater tubes in small coal-fired boilers and on furnace-wall tubes in the vicinity of burners of large boilers. However, the chief chlorine compound in the flue gas is HCl.

Mayer and Manolescu (1980) have demonstrated that there was a marked increase in the corrosion rate of boiler steel specimens at 815 K when the HCl concentration of the synthetic flue gas increased from zero to 2 percent by volume as shown in Fig. 17.12. The authors have related the high rate of corrosion to the increasing HCl concentration to changes in the morphology of oxide layers on the metal surface. In the absence of HCl, the steel specimen exposed to the synthetic flue gas—81.5 N_2, 14.1 CO_2, 4.0 O_2, 0.21 CO, and 0.18 SO_2, expressed in volume percent—acquired a continuous, nonporous scale of iron oxides:

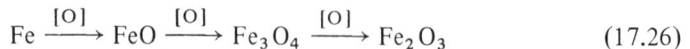

$$Fe \xrightarrow{[O]} FeO \xrightarrow{[O]} Fe_3O_4 \xrightarrow{[O]} Fe_2O_3 \qquad (17.26)$$

where the rate-controlling step was the diffusion of oxygen through the outer hematite layer. The scale formed on the specimens exposed to the gas mix containing 0.1 percent by volume HCl was largely intact, but the hematite layer contained some blisters and cracks, and the Fe_2O_3 structure became porous and

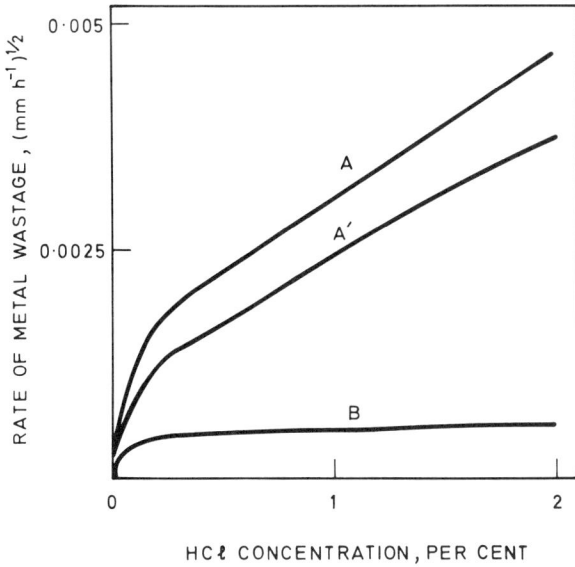

Figure 17.12 Corrosion of boiler-tube steels at 813 K in synthetic flue gas at 1333 K: A,A': low-chromium ferritic steels; B, austenitic steels.

discontinuous when the HCl concentration was increased to 0.2 percent. A further increase of HCl content of 0.8 percent caused complete disintegration of the hematite layer.

Hirsch and Rasch (1968) have calculated the thermodynamic free energy changes for a number of possible reactions between HCl and iron oxides. They concluded that in a consistently oxidizing atmosphere there should be little damage by chloride to the protective oxide scale of Fe_3O_4 and Fe_2O_3. Simons et al. (1955) have, however, suggested that an autocatalytic reaction may take place once initial reduction of sulfate by metal to sulfide has occurred. Cutler and Raask (1981) have stated that addition of 1 percent chloride to the eutectic melt of alkali-metal sulfates under a dry gas atmosphere of O_2, SO_2, SO_3, and N_2 did not significantly increase the rate of corrosion as measured by the electrochemical method described by Cutler et al. (1970). An increase in the corrosion rate was, however, observed when water vapor was added to the HCl-containing gas mix sweeping over the sulfate melt. Water vapor has an appreciable solubility in the sulfate melt as a result of formation of bisulfate ions and thus influences the corrosion reaction, since it provides an additional oxidant for the reaction with metal:

$$SO_4{}^{2-} + H_2O + SO_3 \rightarrow 2HSO_4{}^- \tag{17.27}$$

The bisulfate ions could be subsequently reduced to give hydrogen:

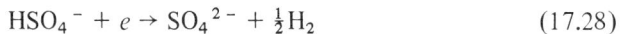

$$HSO_4{}^- + e \rightarrow SO_4{}^{2-} + \tfrac{1}{2}H_2 \tag{17.28}$$

The evolution of hydrogen could be observed at the metal–molten sulfate interface, and the hydrogen, together with HCl, constitutes an aggressive mixture for degradation of the diffusion resistant oxides:

$$Fe_2O_3 + 2HCl + H_2 \rightarrow FeO + FeCl_2 + 2H_2O \qquad (17.29)$$

$$Fe_3O_4 + 2HCl + H_2 \rightarrow 2FeO + FeCl_2 + 2H_2O \qquad (17.30)$$

In addition, the flue gas in coal-fired boilers usually contains an average of 100 ppm CO with excursions up to 500 ppm, as shown by Ormerod (1981). These concentrations may be sufficient to catalyze attack on iron oxides by chloride:

$$Fe_2O_3 + 2HCl + CO \rightarrow FeO + FeCl_2 + H_2O + CO_2 \qquad (17.31)$$

$$Fe_3O_4 + 2HCl + CO \rightarrow 2FeO + FeCl + H_2O + CO_2 \qquad (17.32)$$

Once the protective oxide layer is rendered porous, HCl together with the oxidants SO_3 and oxygen will diffuse to the metal surface, resulting in a corrosion-product mixture of sulfide, chloride, and oxide:

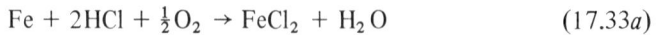

$$Fe + 2HCl + \tfrac{1}{2}O_2 \rightarrow FeCl_2 + H_2O \qquad (17.33a)$$

$$4Fe + SO_3 \rightarrow 3FeO + FeS \qquad (17.33b)$$

Figure 17.13, based on data published by Fichte (1953) and Hirsch and Rasch (1968), shows that $FeCl_2$ has a high saturation pressure at superheater metal temperatures of 800–950 K. Consequently, no significant quantity of iron chloride ($FeCl_2$ or $FeCl_3$) is likely to remain in the scale on corroded superheater tubes, but this may occur in furnace-wall deposits. Mortimer and Raask (1968) have found up to 10 percent chloride in corrosion pits of mild-steel furnace tubes taken from a pulverized-coal-fired boiler. The estimated metal temperature was between 675 and 700 K, which was sufficiently low for sodium or iron chloride to be present. The analysis by electron microprobe showed no distribution correlation between sodium and chloride, so it was concluded that iron chloride, $FeCl_2$, was the chief halide species in the corrosion product.

Figure 17.14 shows a typical failure of furnace-wall tube as a result of

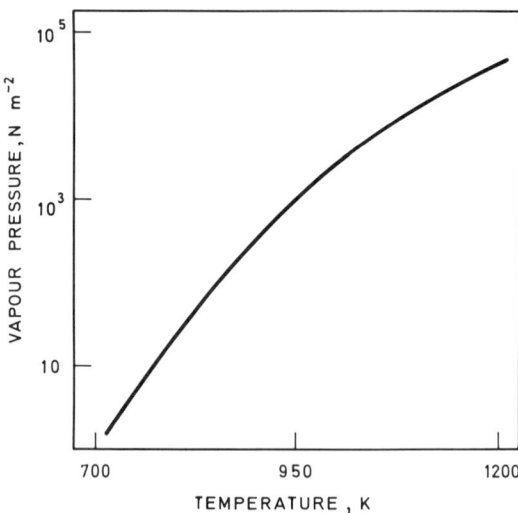

Figure 17.13 Vapor pressure of ferric chloride.

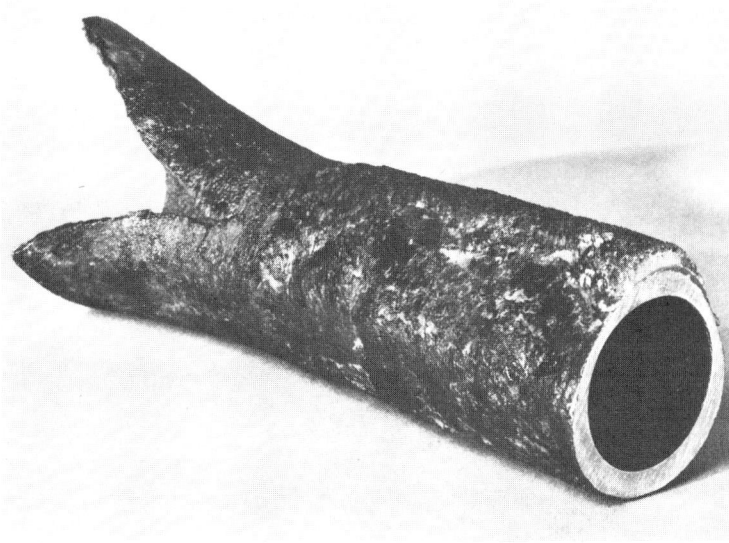

Figure 17.14 Corrosion failure of boiler tube.

high-temperature corrosion, and the topographical and sectional views of the oxide layer on corroded tube are given in Fig. 17.15, *a* and *b*. The oxide is cracked and holed, and thus it is ineffective in providing a sufficient degree of protection to the underlying metal. Figure 17.15*c* shows the sectional views of the layer structure of the protective oxide cover on a noncorroded section of the boiler. It appears that a thin layer of sulfide, FeS, at the metal–oxide interface is not necessarily associated with a high rate corrosion. However, a massive layer of the corrosion products and

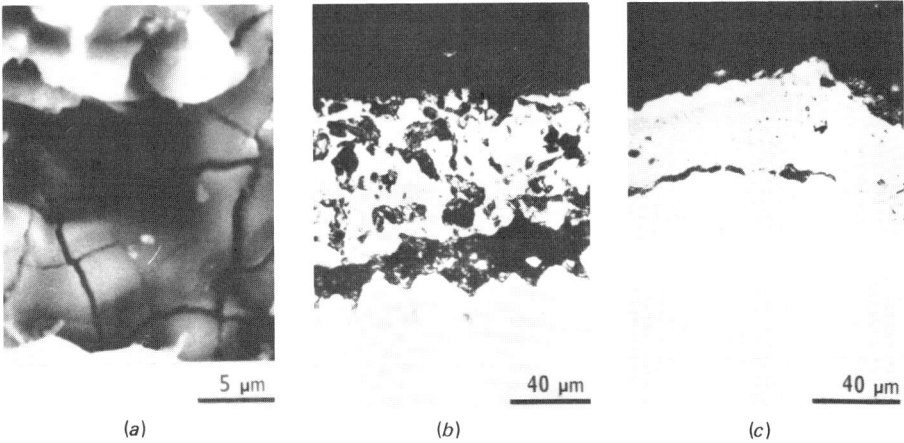

5 μm	40 μm	40 μm
(a)	(b)	(c)

Figure 17.15 Oxide scale on corroded and noncorroded boiler tubes. (*a*) Plan view of scale on corroded tube. (*b*) Intergranular attack on corroded tube. (*c*) Protective scale on noncorroded tube.

deposit constituents at the interface is an indication of a severe attack by the sulfur and chlorine impurities. The risk of high-temperature corrosion with coals of differing chlorine content is discussed in Sect. 17.7.

The high-temperature corrosion of nickel-based alloys and the protective coatings enhanced by sodium chloride vapor have been discussed by Smeggil et al. (1977). A number of these alloys contain aluminum, and an alumina-rich layer forms at high temperatures, which protects the alloy from corrosion by sulfur impurities. However, it has been observed that in the presence of NaCl vapor a volatile compound was formed that effected the transport of aluminum from the metal–oxide interface, depositing it at the surface of sodium sulfate layer. That is, there was enhanced migration of aluminum from the region of low to high oxygen potential. The protective oxide layer was thus rendered porous for enhanced sulfidation, as discussed previously in Sect. 17.4.

The survey on chloride enhanced corrosion can be summarized as follows.

1. In laboratory experiments, addition of small amounts of chloride to sulfate coatings on alloy specimens can produce startlingly high corrosion rates.
2. It is, however, more difficult to establish unequivocally the role of direct participation of chlorine impurities, chiefly HCl and sodium chloride, in the high-temperature corrosion of boiler tubes in the coal-fired plant. This corrosion is more likely to occur in the combustion chamber, where deposition of chloride on furnace tubes can occur and where the flame atmosphere has variable oxygen potential.
3. Initiation of the destructive reactions by chlorides on the protective oxide layer on boiler tubes takes place in the locality of a low oxygen potential, e.g., in the presence of CO or hydrogen in the flue gas, or at the metal–oxide interface.
4. The chloride reaction products—e.g., $FeCl_2$—have a relatively high vapor pressure and thus are swept away from the reaction site by the flue gas. This makes it difficult to establish a creditable mechanism for chloride-enhanced corrosion.

17.6 ENRICHMENT OF COAL TRACE ELEMENTS IN CORROSIVE DEPOSITS

In addition to the major constituent elements that make up the mineral impurities in coal there is a large number of trace elements present, usually in concentrations less than 100 ppm. Bethell (1963) and Gluskoter (1975) have published informative surveys on concentration and the mode of occurrence of trace elements in coal, and these surveys serve as a useful introduction to the subject. Further data on the trace elements in coals mined in the United States can be found in the work published by O'Gorman and Walker (1972) and Swanson et al. (1976).

Numerous studies have been made on the distribution of trace elements in coal-fired boiler ash captured in the electrical precipitators and in the chimney-emitted solids. A representative selection of the vast literature on this subject is the work published by Davison et al. (1974), Abel and Rancitelli (1975), Kaakinen et al. (1975), Klein et al. (1975), Gladney et al. (1976), Ondov et al. (1979), and Raask and Geotz (1981). The research on trace elements has been directed chiefly toward the chimney emission and environmental impact, whereas the possible role of coal minor constituents in the formation of corrosive deposits has been only cursorily investigated.

The coal trace elements can be placed in three categories according to their behavior in pulverized-fuel-fired and cyclone-fired boilers.

1. The flame-involatile, lithophilic elements associated with silicates remain largely trapped in boiler deposits and the ash captured in the electrical precipitators.
2. Chalcophilic elements, chiefly associated with pyrites in coal, volatilize to a different degree in the flame depending on atmosphere and temperature of the flame. The volatilized species are partly captured by fused ash particles, deposited on boiler, or may exist as discrete particles of sulfate. That is, the distribution of trace metals—e.g., Cu, Pb—may show a pattern similar to that of sodium, as discussed in Chap. 7.
3. The flame-volatile but subsequently noncondensible coal impurities consist chiefly of halides that are hydrogenated in the flue gas—i.e., HCl, HF, and HBr are formed. Trace amounts of elemental iodine may be present in the flue gas; the same applies to mercury.

The following discussion on the possible role of coal trace constituents in the formation of corrosive deposits are limited to two elements, namely arsenic and lead. There are other elements, such as lithium, that may play a significant role in the formation of aggressive deposits, but there is a lack of published data for an appraisal. The possible role of arsenic in the form of arsenates in corrosive superheater deposits have been discussed by Kirsch et al. (1968), and above 1 percent concentrations of lead, chiefly as sulfide, have been found in aggressive furnace wall deposits by Raask et al. (1970).

Monkhouse (1950) and Duck and Himus (1951) have studied the mechanism of volatilization of arsenic in the coal coking process. Their results showed that about 50 percent of coal arsenic was volatilized in the reducing atmosphere at 1000 K and the volatile species was sulfide, but some arsine, H_3As, might also have been formed. From the calcium-rich coal only about 10 percent arsenic was volatilized, probably because the trace element was present as involatile arsenate rather than the usual sulfide species in coal. Calcium arsenate was, however, decomposed when chloride was added to the coal, resulting in the formation of a highly volatile arsenic compound, probably chloride.

Significant quantities of arsenic have been found in the phosphate-rich deposits formed in stoker-fired boilers, as discussed by Crossley (1952) and Brown and Swaine (1964). Swaine and Taylor (1970) have carried out a detailed X-ray diffraction analysis of the phosphorus, boron and arsenic compounds in deposits taken from a stoker-fired boiler burning an Australian coal. The coal contained 19.3 percent ash, 400 ppm phosphorus, 6 ppm boron, and <1 ppm arsenic. Two deposit samples contained, respectively, 16 and 15 percent phosphorus, 7 and 3 percent

Table 17.4 Composition of arsenic-rich deposit

Constituent oxides	Deposit layers: Content (wt.%)		
	Next to boiler tube	Middle	Flue gas side
SiO$_2$	28.1	33.2	32.8
Al$_2$O$_3$	19.4	7.2	11.4
Fe$_2$O$_3$	14.4	29.0	21.4
CaO	Trace	2.6	3.3
MgO	Trace	Trace	Trace
K$_2$O	3.0	1.9	1.6
Na$_2$O	1.3	1.2	1.3
As$_2$O$_3$	24.5	14.0	17.4
P$_2$O$_5$	1.9	2.1	2.0
SO$_3$	9.0	8.6	8.5

boron, and 2 and 0.8 percent arsenic, as determined by spectrographic (elemental) analysis. Swaine and Taylor (1970) have identified boron arsenate, BAsO$_4$, in solid solution in boron phosphate, BPO$_4$, as being a major species in the arsenic-rich deposits. They suggest that a copious volatilization of phosphorus, boron, and arsenic in the form of elements from the hot and reducing fuel bed can lead to the subsequent formation of the duplex compound of phosphate and arsenic.

The mode of volatilization of arsenic in oxidizing coal flames has not been systematically studied. Kirsch et al. (1968) have found that the slag discharged from cyclone-fired boilers was practically free from arsenic. They concluded that the high flame temperature and long residence time of molten slag in the system resulted in an extensive volatilization of arsenic, probably in the form of oxide. It is interesting to note that the instances of corrosive deposits with a high arsenic content were formed in cyclone-fired boilers, as reported by Schneider (1962), Lorenz and Kranz (1966), and Kirsch et al. (1968). This of course does not preclude the possibility of arsenic-rich deposit being formed in pulverized-coal-fired boilers.

Borio et al. (1968) determined the concentration of a number of coal trace elements, arsenic, boron, copper, lead, lithium, manganese, phosphorus, vanadium, and zinc in corrosion probe deposits. They found no significant enrichment of the trace elements in probe deposits, and there was no obvious correlation between the presence of any of the above element constituents and the rate of probe metal corrosion. It may be that the combustion and deposition conditions were not favorable for an enrichment of trace elements to occur, or that the deposit collecting time of 300 h was insufficient for the build-up of layer-structured deposits in which diffusion of trace-element species would take place in a manner similar to that of alkali metals.

Table 17.4 shows the composition of arsenic-rich, corrosive superheater deposits in a cyclone-fired boiler investigated by Kirsch et al. (1968).

The data in Table 17.4 show that there was a remarkably high concentration of arsenic in boiler deposit, in particular in its inner layer next to tube-metal scale. The nonenriched coal ash probably contained between 10 and 100 ppm arsenic, expressed as As_2O_3: thus there was over 100 times enrichment in boilers. It appears that the mechanism of formation of arsenic-rich deposit layer is similar to the mechanism that leads to the build-up of alkali-metal sulfate deposit, as described in Sect. 17.2. That is, deposition of a flame-volatile arsenic compound, probably oxide, is followed by diffusion along the temperature gradient across the layer of deposited ash. The silicate-rich, less adhesive layer of ash is repeatedly removed by soot-blowing, leaving behind an arsenic-rich ash.

Kirsch et al. (1967, 1968) were able to show by X-ray diffraction analysis that the principal arsenic species in the deposit layers (Table 17.4) was aluminum arsenate, $AlAsO_4$, which occurred together with $KAl(SO_4)_2$, potassium aluminum sulfate. The authors found also some iron arsenide, FeAs, at the deposit–metal oxide and the oxide–metal interfaces. Based on these findings they proposed a corrosion mechanism analogous to that occurring in molten sulfates. That is, the protective oxide layer on metal (e.g., Fe_2O_3) is first partially dissolved by the arsenate-rich deposit, followed by the formation of iron arsenide. Additional evidence for the presence of a molten arsenate phase has been provided by Lorenz and Kranz (1966), who found that the softening temperature of arsenic-rich deposits was between 825 and 900 K.

Lead is another coal trace metal that has been found in some boiler deposits in concentrations over 100 times higher than the average value in ash. Lead content of ash is usually between 50 and 200 ppm, as evident from the data published by O'Gorman and Walker (1972), and is associated in coal with pyrite impurity fraction as discussed by Bethell (1963) and Gluskoter (1975). Thus, high-sulfur coals tend to have correspondingly increased amounts of the trace metal. Sometimes separate nodules of galena, PbS, can be found in coal, and the lead content of ash can reach 1000 ppm (0.1 percent), but these occurrences are rare.

The behaviour of galena (PbS) mineral on heating has been investigated by Raask (1982, unpublished work); Fig. 17.16 shows (a) the initial deformation at 1095 K, (b) the half-sphere (fusion) stage at 1100 K, and (c) the flow or melting at 1135 K. The latter was closer to the melting point of lead oxide (PbO) at 1195 K than to that of pure PbS at 1387 K (Weast, 1977). This was probably as a result of partial

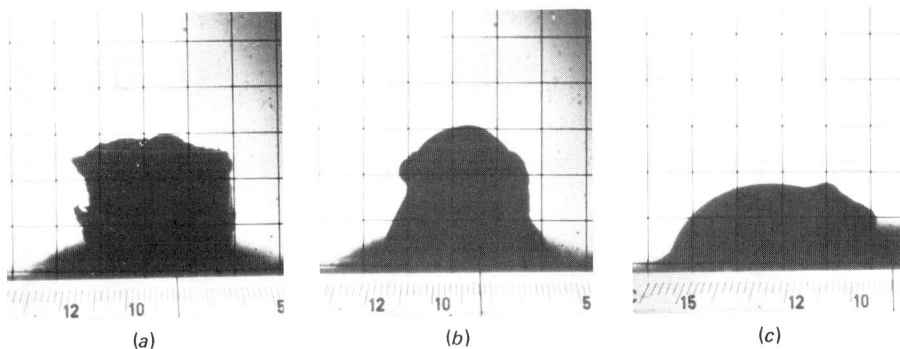

Figure 17.16 The behavior of lead sulfide mineral in heating microscope, 0.5 mm grid: (a) Initial deformation. (b) Fusion. (c) Flow.

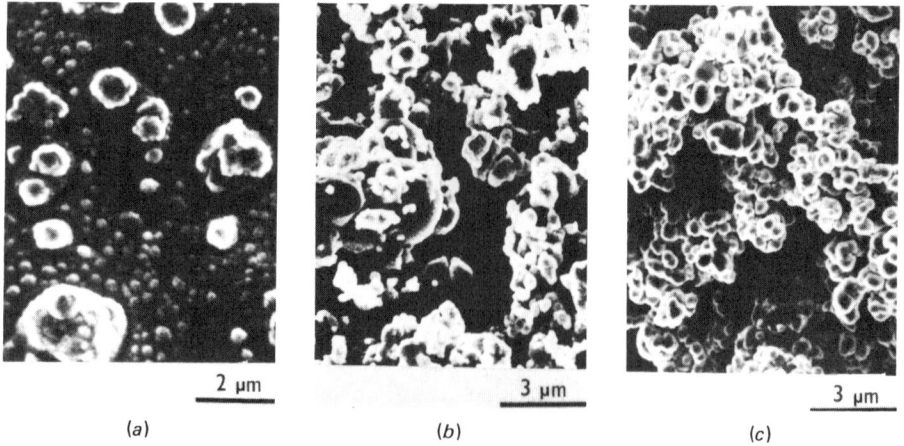

Figure 17.17 Lead compound deposits in laboratory and refinery furnaces. (a) Lead sulfide fume in heating microscope. (b) Lead sulfate deposit in refinery furnace. (c) Lead chloride and sulfide in refinery furnace.

oxidation of the PbS grains in the microscope furnace. Above 1500 K, copious volatilization of PbS in the nitrogen stream took place, and Fig. 17.17a shows the fume particles captured on a target at 500 K. The majority of fume particles were in the size range of 0.1–0.3 μm with aggregated clusters between 0.5 and 1 μm in size.

The concentration of lead in the ash material initially deposited on boiler tubes is too low for determining the mode of deposition of lead species. It was therefore considered that a study of the behavior of lead fume species excaping the lead refinery furnace may give some useful and relevant information. Table 17.5 shows the composition of fume deposited and collected at different temperatures in the duct of a lead refinery plant.

The results in Table 17.5 show that the fume deposit on a hot surface (Fig. 17.17b) consisted largely of lead sulfate, and the sinter and fusion temperatures as determined in a heating microscope (Raask, 1982, unpublished work) were comparatively high, above 1100 K. In contrast the sulfide-, chlorine-, and oxide-rich material

Table 17.5 Composition of fume in lead refinery furnace

Sample	Collecting surface temperature (K)	Constituents (percent by weight)					Sinter temp. (K)	Fusion temp. (K)
		$PbSO_4$	PbS	$PbCl_2$	PbO	Balance		
A	Hot surface, 850	70.8	6.7	2.8	12.5	7.2	1125	1195
B	Cold surface, 600	45.3	13.4	16.9	15.4	9.0	700	735
C	Electrofilter, 450	28.3	13.4	19.6	32.8	5.9	720	745

deposited at lower temperatures (Fig. 17.17c) had corresponding sinter and fusion temperatures below 750 K.

Lead metal and its compounds, other than silicates, have comparatively high vapor pressure, and thus it can be assumed that a significant fraction of lead in coal will volatilize in boiler flames above 1500 K. The concentration of lead species in the flame (chiefly PbS, Pb, PbO, $PbCl_2$) can be calculated from the following formula, assuming that all coal lead is volatilized:

$$PbX_c = CO_2 \times 10^4 \left(\frac{Pb_c}{C_c \times 10^4} \times \frac{MW_C}{MW_{Pb}} \right) \qquad (17.34)$$

where PbX_c denotes the sum of volatilized lead species in ppm by volume, Pb_c and MW_{Pb} are the concentration of lead in coal in ppm by weight and the atomic weight of lead, C_c and MW_C are the weight percent of carbon in coal and atomic weight of carbon, and the CO_2 concentration of the flue gas is given in volume percent. Typical values of carbon in power-station coals and CO_2 in the flue gas, respectively, are 65 and 14 percent; thus Eq. 17.34 becomes

$$PbX_c = 0.012 \ Pb_c \qquad (17.35)$$

An average value of lead in coal may be taken to be 25 ppm, according to data published by O'Gorman and Walker (1972), and thus the total concentration of flame-volatilized lead species, assuming complete volatilization, is 0.3 ppm.

The relative concentration of different volatile species depends on the oxygen concentration and the temperature of the flue gas, as shown in Fig. 17.18. Calculations were based on the thermodynamic data published by Kelley (1960), Kelley and King (1961), JANAF (1963), and Kubaschewski et al. (1967). It was assumed that the thermodynamic equilibrium was established at all temperatures. Figure 17.18 shows that in a reducing atmosphere lead sulfide would be deposited at about 875 K, whereas in an oxidizing atmosphere the oxide is likely to be deposited, which would be converted to sulfate in the presence of sulfur gases. Lead sulfate, $PbSO_4$, is unlikely to exist in the vapor phase. The composition of lead fume deposits taken from a refinery duct (Table 17.5) was in accordance with the thermo-dynamic calculations. That is, the high-temperature deposit consisted largely of sulfate, whereas in low-temperature deposits a significant amount of chloride was present.

The amount of volatilized lead compounds in coal-fired flue gas is likely to be significantly less than that estimated from Eq. (17.35) and from the curves given in Fig. 17.18. The preliminary results of recent work have shown that the magnetic particles of iron oxide in ash contain some lead that had not been volatilized in the coal flame. Also, fused aluminosilicate ash particles are likely to absorb some lead by a mechanism similar to that of alkali-metal capture:

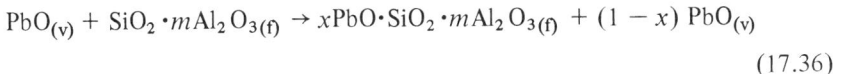

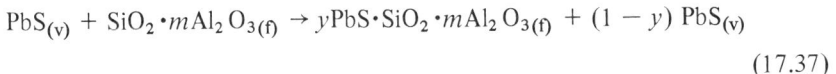

$$PbO_{(v)} + SiO_2 \cdot mAl_2O_{3(f)} \rightarrow xPbO \cdot SiO_2 \cdot mAl_2O_{3(f)} + (1-x) \ PbO_{(v)}$$

$$(17.36)$$

$$PbS_{(v)} + SiO_2 \cdot mAl_2O_{3(f)} \rightarrow yPbS \cdot SiO_2 \cdot mAl_2O_{3(f)} + (1-y) \ PbS_{(v)}$$

$$(17.37)$$

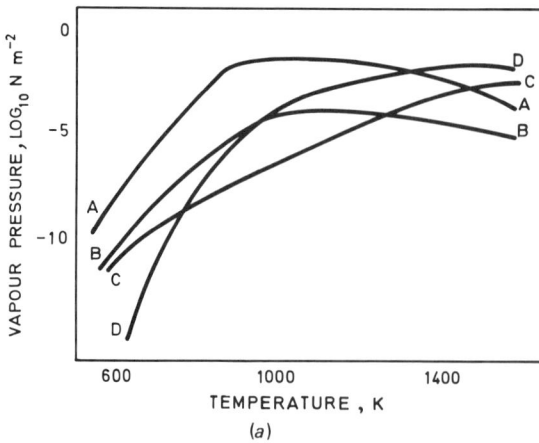

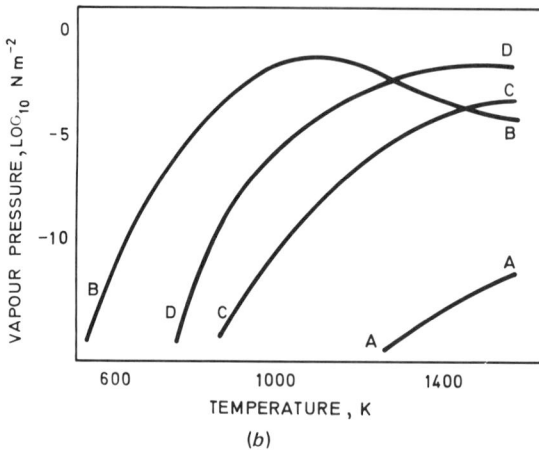

Figure 17.18 Calculated equilibrium concentration of volatile lead species in boiler flue gas. (a) 2 percent oxygen deficient. (b) 2 percent oxygen excess. A, Pbs; B, PbCl$_2$; C, Pb; D, PbO.

where (v) and (f) denote, respectively, the volatile and fused states. Equation (17.36) refers to the capture of lead oxide vapor by fused aluminosilicate ash in an oxidizing atmosphere, and Eq. (17.37) refers to the capture of PbS vapor in a reducing atmosphere. The solubility of lead oxide in fused silicate is higher than that of the sulfide, and thus a significant fraction of the volatilized lead is likely to be absorbed in an oxidizing atmosphere. In contrast, in a reducing atmosphere, lead in the form of sulfide would deposit from the flue gas below 900 K, and this may be a reason for an enrichment of lead in the form of PbS in the deposits on furnace-wall tubes of coal-fired boilers in localities where reducing conditions prevail, as discussed by Brown and Read (1969) and Raask et al. (1970).

The presence of a significant quantity of lead sulfate in refuse-fired boiler deposits has been found by Kerekes et al. (1978). Lead sulfate occurred together with the alkali-metal and zinc sulfates, and in some samples there was also a substantial amount of chloride, as shown in the analysis results given in Table 17.6.

The authors have suggested that there are probably several mechanisms for corrosive attack of furnace-wall tubes in refuse-fired boilers. Lead sulfate did not appear to contribute to the formation of low-temperature melt, but a mixture of lead chloride and oxide can constitute a corrosive medium. This is consistent with the results in Table 17.5, which show that the chloride-rich deposits taken from a lead smelting furnace had low sintering and fusion temperatures between 700 and 745 K. In contrast, May et al. (1975) have shown that lead sulfate, as well as lead oxide in the presence of sodium, constituted a highly corrosive medium in the temperature range of 1075–1225 K. The highly corrosive nature of some of the lead-containing deposits has been further discussed by Sawyer (1961) and Zetlmeisl et al. (1975).

The relevance of a high concentration of lead in deposits to corrosion of furnace-wall tubes in pulverized-coal-fired boilers has been discussed by Brown and Read (1969) and Raask et al. (1970). A high concentration of lead sulfide in boiler deposit can be taken as an indication of unsatisfactory combustion regime where a reducing atmosphere prevails that enhances the chloride-induced corrosion, as discussed in Sect. 17.5. Further, lead sulfide together with HCl may cause some damage to the protective oxide layer on tube metal as a result of reduction reaction.

There may be a number of other coal trace elements that could enhance high-temperature corrosion. Kerekes et al. (1978) have found up to 7.5 percent zinc (ZnO in Table 17.6) in boiler deposits taken from refuse-fired systems. They have used the results of chemical, X-ray, and thermal analysis of the deposit samples and synthetic mixtures to show that zinc sulfate together with those of sodium and potassium can constitute a highly corrosive medium. The molten phases occurred in the low temperature range of 675–725 K as a result of the formation of complex systems of $Na_2SO_4 \cdot ZnSO_4$ and $Na_2SO_4 \cdot K_2SO_4 \cdot ZnSO_4$. Clearly, further research is required to elucidate the mechanisms of enrichment of coal trace elements in boiler deposits and their possible role in enhancing high-temperature corrosion.

Table 17.6 Sulfate and chloride constituents of lead-rich deposits in refuse-fired boilers

	Deposit samples: Constituents (wt. % of deposit)					
	Plant in Germany			Spain	United States	
Constituent	1	2	3	4	5	6
Na_2O	4.9	5.1	4.3	10.7	10.8	5.3
K_2O	11.6	11.2	6.7	12.4	7.1	7.4
PbO	4.2	1.3	10.6	2.4	0.6	14.5
ZnO	0.5	7.2	7.5	5.6	4.8	5.5
Cl	0.6	0.6	0.6	12.2	2.2	0.3
SO_3	34.8	30.8	37.0	15.9	29.0	33.5

17.7 ASSESSMENT OF CORROSION PROPENSITY OF COALS

Discussions in the preceding sections have established that potassium-rich molten sulfates in boiler deposit constitute an aggressive medium for rapid corrosion of boiler tubes. Further, it was inferred that coal chlorine impurities, chiefly HCl and sodium chloride, can enhance corrosion in an atmosphere of changeable oxygen potential.

Potassium occurs in coal chiefly as the aluminosilicate minerals muscovites and illites, and in the absence of flame-volatilized chloride the alkali metal is retained in silicate ash and deposits, as discussed in Chaps. 4, 6, and 12. The potassium retained in the silicates is not readily available for the formation of sulfate, and the silicate ash deposits can be regarded as noncorrosive. As a result of reaction between sodium chloride and silicates, a 5–40 percent fraction of potassium is released that is subsequently sulfated and may lead to the formation of corrosive deposits, as discussed in Chaps. 6 and 7 and in Sect. 17.2. Thus it appears that potassium fraction transferred from the silicate to sulfate phase is an important criterion for an assessment of corrosion propensity of coals.

The potassium in coal minerals is usually determined by a method where the silicates are dissolved in a boiling solution of HF (hydrofluoric acid) or in a borate melt. The total potassium content thus determined does not give any indication of the potentially volatile fraction of the alkali metal and is of no significant value for assessment of the corrosive propensity of coal. Reese et al. (1961) have suggested that the water-soluble, rather than the total, potassium content of coal was an indication for the formation of corrosive sulfate deposits on boiler tubes. They found that there was a high rate of corrosion of boiler-tube steels in coal-fired boilers when

$$K_{SOL}/K_{TOT} > 0.05 \qquad (17.38)$$

where K_{SOL} and K_{TOT} denote, respectively, the water-soluble and total potassium of coal.

Borio et al. (1968) and Borio and Hensel (1972) have suggested that the acid-soluble alkali in coal is a useful gide for the corrosion propensity of boiler deposits. They have proposed a corrosiveness index I_x of coal based on their results in an experimental combustor:

$$I_x = (Na_{Ac}/K_{Ac}) - (Ca_{Ac} + Mg_{Ac}) \qquad (17.39)$$

where Ac denotes the acid-soluble fraction of the four coal mineral constituents. The corrosion assessment index as expressed by Eq. (17.39) has not found a great deal of application. This is probably because it was largely obtained from the experimental results by addition of caustic soda (NaOH), which does not occur naturally in coal. The high sodium-to-potassium ratio thus obtained can be misleading. Addition of a volatile sodium compound to coal increases the release of potassium from the silicate in boiler flames, as discussed in Chap. 6, and this is an important criterion for high-temperature corrosion rather than the ratio of sodium to acid-soluble potassium in the original fuel.

A convenient method of estimating the quantity of alkali metals volatilized in a coal flame is to determine the water-soluble sodium and potassium in the residue from bomb combustion as used for measurements of coal calorific value; the method is described in Chap. 6. The release of potassium is dependent on the volatile sodium content of coal, as shown in Fig. 6.17, but there is some recapture of alkali metals by fused silicate ash, as evident in Table 6.3. The ash content of coals was about 20 percent, and the results in Table 6.3 give the flame-volatilized alkalis (Na + K) as between 0.45 to 1.75 percent of ash. The experience on high-temperature corrosion of boiler tubes in coal-fired boilers burning coals of different chlorine content—i.e., volatilized alkali metals—has established a tentative assessment guide (Table 17.7).

The corrosive index I_c of coals as proposed by Borio et al. (1968) could be modified to

$$I_c = (Na + K)_{VOL} - B(Ca + Mg)_{SOL} \qquad (17.40)$$

where $(Na + K)_{VOL}$ denotes the flame-volatilized (i.e., water-soluble) sodium and potassium in the bomb combustion residue, and $(Ca + Mg)_{SOL}$ is the soluble calcium and magnesium content of the residue. This index based on the flame-volatile alkalis gives a guide to the relative amounts of potentially corrosive sulfates in the material initially deposited on boiler tubes. It should be emphasized that the formation of corrosive deposits on boilers is essentially a long-term phenomenon, where the temperature gradient across the deposit layer and the flue-gas atmosphere have a major influence. It is thus obvious that an assessment of corrosion propensity of coal cannot predict the exact rate of corrosion of boiler tubes. The evaluation can, however, be used to place coals in an "order of merit" table according to their corrosion propensity.

For British coals the chlorine content of bituminous fuels has been found to give a useful indication for the possible risk of high-temperature corrosion. The chlorine content of coal is an indirect indicator of the flame-volatile sodium:

$$Na_c = m \, \frac{MW_{Na}}{MW_{Cl}} \, Cl_c \qquad (17.41)$$

where Na_c, Cl_c = flame-volatile sodium and chlorine content, respectively of coal (weight percent)

MW_{Na}, MW_{Cl} = molecular weights of sodium and chlorine

m = molar ratio of the flame-volatile sodium to chlorine in coal

Table 17.7 Corrosion risk assessment

Flame-volatile (Na + K) of ash (wt. %)	Risk of high-temperature corrosion
<0.5	Low
0.5–1.0	Medium
>1.0	High

For British coals the average value of m is 0.6, so the weight ratio of the flame-volatile sodium to chlorine is

$$Na_c = 0.4 \, Cl_c \qquad (17.42)$$

Baker et al. (1977) have carried out an assessment of high-temperature corrosion of boiler-tube steels in pulverized-fuel-fired boilers burning coals of differing chlorine content. The corrosion rate data were obtained from the trials of superheater test loops (Edwards et al., 1962a,b) and long-term corrosion probes (Cutler et al., 1978), and by monitoring the rate of metal wastage of furnace wall and superheater tubes in coal-fired boilers. Baker et al. (1977) have used the classification of coal corrosion propensity given in Table 17.8.

Table 17.9 gives the observed corrosion rates of different boiler steels with low-, moderate-, and high-chlorine coals.

In boiler design it is usual to set a target minimum of 100,000 h for the operational life of boiler tubes, but most large boilers require replacement of some furnace-wall, superheater, and reheater tubes as a result of high-temperature corrosion. Table 17.9 shows that the higher the chlorine content of coal, the higher is the risk of severe corrosion, and the rate of metal wastage increases with temperature. That is, there is no evidence that the rate of corrosion will decrease above 900 K, as predicted by the corrosion model based on the premise that the aggressive sulfate deposits have an upper temperature stability limit, shown in Fig. 17.11.

Figure 17.19 shows that there is a close similarity in the curves depicting the rate of sulfate deposition and the furnace-wall tube corrosion in pulverized-coal-fired boilers burning fuels of low, medium, and high chlorine content. Both the rate of sulfate deposition and rate of corrosion remained low when the chlorine content of coal was below 0.3 percent, corresponding to below 0.1 percent flame-volatilized alkalis [Eq. (17.42)] but increased rapidly when the chlorine content exceeded this amount. Thus it appears that the chlorine content of coal can be taken as an indicator for the formation of corrosive deposits, chiefly alkali-metal sulfates. However, severe corrosion of the furnace-wall and superheater tubes can occur with low-chlorine coals, and thus the coal chlorine species alone do not determine the corrosiveness of deposits and the flue gas.

The sulfur content of coal also has been used to assess the corrosion propensity of boiler deposits and the flue gas, as discussed by Crossley (1963). He placed coals into four categories according to the sulfur content, namely <1.2, 1.2–18., 1.9–2.5,

Table 17.8 Coal corrosion propensity

Chlorine content of coal (wt. %)	Risk of high-temperature corrosion
<0.15	Slight
0.15–0.35	Moderate
>0.35	High

Table 17.9 Corrosion rates[a] of boiler steels with coals of differing chlorine content

| Metal temperature (K) | Chlorine content of coal (wt. %) | | | | | | | | |
| | Low-alloy ferritic | | | High-alloy ferritic | | | Austenitic | | |
	<0.15	0.15–0.35	>0.35	<0.15	0.15–0.35	>0.35	<0.15	0.15–0.35	>0.35
Gas temperature 1250 K									
815	I	I	M	I	I	I	I	I	I
855	M	M	H	I	I	M	I	I	M
895	S	S	S	I	M	H	M	M	H
935	A	A	A	M	H	S	H	H	S
975	A	A	A	A	A	A	H	S	S
Gas temperature 1325 K									
815	I	I	M	I	I	I	I	I	I
855	M	M	H	I	I	M	I	I	M
895	S	S	S	I	M	S	I	M	H
935	A	A	A	S	S	A	H	S	A
975	A	A	A	A	A	A	H	S	A
Gas temperature 1425 K									
815	I	M	M	M	I	I	I	I	M
855	M	H	H	I	M	M	I	M	H
895	S	S	A	M	S	H	M	H	S
935	A	A	A	S	A	A	S	A	A
975	A	A	A	A	A	A	S	A	A

[a]Code: I, insignificant; M, medium; H, high; S, severe; A, acute. Corrosion rate, nm h^{-1}: I, <25; M, 25–50; H, 50–100; S, 100–200; A, >200;

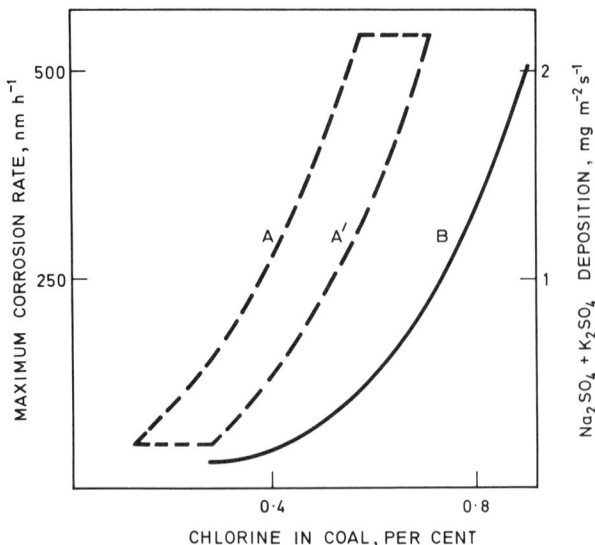

Figure 17.19 Dependence of furnace-wall tube corrosion (A, A') and sulfate deposition (B) on chlorine content of coal.

and >2.5 percent by weight. He suggested that the higher the sulfur content of coal, the higher is the risk of high-temperature corrosion as a result of increased amounts of SO_3 in the flue gas. There is, however, no evidence that the sulfur content of coal is a limiting factor in the formation of corrosion, with the exception of the low-sulfur (0.5 percent), and high-calcium sub-bituminous coals described by Sondreal et al. (1968). With these fuels the SO_3 content of the flue gas is below 1 ppm and the boiler deposit consists largely of comparatively noncorrosive calcium sulfate. Most bituminous coals have sufficient sulfur to form corrosive deposits depending on the presence of other fuel impurities, chiefly alkali-metal compounds, and on the temperature and flue-gas composition.

The high-sulfur coals (i.e., fuels with a large pyrites fraction) constitute a possibility that the FeS residue from pyrites may be deposited and incorporated in the oxide scale on boiler tubes. Also, the coals of high pyrite content are likely to contain relatively large amounts of chalcophilic trace elements, e.g., arsenic and lead, which can concentrate in boiler deposits, thus enhancing corrosion of boiler tubes as discussed in Sect. 17.6.

17.8 THE SCOPE FOR REDUCING HIGH–TEMPERATURE CORROSION BY COAL CLEANING AND BLENDING

The potential benefits of reduced boiler fouling and slagging resulting from coal cleaning are discussed in Chap. 14. The aim is to reduce the ash content as well as the pyrite content of coal; the preferential removal of the latter will also significantly reduce the chimney SO_2 emission. A reduction of sulfur in coal also will decrease marginally the risk of severe high-temperature corrosion. However, the main criterion for the corrosion propensity of coal is the amount of flame-volatile alkali metals in the fuel, for which the chlorine content is an indirect indicator. The corrosion propensity of coal (C_{COR}) could thus be expressed in the form

$$C_{COR} = f(Cl, Cl/A) \qquad (17.43)$$

where Cl denotes the chlorine content of coal and Cl/A denotes the ratio of chlorine to ash in coal.

The relevance of the chlorine to ash ratio to high-temperature corrosion has been discussed in Sect. 17.5. The sulfate deposition and corrosion curves in Fig. 17.19 show that the aim of coal cleaning and blending should be to reduce the chlorine content of power station fuel to a value between 0.25 and 0.3. Additional reduction in the chlorine content would not result in a significant benefit of decreased boiler corrosion. The results in Fig. 17.19 were obtained with coals of an ash content between 15 and 18 percent, giving the ratio of chlorine to ash of 0.018 for the marginal corrosion propensity.

The coal chlorine compounds are largely water-soluble, and it may come as a surprise that there is no significant difference in the chlorine content of washed and untreated coals received at power stations. This is because chlorine is intimately associated with the coal substance, as discussed in Chaps. 3 and 4, and also because the residence time of coal in a conventional coal washing plant is a few minutes, which is too short for removal of a significant fraction of chloride.

A linear regression analysis was made of the ash and chlorine data of some 40 coals, washed and untreated, received at a 2000-MW power station. The analysis showed no significant correlation between the ash and chlorine contents, as summarized in Table 17.10.

The data in Table 17.10 show that although the average chlorine content of high- and low-ash coals was practically the same, the latter have a higher corrosion propensity. This is because of the high chlorine-to-ash ratio, which indicates that a comparatively small fraction of the flame-volatilized alkalis would be captured by silicate ash; the remainder will be converted to potentially corrosive sulfates. This risk of severe corrosion can be minimized by blending the low-ash coals with high-ash fuel and also by combustion control; the latter is discussed in Sect. 17.10.

In order to reduce the chlorine—i.e., the soluble sodium—content of coal it is

Table 17.10 Ash and chlorine in power-station coals

Ash (%)	Average chlorine (%)	Ash-to-chlorine ratio
15–29	0.28	0.015
10–15	0.32	0.025
4–10	0.33	0.040

necessary to adopt special washing procedures where the coal particle size is reduced and the contact time with water is increased. Daybell and Pringle (1958) have studied the aqueous extraction of chloride and showed that 10–30 percent of the total chlorine was retained even after grinding the coals in a ball mill for 24 h followed by extraction in boiling water for 90 min.

Daybell and Gillham (1959) have investigated the possibility of removal of chlorine from power station coal with top size of 13 mm. They concluded that up to 40 percent of chlorine removal could be achieved on prolonged washing. Neavel et al. (1977) have reported on the extraction of sodium from Illinois coals. They found that after soaking coal for 1 h, about 60 percent of the sodium was extracted from coal below 200 μm in size, but is decreased to 20 percent with 13-mm top size coal.

Bettelheim and Hann (1980) have shown that between 15 and 40 percent of total chloride was extractable from 3- to 6-mm coals after a prolonged soaking in water. The authors have developed the concept of countercurrent flow, previously suggested by Neavel et al. (1977) for removal of a significant fraction of coal chloride. The essential feature of the concept is the use of lowest possible ratio of water to coal in order to keep the volume of the effluent to a minimum. The discharge of saline effluent is unlikely to be acceptable on environmental grounds, and it would be thus necessary to recover the salt extract in the form of solid for concentrated solution.

It is evident from the above cursory discussion that an elaborate coal washing system would be required to remove a significant fraction of the extractable chloride. It is therefore imperative to consider carefully the potential gains in boiler operation with coal of reduced chlorine content against the costs of cleaning. With exceptionally high chlorine coals, above 0.5 percent by weight, a case could probably be made for a fuel cleaning system to remove a substantial fraction of the extractable chloride. The risk of severe boiler corrosion with medium–high-chlorine coals (0.3–0.5 percent) can be significantly reduced by fuel blending, by application of the corrosion resistant materials, and by combustion control, as discussed in the next two sections.

17.9 HIGH–CHROMIUM STEELS, COEXTRUDED TUBES, SHIELDS, AND COATINGS FOR COMBATING HIGH–TEMPERATURE CORROSION

The quest for "material" solutions to high-temperature corrosion in the fossil-fuel-fired plant has been ardently pursued for several decades by metallurgists and corrosion research workers. The results of the endeavour have been summarized by Brister and Bressler (1963), who have published the data on the maximum recommended temperatures for different steel alloys in an oxidizing atmosphere and also in a corrosion atmosphere as shown in Table 17.11.

Cutler et al. (1978) have published a similar list of boiler steels with temperature limitations as shown in Table 17.12.

Table 17.11 The oxidation- and corrosion-limiting temperatures of boiler steels

Material	ASTM designation		Maximum operating temperature (K)	
			Oxidation	Corrosion
Carbon steel	A178C, A210		840	780
Carbon-$\frac{1}{2}$Mo	A209	T1a	840	795
$\frac{1}{2}$Cr-$\frac{1}{2}$Mo	A213	T2	850	795
$1\frac{1}{4}$Cr-$\frac{1}{2}$Mo	A213	T11	865	840
2Cr-$\frac{1}{2}$Mo	A213	T3b	895	840
$2\frac{1}{4}$Cr-Mo	A213	T22	910	855
5Cr-$\frac{1}{2}$Mo	A213	T5	920	880
9Cr-1Mo	A213	T9	980	920
18Cr-8Ni	A213	TP304	1145	1035
16Cr-13Ni-3Mo	A213	TP316	1145	1035

The recommended maximum operating temperatures for some steels given in Table 17.12 are lower than those suggested earlier by Brister and Bressler (1963) in Table 17.11. The more cautious approach is necessary with medium,- and high-chlorine coals in order to avoid severe high-temperature corrosion. Partly it is also due to a longer life expectancy of the large coal-fired plant. Previously there were rapid development changes in the design of coal-fired boilers with increased fuel utilization efficiency, so that the plant was obsolete after about 100,000 h of operation. More recently the design of the coal utility plant, chiefly of the pulverized-fuel-firing concept, has become "standardized." That is, the unit size is usually between 500 and 1000 MW output, and the final superheater steam temperature is set at 850 ± 25 K depending on the corrosion propensity of coal, as discussed in the preceding section. It is therefore appropriate to consider the life of

Table 17.12 Alloy composition of boiler-tube steels

Material	Composition (percent; balance Fe)							Maximum operating temperature (K)
	C	Si	Mn	Cr	Ni	Mo	Other	
Mild steel	0.10	0.15	0.5	–	–	–	–	750
1% Cr, $\frac{1}{2}$% Mo	0.12	0.20	0.5	1.2	–	0.5	–	840
$2\frac{1}{4}$% Cr, 1% Mo	0.10	0.25	0.5	2.3	–	1.0	–	855
12% Cr, 1% Mo	0.20	0.4	0.55	11.5	1.0	1.0	–	890
Type 316	–	0.6	1.7	17.5	13.5	2.7	–	955
Esshete 1250	0.1	0.5	6.0	15.0	10.0	1.0	1.4	975
Type 310	–	1.2	1.8	24.5	20.5	–	–	>975[a]

[a]Used for uncooled spacers between steam tube.

the generating plant of such huge capital investment to be extended to 150,000 or 200,000 h operation.

From Tables 17.11 and 17.12 it is evident that increased amount of chromium in an alloy enhances markedly the corrosion-resistent property. This is because the chromium-rich oxide on parent metal constitutes a nonporous barrier to diffusion of metal and corrodant species, and chromium oxide, Cr_2O_3, has a lower rate of dissolution in molten sulfates than that of iron oxides, as discussed in Sect. 17.4. The high-chromium steels can, however, suffer from rapid corrosion in particularly aggressive deposits (Tables 17.1 and 17.3), as evident from the results given in Fig. 17.20 (Raask, 1963c). In these cases it may be necessary to resort to the use of corrosion resistant alloys in the form of coextruded tubes, shields, and coatings.

Flatley et al. (1981) have described a variety of coextruded tubes, also known as duplex tubes, for use in severely corrosive localities of furnace-wall, superheater, and reheaters. A coextruded tube has the usual steel alloy for its inner section sleeved in an outer tube section made from corrosion-resistant alloy. The requirements of the inner face are high creep and rupture strengths, and resistance to corrosion by water or steam impurities. The function of the outer face is to provide resistance to corrosion by molten sulfates and flue-gas impurities. Table 17.13 gives the composition of the corrosion-resistant alloys used in coextruded tubes, together with the wall thickness.

Trials with the 25Cr20Ni coextruded tubes in Britain commenced in 1969, and the results have proved that the alloy has a low rate of corrosion, as discussed by King and Robinson (1973) and Flatley et al. (1981). Rahoi and DeLong (1980) have shown that a 50Cr50Ni alloy has excellent corrosion resistance; there was no significant corrosion after 9 yr in a coal-fired boiler. It is therefore evident that the high-temperature corrosion of the superheater and reheater corrosion can be

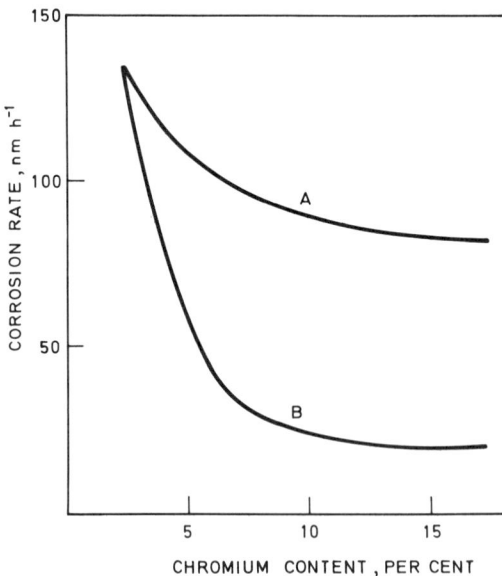

Figure 17.20 Corrosion of superheater steel specimens: A, exposed to flame radiation; B, in sheltered position.

Table 17.13 Coextruded furnace-wall, superheater, and reheater tubes

	Inner tube face		Outer tube face	
Tube location	Material[a]	Wall thickness (mm)	Material[a]	Wall thickness (mm)
Furnace wall	Mild steel	6.5	18Cr10Ni	2
Superheater	15Cr10Ni	3	25Cr20Ni	3.5
Reheater	15Cr10Ni	1.5	25Cr20Ni	2.5

[a]The figures before Cr and Ni refer respectively to approximate chromium and nickel contents by weight percent of the alloys.

combated in arduous conditions by application of high-chromium alloys in the form of coextruded tubes.

In contrast, combating corrosion of furnace-wall tubes made of mild steel has been usually considered by means of the combustion control, as discussed in Sect. 17.10. However, it has been realized that in some cases the use of corrosion-resistant materials is necessary. Figure 17.21 shows the localized areas of high heat flux where the furnace-wall tube corrosion is likely to occur, as discussed by Mortimer and Raask (1968). Trenkler (1967) has described a boiler trial where sections of the corroded mild-steel tubes in a cyclone-fired boiler were replaced by alloys of differing chromium content. It was found that the rate of corrosion of 5, 9, and 11 percent chromium alloys was the same as that of mild steel, whereas 18Cr14Ni alloy had a corrosion rate 5 times lower. Similar results have been reported by Flatley et al. (1981), who found that the coextruded tubes with 18Cr9Ni outer surface in a pulverized-coal-fired boiler had a corrosion rate 3 to 4 times lower than that of mild steel.

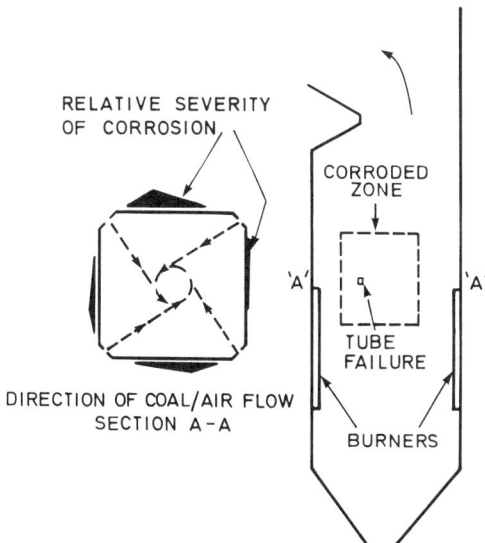

Figure 17.21 The locality of severe furnace-wall corrosion in coal-fired boiler.

The coextruded tubes with a corrosion-resistant outer surface are likely to be widely used in the future to combat the high-temperature corrosion of furnace-wall, superheater, and reheater tubes. The high-chromium alloy, coextruded tubes are expensive compared with conventional tubes, but the sections of furnace-wall tubes (Fig. 17.21) and the number of superheater and reheater tubes subjected to severe corrosions are usually limited. That is, well over 90 percent of the total mileage lengths of water and steam tubes do not suffer significantly from high-temperature corrosion. The main difficulty is to predict accurately the localities in a new boiler where severe corrosion is likely to occur, so that the corrosion-resistant coextruded tubes can be incorporated in the design stage. The severe corrosion of superheater and reheater tubes is usually limited to bottom bends of pendant superheaters (Fig. 14.3) and the first two or three tubes of the high-temperature steam elements, which take full impact of deposition.

A combination of high tube-metal temperature and unhindered deposition should be avoided in the boiler design stage. This can be done by replacing the first three tubes facing the flue gas flow in the high-temperature elements by low-temperature steam or water tubes. This measure of protecting the high-temperature tubes by cooler tubes will be effective only when the edge-view alignment of the tubes in the element remains undistorted. Once the high-temperature tubes leave the shadow of the protecting tubes, the result will be unhindered deposition and high-temperature corrosion. It should thus be possible to limit superheater tube failures as a result of high-temperature corrosion by a judicious layout of pendant superheaters and by a limited use of duplex, coextruded tubes of 25Cr20Ni outer surfaces at exposed positions.

It is more difficult to predict the rate of corrosion of furnace-wall tubes and to pinpoint the exact locality where it is likely to occur in the boiler design stage. However, an assessment of the risk of severe tube wastage can be made based on the corrosive propensity of coal (Sect. 17.7), the heat release rates in the burner zone (Chaps. 13 and 14), and the previous corrosion experience in boilers of similar design. Once the high-risk areas, usually in and immediately above the burner zone, have been established, a decision can be made as to whether the use of corrosion-resistant coextruded tubes would be warranted. The limited corrosion experience of mild-steel tubes coextruded with an austenitic 18Cr11Ni alloy outer surface, 2 mm thick, has shown that these tubes would have double the operation life of monoextruded tubes. An additional advantage resulting from the use of austenitic surface tubes would be that a slag-free zone would be created in this section of the combustion chamber. This is because the adhesive bond of ash deposits on an austenitic steel surface is markedly weaker than the ash bonding on mild-steel and ferritic alloy tubes, as discussed in Chaps. 11 and 14.

Other measures for combating high-temperature corrosion of superheater and reheater tubes include a variety of shields and wrappings, as discussed by Cutler et al. (1974) and Menzo and Rahoi (1979). Since the shields do not have to stand the internal fluid pressure, these can be made of highly corrosion-resistant but less ductile alloys. The protective shields are usually tack-welded or helically wrapped around the tubes, and as a result of poor thermal contact to the tubes the shield metal temperature will be high, above 950 K. That is, the temperature is higher than that of the thermal stability limit of corrosive sulfates, which may result in a reduced rate of corrosion of high-chromium shield material, as discussed in Sect. 17.3. However, this decrease in the corrosion rate above 950 K may not occur in the presence of high concentrations of chloride (HCl) and CO in the flue gas.

The enthusiasm for use of the corrosion-resistant shields and wrappings is on the wane. Prefabricated and attached shields on the superheater tubes can be easily damaged in transport and during installation; attachment of shields *in situ* can be difficult and time-consuming. Further, there have been cases of lost or displaced shields as a result of thermal distortion and sootblower action, thus exposing unprotected sections of the steam tubes. As a result of these unsatisfactory aspects in the application of corrosion-resistant shields, there has been renewed interest in the use of protective coatings, as discussed by Rehn (1980).

Metal coatings, deposited by flame spraying or by electric or plasma arc, have been used in the United States for protection of furnace-wall tubes in cyclone-fired utility boilers and wood pulp liquor recovery boilers. Plumley and Roczniak (1975) have reported on the successful application of a multilayer coating applied by flame spraying. The layers were, 0.05–0.13 mm nickel–aluminum undercoat, 0.18 and 0.22 mm 18Cr8Ni middle coating, and 0.08–0.1 mm aluminum top coating. The authors emphasize that the tubes must be thoroughly cleaned by steel shot or coarse grit blasting, rather than by sand blasting which may leave a residue of fractured particles embedded in the tube metal. In contrast, Flatley et al. (1975) have reported the results of unsuccessful trials with coatings of 13Cr steel, 13Cr steel–aluminum, and nickel–chromium/aluminum on mild-steel tubes. The coatings were applied on cleaned new tubes by arc spraying, but all the coatings had disappeared within 15,000 h of operation without any evidence of protection.

Lees (1978) has suggested that the most likely reason for failure of flame- and arc-sprayed coatings was the presence of interconnected pores, which do not heal in service because the metal particles are preoxidized in the flame or arc. The porosity remains high and in time the coating is lifted off by a layer of tube-metal oxide and deposit constituents formed at the metal/coating interface. Bucklow (1975) has suggested that the plasma-deposited metal coatings have a low porosity and are potentially highly corrosion-resistant. Levine (1978) has discussed the application of corrosion protective alloy coatings of the type MCrAlX, where M denotes Ni, Co or Fe, or a combination of these, and X denotes a constituent metal added to enhance the adherence of oxide scale. Levine has evaluated the oxidation and corrosion resistance and the thermal fatigue performance of a selection of MCrAlX coatings. The applied coatings included aluminides formed by pack cementation and more sophisticated coatings from vapor deposition, overlay coatings, and development of thermal barrier coatings by plasma spray deposition. This last mode of application of the corrosion-resistant coatings is being actively pursued both as a factory-based method and also as an *in situ* spraying technique using mobile units, as discussed by Meringolo (1981). Gal-Or (1980) has suggested that further studies are needed to characterize the chemical reactions and physical changes that occur in metal spray processes.

A trial of combined metal and ceramic coatings on mild-steel furnace-wall tubes was carried out by Raask (1969, unpublished results). A 0.5-mm-thick coating of flame-sprayed alumina was overlayed with 0.5-mm-thick alumina, also flame-sprayed. After 5000 h operation in a 200-MW pulverized-fuel-fired boiler burning high-chlorine coal, there was no evidence of the coating or that tubes had been protected. This was attributed to high porosity of the Al_2O_3 layer, the inter-connecting nonhealing pores in the aluminum coating, and its poor adhesion to boiler-tube metal. Similarly abortive results with a phosphate-bonded silicon carbide material have been reported by Lees (1978). However, the results of extensive trials reported by Rehn (1980) showed that magnesium zirconate ($MgO \cdot ZrO_2$) formed a nonporous, corrosion-resistant coating.

Bishop and Holland (1964) have described a borosilicate glass coating formed at 1225 K. The coating was highly corrosion-resistant but had a poor resistance to thermal and mechanical shocks. It is likely that any other refractory coating would suffer from the same disadvantage. Further, a refractory surface at substantially higher temperature than that of tube metal would enhance bonding of ash deposits, resulting in a stress at the coating–metal interface.

Instead of the external coatings, diffusion "coatings" formed in the tube metal have been considered for combating high-temperature corrosion. In this method a corrosion-resistant component is allowed to diffuse into the parent metal to a depth between 50 and 500 μm. The principal methods of application of diffusion coatings are pack cementation, gas-phase deposition, and molten bath deposition, as described in the book edited by Shreir (1976). Foster and Toft (1962) have reported that chromizing MoVTi (Colmo 1100) steel reduced its corrosion rate by half in coal-fired boilers at 895 K. McGill and Weinbaum (1978) have reported that a diffusion treatment with chromium and aluminum improved significantly the oxidation and wear resistance of boiler steels. Dolinskaya and Grisenko (1980), and Tyul'pin et al. (1980) have claimed excellent results in reducing the rate of corrosion by chromizing 12Cr1MoV superheater tubes.

A variety of surface treatments for steels–e.g., plating, hot dipping, vapor deposition, boronizing, and nitriding–have been discussed in a publication edited by Evans (1978). Lees (1978) has referred to a nitriding process carried out in the atmosphere of ammonium and hydrogen. It is claimed that the nitrided steel had a corrosion rate an order of magnitude less than that of untreated steel. The improved protection results from grain refinement, which enhances oxide adhesion in a manner analogous to the effect produced by rare-earth metals added to high-temperature alloys. The biggest obstacle to overcome in the application of this or any other diffusion method is to ensure that long lengths of boiler tubes can be treated in a process that would result in a layer of uniform thickness and microstructural integrity.

Manolescu and Mayer (1978) have made a critical review of the available methods of application of metal and ceramic coatings for protection of boiler tubes against corrosion and erosion. They concluded that there are several materials and methods of application that will satisfy the requirements for use in the coal-fired boiler. In contrast, Lees (1978), after carrying out a similar survey, was more cautious on the efficacy of corrosion resistant coatings when applied to combat high-temperature corrosion. He found that the results of numerous trials with metal and ceramic-sprayed coatings were inconclusive. This may have been partly due to inexperience in application, but there appears to be a fundamental weakness in that it is difficult to achieve a nonporous layer of metal or ceramic material on boiler tubes by any method other than in the form of a coextruded tube, as discussed by Rehn (1980).

Overview on the Use of Coextruded Tubes, Shields, and Coatings

Menzo and Rahoi (1979) have applied all three methods to combat the high-temperature corrosion of superheater and reheater tubes in cyclone-fired boilers. They found the sprayed metal coatings to be ineffective. The shields protected the tubes from corrosion, but resulted in additional problems of maintenance: the shields fell off repeatedly as a result of thermal stress, boiler vibration, and the action of sootblowers. Further, shields provided the anchorage sites for accumulation of ash deposit. Their solution to the corrosion problem was the use of coextruded tubes with an Inconel alloy, 50Cr50Ni, outer surface.

The findings of Menzo and Rahoi (1979) reflect the view of many research workers and boiler-plant design and operation engineers. That is, the method of coextruded tubes is the most reliable method to protect the superheater, reheater, and furnace-wall tubes against high temperature corrosion. The corrosion-resistant alloys can be used also in the form of shields and bandages, but frequent attachment failures occur, and periodic inspection and replacement of dislodged sections are required.

The results of application of the corrosion-resistant metal and ceramic coatings by flame spraying have been disappointing. The high porosity and poor adhesion of the coatings have been the chief causes for failure. The more recent application of high-temperature plasma spraying is likely to result in improved corrosion-resistant characteristics of the protective coatings. There remains, however, a problem of quality control in application to ensure the unequivocal integrity of several hundred boiler tubes. There is a need for development of a nondestructive technique to measure accurately the layer thickness of plasma-sprayed material. Other measurements are required to monitor the strength of cohesive and adhesive bonds and porosity of the coatings.

The choice of coating and coextruded tube materials is chiefly based on their corrosion-resistant characteristics. Another desirable property is that the surface of coatings and coextruded tubes should have a nonadhesive tendency for bonding of coal-ash deposits. A strong deposit-to-coating bond may result in the protective layer being removed with the sintered or slagged ash. In contrast, a poor adhesion of ash deposits on the coated or coextruded tubes could be utilized to prevent excessive boiler slagging, as discussed in Chap. 14. In particular, there is a need to prevent the build-up of ash deposit on the combustion-chamber walls of boilers converted from oil to coal firing, as discussed in Chap. 20.

17.10 BOILER DESIGN AND COMBUSTION CONTROL MEASURES TO ALLEVIATE FURNACE–WALL AND SUPERHEATER CORROSION

The thermal design regime of pulverized-coal-fired boiler should be consistent with the corrosion propensity of coals in a manner analogous to that of minimized slagging discussed in Chap. 14. That is, the volumetric energy input in the burner zone should be between 450 and 600 kW m^{-3}, depending on the corrosion propensity of coals. Table 14.1 gave the corresponding data for coals of different slagging propensity. Another thermal regime parameter is that the temperature of the flue gas leaving the combustion chamber should be consistent with the corrosion propensity of coals, as suggested in Table 17.14.

Tables 17.11 and 17.12 in Sect. 17.9 gave the choice of suitable steel alloys for superheater and reheater tubes, together with the recommended maximum metal temperatures consistent with the corrosion propensity of different coals. Maximum possible advantage should be taken to protect the high-temperature superheater and reheater elements from direct thermal and deposit impact. That is, the first two or three tubes in the elements facing the flue-gas flow should be replaced by low-temperature steam or water tubes.

Table 14.2 listed the boiler operation measures that can be applied in order to combat boiler slagging. The same measures may be also necessary to alleviate the high-temperature corrosion of furnace-wall and superheater tubes. These include consistently fine coal milling, increase in the amount of combustion air and improved air/coal mixing, decrease in the temperature of preheated combustion air, flue gas circulation, and reduction in boiler load. One of the most sensitive methods of monitoring combustion is to measure the CO content of the flue gas, as discussed in Chap. 14. Figure 14.7 showed that when the excess oxygen in the flue gas drops below 3 percent there may be a significant rise in the CO concentration well above 100 ppm. The high amounts of CO in the flue gas should be avoided in order to reduce the risk of severe high-temperature corrosion as well as boiler slagging. Hadrill (1979) has described in some detail the operation-improving measures that were necessary to minimize the corrosion and slagging problems in 500-MW pulverized-coal-fired boilers.

Boiler operation experience has shown that in many cases the root cause for

Table 17.14 The design temperature of furnace exit flue gas

Corrosion propensity of coal	Furnace exit temperature (K)
Low	1300–1350
Medium	1250–1300
High	1200–1250

severe superheater and reheater corrosion was excessive slagging of the combustion chamber. As a result of build-up of ash deposit on furnace walls, the radiative heat transfer from the flame to water tubes can be drastically reduced, and thus the temperature of the flue gas entering the superheater duct will exceed the design temperature by a margin of 100–250 K. An allowance is made in the boiler design stage for the reduced heat flow through coal ash deposits, but the heat barrier effect may be higher than anticipated. The projected surface area of the combustion chamber of a pulverized-coal-fired boiler is 40 percent larger than that of an oil-fired boiler of the same thermal capacity (Godridge and Read, 1976). The boiler slagging experience has, however, shown that in many coal-fired furnaces this extra area for the radiative heat transfer is not sufficient to prevent the flue-gas temperature at the entry to superheaters exceeding the design limit. Curve A_1 in Fig. 13.11 showed that a layer of loose ash deposit, 3 mm thick, or semifused ash, about 5 mm thick, is sufficient to reduce heat transfer by 40 percent. Each additional 3-mm-thick layer of ash deposit on all furnace-wall tubes will increase the temperature of the flue gas leaving the combustion chamber by about 100 K.

As a result of the high flue-gas temperature, the superheater and reheater metal temperature may exceed the design limit by a large margin. This is particularly the case with reheater tubes, where the low pressure (density) steam is unable to absorb additional heat from the high-temperature flue gas. There will also be a high temperature gradient across the layer of deposit on steam tubes, which enhances diffusion of aggressive constituents towards the surface of tubes as discussed in Sect. 17.3. This could result in an exponential increase in the rate of corrosion of superheater and reheater tubes with tube metal temperature. That is, severe slagging of the combustion chamber may result in an increased rate of corrosion of superheater and reheater tubes, which could be as much as an order of magnitude higher than that occurring when the furnace exit temperature is kept to design value. In order to prevent this occurring the combustion chamber of coal-fired boiler must have an adequate provision of sootblowers. With exceptionally slagging or corrosive coals it may be necessary to augment the usual steam or air blowers with water-jet blowers, as discussed in Chap. 14. Inadequate sootblowing of the combustion chamber usually results in the slagging problem being extended to the pendant superheaters, togethers with a risk of severe corrosion of the superheater and reheater tubes.

17.11 QUEST FOR CORROSION–PREVENTING ADDITIVES IN COAL–FIRED BOILERS

The discussions in Sects. 17.2 and 17.4 have established that gaseous and dissolved SO_3 can play a significant role in the formation of aggressive deposits and also in the mechanism of corrosion of boiler-tube metals. In order to suppress the formation of SO_3 in high concentrations at boiler deposit temperatures, Nelson and Lisle (1964) have introduced the concept of catalyst poisoning. The authors have carried out corrosion experiments where samples of austenitic steel, 18Cr9Ni grade, in a synthetic corrosive mix of ash, alkali-metal sulfate, and iron oxide were exposed to a synthetic flue gas containing 0.25 percent by volume of SO_2. The corrosiveness of the mix was assessed by the weight gain after 5 to 10 d exposure in the temperature range of 750–920 K. Over 80 compounds, chiefly those of group 5A and 6A elements in the Periodic Table, were added to the corrosive mix. The results showed that antimony oxide, Sb_2O_3, was effective in preventing the formation of corrosive deposits. Sodium antimonate was similarly effective, but all other compounds were found either marginally effective or noneffective. The catalysis "poisoning" property ceased to be effective above 920 K.

The authors have explained the efficacy of antimony oxide in reducing the formation of SO_3 by assuming that it formed a chemiadsorptive bond with the catalyst—i.e., iron oxide—which inhibited the SO_2 oxidation at the active site. They further suggested that Sb_2O_3 has a sufficiently high vapor pressure at the temperature range of 750–920 K for distribution on the surface of iron oxide. Above 920 K the adsorptive bond between the catalyst and antimony breaks down and corrosion proceeds unhindered.

The concept of catalysis poisoning in the mechanism of high temperature corrosion is an attractive idea, but so far there have not been any successful boiler trials reported with antimony or any other "anticatalysis" additives. Antimony is present in coal in concentrations between 1 and 3 ppm, and thus the amount of antimony in ash is likely to be around 10 ppm. In Sect. 17.6 it was shown that arsenic and lead can be found in boiler deposits at well over 1 percent concentration, equivalent to over 100 times enrichment compared with the amount found in ash. Some enrichment of coal antimony may occur in boiler deposits, and further research is required to study deposition of trace-element compounds on the steam and water tubes in coal-fired boilers.

A more conventional approach for reducing the SO_3 content of the flue gas is to use basic additives, chiefly magnesium oxide or hydroxide. Levy and Merryman (1966) have suggested that MgO additive reacts with SO_3 in the flue gas, and when deposited on boiler tubes it may mask the catalytic surface of iron oxide. Rahmel (1963) and Borio and Hensel (1972) have suggested that MgO can form complex systems with potassium and sodium sulfates, e.g.:

$$2MgO + K_2SO_4 + 2SO_3 \rightarrow K_2Mg_2(SO_4)_3 \qquad (17.44)$$

Thus the main function of MgO additive is first to reduce the SO_3 partial pressure and secondly to combine with alkali-metal sulfates. Radway and Boyce (1978) have claimed some success with magnesium as an additive when injected into cyclone-fired boilers. The strongly bonded superheater deposits were changed to a more friable form, but the authors were unable to show any significant enrichment of magnesium in superheater probe deposits.

So far there are no unequivocal results to show that basic additives, chiefly magnesium and calcium compounds, could significantly reduce the rate of high-temperature corrosion. In the presence of relatively large amounts of siliceous ash, much of the basic material could be captured in silicates when added to coal or injected into the combustion chamber. This could enhance furnace slagging, which in turn would result in increased superheater corrosion as discussed in the preceding section. The reaction between basic additives and silicate ash could be minimized where magnesium or calcium compounds were to be introduced in the superheater chamber and blown directly to the steam tubes. However, this would be difficult to achieve in practice.

The use of siliceous additives, suggested by Nelson and Lisle (1965), for combating high-temperature corrosion in coal-fired boilers seems at first sight to be a most improbable suggestion. There is such a large quantity of silicate ash in the system that an additive would not make any significant difference. However, the authors have carried out a series of laboratory experiments with different siliceous materials and the results showed that kaolin diatomaceous earth and synthetic silica markedly reduced corrosion of boiler-tube steels in molten sulfates. The corrosive mix consisted of alkali-metal sulfates and iron oxide; the molar ratio of $K_3 Fe(SO_4)_3$, $Na_3 Fe(SO_4)_3$, and $Fe_2 O_3$ was 1.5:1.5:1. It has been claimed that the siliceous additives reduced the sulfate content of deposit by about 50 percent and decreased the corrosion rate by a factor of 4. The authors have suggested that for a siliceous additive to be effective it must have large specific surface (surface area to weight ratio), above 15,000 m^2 kg^{-1}. They claimed that the reduction in corrosion was a result of adsorption of molten sulfates by the highly porous silicate material and that there was no sigificant chemical reaction between the additive and alkali-metal sulfates.

The authors do not recommend injection of the siliceous corrosion inhibiting additives into the combustion chamber. The high temperature would cause a vitrification of the silicate particles, resulting in a loss of the surface area, and hence the effectiveness for subsequent adsorption of the corrosive sulfates would be markedly reduced. They suggest that the additives should be introduced in the locality where high-temperature corrosion occurs. This is, however, difficult to put into practice; thus there are no reports of the siliceous additives being used for combating superheater corrosion in coal-fired boilers, as suggested by Nelson and Lisle (1965).

NOMENCLATURE

A, A_1, A_0	temperature-dependent rate coefficient
$(Al_2 O_3)_{SOL}$	soluble aluminum (sulfate) content of deposit
a	rate constant
b	rate constant
b_a	anodic charge transfer
b_c	cathodic charge transfer
C_c	weight percent carbon in coal
C_{COR}	corrosion propensity of coal
C_r	corrosion rate
C_s	corrosion propensity of deposit
CO_2	carbon dioxide content of flue gas

Ca_{Ac}	acid-soluble calcium content of coal
$(Ca + Mg)_{SOL}$	water-soluble calcium plus magnesium content of coal
Cl	chlorine content of coal
Cl/A	ratio of chlorine to ash in coal
D_c	deposit composition
d	depth of corrosion
E	exponential coefficient of corrosion
E_a	applied voltage
E_c	corrosion potential
$(Fe_2O_3)_{SOL}$	soluble iron (sulfate) content of deposit
I	measured current
I_c	corrosion current
I_x	corrosion index
K_{AC}	acid-soluble potassium content of coal
K_{SOL}	water-solbule potassium content of coal
$(K_2O)_{SOL}$	soluble potassium (sulfate) content of deposit
k_1, k_2	rate constants
Mg_{AC}	acid-soluble magnesium content of coal
M_{ox}	metal oxide composition
MW_C	atomic weight of carbon
MW_{Cl}	atomic weight of chlorine
MW_{Na}	atomic weight of sodium
MW_{Pb}	atomic weight of lead
Na_{AC}	acid-soluble sodium content of coal
Na_c	flame-volatile sodium content of coal
$(Na + K)_{VOL}$	volatile alkali-metal content of coal
$(Na_2O)_{SOL}$	soluble sodium (sulfate) content of deposit
n	rate coefficient
Pb_c	lead content of coal
PX_c	concentration of volatile lead species in flame
R	thermodynamice gas content
T	temperature
T_e	exposure temperature
T_m	tube metal temperature
ΔT	temperature difference
W	rate of weight loss
Λ	electrical conductance
ρ_m	metal density

EIGHTEEN

ASH IMPACTION EROSION WEAR

18.1 CHANGES IN THE ABRASIVE CHARACTERISTICS OF PARTICULATE COAL MINERALS IN THE BOILER FLAME

The abrasive sliding and particle impactation erosion of coal in a handling plant has been discussed in Chap. 5, where it was established that the abrasive quartz and pyrite minerals were chiefly responsible for the damage. The flame imprinted changes in the characteristics of different coal mineral species have been discussed in Chaps. 6–8 and also in papers published by Raask (1981b, c). The more significant changes relevant to the abrasive characteristics of ash are reiterated here.

Vitrification of Aluminosilicates

The aluminosilicate clay species constitute the bulk of the ash in most coals, and the species kaolin, illite, and muscovite are soft minerals occurring in coal chiefly as a fine sediment of particle size below 5 μm, as shown in Fig. 5.7. The nonabrasive aluminosilicate particles vitrify, agglomerate, and spheroidize in the boiler flame, as discussed in Chaps. 6 and 8. A typical spherical particle of aluminosilicate ash is shown in Fig. 18.1a, where it appears to carry some smaller (0.5–1 μm) particles of the same composition, and a large number of sulfate fume particles, 0.1–0.3 μm in size. For the purpose of assessing their abrasive characteristics, the aluminosilicate particles can be regarded as consisting of a glassy material of Vickers hardness

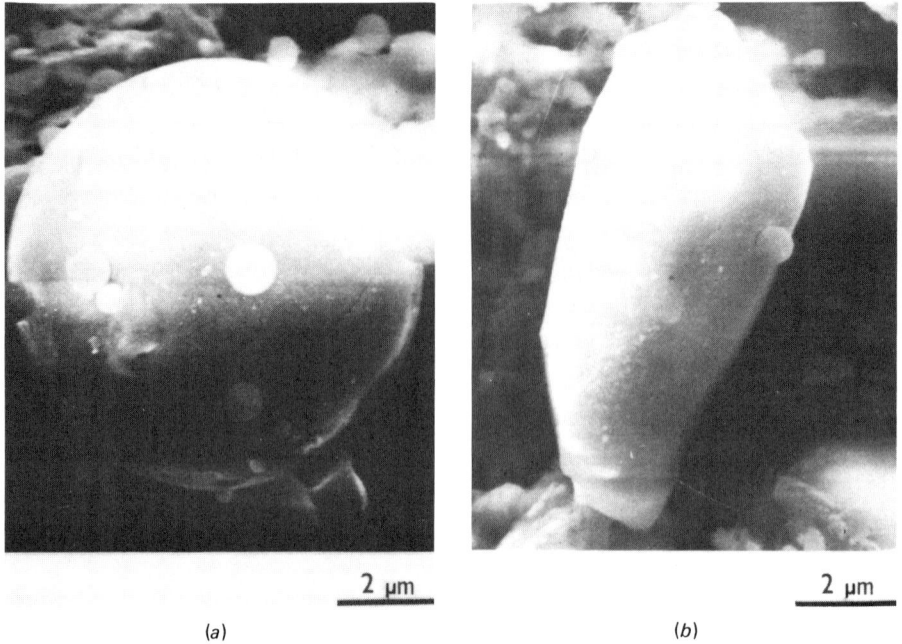

2 μm

(a)

2 μm

(b)

Figure 18.1 The surface characteristics of spherodized and unfused particles in pulverized ash. (a) Aluminosilicate particle. (b) Quartz particles.

number 550–600. The particles may contain encapsulated harder crystalline species, as discussed in Chap. 12 (Fig. 12.2b), but the crystalline needles embedded in the glassy matrix do not come into contact with the impacted surface. Nevertheless, the aluminosilicate particles of flame-heated ash are moderately abrasive whereas the clay mineral species in coal before the heat treatment are nonabrasive. It therefore follows that the flame-heated constituents of silicate ash are more abrasive than the original clay mineral species in coal. Also, as a result of coalescence of the fine clay sediment (inherent ash) in the pulverized-coal flame (Chap. 8), the particle size increases, which will lead to higher rates of erosion wear as discussed in the following section.

Partial Vitrification of Quartz Particles

Quartz mineral in coal is largely responsible for severe erosion wear in fuel handling and milling plant, as discussed in Chap. 5. Quartz is the hardest mineral species usually found in coal, and it occurs in comparatively large particle sizes and requires a high temperature for vitrification. The pure quartz particles are infusible below 2000 K, but the particles in coal are contaminated with aluminum, iron, and alkalis and thus require a lower temperature for vitrification and are partially spheroidized in the pulverized-coal flame, as discussed in Chap. 6. As a result of vitrification, the amount of quartz found in the flame-heated ash is usually less than that present in low-temperature ash, as shown by the results in Table 18.1 (Raask, 1980).

Table 18.1 Quartz content of low-temperature and flame-heated ashes

	Low-temperature ash sample		Flame-heated ash sample		
Content	A	B	C	D	E
Quartz (wt. %)	21	23	5.6	5.2	2.1

The amount of quartz in pulverized-coal ash is usually determined by X-ray analysis and varies between 1 and 10 percent as shown by the data published by Simons and Jeffery (1960) and Watt and Thorne (1965). It occurs partly as microcrystals in glassy particles, as discussed in Chap. 12 and shown in Fig. 12.2a, and partly as slightly rounded particles as shown in Fig. 18.1b. The amount and shape of quartz particles in pulverized coal is largely influenced by the flame temperature, and as a result of the high temperature, the ash from cyclone-fired boilers does not contain any significant quantity of angular particles of quartz. In contrast, some sub-bituminous coal ashes from pulverized-coal-fired boilers contain a large fraction of quartz (Burger, 1961). The coarse mineral particles do not vitrify in the flame of comparatively low temperature, and the ash is thus highly abrasive.

Transformation of Angular Pyrite Particles to Spheres of Iron Oxide

The pyrite particles dissociate rapidly in the flame, with the residues melting at around 1350 K, and the spherical particles thus formed are subsequently oxidized to magnetite (Fe_3O_4), as discussed in Chap. 7. The magnetite particles have a characteristic "orange peel" surface when viewed under the microscope, as shown in Fig. 8.5c. There is a decrease in hardness as pyrite is transformed to iron oxide, and thus the magnetite particles in ash are less abrasive than the pyrite particles in pulverized coal before combustion.

Dissociation of Carbonates

The carbonate particles dissociate in the coal flame and are partly captured by fused silicate ash; the remainder are converted to sulfates, chiefly calcium and magnesium sulfates (Chap. 7). The latter consist largely of fume particles, below 1 μm in size, which do not cause any impact erosion wear.

General Appearance of Pulverized-Fuel Ash Particles

The majority of flame-heated particles of pulverized-coal ash are spherical in shape but are not smooth spheres when viewed under the scanning electron microscope. Figure 18.2a shows the appearance of a typical sample of the spherical particles in

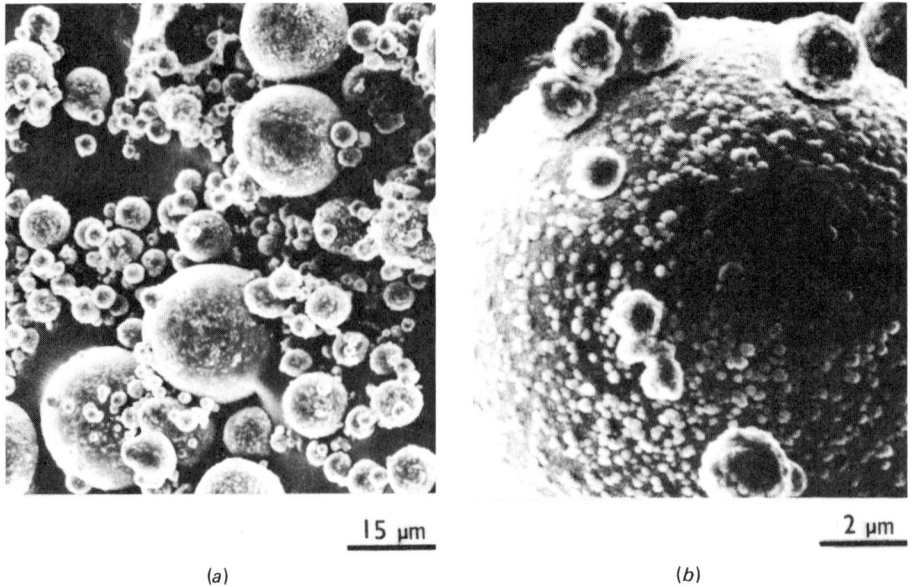

15 μm   2 μm

(a)   (b)

Figure 18.2 Typical pulverized-coal ash. (a) General view. (b) Surface detailed view.

pulverized-coal ash. When viewed at a higher magnification, the surface of the ash particle appears to be largely covered by small particles 0.1–0.3 μm in diameter, as shown in Fig. 18.2b. The results of diagnostic tests given in Chap. 8 and also in the paper of Raask and Goetz (1981) showed that the submicron-size particles consist chiefly of sulfates but include some insoluble crystalloids of silicates and iron oxide. The sulfate particles of low hardness are unlikely to enhance erosion on impaction of the host particles, but the surface crystalloids may increase the rate of erosion. It is therefore to be expected that the ash particles are likely to be more abrasive than the smooth glassy spheres of the same size and hardness.

The gas-borne ash in pulverized-coal-fired boilers usually contains a small amount of grit particles 100–500 μm in size. These particles consist largely of debris of sintered ash deposits dislodged from the furnace-wall and superheater tubes by sootblowing. Figure 8.6 showed that the large grit particles are nonspherical in shape and can be highly abrasive. The changes in hardness of coal mineral species in the flame are summarized in Table 18.2.

The hardness number of glassy silicate particles was assumed to be the same as that of bulk material. The microhardness of large (above 20 μm) particles can be determined by dispersing the ash in a resin and by carrying out diamond stylus indentation measurements, as described by Goodwin et al. (1969). The hardness values of the ash constituents given in Table 18.2 show that, with the exception of a small amount of sulfate (2–5%) and approximately the same amount of carbon residue, the ash particles have a significantly higher hardness than that of boiler steels, which have a Vickers hardness number of about 200 (Fig. 5.11).

The Vickers hardness number of coke residue in pulverized-fuel ash is probably below 200, but the comparatively large particles have a sharp-edged contour and usually have some silicate ash globules embedded in the surface layer as shown in Fig. 8.4*a*. It is therefore assumed that the abrasive characteristics of the coke residue particles are approximately the same as those of glassy silicate ash as discussed in the next section.

18.2 ABRASIVE CHARACTERISTICS OF PULVERIZED–COAL ASH

It is a more difficult task to assess the abrasive characteristics of flame-heated ash than pulverized-coal minerals discussed in Chap. 5. The latter can be considered to consist of a mixture of soft clay mineral particles and of hard quartz and pyrite particles. Thus, the abrasive property of pulverized coal depends largely on the amount of quartz and pyrite in the fuel. In contrast, the flame-heated silicate ash particles have a Vickers hardness number above 500, as shown in Table 18.2. The Vickers hardness number of boiler steels is around 200, and it is thus evident that the ash particles can cause some erosion wear damage depending on the size and shape of the impacting particles. Goodwin et al. (1969) have suggested that there exists a power-law relationship between the rate of erosion wear E_w and the hardness number H of impacting quartz particles:

$$E_w = BH^{2.3} \tag{18.1}$$

where B is a constant. In more general terms, the erosion wear–hardness relationship

Table 18.2 Hardness of coal mineral species and ash constituents of typical bituminous coal

Species	Amount (wt. %)	Hardness Mohs	Hardness Vickers (kg mm^{-2})
Coal minerals			
Aluminosilicates	70	2–2.5	20–80
Quartz	15	7	1200–1300
Pyrite	10	6–7	1100–1300
Carbonate	5	3–4	150–450
Pulverized-fuel ash			
Glassy spheres with embedded mullite needles and quartz crystalloids	80	5	550–600
Small glassy and large nonspherical quartz particles	10	6–7	600–1200
Spherical particles of iron oxide	5	5–6	500–1100
Fume particles of sulfate	2	Nonabrasive	
Coke particles with inherent and surface ash	3	3–5	100–500

may be written in the form

$$E_w = B_1(H - H_1)^n \qquad (18.2)$$

where H_1 is the hardness of material being eroded and B_1 is a constant.

The erosion wear is markedly influenced by the particle size of impacting material, as shown by the results published by Goodwin et al. (1969) and Kotwal and Tabakoff (1980). The experiments were carried out with quartz sand of different particle sizes and found that the rate of erosion was insignificant when the impacting particles were below 5 μm in diameter. This is because the small particles have insufficient amounts of the inertial momentum to impact the target and are carried around the object by the gas flow, as shown in Fig. 12.3. As the particle diameter was changed from 5 to 45 μm, the erosion rate increased correspondingly and remained approximately constant for larger particles, as shown in Fig. 18.3. It is therefore expedient to separate ash into three size fractions for the purpose of the abrasion property characterization: <5 μm, 5–45 μm, and >45 μm. Assuming that the size fraction above 45 μm has an abrasive index I_{1a}, the relative abrasion index of other fractions would be:

Particle diameter (μm)	>45	5–45	<5
Relative abrasive index	I_{1a}	$0.5 I_{1a}$	0

Assuming that the >45-μm size cut consists of a mixture of moderately abrasive glassy particles of index I_{1g} and highly abrasive quartz particles of index I_{1q}, an expression can be written for the abrasive index I_{1a} of this fraction:

$$I_{1a} = x_1[(1 - l_1)I_{1g} + l_1 I_{1q}] \qquad (18.3)$$

where x_1 is the weight fraction of >45-μm size cut of the ash and l_1 is the weight fraction of quartz of the >45-μm size cut. A similar equation can be written for the

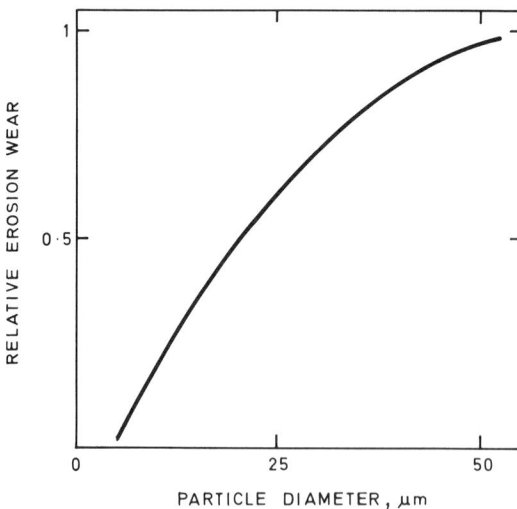

Figure 18.3 Dependence of erosion wear on the size of impacting quartz particles.

abrasive index I_{2a} for the 5–45-μm size cut:

$$I_{2a} = x_2 [(1 - l_2)I_{2g} + l_2 I_{2q}] \tag{18.4}$$

where the corresponding notations are as for Eq. (18.3). In order to simplify the expression for an abrasive index covering both fractions of ash, it is assumed that I_{2g} equals 0.5 I_{1g} and I_{2q} equals 0.5 I_{1q} (Fig. 18.3). That is, the erosion damage caused by the 5–45-μm size cut of glassy or quartz particles is approximately 50 percent of that caused by the >45-μm size cuts. By combining Eq. (18.3) and (18.4), an expression for the abrasive index I_a of ash can be obtained:

$$I_a = [x_1(1 - l_1) + 0.5x_2(1 - l_2)]I_{1g} + (x_1 l_1 + 0.5x_2 l_2)I_{1q} \tag{18.5}$$

where I_{1g} and I_{1q} denote respectively the abrasive index of the >45-μm size cut of glassy and quartz particles. When the weight fractions of quartz are comparatively small—i.e., when $l_1 < 0.1$ and $l_2 < 0.1$—Eq. (18.5) can be simplified:

$$I_a = (x_1 + 0.5x_2)I_g + (x_1 l_1 + 0.5x_2 l_2)I_q \tag{18.6}$$

The experimental work carried out on the particle erosion wear characteristics by Raask (1969a), Sage (1969), Bratchikov (1972), and Tabakoff et al. (1979) suggests that the abrasive index of the spherical ash particles with a nonsmooth surface (Fig. 18.2a) is 0.4 times that of quartz, i.e., $I_{1g} = 0.4I_{1q}$. Equation (18.6) can be thus written in the form

$$I_a = [x_1(l_1 + 0.4) + x_2(0.5l_2 + 0.2)]I_{1q} \tag{18.7}$$

where, reiterating the notations, I_a is the abrasive index of ash, x_1 and x_2 are, respectively, the weight fractions of the >45-μm and 5–45-μm size cuts, l_1 and l_2 are, respectively, the weight fractions of quartz in the two size cuts, and I_q is the abrasive index of the >45-μm-size quartz. A selection of ash samples from typical British (bituminous) coals were size-fractionated and analyzed for the quartz content (Table 18.3). Figure 8.7 showed the particle-size distribution for a typical ash.

It is evident from Table 18.3 that typical pulverized-fuel ash is more abrasive than that of the parent coal (Table 5.2), where the wear-causing minerals, chiefly quartz and pyrite, are diluted by comparatively nonabrasive coal substance and clay species.

It should be noted that the abrasive index as defined by Eq. (18.7) is different from that suggested by Kotwal and Tabakoff (1980). They assumed that chemically analyzed silica and alumina contents can be taken as an indicator for the abrasiveness of ash. Table 5.1 shows that the "free" crystalline alumina mineral is hard and highly abrasive, and accordingly the rate of impaction erosion wear is approximately double that of the same size quartz particles. In the majority of pulverized-coal ashes, however, alumina is present in a "combined" form of aluminosilicate glass with some mullite needles dispersed in the glassy matrix, as shown in Fig. 12.2b. There may be present small amounts of the highly abrasive alumina particles in ashes that have a total Al_2O_3 content above 35 percent. However, the usual indications for an abrasive ash are high silicate content, above

Table 18.3 Size fractions, quartz content, and the abrasive index of typical bituminous coal ashes

Ash number	Fraction of total ash in size cut (wt. %)			Quartz content of fraction (wt. %)		Abrasive[a] index of ash, I_a
	>45 μm	5–45 μm	<5 μm	>45 μm	5–45 μm	
1	25.5	64.7	9.8	4.5	3.3	0.25
2	7.5	69.5	23.0	5.1	4.0	0.18
3	27.0	64.0	9.0	5.3	4.7	0.27
4	12.8	68.2	19.0	5.7	4.8	0.21
5	16.2	69.5	14.3	6.7	5.3	0.23
6	24.0	65.6	10.4	11.0	9.7	0.28

[a]The abrasive indices of ashes were calculated from Eq. (18.7) where the abrasive index of quartz >45 μm in size was taken to be unity; for quartz and glass particles <5 μm, the index was considered to be zero.

50 percent, and a high ratio of silicate to alumina. The ratio above 1.5 indicates the presence of abrasive quartz mineral in coal (Chap. 5) and also in the flame-heated ash.

Figure 18.4 shows the change in relative abrasiveness of high, medium and low quartz ashes when heated to different temperatures. The composition of the mineral matter in typical medium quartz coal (curve B) was taken to be 75 percent clay species, 10 percent quartz, 10 percent pyrites, and the balance carbonates and chlorides. First there is a slight decrease in the abrasive property as sharp-edged pyrite particles change to spherical iron oxide particles at about 1200 K, as was shown in Fig. 7.3. Subsequently there is an increase in abrasiveness when the ash is

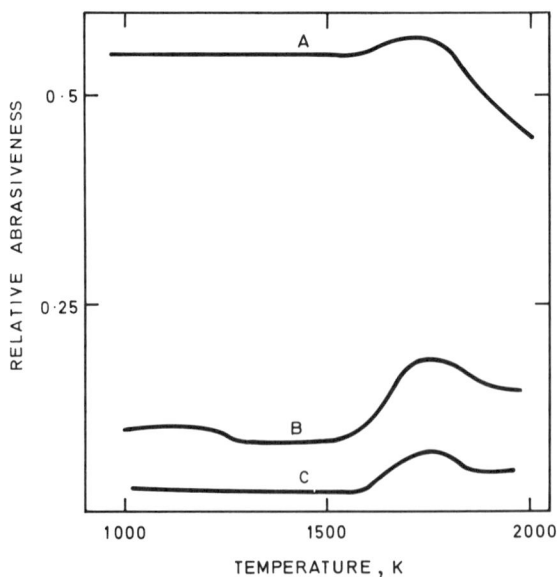

Figure 18.4 The effect of temperature on abrasiveness of pulverized-fuel ash: A, high-quartz ash; B, medium-quartz ash; C, low-quartz ash.

heated from 1200 to 1750 K, as a result of the soft clay mineral particles being changed to glassy particles. Above 1700 K the hard, sharp-edged quartz particles commence to change to spherical shapes, as was shown in Fig. 6.5, resulting in a less abrasive ash. The flame temperature of cyclone-fired boilers is sufficiently high to vitrify the quartz particles, and the particle size of the gas-borne ash is low (Fig. 8.7). Thus the fly ash from cyclone-fired boilers has a low abrasive index. In contrast, the flame temperature in pulverized-coal-fired boilers is not sufficiently high to vitrify the quartz particles (Fig. 6.6), and quartz-rich coals will produce an abrasive ash, as discussed by Burger (1961).

An extensive program of accelerated erosion wear tests with a selection of pulverized-ash samples has been carried out for the EPRI (Stringer, 1981) at the Westinghouse Research Laboratories at Pittsburgh, Pa. The measurements were carried out at an impact velocity of 191 m s^{-1}, which is relevant to the flow velocity in the induced draft fan where severe erosion wear may occur. The results were used to assess the abrasive characteristics of ashes based on the chemical analysis, with conclusions presented in Table 18.4.

A large number of bituminous coals have an ash of silica content between 40 and 50 percent and thus are moderatly abrasive. When the silicate content of ashes is between 50 and 60 percent, the abrasive propensity shows a large variation depending on the quartz content and particle size of the ash, as defined by Eq. (18.7). The sub-bituminous and lignite coals can have either a nonabrasive ash or highly abrasive ash, depending on the silica content. The chief constituents of the nonabrasive ashes are calcium and sodium sulfates, below 5 μm in particle size, and do not contain any quartz. In contrast, a number of sub-bituminous coal ashes and a few bituminous coal ashes have an exceptionally high quartz content, above 60 percent of the total. These ashes are highly abrasive, and utilization of the quartz rich coals has led to the development of boilers of low flue-gas velocity, as discussed in Sect. 18.4.

18.3 TEMPERATURE REGIME AND VELOCITY PARAMETER GOVERNING BOILER–TUBE EROSION BY ASH IMPACTION

The combustion chamber and the high-temperature superheater and reheater sections of coal fired boilers do not usually suffer from tube erosion by ash impaction.

Table 18.4 Abrasive characteristics of ashes

Silica (SiO$_2$) content of ash (wt. %)	Erosion wear propensity
<40	Low, <0.02 wear units
40–50	Medium, 0.02–0.08 wear units
50–60	Medium to high, 0.04–0.28 wear units

That is, the erosion wear cannot occur in the high-temperature regions where adhesive deposits attach to the furnace-wall and superheater tubes, since a layer of bonded ash or molten sulfate will absorb kinetic energy of the impacting particles. The formation of bonded ash deposits ceases to be a problem in the convection heat-transfer passages where the flue-gas temperature has fallen below 1150 K and the tube-metal temperature is below 750 K, as discussed in Chap. 12.

These temperatures can be taken as threshold limits below which the erosion wear of boiler tubes by ash impaction may occur. That is, the erosion wear damage usually occurs in the low-temperature (primary) superheaters and reheaters, and in the water preheaters (economizers) of pulverized-coal-fired boilers as shown in Figs. 18.5 and 18.6. It can be said that the characteristics of the flue-gas-borne ash particles change from being adhesive to abrasive in the temperature range of 1100–1300 K, depending on the composition of the ash. The changeover temperature, from the captive impaction resulting in deposition formation to noncaptive impaction with a risk of erosion wear, is particularly important in high-velocity systems, i.e., coal-fired gas turbines, as discussed in Sect. 18.9.

It was already shown that both boiler slagging and furnace-wall and superheater corrosion increase exponentially with temperature. That is, the metal and flue-gas

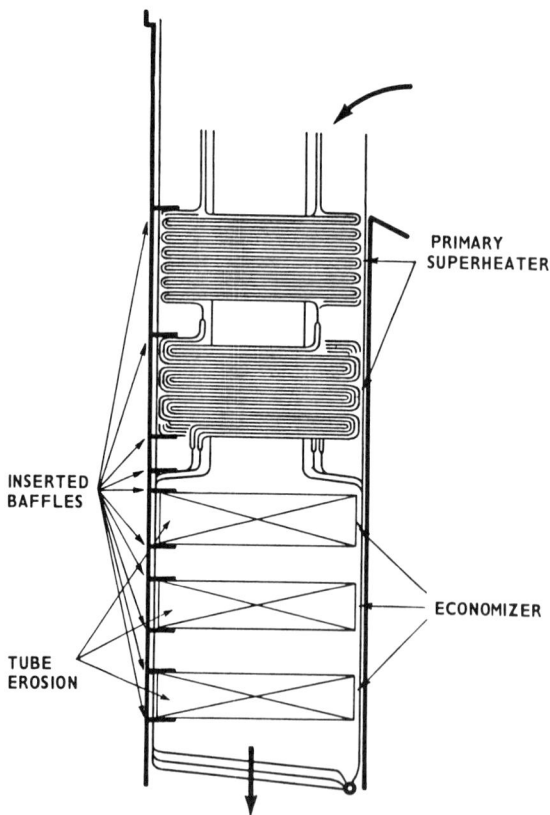

PRIMARY
SUPERHEATER

INSERTED
BAFFLES

ECONOMIZER

TUBE
EROSION

Figure 18.5 Economizer-tube erosion and baffle arrangement to reduce gas velocity in the gap.

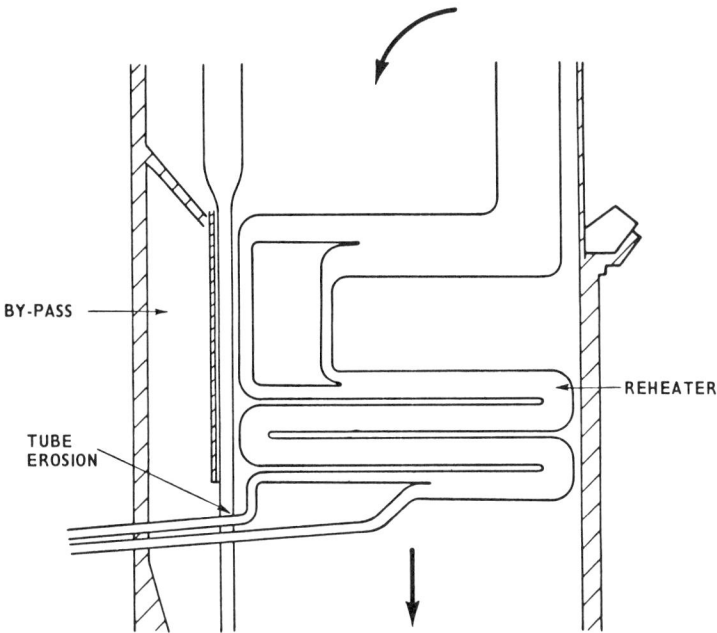

Figure 18.6 Tube erosion underneath reheater bypass.

temperatures are the rate-controlling parameters. In contrast, in the case of erosion wear of boiler tubes, the rate-controlling parameter is the velocity of the impacting particles, and the rate of metal loss increases according to the equation

$$E_w = Au^n \tag{18.8}$$

where E_w = rate of weight loss from the eroded target per unit weight of impacting material

u = velocity of impacting particles

n = velocity index

A = constant

The boiler section where ash erosion wear frequently occurs, e.g., in the economizer, is fabricated from mild steel. It is therefore appropriate to consider first the results of erosion wear measurements made of this grade steel. Particulate quartz is frequently used as the impacting material in the measurements, and the results are relevant to boiler-tube erosion because an abrasive ash contains a significant quantity of quartz. An expression for erosion wear of mild steel when impacted with quartz particles, 76–105 μm size cut, was obtained from the data published by Tilly (1969), Fehndrich (1969), and Raask (1979c):

$$E_w = 7.58 \ W(10^{-3})u^{2.3} \tag{18.9}$$

where E_w = wear (mg kg^{-1})

W = weight (kg) of impacting quartz

u = velocity of the impacting particles (m s^{-1})

It should be noted that Eq. (18.9) applies only for mild steel of the boiler-tube grade ("killed" steel) when impacted with a material that contains 97 percent by weight quartz as determined by X-ray diffraction and has a sieved size cut of 76–105 μm with no undersize material. The angle of impaction is set to give the maximum erosion wear that occurs when the angle between the impacting particle stream and the target is 0.61 rad (35°C), as shown in Fig. 5.10.

The rate of erosion wear can be expressed in the units of metal thickness lost in unit time when the concentration of impacting material in the carrier gas is given in weight/volume units. The rate of impaction of the erodant material per unit area (W_i) with impaction angle of $\pi/2$ rad is given by

$$W_i = C_p u \tag{18.10}$$

where C_p is the concentration of impacting material in the carrier gas and u is the velocity of the impacting particles. The weight loss of eroding metal E_w per unit area is given by

$$E_w = dD \tag{18.11}$$

where d is the thickness of eroded metal and D is the metal density. The rate of erosion E_d is obtained from Eqs. (18.9–18.11):

$$E_d = \frac{C_p u^{3.3}}{D} \tag{18.12}$$

The density D of mild steel is 7800 kg m^{-3}, and thus the maximum erosion is

$$E_d = (9.7 \times 10^{-7}) C_p u^{3.3} \tag{18.13}$$

where E_d = erosion wear rate of mild steel (μm s^{-1})
  C_p = concentration of quartz particles in the carrier gas (kg m^{-3})
  u = impaction velocity (m s^{-1})
The rate of tube-metal loss in power station practice is frequently expressed in nm h^{-1}, nanometer per hours units (1 nm = 10^{-9} m), and the concentration of gas-borne particles is expressed in g m^{-3}; thus Eq. (18.13) can be written

$$E_d = (3.49 \times 10^{-3}) C_p u^{3.3} \tag{18.14}$$

The maximum wear rate of mild steel occurs at the impaction angle of 0.61 rad, as shown in Fig. 5.10, but the particle impaction rate at the angle of $\pi/2$ rad is retained in Eqs. (18.13) and (18.14). This is because the particles in a stream approaching the tubular target will deviate from the central direction line, resulting in an enhanced impaction at more oblique angles, as was shown in Fig. 12.3.

The rate of erosion wear of metals and alloys may increase or decrease with the increase in metal temperature, depending on the mechanism of erosion, as discussed by Gat and Tabakoff (1980). The transition temperature from brittle to ductile behavior of the metals and the annealing temperature are of particular interest. Bitter (1963) has suggested that ductile metals and alloys should show a reduction in the erosion wear at higher temperatures as a result of the annealing recovery of deformation defects caused by particle impaction. Tilly (1969) and Raask (1979c) have shown that the erosion wear rate of mild steel decreased with the increase in metal temperature.

Fehndrich (1969) has shown, however, that the thermal erosion behavior of steels depends on the particle size of impacting particles. With small particles the erosion rate decreased with the increase in temperature as a result of formation of a layer of oxide, which had a higher resistance to erosion than that of the parent metal. He claims, however, that larger particles in an impacting stream were able to penetrate the oxide layer, thus exposing metal to erosion, and there was no significant reduction in the wear rate with increased temperature. The pulverized-coal ash contains some large gritty particles that may cause damage to the protection oxide. It was therefore considered to be prudent to use the comparatively high erosion wear rate of mild steel measured at room temperature rather than the lower rates obtained in laboratory furnace measurements (Raask, 1979c) in subsequent calculation of wear rates of boiler tubes.

The rate of erosion wear of boiler tubes can be estimated in terms of the amount and abrasiveness of ash and the velocity and temperature of the ash-laden flue gas. It is assumed that the velocity of the impacting ash particles is the same as that of the flue gas. The volume V of flue gas produced on burning 1 kg coal is

$$V = \frac{22.4 \, W_C T_g}{(12 \times 273) V_{CO_2}} = \frac{0.00684 \, W_C T_g}{V_{CO_2}} \tag{18.15}$$

where W_C is the weight fraction of carbon in coal; 22.4 is the CO_2 molar volume at the atmospheric pressure and when T_g, the gas temperature, is 273 K; 12 is the atomic weight of carbon; and V_{CO_2} is the volume fraction of CO_2 in the flue gas.

The ash burden C_A of the flue gas is given by

$$C_A = \frac{W_A}{V} = \frac{(1.46 \times 10^5) R_{A/C} V_{CO_2}}{T_g} \tag{18.16}$$

where W_A = weight fraction of ash in coal

$R_{A/C}$ = ash to carbon ratio in coal

V_{CO_2} = volume fraction of the flue gas at temperature T_g and atmospheric pressure

Equation (18.16) applies when the total ash is entrained in the flue gas. In pulverized-coal-fired boilers the entrainment factor is about 0.8—that is, 20% of ash is deposited on the furnace walls and discharged as slag and clinker. Thus the ash burden of the flue gas (C_{A_1}) is

$$C_{A_1} = \frac{(1.17 \times 10^5) R_{A/C} V_{CO_2}}{T_g} \tag{18.17}$$

A typical value of the ash to carbon ratio in coal is 0.25, and the volume fraction of CO_2 in the flue gas is 0.14, which gives an ash burden at 1000 K as follows:

$$C_{A_1} = \frac{1.17 \times 10^5 \times 0.25 \times 0.14}{10^3} = 5.8 \text{ g m}^{-3} \tag{18.18}$$

In cyclone-fired boilers, about 75% of ash is discharged from the furnace as slag and the ash burden of the flue gas is reduced correspondingly:

$$C_{A_2} = \frac{(3.65 \times 10^4) R_{A/C} V_{CO_2}}{T_g} \qquad (18.19)$$

An equation for calculating the rates of erosion wear of mild-steel boiler tubes in pulverized-coal-fired boilers can be obtained by replacing the $C_{W/V}$ term in Eq. (18.13) by the C_{A_1} term from Eq. (18.17) and by introducing the abrasive index of ash, I_a:

$$E_m = \frac{408 \, I_a R_{A/C} \, V_{CO_2} u^{3.3}}{T_g} \qquad (18.20)$$

where E_m is the erosion wear of mild steel (nm h^{-1}) and $I_a < 1$ [Eq. (19.8)]. The volume fraction of CO_2 in the flue gas of pulverized-coal-fired boilers is usually 0.14, and thus Eq. (18.20) can be written as

$$E_m = \frac{57.2 \, I_a R_{A/C} u^{3.3}}{T_g} \qquad (18.21)$$

Table 18.3 in Sect. 18.2 shows that the abrasive index of typical pulverized (bituminous) coal ashes was between 0.18 and 0.28 when the abrasive index of 76–105-μm quartz was taken as 1.0. The range of ash abrasiveness was extended for the erosion wear rate calculations to include exceptionally abrasive ashes and nonabrasive ashes, as shown in Table 18.5. Type 1 ash was taken to contain 60 percent quartz and thus it would be exceptionally abrasive. Types 2 and 3 ashes are those of typical bituminous coals, and type 4 represents a low silica, nonabrasive ash from a sub-bituminous coal (O'Gorman and Walker, 1972).

The last column of Table 18.5 shows the relative erosion wear rate of economizer tubes in pulverized-fuel-fired boiler burning coals of different quality. An erosion wear rate of 50 nm h^{-1} represents the tube thickness loss of 5 mm in 100,000 h operation. Clearly, the erosion wear rate, 150 nm h^{-1}, with type 1 coal would be too high when the boiler flue gas velocity is 15 m s^{-1}. Type 2 coal gives the erosion wear rate of 50 nm h^{-1}, and coals 3 and 4 have the rate well below this value with the impaction velocity of 15 m s^{-1}. Figure 18.7 shows the erosion wear/velocity relationship for the four types of coal ashes.

The ratio of ash to carbon in coal ($C_{A/C}$) and the abrasive index I_a of ash can

Table 18.5 Erosion wear rates of mild-steel boiler tubes

Coal type	Ash (wt. % of coal)	Abrasive index	Boiler flue gas Velocity (m s^{-1})	Temperature (K)	Erosion wear (nm h^{-1})
1	25	0.6	15	800	148
2	15	0.4	15	800	50
3	10	0.2	15	800	16
4	5	0.1	15	800	5

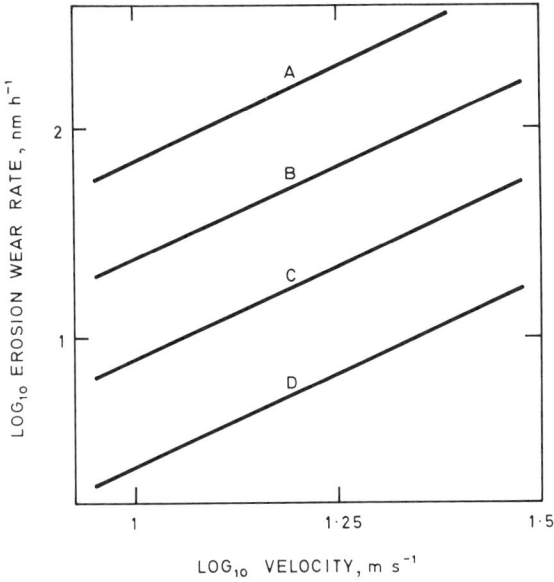

Figure 18.7 Erosion wear of boiler tubes with different ashes: A, type 1; B, type 2; C, type 3; D, type 4 (Table 18.5).

be combined in an erosion parameter of ash P_a:

$$P_a = C_{A/C} \times I_a \tag{18.22}$$

and thus Eq. (18.20) becomes:

$$E_m = \frac{57.2\, P_a u^{3.3}}{T_g} \tag{18.23}$$

where E_m = metal erosion rate (nm h^{-1})

P_a = ash erosion parameter as already defined

u = flue gas velocity (m s^{-1})

T_g = gas temperature

Figure 18.8 shows the maximum velocity of the flue gas in the superheater, reheaters, and economizer consistent with a minimum tube life of 100,000 h operation. It was assumed that the high-pressure superheater economizer tubes have a wall thickness of 6 mm allowing for 5 mm erosion wear, and the low-pressure reheater tubes have 3-mm-thick wall allowing for 2.5 mm erosion wear. The temperature of flue gas in the superheaters and reheaters was taken to be 900 K, and that in the economizer was 650 K. Most bituminous coals have an ash of erosion parameter between 0.05 and 0.1 as defined by Eq. (18.22), so the velocity of the flue gas in the primary superheaters, reheaters, and economizers of pulverized-coal-fired boilers should be between 12 and 20 m s^{-1} in order to prevent excessive erosion wear by ash impaction.

The ash burden of the flue gas in cyclone-fired boilers is about four times less than that of the pulverized-coal-fired flue gas. Moreover, the ash in the cyclone boiler flue gas is comparatively nonabrasive because of the small particle size and low quartz content as discussed in Sect. 18.1. The approximate ratio value of the

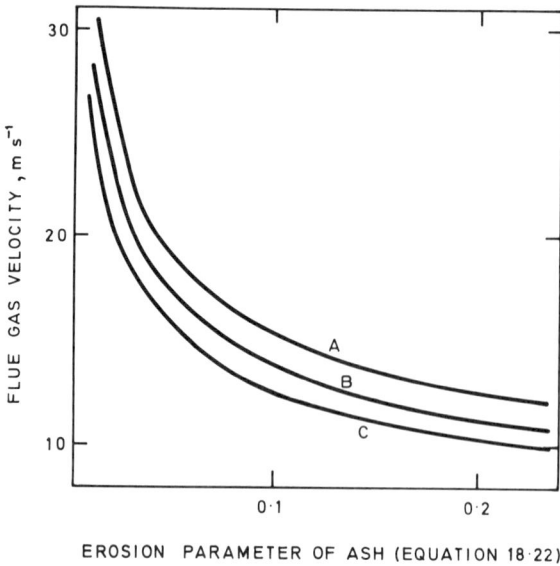

Figure 18.8 The relationship between ash abrasive characteristics and maximum permissible flue-gas velocity: A, superheater; B, economizer; C, reheater. Erosion wear rate limit: superheater and economizer, 50 nm h^{-1}; reheater, 25 nm h^{-1}.

erosion wear parameter P_{a_1} of cyclone boiler ash is given by

$$P_{a_1} = 0.2 P_a \tag{18.24}$$

where P_a is the erosion wear parameter of pulverized fuel ash from the same coal. The corresponding velocity ratio for the same rate of erosion wear is

$$\frac{u}{u_1} = \left(\frac{P_{a_1}}{P_a}\right)^{0.3} = 0.61 \tag{18.25}$$

where u and u_1 are respectively the velocity of the flue gas in pulverized-coal-fired and cyclone-fired boilers.

The slag retention in a vertical cyclone or slag-tap boiler amounts to about 60 percent of the total ash and the flue-gas-borne ash is slightly less abrasive than that in pulverized-coal-fired boilers. The erosion wear parameter P_{a_2} of slag-tap boiler ash is

$$P_{a_2} = 0.45 P_a \tag{18.26}$$

and the corresponding velocity ratio for the same rate of erosion wear is

$$\frac{u}{u_2} = \left(\frac{P_{a_2}}{P_a}\right)^{0.3} = 0.78 \tag{18.27}$$

where u_2 is the flue-gas velocity in the slag-tap boiler.

Figure 18.9 shows the maximum permissible velocity of the economizer flue gas in cyclone, slag-tap, and pulverized-fuel-fired boilers burning different quality coals. The limiting value for the erosion wear rate was taken to be 50 nm h^{-1} (5 mm in 100,000 h operation). The plots in Fig. 18.9 show that it would be a significant advantage in terms of a higher permissible flue-gas velocity to utilize coals with an

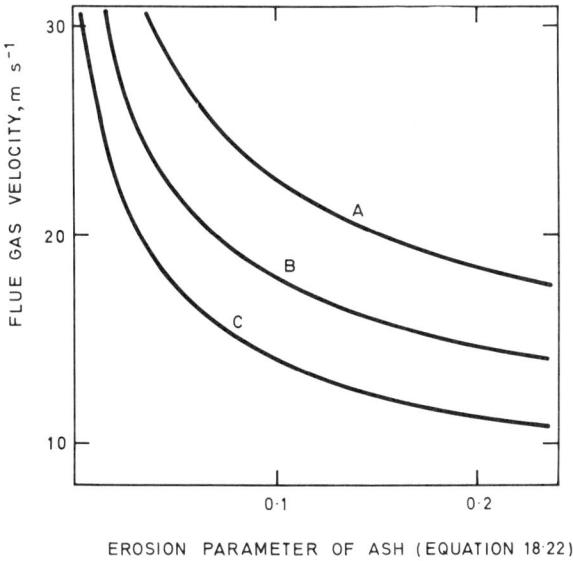

Figure 18.9 Maximum permissible flue-gas velocity in boilers of different design. Erosion wear limit of economizer tubes, 50 nm h⁻¹. A, cyclone firing, 80% slag retention; B, slag-tap firing, 60% slag retention; C, pulverized-fuel firing, 20% slag retention.

abrasive ash in a cyclone-fired boiler. However, these coals usually have an infusible ash and hence are unsuitable for the use in cyclone-fired boilers.

18.4 EROSION WEAR DAMAGE IN COAL-FIRED BOILERS

The erosion wear experience in the coal-fired plant is in general agreement with the calculated wear rates shown in Table 18.5 and in Figs. 18.8 and 18.9. Most power-station coals produce moderate abrasive ash in pulverized-fuel-fired boilers, and the tube failures as a result of ash erosion wear do not constitute a major problem in the first 10 yr of operation. The design flue-gas velocity is between 15 and 18 m s⁻¹, and with typical bituminous coals (ash content of coal <20 percent, quartz content of ash <15 percent) there are few erosion wear tube failures during the first 50,000 h of operation. However, the tube failures become more frequent in the operation period of 50,000–75,000 h, corresponding to the erosion wear rate of 75–100 nm h⁻¹.

The tube failures occur frequently at the bends near the duct wall (Fig. 18.5), and the tubes situated underneath the flue-gas bypass (Fig. 18.6) also suffer from erosion wear. The erosion damage is usually limited to a short length on each tube, as illustrated by the wear pattern of failed reheater tube in Fig. 18.10. There was no significant wear damage at a distance of 50 mm from the erosion hole. Examination of the eroded tube surface under the scanning electron microscope revealed skid marks, 1–5 μm wide and up to 20 μm in length, as shown in Fig. 18.11a. This would suggest that the erosion wear damage was caused by comparatively large particles. Another interesting feature was that the section of 1Cr0.5Mo

Figure 18.10 Erosion-failed reheater tube.

steel tube just outside the eroded area had a pore-free layer of oxide (Fig. 18.11b) strongly adhering to the parent metal. It appears that an oxide layer formed on boiler steel, under the conditions of continuous impaction of ash with velocities below which severe erosion wear occurs, is highly resistant to further oxidation and corrosion. However, once this threshold impaction velocity is exceeded, the erosion wear increases according to the power-law relationship [Eq. (18.23)]. That is, a 20

5 μm

(a)

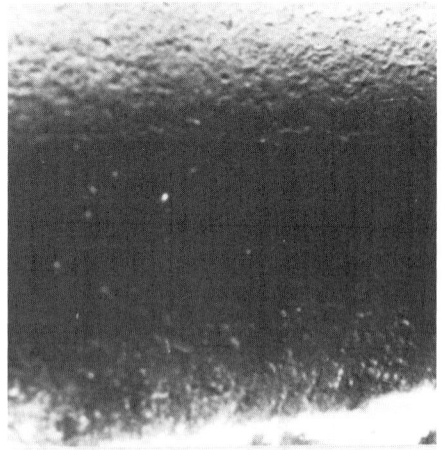

25 μm

(b)

Figure 18.11 Surface views of ash-impacted reheater tubes. (a) Particle skid marks on eroded tube. (b) Magnetite on noneroded tube.

percent rise in the impact velocity would increase the rate of erosion wear by 80 percent, and doubling the velocity of the ash-laden flue gas would increase the wear rate by an order of magnitude.

It is interesting to note that erosion wear damage frequently occurs at the rear wall of the primary superheater or economizer duct, as shown in Fig. 18.5. This suggests that when the flue gas flow changes its direction through an angle of π rad in the pendant superheater section there is a tendency for large and abrasive particles of ash and debris of sintered deposits to concentrate in the stream near the rear wall as a result of inertial momentum of the particles. This phenomenon was found to occur in the earlier design of pulverized-coal-fired boilers burning sand-rich sub-bituminous coals, as discussed by Herrmann and Krause (1965).

The utilization of sub-bituminous coals rich in quartz sand, mined in Germany, for power generation has resulted in severe erosion wear of boiler tubes, as discussed by Beer (1961) and Burger (1961). The coals as mined contained about 50 percent water, and the ash content varied between 3 and 20 percent. The inherent ash was free from quartz, but it amounted to only about 3 percent of coal; the extraneous mineral matter consisted largely of highly abrasive quartz [Eq. (5.21)]. Thus the abrasive index of ash as defined by Eq. (18.7) was between 0.5 and 0.7, and the corresponding erosion wear parameter [Eq. (18.22)] was around 0.3.

The severe erosion wear damage of boiler tubes had led to extensive aerodynamic studies on flow visualization of the airborne quartz particles in model assemblies representing large coal-fired boilers, as discussed by Herrmann and Krause (1965), Recknagel (1967), and Buxmann and Schwab (1973). It was established that in a conventional design of a two-passage boiler—upward flow in the combustion chamber and downward flow in the convection heat-transfer section—there was a nonuniform distribution of large quartz particles in the flue gas after the change in the flow direction through an angle of π rad at the top of the combustion chamber. The large, highly abrasive quartz particles were found in high concentration near the walls as a result of the centrifugal force acting on the particles. A combination of highly abrasive characteristics and enhanced concentration of ash in the zones of erosion wear resulted in rapid tube failures.

The results of the flow model studies and the erosion wear tests by Fehndrich (1969) led to development of the single-pass (Einzug) boiler design, as described by Buxmann and Schwab (1973), and Sur (1979). The noticeable features of this boiler design are that the convection heat-exchange banks are stacked horizontally at the top of the combustion chamber and that in the downflow duct there are no boiler-tube banks. The velocity of flue gas in the single-pass boilers is between 8 and 12 m s^{-1}, which is consistent with the erosion wear characteristics of quartz-rich coals as shown in Figs. 18.7 and 18.8. Some tube failures as a result of ash impaction erosion occur at these low design velocities, as it is difficult to ensure that there are no higher velocity-zone gaps between the tube banks and boiler duct wall.

It has to be accepted that tube failures caused by the ash impaction are likely to occur in a large number of coal-fired boilers. In particular, this will be the case when the 500- to 900-MW units will be earmarked for an extended life, longer than the originally designed operation of 120,000–150,000 h. The tube failures frequently occur in awkward places, deep in the superheater, reheater, and economizer banks, and this makes it difficult to locate the erosion-damaged tubes after the boiler is taken off-load. Much time could be saved if the position of holed tubes could be pinpointed accurately before the unit is shut down for repairs. There are several acoustic devices for locating the noise source of a high-pressure steam jet escaping from holed tubes, and one such electroacoustic probe has been described by Kuper and Werker (1981).

18.5 MAXIMUM FLUE–GAS VELOCITY COMPATIBLE WITH ASH EROSION PROPENSITY

The erosion wear experience with the sub-bituminous coals with an abrasive ash has had a marked influence on the boiler design, as discussed in previous sections. That is, the maximum velocity limit of the flue gas is adjusted to the erosion wear characteristics of the ash assessed by previous experience with similar coals. A similar care has not always been exercised in the design of pulverized-fuel-fired boilers intended for burning bituminous coals with a moderately abrasive ash. The design engineers have been handicapped by the lack of data on the abrasive characteristics of different coal ashes and on the erosion wear rate–impact velocity relationship, as discussed in Sects. 18.1–18.3. The appropriate design velocity of the flue gas in pulverized-coal-fired boilers is arrived at by the results of past experience, but for new coals the erosion wear propensity must be assessed from the results of chemical and mineralogical analysis of laboratory-prepared ash. Appropriate allowances should be made for changes in the hardness, size, and shape of the coal mineral species in boiler flame, as summarized in Tables 18.1 and 18.2.

Once the erosion wear propensity of ash has been established, the maximum permissible velocity of the flue gas for a given limiting wear rate–e.g., 50 nm h^{-1} or 25 nm h^{-1}–can be obtained from Eq. (18.23), and Figs. 18.7 and 18.8. The limiting velocity thus obtained should err on the side of being conservative, and in operation a 10 percent increase in the flue-gas velocity above the design value has been allowed in the calculated erosion wear rates in Figs. 18.7 and 18.8. This is because Eq. (18.23) was derived on the assumption that the rate of erosion of boiler steels does not decrease with the increase in metal temperature. This built-in safety margin cannot, however, account for 50–100 percent increase in flue-gas velocity, which may occur in the gaps between the tube banks and boiler casing and in and underneath the bypass ducts. In order to avoid severe erosion wear damage caused by the flue-gas stream emerging from a bypass duct, there should not be any boiler tubes placed at a distance of 2 m from the duct exit. That is, the design detail as shown in Fig. 18.6 is unacceptable. The reheater tubes of 4 mm wall thickness in this position failed after 16,000 h operation, giving an average wear rate of 250 nm h^{-1}. The erosion wear parameter of ash was 0.08 as defined by Eq. (18.22) and the flue gas temperature was 950 K; thus from Eq. (18.23) the velocity of the flue gas at the locality of maximum erosion wear was 26 m s^{-1}. The design velocity of the flue gas in the main passage was 15 m s^{-1}; thus there was over 60 percent increase in the velocity above the design value.

The erosion wear problem in the high-velocity zone gaps between the tube banks and boiler casing (Fig. 18.5) is more difficult to resolve. These gaps are necessary to accommodate thermal expansion of the water and steam tubes, which amounts to over 0.1 m for 25-m-long tubes when heated from ambient temperature to 400 K. A variety of baffle-plate arrangements have been applied to prevent a high-velocity flue-gas stream emerging from the gaps, as shown in Fig. 18.5. The problem may arise that a plate will create a reduced pressure zone underneath the plate, thus accelerating the flue-gas stream around the edge of the plate. Thus the problem of erosion wear is not solved but is merely displaced to another locality. A multistage baffling arrangement, as shown in Fig. 18.5, should avoid the problem of slipstream acceleration. Another solution is to use an "aerodynamically balanced" baffle system, where perforated plates are set at an angle to the direction of flue-gas flow. The idea is that the perforated plates should offer the same degree of resistance to the flue-gas flow as that caused by tube banks. The oblique angle setting of the baffles would prevent ash accumulation on the plates causing blockage of the gas flow holes. The design and arrangement of the flow equalizing baffles for coal-fired boilers where some erosion wear damage is to be expected requires a flow characterization study in a rig assembly. Otherwise, it would be difficult to predict the exact locality of high-velocity flue-gas flow in a large boiler duct.

The boiler-plant design engineers, when considering the problem of high-temperature corrosion of superheater and reheater tubes, have a choice of corrosion-resistant alloys, including coextruded tubes, as discussed in Chap. 17. A similar choice of erosion-resistant materials is not available to combat the erosion wear of the primary superheaters, reheaters, and economizers. In order for a metal alloy to have a significantly higher erosion wear resistance than that of mild steel, it must have a Vickers hardness number above 400, as shown in Fig. 5.11. These types of alloys are brittle and are not suitable for fabrication of pressure tubes. The erosion wear resistance of austenitic steel is hardly better than mild steel (Fig. 5.11), and Fehndrich (1969) has shown that the rates of erosion wear of different steels tend to converge at boiler-tube temperatures. It would be possible to deposit by welding or other techniques a layer of erosion-resistant alloy on boiler tubes, but it is not a practicable measure to be recommended in the design stage. It would require pinpointing exact locations of the erosion wear, and this can be acquired only by operation experience.

The boiler design engineers have a choice of the alternative design (namely, cyclone firing) for coals of highly slagging ash, which are difficult to burn in pulverized-coal-fired boilers with "dry" ash removal. Figure 18.9 shows that there would be a significant reduction in the erosion wear when burning the coals of highly abrasive ash in cyclone-fired boilers, as a result of low ash burden of the flue gas. However, this option is not available for coals of a high quartz content ash, as the ash is infusible and requires exceptionally high temperatures above 1800 K for the slag to flow. These coals are not therefore suitable for burning in cyclone-fired boilers.

18.6 MEASURES TO COMBAT ASH IMPACTION EROSION WEAR OF BOILER TUBES

There are several measures listed in Table 18.6 that can be considered when the problem of erosion wear caused by ash impaction becomes apparent.

A marked change in the quality of coals delivered to power stations can cause excessive erosion wear of boiler tubes. Frequently the average ash content of coals is significantly above the value taken in the boiler design stage to specify the maximum velocity of flue gas in the convection heat-transfer passages. The fuel suppliers are invariably optimistic in quoting the maximum ash content of coal for future deliveries. Also, low-grade, ash-rich coals are frequently offered at attractive prices based on the calorific value. In many cases an increase in the average ash content from around 15 percent to over 20 percent could significantly increase the rate of erosion wear. The point to note is that, usually, the higher the ash content of coal, the higher the quartz content of ash. It is thus possible that a 25 percent increase in the ash of coal may double the erosion wear propensity of ash, as defined by Eq. (18.22) in Sect. 18.3. That is, the erosion wear propensity and other properties—e.g., slagging and sulfur content—of ash of low-grade "bargain" coals should be carefully considered before these are purchased for use in boilers that have not been designed to burn this type of fuel.

An inadequate performance of sootblowers in preventing the formation of an excessive amount of deposit can cause a significant increase in the tube erosion

Table 18.6 Measures to alleviate ash impaction erosion wear

Measure	Effect on erosion	Comment
Decrease of ash content of coal	Could have a significant effect as a result of reduced dust burden and less abrasive ash	May not be practicable and will increase the cost of fuel
Improved soot-blowing	Reduces temperature, hence the velocity of the flue gas; prevents increase in velocity due to partial blockage of the passages	Improves the heat transfer, hence the performance of the boiler
Removal of ash from superheater and economizer hoppers	Prevents reentrainment of coarse and abrasive ash from fuel hoppers	Many boilers do not have the hoppers
Reduction of combustion air	Would reduce the volume, hence the velocity, of flue gas	This is not usually a practical measure; it may increase boiler slagging and corrosion
Baffles for high-velocity gaps	Can significantly reduce localized erosion wear	Should be applied as soon as the erosion wear pattern becomes evident
Application of weld deposits and shields	Will extend the life of tube when erosion wear is localized	Should be applied before the damage becomes extensive

wear. As a result of reduced heat transfer, a badly slagged combustion chamber will cause a 15 percent increase in the temperature of the combustion chamber at furnace exit. Consequently the temperature and velocity of the ash-laden flue gas in the superheater, reheater, and economizer sections are increased by the same margin. If coupled with the furnace slagging, there is a partial blockage of the convection heat-transfer passages by ash deposit, say 10 percent of the total, then the combined effect would be to increase the flue-gas velocity by 25 percent. This would result in doubling the rate of erosion wear of superheater, reheater, and economizer tubes. Clearly, inadequate performance of sootblowers can be a significant factor when excessive ash impaction erosion wear occurs in pulverized-coal-fired boilers.

It is surprising to note that although the flue-gas velocity is the erosion wear rate-controlling parameter and is highly sensitive to boiler operation conditions, there are hardly any coal-fired boilers where the velocity is monitored. The technique of velocity measurements by differential static and dynamic pressure tubes is well established and, when coupled to recording micromanometers, should give reliable data. There may be some problems of blockage of the measuring heads by ash and changes in the direction of the flue-gas flow. These could probably be overcome by the use of multihead or rotating sensors kept clean by intermittent air purges. The lack of data on flue-gas velocities in the existing boilers makes it difficult to establish an accurate relationship between the erosion rate of boiler tubes, the velocity of the ash-laden flue gas, and the abrasiveness of ash.

A steam or air-jet sootblower in the wrong position can cause severe erosion wear of boiler tubes. The expanding sootblower jet mixes with approximately an equal volume of the flue gas for every distance of the jet diameter it travels, as discussed in Chap. 14. Thus by the time the jet impacts boiler tubes it consists largely of the ash-laden flue gas. Sootblowing is usually required once every 8 h, and the jet action on a boiler tube may last 10 to 20 s, but this short duration may be sufficient to cause severe erosion wear by the entrained ash particles at sufficiently high velocities. Figure 18.12 shows the erosion wear–velocity relationship during sootblowing for a weakly abrasive ash (curve A) and for a moderately abrasive ash (curve B). The data should be useful for locating the sootblowers in boiler plant so that a minimum permissible distance between jet nozzle and boiler tubes will be maintained.

It is fortunate that highly slagging ashes are usually low in quartz and thus are only moderately abrasive. This will permit a closer positioning of sootblowers to boiler tubes for a more effective action without a risk of severe erosion wear. Some sub-bituminous coals of high quartz, calcium, and sodium contents constitute an exception, in that the ash is highly abrasive (Section 18.3) and has a high deposit-forming propensity as well, as discussed in Chaps. 10 and 15.

Severe damage can be caused by sootblowers placed near the sloped floor underneath the pendant superheaters or situated immediately above the ash hoppers. The minimum distance should be at least 1.5 m, otherwise the high velocity jet will entrain a large quantity of coarse and abrasive ash from the settled material. The ash hoppers, when provided in the primary superheater and reheater economizer

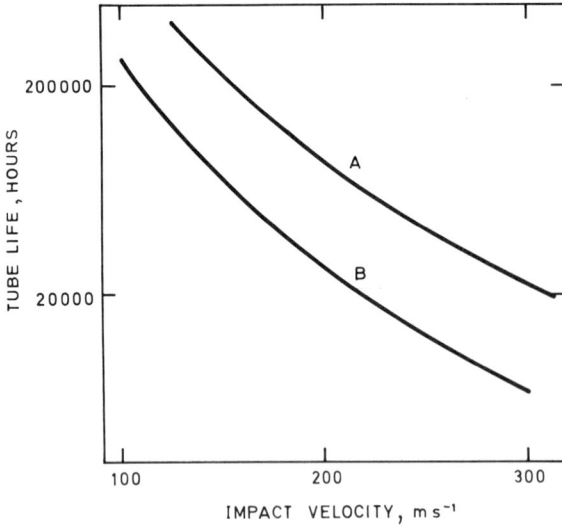

Figure 18.12 Erosion wear of 5-mm-wall tubes during soot-blowing: A, weakly abrasive ash; B, moderately abrasive ash.

sections, should not be allowed to overflow with ash. The coarse ash (Fig. 8.6) is highly abrasive, and if it is not removed from the duct some of it will be reentrained in the flue gas. Also, it is recommended that the opportunity should be taken on each boiler outage to remove the coarse ash settled in the superheater, preheater, and economizer sections.

The large coal-fired boilers are operated with excess oxygen in the flue gas, usually in the range of 2–4 percent, which corresponds to an excess of combustion air between 10 and 20 percent. The excess combustion air in smaller industrial boilers may be higher. A reduction of combustion air would decrease the velocity of the flue gas, but in power-station boilers the amount of excess combustion air is usually close to the optimum limit. Any reduction would result in a loss of combustibles and in enhanced boiler slagging and corrosion, as discussed in Chaps. 14 and 17. There may be some scope in decreasing the excess air in the flue gas of some industrial boilers to alleviate the erosion wear problem.

A reduction of boiler load is another possible measure for reducing the rate of erosion wear of boiler tubes, but with high-merit base-load boilers this course of action would not be justified on the grounds of economy. Burger (1961) has reported a case where the tube erosion was significantly reduced by a temporary measure of decreasing the boiler load by 20 percent. The boiler burned low-grade sub-bituminous coals with a highly abrasive ash, and the flue gas velocity in the zone of excessive erosion wear was above 15 m s^{-1}. Subsequently the flue gas passage was changed to reduce the flue gas velocity to 10 m s^{-1} in the erosion zone on full-load operation.

A variety of ingeneous arrangements of baffle plates have been devised to equalize the flow velocities; some of these are discussed in Sect. 18.5. It is unfortunate that there are usually no records of the flow conditions—e.g., the velocity and the dust burden of the flue gas before and after insertion of baffles—required for an assessment of the effect of baffle arrangements.

The partially eroded and new replaced tubes can be protected by shields fabricated from hardened alloys that have a better erosion wear resistance than that of mild steel (Fig. 5.11). The problem is that severe erosion wear occurs frequently on tube bends (Figs. 18.5 and 18.6) and thus the shields have to be tailor-made to fit the tubes. Eroded areas of boiler tubes can be made good by weld deposit, but the repair work is time-consuming and the locality of severe erosion is frequently inaccessible.

The application of erosion-resistant materials, chiefly alumina and silicon carbide, has been considered for protection of boiler tubes. The results in Table 5.3 showed that these materials can have a low rate of erosion wear, but there are many practical difficulties. An erosion-resistant refractory material must have good grain-to-grain bonding and low porosity, which can be achieved in manufacturing processes but is difficult to accomplish with *in situ* application. Chemically bonded and flame-sprayed coatings are unlikely to satisfy the stringent adhesion and low porosity requirements. The development of *in situ* application of plasma-sprayed ceramic coatings, as discussed by Meringolo (1981), may lead to a useful measure for combating boiler-tube erosion by ash impaction. The plasma temperature is sufficiently high to accomplish a firm grain-to-grain bonding, resulting in a material of low porosity and resistant to erosion.

18.7 COMBATING WEAR DAMAGE OF ASH TRANSPORT LINES

The severity of erosion wear in the pneumatic transport lines depends largely on the amount of gritty material in the ash, e.g., the material from economizer hoppers. Severe abrasive wear also occurs in the hydraulic pipe lines transporting the abrasive clinker material (furnace-bottom ash) to disposal sites. The erosion wear of dry-ash pipelines occurs at bends and elbows and can be reduced by the same measures used to combat the wear damage in the pulverized-coal pipes, as discussed in Chap. 5. These measures include the design of pipeline systems where sharp bends are avoided and the flow velocity is kept to the minimum necessary to prevent ash settlement, and the use of erosion wear-resistant sleeve or lining materials. When sharp bends and elbows in the pneumatic ash transport line are unavoidable, these curved sections should be accessible and designed for easy replacement by new and repaired sections.

The abrasion wear in the hydraulic ash transport lines is caused by water dragging the abrasive clinker along the bottom wall of pipes. As a result the abrasion wear can be extensive: the straight-line sections as well as bends and curved sections will suffer wear damage. The factors affecting the abrasion wear in hydraulic systems—e.g., the hardness of pipe material and abrading particles, and the flow velocity—have been discussed by Truscott (1970). Stauffer (1958) has carried out wet and abrasion tests with different hardened steels, alloys, and nonmetalic materials. The relative wear rate and the hardness of abraded material show the same relationship as that shown in Fig. 5.11. That is, hardened steel alloys that have a Vickers hardness number between 600 and 800 will have the wear rate 3 to 4 times less than that of mild steel when subjected to sliding wear in a hydraulic transport system.

A variety of pipe materials have been tested for abrasion wear in the hydraulic systems transporting ash clinker from coal-fired power stations to disposal sites. These include cast iron and a variety of hardened steels, cast basalt, high-alumina concretes and refractory materials, and heavy-duty plastics. In many applications high-alumina concrete and refractory pipes and polyurethane pipes have given satisfactory service. Paterson and Watson (1979) have arrived at similar conclusions after testing several materials in a pilot installation of hydraulic transport of colliery waste.

18.8 EROSION WEAR IN OIL–TO–COAL CONVERTED BOILERS

The changeable situation in the primary fuel supply to power stations dictates that from time to time fossil-fuel-fired boilers designed to burn coal, oil, or gas need to be converted to another fuel. For example, in the period from 1950 to 1960, which heralded the epoch of two decades of plentiful and cheap oil, a large number of coal-fired boilers were converted to oil firing. Some 20 yr later the situation is the reverse: oil-fired boilers are now being converted to coal firing, only this time, the problems on conversion are much more formidable, as discussed in EPRI Seminar Proceedings (Drenker, 1981) and by Richards (1978). The boiler units to be converted are of 150 to 600 MW capacity, instead of the 15- to 60-MW output boilers two decades ago, and it is a much more difficult task to convert an oil-fired boiler to coal firing than to carry out the change from coal to oil burning. In many oil-fired power stations there is the site restriction for adequate coal and ash handling facilities; the combustion chamber of an oil-fired boiler is far too small for a coal-fired boiler of the same output capacity; and there may be severe erosion wear when coal ash is introduced in a boiler designed for oil firing. The mode of introducing fuel into converted boilers may be coal/oil or coal/water slurry, or pulverized fuel alone. An assessment of the possible erosion wear problems is made on these modes of firing.

The choice of coals for coal/oil slurries will probably be limited to bituminous fuels of comparatively high calorific value (Table 18.7).

For the abrasive index of ashes it was assumed that the solid fuel in coal/oil slurries will be milled to a fine size, e.g., 90 percent by weight passing through a 76-μm (200-mesh) sieve. Finely ground coal will produce a correspondingly fine ash (sample 2 in Table 18.5). It was further assumed that the lower the ash content of coal, the lower is the ash abrasive index, since the inherent ash contains usually only a small amount of quartz. The erosion parameter values were calculated from Eq. (18.22).

The combustion chamber of an oil-fired boiler is significantly smaller than that of a pulverized-coal-fired boiler of the same thermal output. For example, a 500-MW oil-fired boiler has the projected surface area of furnace walls of 2000 m^2,

Table 18.7 Abrasiveness of coal/oil slurry ashes

Coal type	Calorific value (MJ kg^{-1})	Percent by weight		Abrasive index, I_a	Erosion parameter, P_a
		Ash	Carbon		
Medium ash	26	12	67	0.18	0.032
Low ash	28	8	70	0.16	0.018
Clean	30	4	73	0.14	0.008

whereas a pulverized-coal-fired boiler of the same output has a furnace-wall area of 3500 m^2 (Godridge and Read, 1976). It is therefore evident that an oil-fired boiler converted to pulverized-coal firing will have a reduced thermal output unless the combustion chamber is rebuilt and enlarged. A reduced boiler load will decrease the flue-gas velocity, but the reduction is not in direct proportion, as shown in Table 18.8. This is because the amount of excess combustion air in coal-fired boilers is higher than that in oil-fired boilers.

The probable rates of erosion wear of superheater, reheater, and economizer tubes in an oil-to-coal converted boiler can be calculated from Eq. (18.23) using the velocity data of Table 18.8 and the erosion wear parameter (P_a) numbers in Table 18.7. For the calculated erosion rates given in Table 18.9 it was assumed that the maximum velocity of flue gas in the original oil-fired boiler was 30 m s^{-1}. For any other velocity value U_o, the erosion rates should be multiplied by the factor $(U_o/30)^{3.3}$. For example, if U_o is 25 m s^{-1}, the multiplication factor is 0.55.

The thermal output of converted pulverized-coal-fired boilers is likely to be between 60 and 70 percent of the original oil-firing design. Correspondingly, the flue-gas velocity will probably be in the range of 20–25 m s^{-1}. It is evident from Table 18.9 that coals low in ash, 4–8 percent by weight, could be burned in the converted boilers without causing severe superheater or economizer tube erosion by ash impaction. When the design velocity of the flue gas in the original oil-fired boiler was less than 30 m s^{-1}, coals with a higher ash content could be utilized after the conversion to solid-fuel firing.

In contrast, when low-ash coals contain comparatively large amount of quartz, the velocity of the flue gas has to be limited to a value below 20 m s^{-1}. Kirkwood and Chatfield (1981) have described conversion of two 200-MW oil-fired boilers to pulverized-fuel firing where the maximum velocity of the flue gas was restricted to 15 m s^{-1}. The fuel supply consists of a mixture of 4 percent and 7 percent coals mined in western Australia. According to the calculated wear rates in Table 18.9,

Table 18.8 Comparison of oil and coal boiler firing

Parameter	Oil	Pulverized coal		
Capacity (% of original)	100	80	70	60
Maximum flue-gas velocity (m s^{-1})	30	28.5	25.0	21.3

Table 18.9 Erosion wear rates of boiler tubes in oil-to-coal converted boilers (flue gas temperature = 800 K)

Ash content of coal (wt. %)	Rate of erosion wear (nm h^{-1})	Life of 5-mm wall thickness tube (h)
80% Output; flue-gas velocity = 28.5 m s^{-1}		
12	145	34,500
8	81	61,500
4	36	139,000
70% Output; flue-gas velocity = 25 m s^{-1}		
12	94	53,000
8	53	94,500
4	23	217,000
60% Output; flue-gas velocity = 21.3 m s^{-1}		
12	55	91,000
8	31	161,000
4	14	361,000

the flue-gas velocity of 20-22 m s^{-1} would be permissible with coals with a moderately abrasive ash. Some of the coals mined in western Australia are, however, known to contain comparatively large amounts of quartz, thus producing a highly abrasive ash.

From the point of view of boiler-tube erosion it would be advantageous to convert an oil-fired boiler to cyclone firing, where up to 80 percent of coal ash is retained and discharged from the furnace in the form of slag. As a result of the reduced ash burden and less abrasive nature of the ash in the cyclone-fired boiler compared with the pulverized-coal-fired boiler, the velocity of the flue gas in cyclone fired boilers can be increased by a factor of 1.6 [Eq. (18.25), Sect. 18.3]. The corresponding multiplication factor for the maximum permissible velocity of the flue gas in slag-tap (vertical cyclone) boilers is 1.3 [Eq. (18.27)]. It thus appears that when an oil-fired boiler is converted to cyclone firing where 75 percent of ash is retained as slag, up to 80 percent of the original output could be obtained by burning pulverized coal of 8-12 percent ash without a serious risk of erosion wear by ash impaction.

The utilization of a mixed fuel of coal and oil in the form of a slurry is a possible alternative to the conversion of boilers to pulverized-coal firing, which is expensive to carry out and results in a reduced thermal output. The results of extensive work in the field of stabilizing the particulate coal suspension in oil have shown that up to 40 percent of oil can be replaced by the solid fuel, as discussed by Cook and Furman (1981). The ash content of oil is usually 0.05-0.1 percent by weight, which is small in comparison with that of coal. It is therefore possible to consider a 60/40 mixture of oil and coal as a fuel where the ash content has been reduced by a factor of 0.4 when compared with that of solid fuel. In order to facilitate the formation of stable suspension of coal in oil, the solid fuel is required to be milled to a significantly smaller particle size than that of pulverized coal. In general, a finely ground coal will produce a less abrasive ash and thus the abrasive index values of ashes given in Table 18.7 were used for calculating the erosion wear rates from Eq. (18.23), and the ash burden was reduced by a factor of 0.4.

Figure 18.13 shows the erosion wear rates of superheater and economizer tubes in boilers designed for oil firing, burning 40/60 fuel mixture operating with flue-gas velocities between 25 and 35 m s^{-1}. The flue-gas temperature was taken to be 800 K. The estimated erosion wear rates suggest that when a coal with ash content below 8 percent is used to produce a coal/oil suspension for burning in oil-fired boilers there should not be any serious tube wear as a result of ash impaction. That is, the wear rate should not exceed 50 nm h^{-1}, corresponding to an operation life of 100,000 h for 6-mm-wall tubes, allowing 5 mm metal loss. High-quartz coals with an abrasive ash are an exception and require the flue-gas velocity on coal/oil mixture firing to be restricted to below 25 m s^{-1}.

This cursory assessment shows that with the majority of low-ash coals there should not be a serious problem of boiler-tube erosion unless an extensive

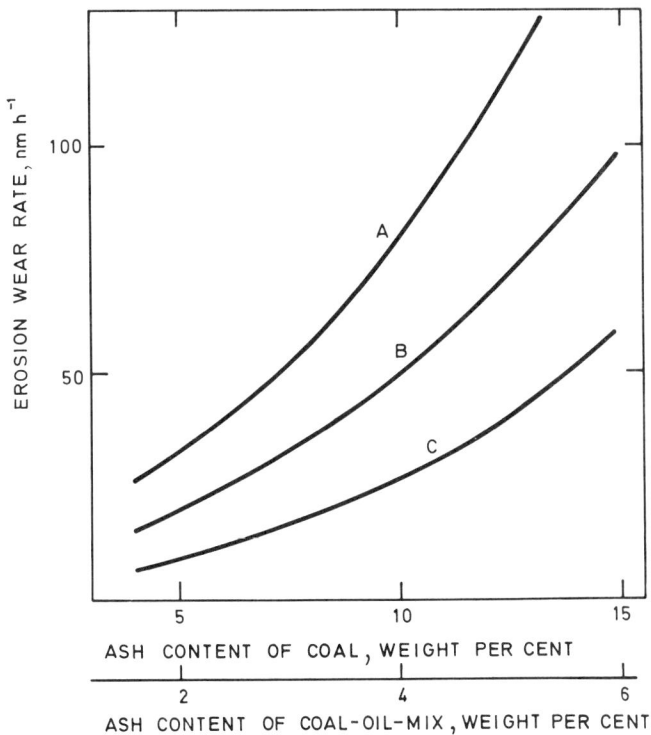

Figure 18.13 Erosion wear of boiler tubes by ash impaction on coal/oil slurry firing. Flue gas velocity: A, 35 m s^{-1}; B, 30 m s^{-1}; C, 25 m s^{-1}.

coalescence of small coal mineral particles takes place in the dual-fuel flame. This may be the case when the small particles of coal mineral matter trapped in oil droplets coalesce by sintering and viscous flow as the devolatilized coke residue burns. An extensive coalescence of the inherent ash takes place when the coal coke residue burns in a pulverized-fuel flame, as discussed in Chap. 8. Sarofim et al. (1977) have concluded from the results of their work on ash coalescence that on average four ash spheres are formed from the inherent mineral matter dispersion in each pulverized-coal particle. A similar degree of coalescence of the "inherent" mineral matter trapped in the mixed coal/oil droplets would lead to a significant increase in the average particle size of ash. It may be, however, that the coal and mineral particles will be ejected from the oil droplets during the devolatilization stage, and thus the degree of coalescence would be less extensive.

There is another possible source for large and abrasive particles in a dual-fuel-fired boiler. The introduction of coal in a boiler designed for oil firing is likely to increase the amount of ash deposited on the furnace wall, and on the high-temperature superheater and reheater tubes. In order to maintain the necessary heat-transfer rates extensive sootblowing would be required. The sootblower action on slightly sintered deposit will produce large (above 100 μm) gritty particles (Fig. 8.6), which are highly abrasive. It is therefore highly likely that the rate of erosion wear in dual-fuel-fired boilers will depend markedly on the degree of agglomeration of the small particles of coal minerals in burning oil droplets and at the surface of boiler tubes. The formation of large and abrasive particles may be sufficiently extensive to double the erosion wear rates plotted in Fig. 18.13.

18.9 EROSION WEAR IN COAL–GAS TURBINES COUPLED TO FLUIDIZED–BED COMBUSTORS

Industrial gas turbines have been extensively used for peak load power generation. The units can be brought to service quickly, but require expensive distillate fuel for operation and will decline unless a cheaper replacement fuel can be found. Natural-gas-fired turbines can be coupled to the Rankine cycle boiler plant giving an overall generating efficiency of 42–43 percent compared with 36–38 percent for the conventional boiler. It is, however, unlikely that many such combined power units fired with natural gas will be built because of the scarcity of the fuel. The possibility of using coal-gas turbines has been examined in the past, but the concept was rejected because of the problems of high-temperature corrosion and erosion by the mineral impurities in coal.

As a result of a favorable price of solid fuel compared with that of oil or natural gas, coal-gas turbine designs are once again being considered, in particular when combined with fluidized-bed combustion boilers as discussed by Mesko (1980). The problem of high-temperature corrosion in fluidized-bed boiler/gas turbine systems can be solved by application of high-chromium steels and corrosion-resistant alloys. The results of extensive materials testing in fluidized combustion systems, reported by Minchener et al. (1980) and Roberts et al. (1980), showed that the rate of corrosion could be kept below 100 nm h^{-1} at temperatures up to 1175 K by a suitable choice of high-chromium steels and nickel-based alloys.

There remains the question of erosion wear in power-generating systems based on the mode of fluidized-bed combustion. There should not be any serious erosion wear of the water and steam tubes located in the fluidized bed. The bed material is coarse, as shown in Fig. 18.14a, and some components (e.g., quartz particles) are highly abrasive, but the particle impact velocity is low, probably between 1 and 3 m s^{-1}. Thus there should not occur frequent incidents of tube erosion in fluidized-bed combustors. Some erosion wear of the bed construction components in the vicinity of the air inlet annulus can occur where the velocity may be above 5 m s^{-1}.

The flue gas from a fluidized-bed combustor has a high ash burden, in particular when crushed limestone or dolomite is added to reduce the SO_2 emission. This may seem surprising as the coal and additive material are much coarser than pulverized-fuel ash and the elutriation velocity is low. However, much of the mineral matter in coal is present as a fine clay dispersion and thus the ash always contains a large weight fraction of 0.1–10-μm particles irrespective of the size of original coal. The inherent mineral matter in large coal particles can form clusters of fine ash in a fluidized-bed combustor, as shown in Fig. 18.14b. The majority of these clusters of small ash particles will break down in a fluidized combustion system, when the temperature (between 1000 and 1300 K) is not sufficiently high to coalesce the ash to large particles as occurs in the pulverized-fuel flame discussed in Chap. 8.

A large amount of small particles (0.1–0.3 μm in size) is formed as a result of sulfation of limestone or dolomite additives at fluidized-bed temperatures. The

500 μm   5 μm

(a)   (b)

Figure 18.14 Fluidized-bed combustion ash: (a) Coarse ash. (b) Agglomerated fines.

submicron-size sulfate particles formed at the surface of $CaCO_3$ grains are removed by the comminution action of other large particles in the fluidized bed, and much of the sulfate fume is entrained in the flue gas. The size and shape of the fume particles of calcium are the same as those of the $CaSO_4$ fume entrained in the flue gas of a pulverized-coal-fired boiler (Fig. 8.3b).

The flue gas leaving the fluidized-bed combustion zone without the additive will usually have a lower ash burden when compared with the dust loading in the flue gas in pulverized-coal-fired boilers. With limestone or dolomite additive the burden of the flue-gas-borne ash in a fluidized combustion system can be exceptionally high. Pillai and Wood (1980) have stated that 80 percent of the coal ash and 30–40 percent of dolomite inerts were elutriated from the fluidized bed of a pressurized system. Clearly, a gas turbine could not be operated in an atmosphere of such a high dust burden, which may be as high as 1 percent by weight of the flue gas and is equivalent to 10 g m^{-3} concentration in a pressurized system.

The flue gas in conventional coal-fired boilers is cleaned in the electrical precipitators with higher than 99 percent efficiency, leaving a solid burden in the chimney flue gas between 50 and 150 mg m^{-3}. A similar degree of ash removal is aimed at for the fluidized combustor flue gas before it is fed to the gas turbine. The high-temperature gas cleaning processes using granular-bed filters and ceramic filters, and multicyclone systems have been discussed by Nutkis et al. (1977) and Hoke and Ernst (1980). The multistage cyclone systems are employed for high-temperature gas cleaning in fluidized test combustors to evaluate the performance of gas turbine materials, as discussed by Roberts et al. (1980).

The abrasive wear propensity of fluidized-bed ash is markedly different from that of the ash captured in the electrical precipitators of a conventional pulverized-coal-fired boiler. The combustion temperature in fluidized-bed systems does not usually exceed 1200 K as a result of rapid heat transfer, whereas the flame temperature in pulverized-coal-fired boilers may exceed 1750 K. Figure 18.4 showed that an ash low in quartz is relatively nonabrasive in the temperatue range of 1000 to 1200 K (curve C). At these temperatures the nonabrasive clay minerals—i.e., the bulk of ash—retain their low erosion wear propensity, whereas the aluminosilicate species vitrify in the conventional boiler flame and the resultant glassy particles are significantly more abrasive. In contrast, a quartz-rich ash retains its highly abrasive characteristics at temperatures up to 1700 K. It is therefore evident that the abrasive characteristics of the ash escaping the high-temperature gas cleaning system in a fluidized-bed combustor will depend chiefly on the amount and particle size of quartz in the ash.

An abrasive index of the nonquartz fraction of a typical fluidized-bed ash captured in the cyclone collector was obtained from the results of erosion measurements reported by Sage (1969). For the >45-μm particles, it is 0.17 that of quartz; that is, the nonquartz fraction of fluidized-bed ash is about 2.5 times less abrasive than that of pulverized-fuel ash. Thus from Eq. (18.7) in Sect. 18.2, an expression for the abrasive index I_{ab} of fluidized-bed ash can be obtained:

$$I_{ab} = [x_1(l_1 + 0.17) + 0.5x_2(l_2 + 0.17)]I_{1q} \qquad (18.28)$$

where x_1, x_2 = weight fractions of the >45-μm and 5-45-μm size cuts, respectively

$\quad l_1, l_2$ = weight fractions of quartz in the two size cuts

$\quad I_{1q}$ = abrasive index of the >45-μm size quartz

Equation (18.28) gives an abrasive index value for fluidized-bed ash elutriated from the combustion zone and captured in a cyclone collector. It does not apply for the ash escaping the collectors, which is unlikely to contain particles >45 μm in size. The results reported by Pillai and Wood (1980) showed that the ash that escaped the three-stage cyclone collectors in a fluidized-bed combustor had the following size distribution:

Particle size, μm	45	25	10	5
Weight percent above the size	0	0.2	1	3

If the above size distribution is typical of the ash escaping the ash-cleaning system in a fluidized-bed combustor, an expression for the abrasive index I_{ab} of ash can be written with the abrasive index of quartz equal to 1.0, i.e., $I_1 = 1$:

$$I_{ab} = mx_2(l_2 + 0.17) \tag{18.29}$$

where x_2 is the >5-μm size cut of ash, l_2 is the quartz weight fraction of the >5-μm size cut, and m is the abrasiveness factor depending on the size distribution of particles in the >5-μm size cut. The results of Pillai and Wood (1980) showed that the particles above 5 μm in size in the cleaned flue gas from a fluidized-bed combustor were chiefly in the range of 5-25 μm, and the relative abrasiveness [value of m in Eq. (18.29)] of this size fraction is approximately 0.2, as shown in Fig. 18.3. Thus, the abrasive index I_{ab} of the residual ash in cleaned fluidized combustor flue gas is given by

$$I_{ab} = 0.2x(l_2 + 0.17) \tag{18.30}$$

where the notations are as for Eq. (18.29).

Table 18.10 shows the abrasive index of typical fluidized-bed ashes escaping and captured in the gas cleaning system, when the abrasive index of >45-μm quartz is 1.0.

Type 1 ash in Table 18.10 has the size characteristics of the residual solids in

Table 18.10 Abrasive index of typical fluidized-bed ashes

Type of ash	Percent by weight particle size distribution (μm)			Quartz content of <5-μm fraction (wt. %)	Abrasive index
	>45	5-45	<5		
1	None	3	Balance	None	0.001
2	None	5	Balance	2	0.002
3	None	10	Balance	5	0.004
4	5	25	Balance	10	0.043

cleaned fluidized combustor flue gas, as reported by Pillai and Wood (1980). The results of chemical analysis of the flue-gas solids showed a low (below 1.5) weight ratio of SiO_2 and Al_2O_3; thus no significant quantity of quartz was present. The small particle size and the absence of quartz resulted in the ash having an exceptionally low abrasive index. With coals of higher quartz content and with different cleaning arrangements, the residual solids may be more abrasive as exemplified with ash types 2 and 3.

An expression can be obtained from Eq. (18.14) (Sect. 18.3) for estimating the erosion wear rate of gas turbine metals in the cleaned flue gas of fluidized coal combustion systems:

$$E_m = (3.49 \times 10^{-6})C_A I_{ab} u^{3.3} \tag{18.31}$$

where E_m = erosion wear rate of gas turbine metal components (nm h^{-1})

$\quad C_A$ = ash concentration of the flue gas (mg m^{-3})

$\quad I_{ab}$ = abrasive index as defined by Eq. (18.30)

$\quad u$ = velocity of the impacting particles

Figure 18.15 shows the estimated erosion wear rates of gas turbine components when impacted with an ash of low abrasion propensity (sample 1 in Table 18.10). Roberts et al. (1980) have reported that there is very little erosion wear of gas turbine metals when tested in the cleaned flue gas of a fluidized coal combustor when the ash burden, after cleaning, was in the range of 50–140 ppm. It was found

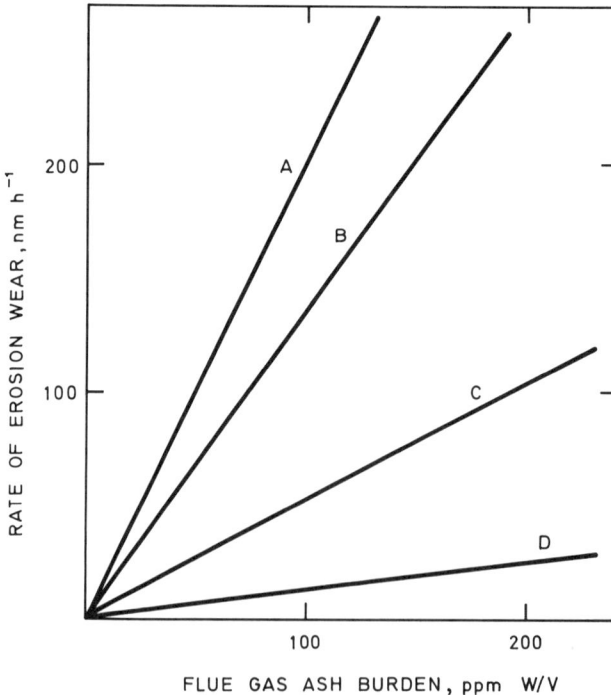

Figure 18.15 Calculated erosion wear rates of turbine components in cleaned fluidized combustor flue gas. Impact velocity: A, 450 m s^{-1}; B, 400 m s^{-1}; C, 300 m s^{-1}; D, 200 m s^{-1}.

that there was no measurable loss in thickness of the test samples in 1000 test runs. Some erosion wear occurred, however, as a result of malfunction of the cyclone ash collectors when the ash concentration of the flue gas increased by about ten times.

The low erosion wear rates, lower than those predicted from Fig. 18.15, in the experimental test assemblies were probably due to the fact that residual solid in the cleaned flue gas had a large weight fraction of nonabrasive calcium and magnesium sulfate formed from the dolomite additive, as evident from the analysis results published by Pillai and Wood (1980). It is possible that the submicron-size particles of sulfate, either captured on ash particles or gas-borne, will have a cushioning effect for more abrasive particles. Thus the erosion wear rates given in Fig. 18.15 are more appropriate for the ash residue in fluidized combustor flue gas when the sulfate-forming minerals (e.g., limestone or dolomite) are not added to the fuel.

Equation (18.29) and the abrasive characteristics of fluidized-bed ash (Table 18.10) can be used to show that the erosion wear of cyclone collectors is likely to be small. The abrasive index of the ash entering a cyclone collector is higher than that of the solids in the flue gas leaving the cleaning systems; nevertheless, a velocity up to 50 m s^{-1} does not cause a significant erosion wear. The high-quartz ashes, above 10 percent, are an exception and can cause severe erosion wear.

NOMENCLATURE

A	constant
B	constant
B_1	constant
C_A	ash burden of flue gas
C_{A_1}	ash burden of pulverized-coal-boiler fired flue gas
C_{A_2}	ash burden of cyclone-fired flue gas
$C_{A/C}$	ratio of ash to carbon in coal
C_P	particle concentration of carrier gas
D	eroded metal density
E_d	erosion rate of thickness wear
E_m	erosion rate of boiler tube metal
E_w	erosion rate by weight loss
H	hardness number of impacting particles
H_1	hardness number of eroded material
I_a	abrasive index of ash
I_{ab}	abrasive index of fluidized combustion ash
I_{1a}	abrasive index of above 45-μm size ash
I_{1g}	abrasive index of above 45-μm glassy particles
I_{1q}	abrasive index of above 45-μm size quartz particles
I_{2a}	abrasive index of 5-45-μm size ash
I_{2g}	abrasive index of 5-45-μm size glassy particles
I_{2q}	abrasive index of 5-45-μm size quartz particles
P_a	erosion parameter of pulverized coal ash
P_{a_1}	erosion parameter of cyclone boiler ash

P_{a_2}	erosion parameter slag-tap boiler ash
$R_{A/C}$	ratio of ash to carbon in coal
T_g	gas temperature
V	flue gas volume
V_{CO}	carbon dioxide content of flue gas
W	weight of impacting quartz particles
W_A	weight fraction of ash in coal
W_C	carbon content of coal
W_i	rate of particle impaction
d	thickness of eroded material
l_1	quartz content of above 45-μm size ash
l_2	quartz content of 5–45-μm size ash
m	relative abrasiveness of ash
n	power index
u	gas velocity
u_1	velocity of cyclone boiler flue gas
u_2	velocity of slag-tap boiler flue gas
x_1	weight fraction of above 45-μm size ash
x_2	weight fraction of 5–45-μm size ash

LOW-TEMPERATURE FOULING
AND CORROSION

19.1 THE TEMPERATURE REGIME
AND DESIGN OF AIR HEATERS

The air heaters constitute the final stage of heat exchangers in coal-fired boilers before the flue gas passes to the electrical precipitators for ash removal and is subsequently discharged through chimneys into the atmosphere. The air heater receives the flue gas at about 600 K and cools it to a temperature between 380 and 420 K, depending on the design and operation performance, i.e., the degree of fouling of the low-temperature heat exchanger. The aim is to extract the maximum amount of heat from the flue gas without causing extensive condensation of acid in the air heaters, in the electrical precipitators, and in the chimney. The incoming air is heated from the ambient temperature to about 550 K and passed to the boiler combustion chamber. Both the flue gas and air are near atmospheric pressure, so the heat transfer takes place from one nonpressurized gaseous medium to another, requiring an extensive surface area.

There are two main types of air heaters: the static recuperative and the rotary regenerative. In the recuperative air heaters, the air and the flue-gas streams are confined to separate channels by an arrangement of tubes or plates. In contrast, the regenerative air heaters rely on the heat-storage capacity of a matrix of corrugated steel plates that is exposed alternately to the hot flue gas and to the cool air. The regenerative air heaters are more compact and cheaper to construct than the recuperative tubular type and are preferred on large boilers. The heat-exchange matrix is usually in the form of a squat cylinder, and alternate exposure of the elements to the air and flue gas streams is arranged by a relative rotation of the matrix and its connecting ducting. In the more conventional Ljungstrom design, the matrix rotates between stationary headers (Fig. 19.1a), while the Rothemuhle design has a stationary matrix with an arrangement of rotating hoods (Fig. 19.1b).

The design of element matrix in the regenerative air heaters has received a great deal of attention, as discussed by Chojnowski and Chew (1978). The requirement is to maximize the rate of heat transfer without causing fouling by ash deposits. All three designs of undulated plates shown in Fig. 19.2 promote turbulence as the flue gas or air passes through the gaps, resulting in a high rate of heat transfer. The arrangements shown in Fig. 19.2, b and c, have an added advantage of causing less air-heater fouling, as discussed in Sect. 19.6.

The temperature of heat storing matrix of the regenerative air heater is continually changing, as depicted in Fig. 19.3. When cooled segments of the heat exchanger leave the air stream and enter the flue-gas flow, the temperature of the heat-storing elements increases rapidly first and more slowly later, as shown by curves T_1 and T_2. During this period condensation of sulfuric acid may take place depending on the extent and duration of metal temperature excursions below the acid dewpoint of the flue gas, as depicted by curve T_2 in Fig. 19.3. The concentrated solution of sulfuric acid thus formed will react with alkaline ash and also with air-heater metal, which may result in severe fouling and corrosion of the flue-gas outlet section, as discussed in Sect. 19.4.

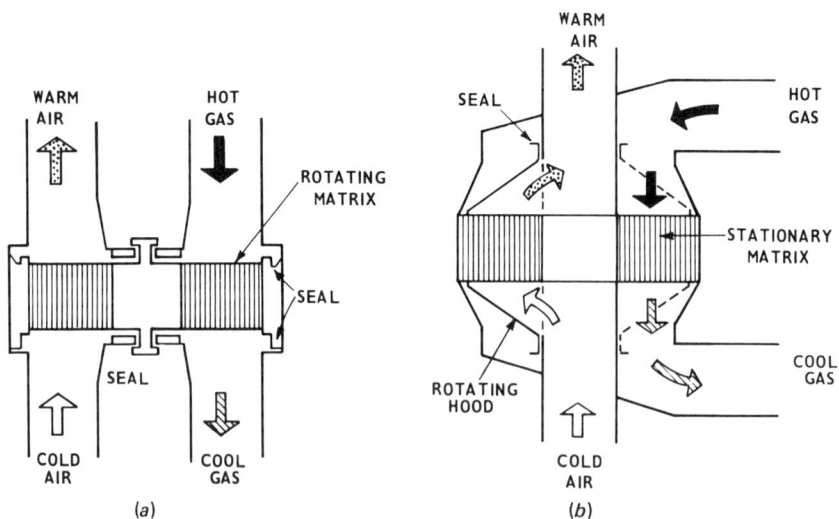

Figure 19.1 Regenerative air heaters. (a) Ljungstrom rotating matrix design. (b) Rothemuhle rotating hood design.

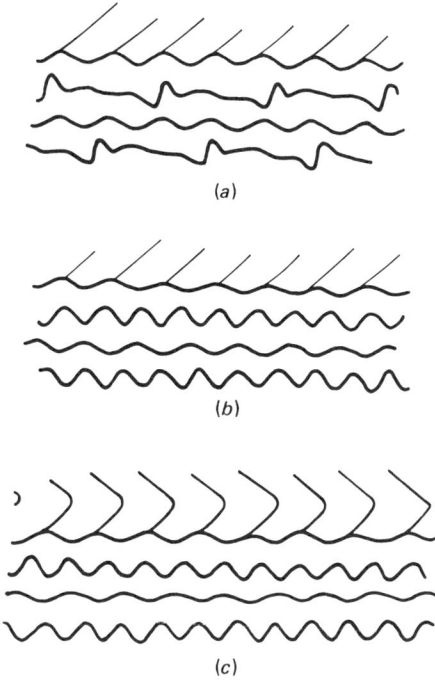

Figure 19.2 Regenerative air-heater matrix designs. (*a*) Double undulated. (*b*) Corrugated undulated. (*c*) Corrugated herringbone.

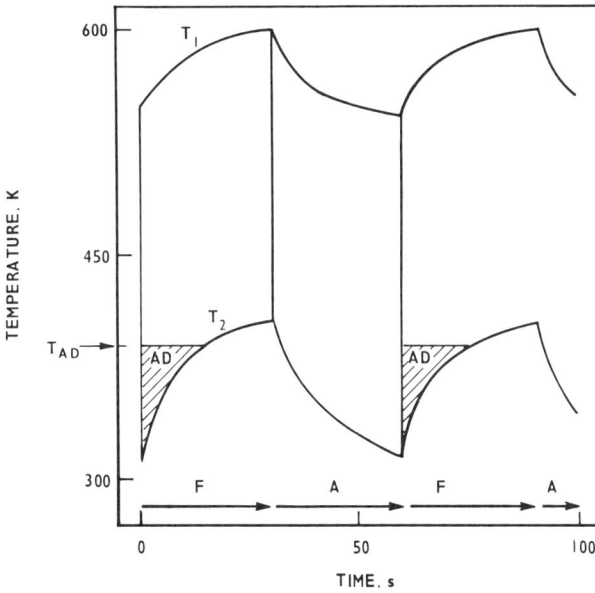

Figure 19.3 Changing metal temperatures and acid dewpoint in rotating regenerative air heater: T_1, air-heater inlet tempeature; T_2, air-heater outlet temperature; F, heating in flue gas; A, cooling in air; T_{AD}, acid dewpoint tempeature; AD, acid deposition.

19.2 ACID DEPOSITION MEASUREMENTS
IN COAL–FIRED BOILERS

Raask (1961a) and Alexander et al. (1962) have measured the rate of deposition of sulfuric acid in pulverized-coal-fired and cyclone-fired boilers. An air-cooled probe described by Alexander et al. (1960) was inserted in the air-heater duct and cooled to temperatures below the acid dewpoint. The amount of acid deposited in 30 min was determined by washing 150-mm-long sections of the probe and titrating the washings with the sodium hydroxide solution.

Figure 19.4 shows typical curves for the high and low rates of acid deposition on the cooled probe inserted in the air heater duct of pulverized-coal-fired boilers. Curve A shows that the dewpoint temperature was around 405 K and the maximum rate of acid deposition was 4.0 mg m^{-2}. This rate of acid deposition corresponds to about 10 ppm SO_3 in the flue gas, assuming the same acid deposition/SO_3 relationship as that evaluated by Raask (1984b) for oil-fired boilers. In another pulverized-coal-fired boiler, the maximum acid deposition rate was only about 0.8 mg m^{-2} (curve B_1, Fig. 19.4), equivalent to around 1 ppm SO_3 in the flue gas. The sulfur content of coal B was higher, 1.5 versus 0.9 percent of boiler A coal. It is thus evident that the sulfur content of fuel by itself cannot be used to predict the rate of acid deposition or the dewpoint temperature in coal-fired boilers. The rate of acid deposition was low with coal B (Fig. 19.4) because the ash was rich in calcium, over 10 percent, and much of the SO_3 formed in the flame had reacted with calcium oxide to form sulfate. A calcium-rich coal burned in the cyclone-fired boiler also resulted in a low rate of acid deposition, as shown by curve B in Fig. 19.5.

The results in Fig. 19.6 show that the rate of acid deposition was influenced by the amount of excess oxygen of the flue gas. A negligible amount of acid was

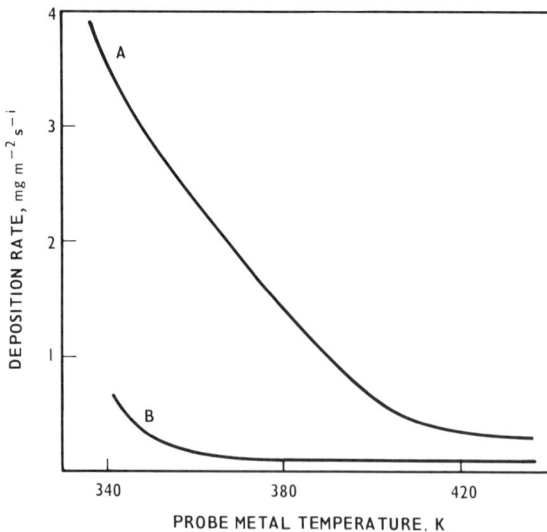

Figure 19.4 The effect of ash composition on acid (SO_3) deposition in pulverized-coal-fired air heater: A, low-calcium ash, 0.9 percent sulfur in coal; B, high-calcium ash, 1.5 percent sulfur in coal.

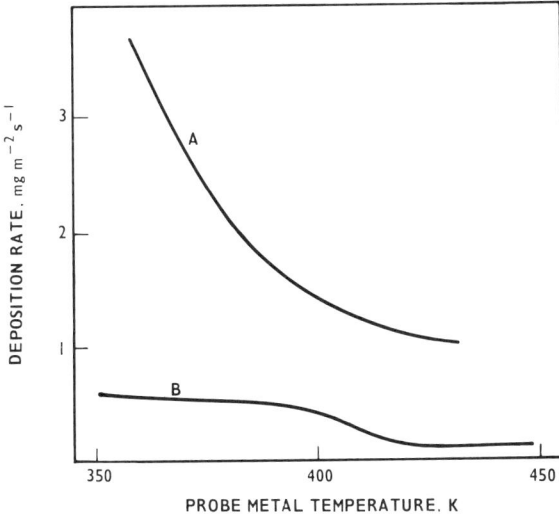

Figure 19.5 The effect of ash composition on acid (SO_3) deposition in cyclone-fired boiler air heater: A, low-calcium ash, 3.2 percent sulfur in coal; B, high-calcium ash, 2.5 percent sulfur in coal.

deposited when the oxygen content was temporarily reduced to 2.1 percent, as shown by curve C. This was probably because the low excess of oxygen produced a high concentration of CO in the flue gas (Fig. 14.7), which suppressed the SO_3 formation. The boiler operation with such a high concentration of CO in the flue gas, over 100 ppm, is highly undesirable because of the loss of combustibles and because of a high risk of furnace slagging and high-temperature corrosion, as discussed in Chaps. 14 and 17.

A realistic optimum level of excess oxygen in the flue gas of pulverized-coal-fired boilers is between 2.5 and 3 percent. With good coal milling and an adequate supply of the combustion air to each burner, this level of excess oxygen should be

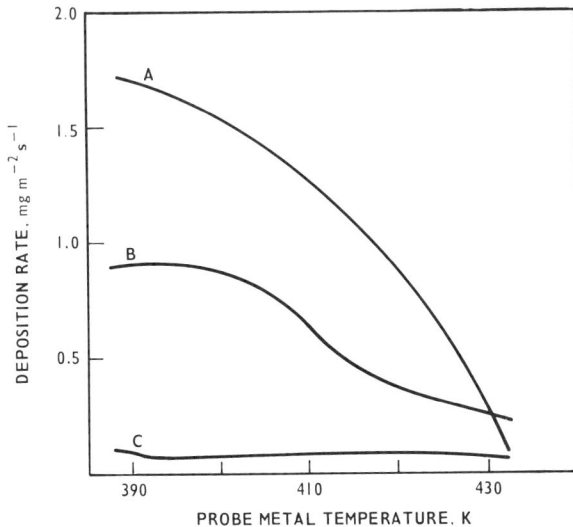

Figure 19.6 The effect of flue-gas oxygen on acid deposition in pulverized-coal-fired air heater: oxygen concentration, A, 3.9 percent; B, 3.1 percent; C, 2.1 percent.

sufficient to prevent severe high-temperature corrosion and slagging. The acid condensation in the air heaters is not completely suppressed, but the rate of deposition may be reduced by 50 percent as a result of decreasing excess oxygen in the flue gas from 4 to 3 percent, as shown by curves A and B in Fig. 19.6.

In order to relate the measured rates of acid deposition to the amount of acid condensed on the air heater plates or tubes it is necessary to estimate an average temperature difference (ΔT):

$$\Delta T = T_a - T_m \tag{19.1}$$

where T_a is the acid dewpoint temperature of the flue gas and T_m is the air-heater metal temperature when exposed to the flue-gas stream, as shown by curve T_2 in Fig. 19.3. Taking 15 K as a value for ΔT gives a rate of acid deposition of 1 mg m^{-2} s^{-1} from curve A in Fig. 19.4 when the SO_3 concentraton of the flue gas is around 10 ppm. The acid condensation usually occurs in a 100–200-mm deep zone measured from the cold end of the air-heater plates, and the deposition rate of 1 mg m^{-2} s^{-1} represents a build-up of acid layer of 1 mm thick in 500 h. This amount is likely to be sufficient to form strongly bonded deposits as a result of reaction between the acid, ash, and air-heater metal, as discussed in Sect. 19.4.

19.3 ACIDITY OF THE FLUE GAS AND COAL QUALITY

There are numerous empirical formulas for assessing the slagging propensity of different coals from the results of ash analysis, ash fusion, and sintering tests (Chap. 10). An estimate can also be made of the high-temperature corrosion propensity of coals based on the chloride, volatile alkali metal, and sulfur contents of the fuel (Chap. 17). The relationship between the coal quality and air-heater fouling is less easily definable. There is a general trend that the SO_3 concentration of the flue gas is high with sulfur-rich coals, resulting in a high acid dewpoint temperature and excessive condensation of sulfuric acid in air heaters, as discussed by Maloney and Benson (1981) and Gray (1981).

However, the results of acid deposition measurements in boilers burning 1–2.5 percent sulfur coals (Figs. 19.4 and 19.5) showed that the rate of acid deposition was influenced more by the alkalinity of ash, chiefly by the CaO content, and less by the sulfur content of coal. The SO_3 content of the air-heater flue gas (C_{SO_3}) is the difference between the total amount formed (C'_{SO_3}) and that reacted to form calcium and alkali-metal sulfates (C''_{SO_3}):

$$C_{SO_3} = C'_{SO_3} - C''_{SO_3} \tag{19.2}$$

The total amount of SO_3 formed in the high-temperature flue gas should bear a relationship to the sulfur content S of coal:

$$C'_{SO_3} = xS \tag{19.3}$$

where x is the fraction of coal sulfur oxidized to SO_3; the remainder is discharged as SO_2 with the flue gas. The degree of oxidation of SO_2 to SO_3; i.e., the value of

x in Eq. (19.3), depends on the flame temperature, the catalytic effectiveness of particulate ash and boiler deposits, and the oxygen content of the flue gas. Raask (1982a) has shown that the fraction of coal sulfur retained in ash as sulfate (C''_{SO_3}) depends on the amount of nonsilicate calcium and sodium compounds, chiefly carbonates and chlorides in coal, and the relationship is

$$C''_{SO_3} = \frac{(0.58\ CO_2 + 0.27\ Cl)(1 - kW^{2/3})}{S} \tag{19.4}$$

where CO_2, Cl, W, and S denote respectively the carbonate, chlorine, ash, and sulfur contents of coal, and k is a constant. The $(1 - kW^{2/3})$ factor represents the fraction of calcium and sodium captured by fused ash at high temperature and is not available for reaction with SO_3. Equation (19.4) can be written in an alternative form:

$$C''_{SO_3} = \frac{(0.8\ Ca_{SOL} + 0.7\ Na_{SOL})(1 - kW^{2/3})}{S} \tag{19.5}$$

where Ca_{SOL} and Na_{SOL} denote, respectively, the weight percentages of dilute-acid-soluble calcium and sodium in coal. Thus Eq. (19.2) for the SO_3 concentration of air-heater flue gas can be written in the form

$$C_{SO_3} = xS - \frac{(0.8\ Ca_{SOL} + 0.7\ Na_{SOL})(1 - kW^{2/3})}{S} \tag{19.6}$$

The maximum rate of acid deposition in air heaters (R_{SO_3}) should be related to the SO_3 concentration of the flue gas in a manner analogous to that in oil-fired boilers (Raask, 1984b) and could be written as

$$R_{SO_3} = m\left\{\frac{xS - [(0.8\ Ca_{SOL} + 0.7\ Na_{SOL})(1 - kW^{2/3})]}{S}\right\}^{3/4} \tag{19.7}$$

where the values of m and x depend on boiler combustion conditions. The rate of acid deposition in the oil-fired boiler air heater can be predicted with a reasonable degree of accuracy from the sulfur content of fuel and the excess oxygen in the flue gas (Raask, 1984b). In contrast, the SO_3 content of the flue gas and hence the rate of acid deposition in the coal-fired boiler air heaters are much more variable as a result of the reaction of SO_3 and alkaline coal ash. The acidity characteristics of the flue gas with 4 percent excess oxygen, given in Table 19.1, should be therefore considered as approximate estimates.

The high-sulfur bituminous coals are usually low in nonsilicate calcium and sodium, and as a result the air-heater flue gas is highly acidic, as shown by the data in Table 19.1. In particular, this is the case when the high-sulfur coals are fired in cyclone boilers where the bulk of calcium and alkali-metal constituents of ash is captured by molten slag. As a result of the capture of calcium and sodium by silicates and the high flame temperature, the concentration of SO_3 in cyclone-fired boilers is likely to be higher than that in pulverized-coal-fired boilers. In contrast, the flue gas in boilers fired with low-sulfur, high-calcium sub-bituminous coals is

Table 19.1 SO_3 concentration, dewpoint temperature, and acid deposition in coal-fired boilers

Type of coal	Sulfur in coal (%)	CaO in ash (%)	SO_3 in flue gas (ppm v/v)	Dewpoint temperature (K)	Maximum acid deposition rate (mg m^{-2})
High sulfur, low calcium	>2.5	2–5	10–25	400–410	5–10
Medium sulfur, low calcium	1–2.5	2–5	5–10	295–400	2.5–5
Medium sulfur, medium calcium	1–2.5	5–10	1–5	285–295	1–2.5
Low sulfur, high calcium	<1	>10	<1	<285	<1

practically nonacidic, i.e., the SO_3 concentration is less than 1 ppm. Apart from these exceptionally high- and low-sulfur fuels, the majority of power-station coals have the sulfur content between 1.0 and 2.5 percent (Chap. 4). According to the results of acid deposition measurements in Figs. 19.4 and 19.5, the medium-sulfur coals will produce flue gas that may be highly acidic, moderately acidic, or nonacidic, depending on the amount of nonsilicate calcium and alkali-metal constituents in coal, and on the level of excess oxygen in the flue gas (Fig. 19.6).

19.4 THE MECHANISM OF AIR–HEATER FOULING AND CORROSION

The flue-gas temperature at the air-heater inlet is about 600 K, which is far too low for the formation of sintered or bonded deposits. It may therefore appear surprising that there are frequent occurrences of blockage of the gas-inlet section of the air heaters by ash plugging in coal-fired boilers. The blockage usually starts with large pieces of sintered deposit debris and bloated slag globules being wedged in the narrow gaps between the plates. Subsequently the gas-flow channels are blocked by compacted ash. The coarse deposit and slag debris material is originally formed on the furnace wall and superheater tubes and falls or is transported by the flue gas to the air heaters.

In a 500-MW coal-fired boiler, about 50 t ash is deposited on the furnace-wall and high-temperature steam tubes in every 8 h, a typical interval between sootblowing. Much of this material is removed from the tubes during the next sootblowing cycle, and the dislodged material is in the form of large pieces of sintered and fused deposits and loose ash. That is, the particle size of dislodged material may range from less than 1 μm to over 100 mm. The large particles fall through the combustion chamber and will be discharged as clinker (furnace-bottom ash), but the flue gas is able to entrain and drag along comparatively large particles depending on the flue-gas velocity and particle density.

The mechanism of formation of the bloated slag globules formed at high temperatures and subsequently transported to the air heater is likely to be the same as that leading to the formation of ash cenospheres as discussed in Chap. 6. The evolution of gas, chiefly CO and CO_2, results from the reaction between coal carbon and slag catalyzed by iron. It is therefore to be expected that formation of the lightweight globules is enhanced by deposition of burning coal particles, i.e., flame impingement on boiler tubes, and with iron-rich ashes.

The blockage of gas-passage channels of air heaters by compacted ash deposit can take place without the carry-over of bloated slag from the furnace. This occurs when ash is wetted by water leaking from a faulty sootblower or from a failed economizer tube. That is, water is acting as a "bonding agent" and the wet particles are brought into intimate contact by the action of the surface tension. The pellets or aggregates thus formed are said to have a "green" or wet strength. The ash contains usually between 2 and 5 percent soluble matter, chiefly calcium and alkali-metal sulfates together with a small amount of leachable silicates (Raask, 1980). On evaporation, a matrix of crystalline material needles and plates (Fig. 8.3c) is formed, and thus an ash compact will have sufficient strength to cause a blockage in narrow channels between the air-heater plates.

The mechanism of formation of the compacted deposits from wetted ash is similar to that utilized in the manufacture of sintered ash aggregates as described by Barber et al. (1972). In the process the moist ash, without any additive, is pelletized to 20 mm in size on a rotating pan, and the pellets thus formed are sintered at temperatures between 1200 and 1400 K. The strength of pellets after drying but before sintering—i.e., in the temperature range of 400–1200 K—depends on formation of the crystal matrix of high-temperature cement as described above. It has been found that the pellet strength is high in the intermediate temperature range with calcium-rich ashes or when lime (1 percent) is added to a low-calcium ash. It is therefore to be expected that calcium-rich ashes are likely to form deposit compacts of a high cohesive strength in air heaters when there is a water leak from a failed economizer tube or faulty sootblower.

The formation of bonded deposits in the flue-gas outlet section of air heaters is intimately related to condensation of sulfuric acid at temperatures below the dewpoint. Table 19.2 gives the analysis of bonded deposits taken from the air heaters of three pulverized-coal-fired boilers (Alexander et al., 1962).

The results of Table 19.2 show that aluminum, iron, and calcium sulfates constituted the bulk of the soluble material in air-heater deposits. Aluminum and calcium sulfates were formed in the ash deposit by the action of hot concentrated solution of sulfuric acid:

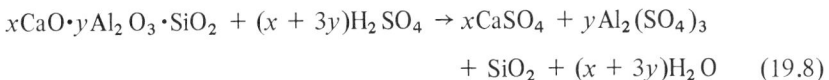

$$x\text{CaO} \cdot y\text{Al}_2\text{O}_3 \cdot \text{SiO}_2 + (x + 3y)\text{H}_2\text{SO}_4 \rightarrow x\text{CaSO}_4 + y\text{Al}_2(\text{SO}_4)_3$$

$$+ \text{SiO}_2 + (x + 3y)\text{H}_2\text{O} \qquad (19.8)$$

Iron in sulfate could have originated from the iron oxide of ash deposit or oxidized air-heater metal, or from a direct reaction with the steel:

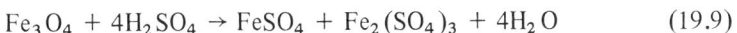

$$\text{Fe}_3\text{O}_4 + 4\text{H}_2\text{SO}_4 \rightarrow \text{FeSO}_4 + \text{Fe}_2(\text{SO}_4)_3 + 4\text{H}_2\text{O} \qquad (19.9)$$

Table 19.2 Analysis of bonded air-heater deposits[a]

Constituent	Sample A		Sample B		Sample C "baked" deposit
	Inner layer (A$_1$)	Outer layer (A$_2$)	Inner layer (B$_1$)	Outer layer (B$_2$)	
Water-soluble[b]					
Total	76.1	47.6	77.7	57.2	8.6
Loss at 385 K	6.6	2.5	17.1	11.2	0.8
Loss at 625 K	9.6	5.5	8.7	5.5	1.8
SiO$_2$	0.1	0.8	<0.1	<0.1	0.3
Al$_2$O$_3$	11.4	8.8	3.6	5.6	0.3
FeO	0.5	0.2	0.4	0.2	0.3
Fe$_2$O$_3$	3.4	1.6	4.4	2.5	0.7
CaO	4.0	3.8	0.3	4.6	0.6
MgO	0.6	1.2	1.8	1.3	0.9
K$_2$O	0.9	0.5	0.9	0.5	<0.1
Na$_2$O	1.5	0.5	1.6	0.7	0.2
SO$_3$	36.5	20.7	32.2	23.1	2.1
Cl	0.1	0.1	<0.1	<0.1	0.2
Water-insoluble					
SiO$_2$	24.0	26.7	11.8	21.0	9.3
Al$_2$O$_3$	4.7	15.1	3.8	9.0	3.3
FeO + Fe$_2$O$_3$	1.2	6.1	1.3	5.5	72.9
CaO	1.5	1.9	0.4	1.6	1.5
MgO	0.8	1.5	0.3	0.4	0.6
K$_2$O	0.5	1.1	0.4	1.1	0.2

[a]Results expressed as percent by weight of sample.
[b]The "water-soluble" extraction was carried out in 0.02 M HClO$_4$ (perchloric acid) solution in order to prevent hydrolysis of aluminum and iron sulfates in the acidic deposits.

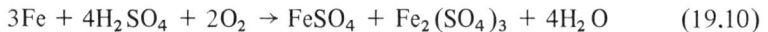

$$3Fe + 4H_2SO_4 + 2O_2 \rightarrow FeSO_4 + Fe_2(SO_4)_3 + 4H_2O \qquad (19.10)$$

The deposit material was hygroscopic: this was indicated by the weight loss at 385 K in Table 19.2. The weight loss at 625 K represents the excess of sulfuric acid present in the air-heater deposits.

It appears that the bonding constituents of the low-temperature deposit in air heaters consisted chiefly of aluminum, iron, and calcium sulfates. The amount of alkali-metal sulfates was comparatively low and seemed to not play any significant role in the bonding mechanism. The results in Table 19.2 show that the sulfate (SO$_3$) content of the acidic air-heater deposit is on average 25 percent by weight. The maximum rate of sulfuric acid (H$_2$SO$_4$) deposition in the air heater is around 1.0 mg s^{-1} (Sect. 19.2), and thus about 1000 h operation would be required to build a 5–6-mm-thick ash deposit.

In the presence of excess acid, for samples A$_1$, A$_2$, B$_1$, and B$_2$ in Table 19.2, the deposits remain soft and can be removed by water washing. However, this will not be the case when the acidic deposit is allowed to "bake" at temperatures above the acid dewpoint. If the flue-gas temperature increases as a result of air-heater fouling, or through any other causes, the acidic deposits previously formed start to lose the excess of acid and will adhere strongly to the air-heater plate metal. Sample C in Table 19.2 was a typical deposit that was strongly adhering to the plate metal, and the result of analysis shows that its sulfate content was less than that of the acidic, "non-baked" deposit samples. The adhesive bond of the "baked" deposit is sufficiently strong to resist the action of sootblowers, and when it is chipped from the air-heater element a layer of corroded metal is removed with the deposit. This is indicated by the high iron content of deposit sample C in Table 19.2.

The composition of the acidic deposits in air heaters of coal-fired boilers (Table 19.2) suggests that there is an extensive reaction between sulfuric acid condensed from the flue gas and the ash captured by the liquid film. For this reason a moderate amount of deposited acid may cause severe fouling, whereas the incidents of severe corrosion of air heater metal are less frequent. However, severe air-heater corrosion can occur when the concentration of SO_3 in the flue gas exceeds 5 ppm, which may be the case with high- and medium-sulfur coals low in calcium (Table 19.1). The corrosion of air heater metal is closely related to the dewpoint and the rate of acid deposition, as shown in Fig. 19.7. The maximum rate of metal corrosion occurs at temperatures some 30–50 K below the acid dewpoint, where the concentrated solution of sulfuric acid, about 80 percent by volume, is highly corrosive. The acid will dissolve both the oxide scale and the air-heater metal [Eqs. (19.9) and (19.10)].

Pollmann (1978) has suggested that chloride in the acidic air-heater deposits may significantly enhance the rate of corrosion of air-heater metal. The chemical and X-ray diffraction analysis established the presence of 3 percent chloride in highly corrosive deposits and also in the corrosion products. The rate of corrosion was exceptionally high; 1-mm-thick mild steel plates in the air-heater cold section had been completely corroded during 4000 h operation. The boiler was fired with

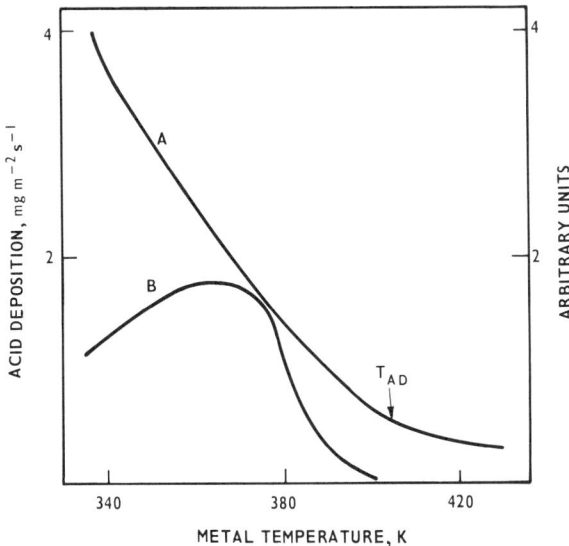

Figure 19.7 The relationship between acid deposition (A) and air-heater metal corrosion (B) (T_{AD}, acid dewpoint temperature).

sub-bituminous coals with a chlorine content between 0.06 and 0.6 percent.

Pollmann (1978) has compared the composition of corrosion products on air-heater plates with that of rusty deposits on steel structures attacked by sea water; the latter had been examined previously by Keller (1969). In both cases the hydrated iron oxide corrosion product had between 2.3 and 6.4 percent chloride ion incorporated in the crystalline structure. It appears that the small amount of chloride in the corrosion product is sufficient to destroy the "passivity" of hydrated iron oxide, which could slow down the rate of corrosion of air-heater steel in a sulfuric acid solution.

19.5 OPERATION EXPERIENCE OF FOULING AND CORROSION OF COAL–FIRED BOILER AIR HEATERS

Alexander et al. (1962) have carried out a limited survey on the problems associated with fouling and corrosion in air heaters of coal fired boilers. Their findings are summarized in Table 19.3.

The results in Table 19.3 show that the increase in the pressure differential across the air heater is directly related to the degree of fouling. That is, the extent of air-heater fouling is defined in terms of increased pressure differential. As discussed in Sect. 19.4, the air-heater fouling is usually in the form of plugging of the element channels by compacted deposit ash at the flue-gas inlet (hot section) or by the acidic, bonded deposit at the air inlet (cold section). That is, the nonplugged sections remain relatively clean and thus there is only a comparatively small reduction in the heat transfer. Figure 19.8 (curve A) shows a typical draught differential–time plot for a regenerative air heater where the resistance to the flue gas and air flow steadily increased in 7500 h operation as a result of the formation of acid-bonded deposit at the flue-gas outlet section. During the time when the pressure differential increased fourfold, there was only a 10 K increase in the flue-gas inlet and outlet temperatures.

As the pressure differential across the air heater increases as a result of fouling, the induced and forced draught fans may not be able to maintain the volume of combustion air necessary for full-load operation. When the air requirement exceeds the fan capacity, boiler load must be decreased; thus air-heater fouling can be one of the causes that limit the output from coal-fired boilers. The corrosion of the air-heater plate elements or tubes in coal-fired boilers is a lesser problem, as shown in Table 19.3. Similar conclusions were reached by Maloney and Benson (1981), who have reported on the survey of air-heater fouling in fossil-fuel-fired boilers. Figure 19.9 shows the frequency of reported difficulties and faults on the regenerative and recuperative air heaters. The survey included 236 air heaters installed in 118 fossil-fuel-fired boiler units of 500 MW and higher outputs in the United States.

The results show clearly that the problems associated directly with ash deposition and acid condensation, namely the channel blockage (plugging) and

Table 19.3 Air-heater fouling and corrosion in coal-fired boilers

Station	Type of air heater	Element spacing[a] (mm)	Air inlet temperature (K)	Flue-gas outlet temperature (K)		Draught differential (N m^{-2})		Observations	
				Clean	Fouled	Clean	Fouled	Fouling	Corrosion
1	Recuperative plate	12.5	308	416	433	81	182	Moderate	Slight
2	Recuperative tubular	62.5	303	416	427	122	212	Slight	None
3	Recuperative tubular	37.5	308	405	427	183	366	Very severe	Severe
4	Rotary regenerative	7.9	298	422	422	142	142	None	None
5	Rotary regenerative	7.9	303	405	427	81	230	Severe	Slight
6	Rotary regenerative	7.9	298	405	405	81	81	None	None
7	Rotary regenerative	7.9	303	411	425	183	466	Severe	Slight
8	Rotary regenerative	9.5	308	405	414	114	147	Moderate	Moderate
9	Rotary regenerative	7.9	293	408	419	142	244	Severe	Slight

[a]Element spacing refers to the maximum distance between the tubes or plates in the flue-gas outlet section of air heaters.

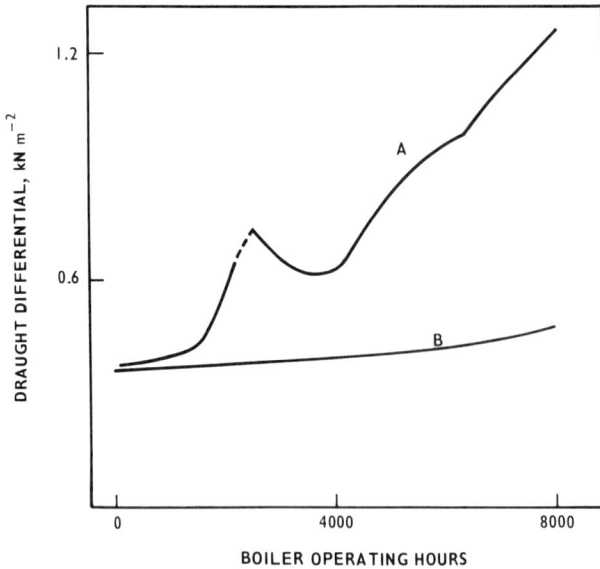

Figure 19.8 The effect of warm air circulation on regenerative air heater fouling: A, without recirculation (severe fouling); B, with recirculation (no fouling).

fouling and corrosion, were the dominant problems limiting the performance of air heaters of both the regenerative and recuperative design. The air-heater blockage and fouling were particularly severe in coal-fired boilers burning high-sulfur bituminous fuel and also calcium rich sub-bituminous fuel. Maloney and Benson (1981) concluded from their survey that air-heater hot end blockage and cold end fouling had frequently restricted boiler output. The load reductions were necessary when the forced and induced draught fans could not maintain the sufficient flow of combustion air through the deposit-fouled air heater.

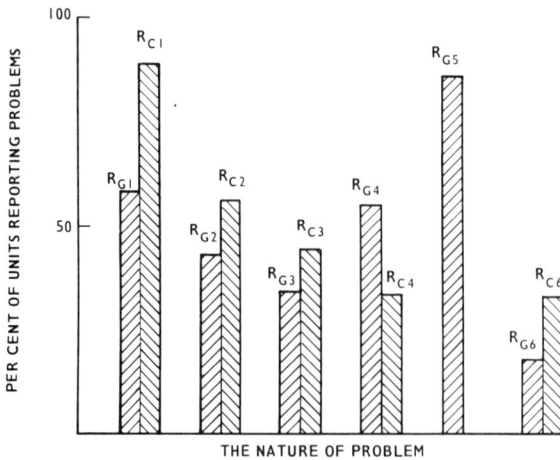

Figure 19.9 Operation experience with regenerative (R_G) and recuperative (R_C) air heaters: R_{G1}, R_{C1}, gas inlet fouling (plugging); R_{G2}, R_{C2}, gas outlet fouling and corrosion; R_{G3}, R_{C3}, load reductions; R_{G4}, R_{C4}, sootblowing and water washing; R_{G5}, rotor drive, bearing, and seals; R_{G6}, R_{C6}, other.

The air-heater fouling in large coal-fired boilers can also result in hidden penalties that may not be apparent as the prime cause for boiler slagging and corrosion. When air-heater fouling reaches a degree where the forced draft and induced draft are unable to maintain the required amount of excess oxygen in the flue gas, the result may be severe boiler slagging as discussed in Chap. 14, or an enhanced rate of corrosion of furnace-wall, superheater, and reheater tubes (Chap. 17). In many instances there is an insidious choice to be taken as a result of severe air-heater fouling of high-merit coal-fired boilers: either to reduce the load and pay the immediate penalty of reduced output, or to keep to the load with the inadequate amount of combustion air and risk later forced outages caused by corrosion of boiler tubes.

The results of a survey reported by Anson (1977) showed that the estimated costs of malfunctioning and failures in the air heaters of fossil-fuel-fired boilers were less than those resulting from the furnace tube failures. The costs of air-heater problems were not, however, insignificant—13 million dollars at mid-1970s prices—considering the relatively small number of boilers covered in the survey. In addition, there may have been hidden costs of increased slagging and corrosion of furnace-wall and superheater tubes due to the lack of combustion air resulting from air-heater fouling, as already discussed.

19.6 MEASURES TO COMBAT AIR–HEATER FOULING AND CORROSION

The severity of fouling in the regenerative air heaters is closely linked with the design of and spacing of the heat-exchange elements. The double-undulated arrangement (Fig. 19.2a) is frequently used with the corrugations of adjacent plates running parallel to each other at an angle of 0.52 rad to the gas flow. The design will ensure a high degree of turbulence in the gas flow, resulting in rapid heat transfer, but the arrangement may lead to a high rate of fouling by ash deposits. The initial build-up occurs in the narrow notched channels (Fig. 19.2a), as discussed by Chojnowski and Chew (1978).

In order to eliminate the narrow channels in the notched double-undulated arrangement a "corrugated undulated" design has been considered, as shown in Fig. 19.2b. The matrix elements consist of an alternative arrangement of undulated plates, with shallow corrugations running at an angle of 0.52 rad to the flow direction, and corrugated plates with deeper furrows running parallel to the flow. The shallow corrugations keep the flow turbulent and thus the heat-transfer performance is equivalent to that of the double-undulated design (Fig. 19.2a). It was found, however, that the shallow corrugations tended to bias the gas flow to one side, leading to localized fouling. A modified arrangement of the corrugations in herringbone fashion, as shown in Fig. 19.2c, corrects the bias toward uneven gas flow and localized fouling.

Table 19.4 summarizes the causes of air-heater fouling and corrosion, together with the corrective measures in the air-heater design and boiler operation.

Table 19.4 Air-heater fouling and corrosion in coal-fired boilers

Cause and effect	Corrective measures
Hot end fouling (plugging): Carry-over of sintered ash and slag deposit debris from the furnace and superheater tubes can block the flue-gas entry section of air heater.	Careful consideration should be given to the design, location, and mode of emptying of the ash hoppers and traps in the superheater, reheater, and economizer sections. The aim should be to prevent the coarse ash and deposit debris reaching the air heater. It may be necessary to install baffles and grid screens in the air-heater entry duct.
Water leakage from pin-holed economizer tubes and faulty sootblowers will result in formation of compacted deposits, as discussed in Sect. 19.3.	Early detection of leaking economizer tubes as a result of ash erosion (Chap. 18) or weld failure is essential. Careful checks should be made for water leaks from the steam and air sootblowers located in superheaters, reheaters, economizers, and air heaters.
Low-temperature fouling and corrosion: Condensed sulfuric acid reacts with deposited ash and air-heater metal to form bonded deposits.	The flue-gas outlet and air inlet temperatures should be consistent with the acidic characteristics of the flue gas. Corrosion and fouling of air heaters can be minimized by the use of corrosion-resistant steels for fabrication of the low-temperature air-heater elements.
Air-heater element and matrix design: Badly designed heat-exchange matrices may cause ash to be trapped in the element gaps, and the fixed-basket installation will make it difficult to replace corroded and badly fouled elements.	The appropriate element design is crucial for minimizing the blockage of gas passages by compacted ash and bonded deposits. The matrix baskets should be designed and installed for quick replacement of corroded and badly fouled elements.
Sootblowing and water washing: Inadequate sootblowing and water washing frequently fail to remove firmly lodged deposits in the hot section and the strongly adhering deposits in the cold section. Inadequate washing may, in fact, result in an increase of fouling by compacted deposits, as described in Sect. 19.3.	Single-jet and retractable multijet nozzles for sootblowers are available for on-load air-heater cleaning. The multijet sootblowers where the nozzles are spaced along the lance are recommended for large air heaters. Water washing is usually carried out by means of a stationary multijet arrangement. For strongly adhering deposits it may be necessary to use high-velocity water jets, pressure up to 700 $MN\ m^{-2}$.

The rate of fouling of the gas inlet section of air heaters depends largely on the formation of coarse ash at high temperatures, and on the locality and the mode of arrangement of the heat exchangers. The bloated slag globules and sintered ash debris formed on the furnace wall and superheater tubes (Sect. 19.3) may be transported to the air heaters in series of "kangaroo hops." First the coarse, low-density material is dislodged from the high-temperature tubes by sootblowing and settles on the slope underneath the pendent superheaters, and in the primary superheater and economizer hoppers. Subsequently, the coarse material is disturbed again by sootblowers in the locality and thus the large particles are transported further downstream, eventually arriving in the air heater. The gaps between the undulated plates may be sufficiently narrow to capture the coarse, lightweight ash, and it takes a comparatively small amount of the furnace-wall and superheater deposit debris to cause severe blockage of the air heater.

It is therefore essential that the coarse ash material that settles in the ash hoppers upstream of the air heater should be removed at frequent intervals. Also, the steam or compressed-air sootblowers should not be placed too near the sloping flow underneath the pendent superheaters, or over the ash hoppers in the primary superheaters, reheaters, and economizer. The removal of coarse ash would reduce the risk of severe fouling of the flue-gas inlet section of air heaters, as well as decrease the rate of erosion wear of the superheater, reheater, and economizer tubes (Chap. 18).

The hot section fouling can be particularly severe when the air heaters are situated immediately below the economizer and the flue-gas flow is downward through the heat exchanger (Fig. 19.10a). The lightweight coarse debris of high-temperature deposits and the ash wetted by a water leak are readily compacted in the narrow channels. Once the deposits are firmly lodged in the undulated gas passages, they will be difficult to remove. The inverted arrangement where the flue gas flows upward (Fig. 19.10b) would reduce the number of incidents of flow-channel blockage by coarse ash as a result of the large particles settling out from the upward-directed gas stream. This arrangement with the cold end top section is likely, however, to increase the depth of fouling by the acidic deposits. The gaps between the plates, 12–15 mm, in the cold section are about three times wider than those in the hot section. The acidic deposit formed in the cold section would fall to

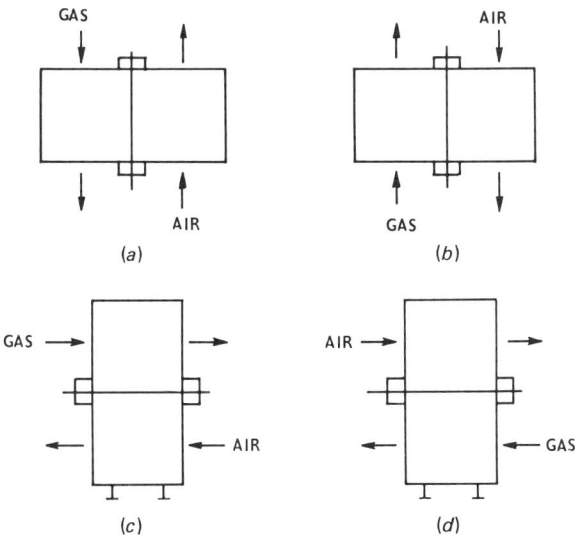

Figure 19.10 Possible arrangements of regenerative rotary air heaters. (a) Vertical standard. (b) Vertical inverted. (c) Horizontal gas/air. (d) Horizontal air/gas.

the middle zone of the air heater where it would "bake" and adhere strongly to the heat exchanger plates.

From the point of view of air-heater fouling, the flue gas and air-flow arrangement with the hot section at the top (Fig. 19.10a) is appropriate for the majority of coal-fired boilers. In an inverted layout with the cold section at the top (Fig. 19.10b), there is a risk of enhanced fouling by the acidic deposits, already as discussed. The low-sulfur and high-calcium non-bituminous coals constitute an exception. The SO_2 content of the flue gas is likely to be below 1 ppm (Gronhovd et al., 1973) and there is a remote possibility of the acidic deposits being formed. With these coals the inverted arrangement (Fig. 19.10b) should reduce the risk of plugging the hot section by the debris of furnace wall and superheater deposits. The regenerative air heaters could be designed to rotate on the horizontal axis (Fig. 19.10, c and d), and the arrangement may have some advantage in minimizing the hot end blockage. The large particles of gas-borne debris of the high-temperature slag and sintered ash deposit will fall toward the duct floor in an horizontal gas stream. It should therefore be comparatively easy to prevent the deposit debris entering the air heater with baffles or a meshed screen.

One of the most effective measures of alleviating air-heater fouling and corrosion is to control the element metal temperature by preheating the incoming air. The object is to prevent the metal temperature in the air inlet section falling too far below the acid dewpoint (Fig. 19.3). Cadrecha (1980) has recommended that the optimum inlet air temperature in each boiler plant should be determined from the acid dewpoint measurements, rather than accepting the set temperature recommended by manufacturers. Rolker (1973) has described a recording dewpoint instrument where the readings are not markedly influenced by the flue-gas solids accumulating on the probe.

The air inlet temperature can be controlled by several methods, including circulation of hot air taken from the intermediate or hot section of the air heater, cold air and flue-gas bypasses, and preheating air by the steam or glycol heat exchangers. Figure 19.8 shows the marked reduction in fouling as measured by the pressure differential across the air heater (curve B) when 10 percent of hot air was recirculated to the cold air inlet. The steam loop and glycol-filled heat exchangers have proved to be a less reliable method of preheating air, as discussed by Maloney and Benson (1981).

The air heaters in large coal-fired boilers are equipped with the steam or compressed-air sootblowers, which are effective in blowing off the loose dust but are less efficient in removing the bonded and compacted deposits. It is essential that the blowing medium is kept dry. A small amount of condensate water in the low-pressure steam or compressed-air jets can result in the formation of a sticky mass that becomes bonded to the surface of air elements when dried and is difficult to remove. Water washing is frequently necessary to remove the bonded and compacted deposits, which are not dislodged by sootblowing, as discussed by Keck (1963). The acid-bonded deposits in the air-heater cold sections are usually removed by water lancing operated with a jet pressure of 35 MN m^{-2}. In contrast, higher pressure water jet (hydrablast) washing is required to remove the "baked" deposits and to unplug the hot section, which is blocked by deposit debris originally formed on the furnace-wall and superheater tubes.

Ostlie and Stemper (1979) have described the experience of water washing of badly plugged air heaters of a 757-MW coal-fired boiler burning low-sulfur coals. The blockage was caused by carry-over of the lightweight ash deposit debris "popcorn slag," which was wedged firmly between the undulated plates. Initially water washing was carried out with a jet pressure of 35 MN m^{-2} for 48 h, and much of the ash material was removed. But when the boiler was put back on load the pressure differential across the air heater returned to its high precleaning values. In another attempt to dislodge the deposit debris in the middle of air heater element matrix, the water pressure was increased to 60 MN m^{-2}. The speed of rotation of the air heater was reduced from 1 to 0.25 rpm to ensure maximum penetration of the water jets. The cleaning process was completed in about 60 h, and when the unit was put back on load the low pressure differential across the air heater confirmed that the heat exchanger was completely free from deposits.

The SO$_3$ formation-suppressing flame additives and the low-temperature basic additives are frequently used to combat air-heater fouling and corrosion in oil-fired boilers, as discussed by Locklin et al. (1980). In contrast, reports on the successful use of additives to combat air-heater fouling and corrosion in coal-fired boilers are rarely found in literature. Gray (1981) has claimed that significantly reduced air-heater fouling was obtained when a volatile organometal magnanese additive was injected with the coal/air mixture into a 525-MW cyclone-fired boiler. The boiler was fired with high (3.8 to 4.2 percent) sulfur coals and thus the SO$_3$ concentration of the air-heater flue gas and the rate of acid deposition was high, as estimated from Table 19.1. The additive, methyl cyclopentadienyl manganese tricarbonyl, was injected at the rate of 20 ppm by weight of coal, and it has been in use for a number of years. It has been claimed that the additive has eliminated the deposition of acidic agglomerates from the chimney plume and reduced fouling and corrosion of the air heaters, as well as slagging in the secondary combustion chamber of the cyclone-fired boilers. The SO$_3$ concentration of the flue gas in most pulverized-coal-fired boilers is not excessive, and additives to control air-heater fouling are not required.

The corrosion resistance of a variety of steel alloys and coating materials have been tested mainly in oil-fired boiler air heaters, but the results should be relevant to the low-temperature heat exchanges in coal-fired boilers (Maloney and Benson, 1981). A number of high-chromium ferritic and austenitic steels have been found to be more corrosion-resistant than mild steel. The justification for using these more expensive materials for fabrication of the air-heater elements in coal-fired boilers depends on the severity of corrosion. There is another advantage in using the corrosion-resistant steels. The acidic deposits on the surface of a noncorroding metal do not form a strong adhesive bond, whereas the chemical reaction between sulfuric acid and mild steel results in a strong adhesive bond being formed.

An enamel and glass coating on mild steel can be highly resistant to the acid corrosion, but the coatings are easily chipped and cracked, resulting in an exposure of metal to acid attack. The same applies to the heat-exchanger elements made entirely from glass or a ceramic material. A variety of corrosion-resistant coatings could be applied on the air-heater elements by flame and plasma spraying, by plating and painting, but these techniques have not been yet proven conclusive in practice.

The erosion wear of air heater element plates and tubes caused by ash impaction is not usually a major problem. However, Rudenko and Spodyryak (1977) and Vasil'ev and Dombrovskii (1978) have described the excessive erosion wear damage with some of the coals mined in the Kazakh District (U.S.S.R.). There was an additional problem that the ash from some other coals in the same district caused severe blockage of the gas inlet section of the regenerative air heaters. In order to overcome the problems it was decided to revert to the recuperative design with wide gaps between the heat-exchange tubes. The rate of heat transfer from the flue gas to air was increased by adopting a design of corrugated tubes where the radius ρ of undulation curvature in relation of tube diameter d was found to be an important parameter. The ratio for maximum heat transfer was

$$\frac{d}{\rho} = 0.255 \pm 0.005 \tag{19.11}$$

Rudenko and Spodyryak (1977) have claimed that the corrugated-tube air heater (air flow inside the tubes) was cheaper by 18 percent to construct, and the running and maintenance costs were significantly lower than those of the corresponding capacity regenerative air heater.

19.7 MONITORING ACID PENETRATION IN CONCRETE STRUCTURES

Some of the electrical precipitator casings and all the tall chimneys of large coal-fired boilers are constructed from Portland cement concrete. The precipitator casings are protected internally by a lining of acid-resistant mortars against the attack by sulfuric acid. The concrete chimney flues are usually protected internally by a lining of acid-resistant bricks, but other materials have also been used. Umpleby (1976) and Lowrey and Taylor (1976) have described the application of a coating material of fluoroelastomer (fluorocarbon) on the three large concrete flues serving a 4000-MW coal-fired power station.

The purpose of an internal lining on the precipitator casings and chimney flues is to prevent the acid in the flue gas reaching the concrete structures. In order to assess the performance of the protective linings, it is expedient to have a monitoring system that would give warning when acid starts to penetrate the concrete structures. Raask (1976) has described a method of monitoring acid penetration in concrete structures and mortar and brick linings. The method is based on the electrical conductance measurements: matured dry concrete has a low conductance, whereas the acid wetted concrete has a much higher electrolytic conductance value.

On hydration of Portland cement, calcium hydroxide $[Ca(OH)_2]$ is liberated:

$$3\text{CaO} \cdot \text{SiO}_2 + (3 + nx - x)\text{H}_2\text{O} = x\text{CaO} \cdot \text{SiO}_2 \cdot n\text{H}_2\text{O} + (3 - x)\text{Ca(OH)}_2$$

$$(19.12)$$

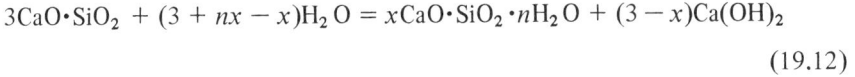

Within a few hours the residual water in the cement paste becomes supersaturated by Ca^{2+} and OH^- ions, and the supersaturation persists for several hours. Commercial cements contain gypsum (hydrated CaSO_4) and alkali-metal impurities, so the water in a cement mix also contains $\text{SO}_4{}^{2-}$, Na^+, and K^+ ions. Figure 19.11 shows that the conductivity—i.e., the ionic concentration in the cement paste— increases during the first hour after mixing. After about 4 h the conductivity starts to decrease as a result of precipitation of ettringite and formation of hydrated calcium silicates, as discussed by Calleja (1952) and Taylor and Arulanandan (1974). In mature concretes, the conductivity cells buried in large structures have been used by Lee and Bryden-Smith (1968) to study the moisture movement in large structures.

Raask (1976) has designed miniature probes for measuring the electrical conductivity of matured concrete at different depths of chimney flues and precipitator casings. Figure 19.12 shows the probes having two concentric nickel tube electrodes supported by a silica tube in a protective steel sheath. Figure 19.12a shows the probe with an external thermocouple; alternatively, a thermocouple can be placed inside the inner electrode, as shown in Fig. 19.12b. The cell "constant" of the conductivity probes was determined in usual manner by dipping in a potassium chloride solution. Subsequently the space between the two electrodes was filled with Portland cement paste and cured in a moist, CO_2-free atmosphere.

The response to sulfuric acid was tested by immersing the dry probes in a concentrated, 80 percent by volume, solution of H_2SO_4 and also by allowing the acid to condense on the probes. In both cases there was a rapid response, as shown in Fig. 19.13 (curve A), where the conductivity increased from 10^{-4} S m^{-1} to 1.15 S m^{-1} in 25 s. The response of the probe in calcium hydroxide and in water was

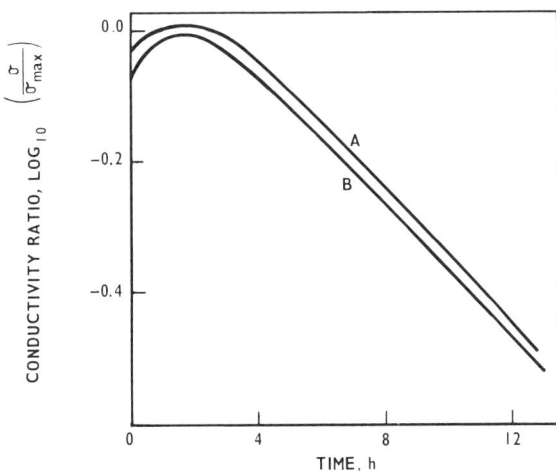

Figure 19.11 Change of conductivity of hydrating cement paste during first 12 h: A, above water; B, immersed in water.

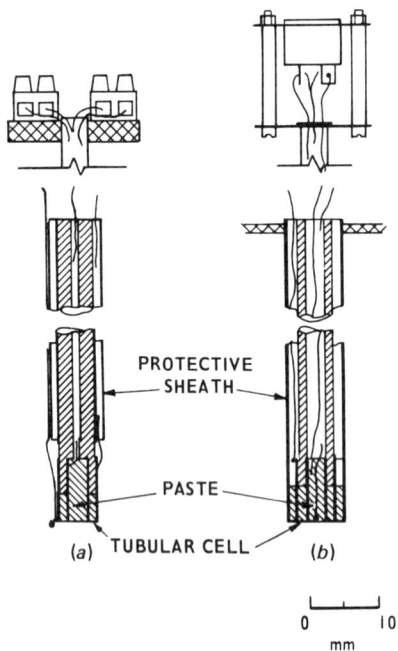

Figure 19.12 Conductivity probes for measuring acid penetartion in concrete structures. (a) Attached thermocouple. (b) Internal thermocouple.

much slower and less extensive, as shown by curves B and C. It should not therefore be difficult to distinguish wetting of the chimney concrete by water or calcium hydroxide solution, which is harmless, from the penetration of an acid solution resulting in a destruction of the cement paste matrix.

The experience with the conductance measuring probes mounted in the concrete chimney flues (Raask, 1976) showed that the readings were as predicted from the laboratory measurements. The probes are comparatively inexpensive and can be mounted immediately after chimney construction or inserted later. There are

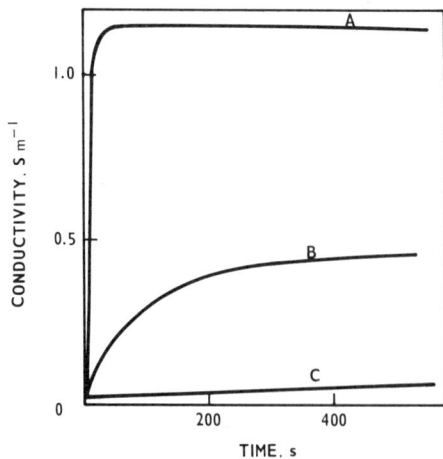

Figure 19.13 The response of conductivity probe: A, in 80 percent H_2SO_4; B, in saturated $Ca(OH)_2$; C, in distilled water.

many other applications where the conductance measuring probes would give an early warning of the penetration of acids, alkalis, or a salt solution into concrete and brick structures. It would also be useful to insert the conductance probes in chimney flues where it is necessary to renew or repair the acid-damaged lining. The probe readings from different depths of the lining and concrete would enable an estimate to be made on the safety and life of the chimney structures.

NOMENCLATURE

Ca_{SOL}	acid-soluble calcium content of coal
Cl	chlorine content of coal
CO_2	carbonate content of coal
C_{SO_3}	sulfur trioxide content of air-heater flue gas
C'_{SO_3}	total amount of sulfur trioxide formed
C''_{SO_3}	sulfur trioxide reacted with basic oxides
d	tube diameter
k	constant
m	proportionality factor
Na_{SOL}	acid soluble sodium content of coal
R_{SO_3}	maximum rate of sulfuric acid deposition
S	sulfur content of coal
T_a	acid dewpoint temperature
T_m	air-heater metal temperature
ΔT	temperature difference
W	ash content of coal
x	fraction of sulfur dioxide converted to trioxide
ρ	radius of curvature of tube bends

TWENTY

COMPARISON OF ASH-RELATED PROBLEMS IN PULVERIZED-FUEL-FIRED AND OTHER COAL-FIRED SYSTEMS

20.1 BOILER OPERATION DIFFICULTIES RELATED TO SPECIFIC COAL MINERAL SPECIES

It would be expedient here to summarize the boiler operation difficulties that can be related to specific mineral species in coal. Table 20.1 is laid out in the form of a balance sheet presenting the adverse and some beneficial effects of different coal mineral species on boiler operation and chimney emission. The adverse or beneficial effects will depend on the design of the coal combustion plant. For example, high-sulfur coals with iron-rich ash are likely to cause severe boiler slagging in pulverized-coal-fired boilers, whereas in cyclone-fired boilers the iron-rich ash does not require an excessively high temperature for a satisfactory slag flow (Chap. 9). Many problems are, however, common to all boiler systems. High-sulfur coals can cause furnace-wall, superheater, and air-heater corrosion both in pulverized-fuel-fired and in cyclone-fired boilers. There is no combustion system that can utilize sulfur-rich coals without releasing large amounts of SO_2. The SO_2 emission can be reduced by high- or low-temperature gas cleaning, but the processes are expensive and may cause problems in disposing the end products, i.e., huge quantities of calcium sulfate-rich ash or slurry.

The ash-related problems in pulverized-coal-fired boilers are summarized in Table 20.1.

Table 20.1 Summary of coal quality and ash-related problems

Coal and ash characteristics	Effect on boiler operation and chimney emission	
	Adverse	Beneficial
Wet (>12% water) bituminous coals	May freeze in transport, and stick to conveyers and bunkers. May limit mill performance, hence boiler output.	
High-moisture and high-volatile sub-bituminous coals	Require large amount of heat for drying and thus need special boilers where the water pre-heater (economizer) is omitted.	Sub-bituminous coals have a high combustible volatile content and thus do not require a fine milling.
Low (<15%) combustible volatile coals	Produce high-carbon residue ash; require extra-fine milling and boilers with long flame path.	
High (>20%) ash coals	Cause abrasion and particle impaction erosion wear of fuel handling plant, mills, burners, boiler tubes and ash pipes; can also cause boiler slagging and impair performance of the electrical precipitators by ash over-loading. There may be problems of ash disposal.	High-temperature and air-heater corrosion rarely occurs with high-ash coals unless chlorine and sulfur contents are high.
Coals with quartz-rich ash (>15% quartz in ash)	Cause abrasion and erosion wear as above.	Boiler slagging and fouling should not be a major problem; some sub-bituminous coals rich in quartz but also rich in sodium constitute an exception and can cause severe fouling.
High-sulfur (>2%) coals with iron-rich (>15% Fe_2O_3) ash	Cause furnace slagging in pulverized-coal-fired boilers, also fouling and corrosion of superheaters and air heaters. The coals are more suitable for cyclone firing. Emission of SO_2 and with pyrite-associated trace elements, As, Cu, Hg, Pb, Se is high.	The ash has comparatively high surface conductance and thus the electrical precipitators should have a high catch efficiency.
Medium-sulfur (1–2%) and medium-ash (12–20%) coals	Can give some slagging, corrosion erosion and precipitation problems depending on the design and operation conditions of coal milling and boiler plant.	
Low-sulfur (<1.0%) coals	Emission of chimney solids could be high. The ash of low surface conductance is not readily captured in electrical precipitators.	There should be no significant high-temperature or acid corrosion. Emission of SO_2 and the chalcophilic trace elements is low.
High (>0.3%) chlorine coals	Cause furnace-wall and superheater corrosion and also fouling; HCl emission is high.	The ash conductance and the performance of electrical precipitators are marginally enhanced.
Coals with calcium-rich ash (>10% CaO in ash)	Can cause superheater and reheater fouling by sintered deposits and air-heater fouling by compacted deposits.	The sulfate retention in ash is high; the SO_2 emission is reduced correspondingly.

422

The summarized overview of boiler operation and chimney emission problems in Table 20.1 gives some indication of the specific difficulties to be expected with fuels of unusual characteristics. There is no "ideal" coal, and the large number of medium-ash, medium-sulfur fuels utilized in the power station can cause slagging, corrosion, erosion, and emission difficulties. The severity of the problems depends on the design of the boiler plant and the operation efficiency of coal mills, burners, sootblowers, induced and forced draft fans, and the electrical precipitators. The problems related to different mineral species in coal can be set out in two categories: intramural (slagging, corrosion, and erosion of boiler plant) and extramural (emission of SO_2, trace elements and chimney solids, and ash disposal). The boiler design engineers together with research back-up teams should be able to anticipate the specific operation problems with "difficult" coals and incorporate the appropriate combative measures in the design stage. Also, it is possible to make a realistic assessment of cost penalties occurring as a result of slagging corrosion and erosion, and set these against expenses of combative measures.

The external nuisance that may be caused by chimney emission and ash disposal is difficult to quantify, but current concern dictates that the environmental impact from the emission and by the disposed ash should be kept to a minimum. One of the most difficult problems facing the fuel suppliers and electricity undertakings is the utilization of high (>2 percent) sulfur coals and also medium (1–2.0 percent) sulfur coals. The sulfur content of coal is a direct measure of the SO_2 emission; some non-emissible sulfur is retained as sulfate in boiler deposits and in the precipitated ash (Raask, 1982a), but often over 90 percent is discharged chiefly as SO_2. The electricity undertakings in many countries are subjected to coercive persuasion and legislative restrictions to limit the SO_2 emission. That is, either the sulfur content of fuel must be reduced by coal cleaning, or the SO_2 content of the flue gas reduced by gas cleaning, or by combination of the two methods. Detailed discussion on different methods of coal cleaning and SO_2 removal are outside the scope of this book, but in the following section an assessment is made of the possible benefits or penalties in boiler operation with cleaned coals.

20.2 BENEFITS AND SOME ADVERSE EFFECTS OF COAL CLEANING ON BOILER OPERATION

The coals intended for utilization in power stations are cleaned for the purpose of reducing the ash content, thus increasing the calorific value of the fuel (Fig. 4.1) or to remove specific deleterious impurity species, e.g., pyrite and chloride. An informative introduction to coal processing methods is given in the book edited by Pitt and Millward (1979). Numerous physical and chemical methods for reducing sulfur in coal have been discussed by Kilgroe et al. (1977) and at the U.S. Environmental Protection Agency Conference, the proceedings edited by Rogers and Lemmon (1979).

The efficacy of a cleaning process is judged by the ratio of ash in raw coal to that in cleaned fuel or by the similar ratio of sulfur in the raw and cleaned coals. The amount of residual mineral matter in cleaned coal depends on the ash content of raw coal, the mode of distribution of the mineral species and on the cleaning processes. For example, it would be unrealistic to expect that a 50 percent ash coal can be readily cleaned to where the fuel would have an ash content of 5 percent. Typically, the ash content of raw coal is reduced by 50-70 percent in gravity washing processes. For example, "as mined" coals in Britain frequently contain 25-35 percent mineral matter, whereas washed coals delivered to power stations contain usually between 8 and 12 percent ash. The lowest ash content of cleaned coals is around 4 percent, and the residual (inherent) mineral matter is not removed by physical coal preparation methods, e.g., density separation or magnetic separation. The cleaned coals could be placed into four categories depending on the ash content:

Grade	1	2	3	4
Ash content (%)	4–6	6–8	8–12	12–15

For utilization in pulverized-fuel-fired and cyclone-fired boilers there are no significant advantages to clean further the coals of an ash content below 12 percent, as discussed in Chap. 14. In contrast, coals with ash content below 8 percent would command a premium for utilization in converted boilers originally designed for oil firing, as discussed in Chap. 18 and also in Sect. 20.3. All coal-cleaning processes incur a penalty of substantial costs and a significant fraction of nonutilized fuel, which is rejected with the mineral matter. These losses must be set against the benefits of a cleaner environment and improvements in the availability and efficiency of boiler plants. Table 20.2 gives a summary of the beneficial consequences, with some adverse effects, in boiler operation resulting from the use of cleaned coals. The scope for alleviating boiler slagging and reducing high-temperature corrosion by coal cleaning has been discussed in Sect. 14.5 and in Sect. 17.8, respectively.

It is evident from Table 20.2 that coal cleaning should result in a significant improvement in the availability and performance of boiler plants. There are, however, some notable exceptions. Suydam and Duzy (1977) have discussed the cost penalties in boiler operation that are likely to occur from the use of cleaned coal mined in Arizona. The sub-bituminous coal is notable for its high flame-volatile sodium content, which is the principal criterion for boiler fouling with this type of coal. Sodium is intimately combined with the coal substance, and it is not removed by water washing, as shown by results of experiments carried out by Suydam and Duzy (1977). The sodium content of washed coal was higher, 2.2 percent versus 1.8 percent in raw coal, whereas the ash content was reduced from 23.6 to 13.7 percent. The high ratio of sodium to ash in washed coal would result in enhanced boiler fouling, and the cost penalty of increased boiler fouling together with that of handling wetter coals outweighed the savings resulting from the reduced wear of coal and boiler plant with the washed, less abrasive fuel.

Ruby and Huettenhain (1981) have reported on the survey of potential

Table 20.2 Consequences of coal cleaning on boiler operation

	Effect on boiler operation	
Nature of problem	Beneficial	Adverse
Abrasion and erosion of coal handling and milling plant	Removal of the extraneous abrasive mineral matter will significantly reduce the coal plant wear (Chap. 5).	
Furnace-wall slagging and superheater fouling	Reduction in the amount of ash in cleaned coal will in general decrease slagging and fouling (Chap. 14).	Exception is the coals that remain after cleaning, rich in flame-volatile sodium (Chap. 15). Special coal cleaning methods are required to remove sodium.
High-temperature corrosion of furnace-wall, superheater, and reheater tubes	Marginal reduction in deposit and flue-gas corrosivity will result from removal of coal pyrites (Chap. 17).	Exception is the high-chlorine coals, which may be more corrosive after cleaning because of the high chlorine-to-ash ratio.
Superheater, reheater, and economizer tube erosion by ash impaction	Low solids burden and nonabrasive ash will significantly reduce the erosion wear with cleaned coals.	
Air-heater fouling and corrosion	Reduction of the sulfur and ash content of coal will marginally decrease air-heater fouling and corrosion.	
Inadequate performance of the electrical precipitators	The performance of the electrical precipitators may be significantly improved as a result of lower ash burden with cleaned coals.	
Ash disposal	The cost and the possible environment impact of ash disposal is directly proportional to the ash content of coal; thus significant improvement results from coal cleaning.	
SO_2 emission	SO_2 emission is directly proportional to the sulfur content of coal. The reduction in the emission is the principal motive for coal cleaning.	
Trace-element emission	The emission of chalcophilic trace elements—As, Cu, Hg, Pb, Se—will be significantly reduced as a result of removal of pyrites in coal cleaning (Ford and Boyer, 1978).	

utilization of the vast reserves of sub-bituminous and lignite fuels in the United States. They recommended that the research and development work for the purpose of up-grading the low-calorific fuels should be concentrated in two main areas. First, a partial removal of sodium probably by an ion-exchange method would reduce significantly boiler fouling. Second, there would be significant benefits in reduced costs of fuel transport and a smaller boiler plant if an economic method of reducing the high moisture content could be found.

The lower ash and sulfur in cleaned bituminous coals should in most cases result in significant benefits of significantly reducing wear of coal plant, boiler slagging, and boiler tube erosion, as discussed in Chaps. 5, 14, and 18. Also, the performance of the electrical precipitators is likely to improve markedly as a result of reduced ash burden of the flue gas from cleaned coals. In particular, the removal of coal seam floor material of fine clay sediment of high resistivity should reduce the chimney-ash emission. However, the reduction of sulfur in cleaned coal to below 1.3 percent may result in a poor performance of the electrical precipitators because of the low SO_3 concentration of the flue gas, and a particle surface-conductance-inducing additive may be necessary (Chap. 16).

The economic assessment of coal and flue-gas cleaning processes and the impact of cleaned coals on boiler-plant performance and chimney emissions have been published by Hoffman and Deurbrouck (1977), Cole (1978), Buder et al. (1979), Phillips and Cole (1979), Clifford et al. (1980), and Vivenzio (1980). These assessments have shown that it is less expensive to clean coal by physical methods—e.g., density washing or magnetic separation—than to clean the flue gas in order to reduce the SO_2 emission. Coal cleaning has an added advantage that the costs can be partially recovered from the improved performance of boiler plant, whereas the flue-gas cleaning does not bring any such benefits.

20.3 ASH-RELATED PROBLEMS IN OIL–TO–COAL CONVERTED BOILERS

The erosion wear of water and steam tubes in coal-converted and mixed-fuel-fired boilers has been discussed in Chap. 18. There are many other problems directly related to the mineral matter in coal, as summarized in Table 20.3.

The synopsis of the mineral-matter-related problems in Table 20.3 shows that the boilers converted from oil to coal firing cannot handle large amounts of ash, and the use of cleaned coals will be necessary. This is because of the comparatively small combustion chamber, the narrow spacing between the tube elements, and a high velocity of the flue gas in the convection sections, superheater, reheater, and economizer. There is also a problem of finding the necessary space to install the electrical precipitators to capture large amounts of ash.

The combustion chamber area of pulverized-coal-fired boilers is approximately 40 percent larger than that of oil-fired boilers of the same output capacity (Godridge and Read, 1976). A large combustion chamber for the coal-fired boiler is necessary because of the reduced heat transfer through the ash deposits on

Table 20.3 Coal mineral impurity-related problems in converted boilers

Problem areas	Effect	Combative measures
Fuel supply system and burners	The abrasive minerals in coal can cause severe erosion wear.	Use of wear-resistant materials.
Combustion chamber	Coal-ash deposits on furnace walls will cut down the rate of heat transfer to water tubes by about 30% compared with that in oil-fired boilers.	Coal cleaning: the use of "anti-slagging" tube alloys and coatings, reduced boiler output, efficient sootblowing.
High-temperature superheater and reheater	Narrow passages in the secondary superheater and reheater banks may become blocked by sintered coal-ash deposits.	As above.
Primary superheater and reheater, and economizer	Severe erosion of the steam and water tubes is likely to occur as a result of coal-ash impaction.	Coal cleaning and reduction in boiler output, as discussed in Chap. 18.
Air heater	The flue-gas inlet section of regenerative air heater may become blocked by deposit debris carried over from furnace walls and high temperature superheaters.	Installation of ash hoppers in the economizer and large-particle screens at the entry to the air heater.
Electrical precipitators	The precipitators require a large area, which may be difficult at oil-fired stations.	Coal cleaning and a compact design of the precipitators.

furnace-wall tubes, as discussed in Chap. 13. It is therefore inevitable that there will be a decrease in the heat transfer for steam-generation tubes, leading to a load restriction of a coal-converted boiler. The load restriction could be kept to a minimum in the boiler converted to pulverized-coal firing by the choice of low-ash coals. Another requirement is that the ash should have a high fusion temperature.

The principal ash-related problems—i.e., the combustion-chamber fouling and tube erosion by ash impaction—in oil-to-coal converted boilers could be minimized by adopting a slag-tap or cyclone system for coal firing. Addition of the high-temperature cyclone furnaces to the existing combustion chamber results in a sufficiently high rate of heat transfer to the steam-generating tubes to maintain the original design load. There will also be a significant reduction in the ash impaction erosion wear of superheater, reheater, and economizer tubes. In cyclone-fired boilers, 75–85 percent of ash is retained and discharged from the furnace in the form of slag, and the residual ash entrained in the flue gas is less abrasive than the fly ash in pulverized-coal-fired boilers. Another method of utilizing coal in boilers designed for oil firing is to burn coal-in-oil slurry, as discussed briefly in Chap. 18. Table 20.4 summarizes the merits and disadvantages of the pulverized-fuel-fired and cyclone-fired systems of utilizing coal in boilers originally designed for oil firing.

It is evident from Table 20.4 that the oil-fired boilers converted to the cyclone or coal/oil slurry method of combustion would have some advantages over those changed to pulverized-fuel firing. Near original output capacity would be retained, and there would be fewer problems of erosion wear of the steam and water tubes caused

Table 20.4 Merits and disadvantages of different coal-converted systems

Mode of firing	Cyclone	Pulverized coal	Coal/oil slurry
	Fuel and boiler plant		
Boiler output, % of the original design	90–100	60–80	90–95
Cost of conversion	High	High	Moderate
Cost of fuel preparation	Moderate	Moderately high	High
	Ash-related problems		
Combustion chamber	Slagging and furnace-wall corrosion are highly sensitive to changes in the coal-ash characteristics and boiler operation.	Moderately sensitive to changes in the fuel and combustion conditions.	As for pulverized-coal firing in Table 20.1.
Superheaters, reheaters, and economizers.	The erosion wear caused by ash impaction should be low.	The erosion wear could be high.	Moderate erosion wear.
Electrical precipitators	A compact design precipitator would be adequate for the low ash burden.	Full-size precipitator would be required.	As for cyclone firing.

by ash impaction, and the amount of ash to be captured in the electrical precipitators would be comparatively small. However, the capital costs of converting an oil-fired boiler to coal cyclone firing would be high, and this mode of combustion is sensitive to changes in the quality of coal, i.e., the mineral matter in coal. Also, there would be added risk of furnace tube failures as a result of high-temperature corrosion. For these reasons most of the oil-fired boilers that have been converted or designated to be converted are based on the system of pulverized-coal firing.

Kirkwood and Chatfield (1981) and Kregg (1981) have described the successful outcome of converting two 200-MW oil-fired boilers to pulverized-coal firing. The coal-converted boilers have the designed capacity of 120 MW each, but both boilers have been successfully operated at loads up to 140 MW. The coal contains around 10 percent ash and 0.5 percent sulfur, and the composition of ash suggests it is nonslagging. However, at loads over 100 MW frequent sootblowing is necessary to maintain adequate heat transfer in the combustion chamber.

The conversion of oil-fired boilers to pulverized-coal firing is expensive and time-consuming, and consequently much research is being devoted to finding alternative methods of utilizing coal, e.g., coal/oil and coal/water slurries. The method of preparation, stability, and combustion of coal/oil slurries have been discussed by Dooher et al. (1978), Demeter et al. (1978), and Rowell et al. (1978), and also at an EPRI Seminar (Drenker, 1981). Slepow and Mendlessohn (1981) have discussed the test results of burning coal/oil slurries in a 400-MW oil-fired boiler. The bituminous coal of high calorific values, 30 MJ kg^{-1} (10 percent ash and 1 percent sulfur), was milled to a consistency of 80 percent through a 200-mesh (76-μm) sieve. The solid-fuel fraction of slurries was increased from 10 percent in steps to a maximum of 50 percent by weight (40 percent by heat content). As expected, coal slurries resulted in the formation of ash deposits on furnace walls, and with 30 percent coal mix the fouling became acute and the unit was shut down for installation of sootblowers in the combustion chamber. Subsequently, full-load operation could be maintained only with extensive sootblowing when the boiler was fired with a 40 percent coal slurry. When the coal fraction was increased to 50 percent, full load could not be maintained as a result of combustion-chamber slagging. This was thought to be due chiefly to the lack of combustion air, because a higher amount of excess oxygen is required to burn coal to prevent boiler slagging, as discussed in Chap. 14.

Slepow and Mendlessohn (1981) state that it is technically feasible to burn a 50 percent coal/oil slurry in oil-fired boilers; above 50 percent coal there will be problems because of the high viscosity of the mixed fuel. However, the high capital and operating cost of the fuel slurry facility, plant modifications, and emission-control equipment will offset much of the economic advantage of using coal. Similar conclusions were reported by Dunn (1981) after the trials with coal/oil slurries in an 80-MW boiler. The author suggests that it may be economically more advantageous to burn coal directly as pulverized coal with some oil, or to introduce coal as a water slurry, thus displacing a greater quantity of oil by the solid fuel. The technical feasibility and economic assessments on burning coal/water slurries in oil-fired boilers have been discussed by Chloupek and McKay (1981), Manfred (1981), Philipp (1981), and Stauffer (1981). Cook and Furman (1981) suggest that almost 100 percent oil could be replaced by a slurry which contains 75 percent by weight of coal. The 25 percent water present in the slurry would decrease the combustion efficiency by about 3 percent. There will occur, of course, the ash-related problems of boiler slagging, superheater fouling, and economizer erosion, as summarized in Table 20.3, when a substantial fraction (over 40 percent) of oil is replaced by coal/water slurry.

The method of burning coal in the form of a water slurry may have a beneficial effect of reducing boiler slagging. The evaporation of water will reduce the peak temperature of ash particles, and the rate of sintering and slagging would be reduced. Also, there is some evidence for enhanced crystallization of the glassy phase of ash deposits in a water-vapor-rich atmosphere resulting in a less extensive slagging (Jackson and Raask, 1957). It is therefore to be expected that a coal/water slurry should cause less extensive boiler slagging than on burning the same coal in the form of an oil slurry.

The difficulties in converting oil-fired boilers to coal and in burning coal slurries can be summarized as follows:

1. The small combustion chamber, where the coal-ash deposit will cut down the heat transfer to steam-generating tubes.
2. The narrow gaps between the pendent superheater and reheater elements, which may be bridged by coal-ash deposits.

3. The high flue-gas velocity in the primary superheaters and reheater and in the economizer, which may result in a high rate of tube erosion wear caused by impaction of the abrasive particles in coal ash.
4. The lack of space to install the coal handling and milling plant, the electrical precipitators, and the ash handling plant.

20.4 CHARACTERISTICS OF FLUIDIZED-BED ASH

Fluidized-bed combustion and gasification systems have been subject to innumerable papers and frequent international conferences, and there are several books published on the topic. An informative introduction to the subject has been given by Dainton (1979). It would therefore suffice here to compare and contrast the combustion conditions in pulverized-fuel-fired and fluidized-bed systems that influence the behavior of ash as set out in Table 20.5.

The striking difference between the thermal regime in a pulverized-coal-fired boiler and that in a fluidized-bed combustor is that the combustion temperature in the latter is some 500–600 K lower. Also, the velocity of flue gas in a fluidized combustor is low. It may therefore be surprising to note from Table 20.5 that over 50 percent of the ash can be entrained in the flue gas leaving the fluidized-bed combustion zone, considering that the bed velocity is low and the coal feedstock is coarse. The reason is that much of the mineral matter in coal is present as a dispersion of fine clay, 0.1–10 μm in diameter, as discussed in Chaps. 3 and 4. The fine particles of inherent ash agglomerate when the coal particles burn in pulverized-fuel flame (Chap. 8), but this does not occur at temperatures below 1200 K, and thus the clay particles retain their original size in a fluidized-bed combustor. The behavior of different coal mineral species in pulverized-coal-fired boilers and fluidized-bed combustors is summarized in Table 20.6.

Table 20.5 Combustion conditions and ash entrainment in pulverized-coal and fluidized-bed systems

Condition	Pulverized-fuel firing	Fluidized-bed combustion
Coal particle size (μm)	<100	<2000
Combustion temperature (K)	1700–1800	1100–1200
Residence time of ash in combustion zone (s)	2–5	10–1000
Flue-gas velocity (m s^{-1})	10–20	1–3
Ash entrainment in flue gas (%)	75–80	40–60
Nature of entrained ash	Chiefly spherical particles, 0.01 to 50 μm in diameter	Chiefly nonspherical particles, 0.01–20 μm in size
Nature of nonentrained ash	Sintered clinker with some fused slag	Unfused, nonspherical particles chiefly 20–2000 μm in size

Table 20.6 Silicate and nonsilicate coal mineral species in the high- and low-temperature combustion systems

Species	Pulverized-coal firing	Fluidized-bed combustion
Aluminosilicate	Dehydration, vitrification, spheroidization, and partial recrystallization; partial flame volatilization of potassium and the lithophilic trace elements; extensive sintering and coalescence on particle-to-particle contact	Dehydration and some crystal structural changes; no significant release of potassium or the trace elements; much-reduced particle sintering
Quartz	Partial vitrification	No change
Pyrite	Dissociation and oxidation to SO_2, SO_3, and Fe_3O_4; extensive volatilization of the chalcophilic trace elements	As in pulverized-coal firing, but the trace-element release is less extensive
Carbonate	Dissociation and partial capture of calcium by silicates, the remainder sulfated to $CaSO_4$	Dissociation and sulfation; there is no significant reaction with silicates, but the sulfate contains significant amount of CaO when lime is added to reduce SO_2 emission
Chloride	Volatilization and partial capture of sodium by silicate with a release of potassium; the released potassium and the residual volatilized sodium sulfated to K_2SO_4 and Na_2SO_4; chlorine present in the flue gas as HCl	Sodium chloride sulfated to Na_2SO_4; no significant reaction of silicates or formation of K_2SO_4; chlorine present in the flue gas as HCl

The nonsilicate mineral species in coal—chiefly pyrites, carbonates, and chlorides—will dissociate, oxidize, and be converted to sulfates in a fluidized-bed combustor, as it occurs in a pulverized-coal-fired boiler. The principal difference is that in the low-temperature (fluidized-bed) ash, there is no significant reaction between silicate ash and the nonsilicate species. That is, at the fluidized-bed temperature of 1100–1200 K there is no significant capture of sodium and calcium by silicate ash, whereas this occurs extensively in pulverized-coal flame, as discussed in Chap. 7. The abrasive wear characteristics of fluidized ash have been discussed in Chap. 18. In the next two sections, the sintering and slagging propensities of fluidized-bed combustion ash are briefly evaluated.

In the absence of a large temperature gradient within the ash deposit on the heat-exchange tubes in a fluidized-bed combustor, the formation of corrosive deposits is less extensive than that in conventional coal-fired boilers, as discussed in Chap. 17. It is considered that high-temperature corrosion of the austenitic steels and high-chromium ferritic steels in fluidized-bed combustion systems does not constitute a major problem, as discussed by Minchener et al. (1980) and Lane et al. (1981).

20.5 SINTERING PROPENSITY OF FLUIDIZED–BED ASH

There is an advantage when sinter-testing the fluidized-bed ash rather than pulverized-fuel ash: the low-temperature combustion ash resembles closely a laboratory-prepared ash with respect to chemical composition. Figure 20.1 shows that the fluidized-bed combustor and laboratory-prepared ashes from the same coal had similar sintering characteristics, as measured by the electrical conductance method described in Chap. 10. In contrast, the ash in a pulverized-coal-fired boiler has a markedly different thermal treatment from that of laboratory-prepared ash. A selective deposition of the reactive alkali-metal constituents takes place on cooled furnace-wall and superheater tubes in pulverized-coal-fired boilers, as discussed in Chap. 12. As a result the ash deposited on the boiler tubes is likely to sinter at lower temperatures than those predicted from the results of ash fusion tests made with laboratory-prepared ashes.

Extensive formation of the sintered ash deposits should not occur in fluidized-bed combustion systems when the ash particle temperature does not exceed 1200 K. In particular, this applies to the heat exchanger tubes immersed in the fluidized bed; continuous impaction of the large particles of bed material will help to keep the tubes free from bonded deposits. Mei et al. (1977, 1978) have reported that with anthracite refuse and Texas lignite fuels the fluidized combustion could be operated at temperatures up to 1365 K without formation of the ash clinker in the bed. Some sintered ash deposit could be formed with sodium-rich coals in stagnant, hot zones of the combustion bed, and also on the heat exchanger tubes above the fluidized bed as a result of deposition of the ash of small particle size.

Introduction of limestone or dolomite additive for SO_2 (sulfate) capture may

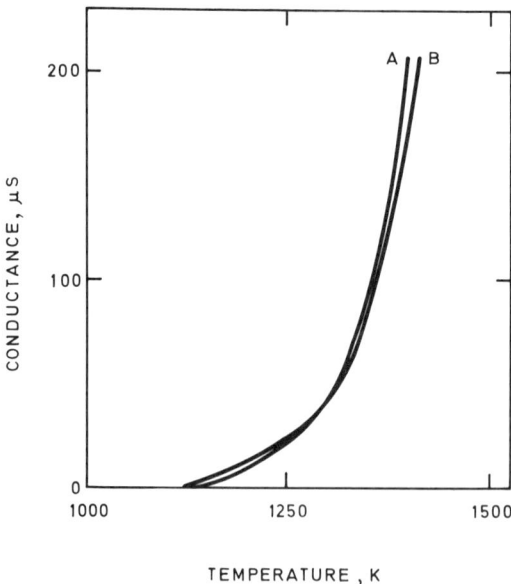

Figure 20.1 Conductance measurements on sintering of (A) fluidized combustor ash and (B) laboratory ash (Illinois high-sulfur coal ash).

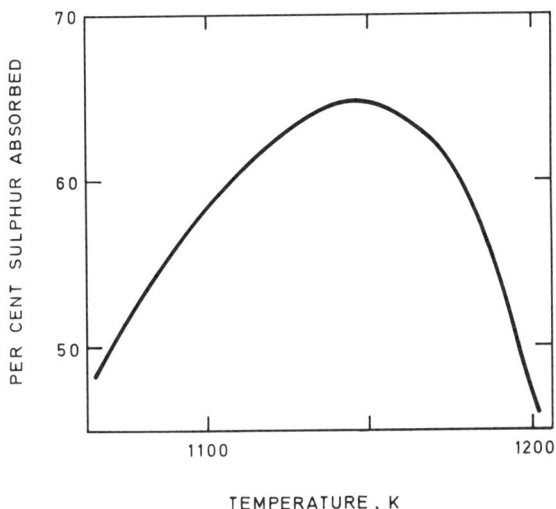

Figure 20.2 Optimum temperature for SO_2 absorption by lime.

lead to some reaction between the calcium oxide formed on dissociation of $CaCO_3$ and aluminosilicates in ash. Mazza et al. (1978) have found some crystalline calcium aluminosilicates in fluidized-bed ash that could act as a bonding agent in the formation of sintered deposits. However, the bed combustion temperature must be limited to 1150 K when the calcium additives are introduced, and thus the rate of ash sintering should be minimal. The temperature limitation is necessary because the amount of calcium sulfate formed, and hence the SO_2 capture, falls off rapidly above 1150 K, as discussed by Moss (1978) and Haque et al. (1979). Figure 20.2 shows the optimum temperature range for the formation of calcium sulfate, and it is evident that there is a sharp decrease in sulfur retention in the form of $CaSO_4$ as the combustion bed temperature increases above 1175 K. At higher temperatures the reactivity of CaO formed on the dissociation of $CaCO_3$ will decrease as a result of change in the structure and porosity of the pseudocrystalline material, as discussed in Chap. 7. Also, at higher temperatures some CaO is "lost" as a result of reaction with aluminosilicate species in the ash:

$$CaSO_4 + xAl_2O_3 \cdot ySiO_2 \rightleftharpoons CaO \cdot xAl_2O_3 \cdot ySiO_2 + SO_2 + \tfrac{1}{2}O_2 \qquad (20.1)$$

Equation (20.1) shows that the thermodynamic conditions for the formation of calcium aluminosilicates, and hence the ash sintering, depends on the partial pressure of SO_2 and oxygen. Shearer et al. (1979) have suggested that the rate of sulfation of limestone is enhanced by the presence of low-temperature-melting salts, e.g., sodium chloride.

The carry-over of large quantities of fine ash (Table 20.5) constitutes a difficult gas-cleaning problem. The ash is rather "unresponsive" to the traditional appliances of flue-gas cleaning, i.e., electrical precipitators and cyclone dust collectors. Figures 16.2–16.5 showed that clay minerals, which make up the bulk of the 0.1–10-μm ash fraction, have a high resistivity when heat-treated in the temperature range of 1000–1400 K. The high resistivity, above 10^{10} Ω m, coupled with the small particle size, results in a low catch efficiency of the electrical precipitator. It may therefore be necessary to resort to the use of low-temperature additives in order to improve the precipitator performance.

There are several high-temperature gas-cleaning methods that have been tested for the capture of the flue-gas-borne ash from fluidized-bed combustors. The cyclone dust collectors, two or more appliances in train, are frequently used, and the dust burden in the flue can be reduced below 150 ppm by weight of the flue gas, as discussed by Pillai and Wood (1980) and Roberts et al. (1980). The cyclone collection efficiency and that of the electrical precipitator would be markedly improved if it were feasible to agglomerate the fine particle ash to above 5 μm size before entering the dust collectors. This could probably be achieved by using coarser coal and increasing the combustion temperature by some 100 or 150 K, where the aim would be to sinter the fine ash at the surface of burning coal particles. The main difficulty is to control the process of particle agglomeration without producing large masses of sintered ash and slag.

The ash-laden flue gas from the fluidized-bed combustion could be passed through another high-temperature fluidized or stationary bed. An additive material would be introduced in order to promote sinter bonding. The aim would be to change the 0.1–10-μm ash to 1–10-mm-diameter pellets with the same density and strength properties comparable to those "Lytag" aggregates shown in Fig. 20.3b and described by Barber et al. (1972). It should be noted, however, that the ash sintering beds have to be operated at temperatures between 1350 and 1500 K: thus there would be some decomposition of sulfate and the release of SO_2 into the flue gas. Clearly, the high-sulfate ashes, i.e., ashes from high-sulfur coals with calcium additive, would be unsuitable for the sintering treatment.

20.6 SLAG AGGLOMERATION IN A FLUIDIZED–BED GASIFIER

A controlled rate of ash agglomeration can probably be more easily attained in a fluidized-bed gasifier, rather than in a combustor. First, the silicate ash can have a markedly lower viscosity range, hence higher sintering and slagging rates in a reducing atmosphere at 1350–1500 K, as discussed in Chaps. 9 and 10. Second, iron and calcium sulfides constitute an effective bonding medium under these conditions. Godel and Cosar (1967), Squires (1970, 1972), and Yerushalmi et al. (1975) have described the gasification systems where the coal was fluidized above an inclined grate. The fluidizing air was injected through the grate at velocities up to 15 m s^{-1}. The bed temperature varied between 1475 and 1675 K, which was sufficiently high to fuse the ash and would lead to the formation of large masses of clinker and slag in a nonfluidized system. However, by manipulating the fluidized air flow the process of ash agglomeration could be controlled and, when grown to a size between 5 and 10 mm, the pellets fell onto the grate and were carried from the gasifier.

Goldberger (1967) and Corder et al. (1973) have described a gasifier system

5 mm

(a)

5 mm

(b)

Figure 20.3 Lightweight ash products for utilization. (a) Slag agglomerates from fluidized-bed gasifier. (b) Sintered aggregates manufactured from pulverized-coal ash.

where the fluidizing velocity was between 2 and 3 m s^{-1} and the bed temperature varied from 1350 to 1420 K. Some ash agglomeration was observed at as low as 1035 K, but this was probably due to the fact that burning coal char particles were significantly hotter than the average bed temperature. A significant degree of ash agglomeration commenced when the temperature exceeded 1300 K. Sandstrom et al. (1976), Rehmat et al. (1978), and Vora et al. (1980) have described pressurized gasification systems operating in the temperature range of 1125-1340 K. The physical properties of the agglomerated ash pellets have been examined by Raask (1978), and the results are summarized in Table 20.7.

The black color of ash pellets indicates that the iron was largely present in the ferrous state. In contrast, the ash pellets of about the same size and density (Fig. 20.3b) manufactured by sintering on a moving grate at 1350-1500 K (Barber et al., 1972) are gray or reddish-gray in color, showing that iron was partly oxidized to the ferric state (Raask and Street, 1978). One unusual property of the ash pellets material from the fluidized-bed gasifier was that it contained over 3 percent combustible material. In contrast, the sintered ash clinker in pulverized-coal-fired boilers and the molten slag discharged from cyclone-fired boilers do not contain any significant quantities of unburned carbon. Curve B of Fig. 20.4 shows that the carbon in fluidized gasifier pellets was significantly less reactive than the coke residue in pulverized-fuel ash when oxidized in air at 650-950 K. It is possible that carbon in the ash pellets was present partly as iron carbide, which is an essential intermediate compound in the coalescence of particulate ash in the formation of porous pellets.

The chemistry of formation of porous slag pellets in the fluidized-bed gasifier and the mode of creation of ash cenospheres in pulverized-coal-fired boilers (Chap. 6) have some features in common. Initially, iron carbides are likely to form at the

Table 20.7 Properties of ash pellets from fluidized-bed gasifier

Appearance	Black
Particle size	3-8 mm
Particle shape	Rounded, some spherical, some ellipsoidal (Fig. 20.3a)
Bulk density	700 kg m^{-3}
Particle density	1200 kg m^{-3}
Water absorption	6.5% by weight
Total porosity	50% by volume (approx.)
"Open" porosity[a]	10% by volume (approx.)
"Closed" porosity	40% by volume (approx.)
Macropore size	50-250 μm
Micropore size	1-10 μm
Combustible content	3.2% by weight
Carbon oxidation on-set	775 K (Fig. 20.4)
Initial softening temperature in air	1400 K
Flow temperature in air	1675 K

[a]The open porosity of ash pellets was defined as the void volume that was filled with water on immersion.

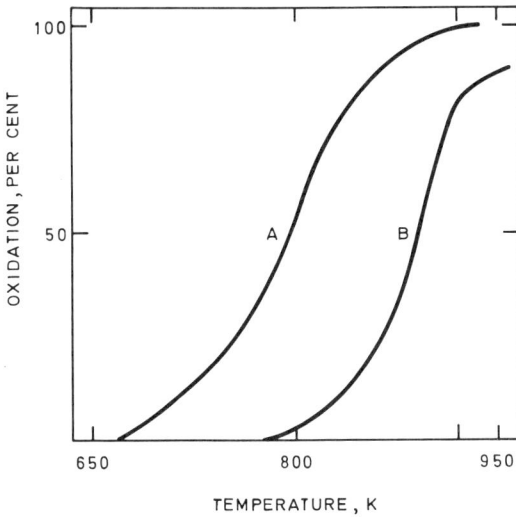

Figure 20.4 Oxidation of carbon residue: A, typical pulverized-coal ash; B, fluidized-bed gasifier slag pellets. Heating rate in air, 0.5 K min^{-1}.

coal/ash interface (Raask, 1966a), which facilitates the adhesion of silicate mineral particles on coal char. Coal pyrite, FeS_2, also is likely to play a significant role in the formation of pelletized slag. The FeS residue formed on dissociation of FeS_2 melts at about 1300 K (Fig. 7.3*b*) and thus acts as a binding agent for silicate ash particles. Some calcium sulfide, CaS, may form at the surface of ash in the reducing atmosphere, thus depressing further the melting temperature of the bonding phase.

Vora et al. (1980) have found some iron sulfide in the slag pellets formed in the fluidized-bed gasifier with high-sulfur coals. However, for the most part the iron and calcium sulfides react with water vapor in a reducing atmosphere after the coal char particles have been oxidized to CO and CO_2. The presence of H_2S in the flue gas indicates that iron and calcium originally of the bonding phase of FeS/CaS mixture had been oxidized to FeO and CaO:

$$FeS + H_2O \rightarrow FeO_{DIS} + H_2S \tag{20.2}$$

$$CaS + H_2O \rightarrow CaO_{DIS} + H_2S \tag{20.3}$$

The change from the sulfide to oxide phase results in the formation of crystalline species, chiefly calcium aluminosilicates and calcium, iron, aluminosilicates, as discussed by Vora et al. (1980). When the crystallization takes place in the surface layer of ash pellets this would result in cessation of further agglomeration. Some trace metals (e.g., copper) may play a significant role in initiating crystallization in the surface layer of slag pellets, which would terminate the slagging and growth stage.

This cursory overview of the slagging mechanism leading to the formation of ash pellets shows that the process is sensitive to changes in the ash composition, slag/coal interreaction, and in the temperature, velocity, and oxygen potential of gasifier flue gas. For successful plant operation of a slag-pellet-producing gasifier it is necessary to define closely the temperature and atmosphere regime consistent with the slagging characteristics of coal ash.

REFERENCES

REFERENCES

Abel, K. H., and Rancitelli, L. A. 1975. Major and Trace Element Composition of Coal and Fly Ash as Determined by Instrumental Neutron Activation Analysis. In *Trace Elements in Fuel,* edited by P. Babu, p. 118. Am. Chem. Soc., Series 141, Washington, D.C.

Abryutin, A. A., and Karasina, E. S. 1970. Emissivity Factor and Absorptivity of Ash Deposits in Boiler Furnaces. *Teploenergetika* (Thermal Engineering) 17, No. 10, English Transl., p. 64.

Adams, A. M., and Raask, E. 1963. Complex sulphates in coal fired boiler plant. *Conf., Mechanism of Corrosion by Fuel Impurities,* p. 196, Marchwood, U.K., Butterworth, London.

Agababov, S. G. 1957. Emissivity of Slags. *Teploenergetika* (Thermal Engineering), 4, No. 8, English Transl., p. 56.

Alekseev, L. S. 1959. Mineral impurities in coal. *Doklady Akad. Nauk, SSSR,* 124, 903.

Alexander, P. A. 1963. Laboratory studies of the effects of sulphates and chlorides on the oxidation of superheater alloys. *Conf., Mechanism of Corrosion by Fuel Impurities,* p. 571, Marchwood, U.K., Butterworth, London.

Alexander, P. A., and Raask, E. 1959. Examination of the external deposits on superheater test element, Skelton Grange Power Station. CERL Report No. 894, Leatherhead, U.K.

Alexander, P. A., Fielder, R. S., Jackson, P. J., and Raask, E. 1960. An air cooled probe for measuring acid deposition in boiler fuel gases. *Jour. Inst. Fuel,* 33, 31.

Alexander, P. A., Jackson, P. J., and Raask, E. 1962. Air heater fouling in pulverized coal fired boilers. CERL Report RD/L/R 1096, Leatherhead, U.K.

Allen, G. C., Honeybourne, C. L., and Mascall, R. A. 1981. Characterization of pulverized fuel ash using Mossbauer spectroscopy and measurements of magnetic susceptibility. *Chem. Ind.,* 18 July, p. 503.

Allen, G. C., Lee, B. J., and Wild, R. K. 1982. Surface analysis of pulverized fuel ash by X-ray photoelectron spectroscopy. *Pow. Ind. Res.,* 2, 35.

Anderson, A. R., and Johns, F. R. 1955. Characteristics of free supersonic jets exhausting into quiescent air. *Jet Propulsion,* 25, 13.

Anderson, B., Baker, B., and Gardiner, W. 1977. The control of slagging and fouling when burning lignite fuel in pulverized fuel generation. Meet., Can. Elec. Assoc., Autumn Session.

Anderson, C. H., and Goddard, C. W. 1968. Equilibrium SO_3 pressures of inner layers of fireside deposits from coal-fired boilers. *Jour. Inst. Fuel,* 41, 357.

Anson, D. 1977. Availability of fossil fired steam power plant. EPRI Report FP-422-SR, Palo Alto, Calif.

Anson, D., Clarke, W. H. N., Cunningham, A. T. S., and Todd, P. 1971. Carbon monoxide as a combustion control parameter. *Jour. Inst. Fuel,* 44, 191.

Anson, D., and Godridge, A. M. 1967. A simple method for measuring heat flux. *Jour. Scient. Instrum.,* 44, 541.

Anson, D., and Godridge, A. M. 1975. Heat flux meter. CEGB Technical Disclosure Bulletin No. 249.

Archer, K., Flint, D., and Jordan, S. 1958. The rapid analysis of coal ash, slag and boiler deposits. *Fuel,* 37, 421.

Archer, W. E. 1977. Fly ash conditioning update. *Power Engr.,* June, p. 76.

Armor, A. F., Parkes, J. B., and Leaver, D. E. 1981. Root cause analysis of fossil power plant equipment failures: The EPRI programme. IEEE Winter Meet., Paper WM 153-6, Atlanta, Georgia.

Asanti, P., and Kohlmeyer, E. J. 1951. Thermal characteristics of cobalt oxides and sulphates. *Zeitschr. Anorg. Chem.,* 265, 90.

ASTM. 1968. Fusibility of coal and coke ash. D 1857-68.

ASTM. 1969. Analysis of coal and coke ash. D2795-69.

ASTM. 1971. Grindability of coal by the hardgrove machine. D 409-71.

ASTM. 1972. Gross calorific value of solid fuels by adiabatic bomb calorimeter. D2015-72.

Attig, R. C., and Barnhart, D. H. 1963. A laboratory method of evaluating factors affecting tube bank fouling in coal fired boilers. *Conf., Mechanism of Corrosion by Fuel Impurities,* p. 183, Marchwood, U.K., 1963, Butterworth, London.

Attig, R. C., and Duzy, A. F. 1969. Coal ash deposition studies and application to boiler design. *Am. Power Conf.,* Chicago, Ill., Illinois Institute of Technology.

Avery, H. S., and Chapin, H. J. 1969. Selection of grinding mill liner alloys for optimum wear resistance. Society of Mining Engineers-AIME Meetings Preprint 69-B6.

Babu, S. P. (ed.). 1975. *Trace Elements in Fuel.* Am. Chem. Soc., Series 141, Washington, D.C.

Bacon, F. J., Hasapis, A. A., and Wholley, J. W. 1960. Viscosity and density of molten silica and high silica content glasses. *Phys. Chem. Glasses,* 1, 90.

Baker, D. W. C., Fountain, M. J., Hart, A. B., Holmes, D. R., Jackson, P. J., Jarman, J., King, C. W., Laxton, J. W., Mortimer, D., and Toft, L. H. 1977. The control of high temperature fireside corrosion. CERL Memorandum RD/L/M 484, CEGB Engineering Documents Unit, London.

Ball, C. G. 1935. Contributions to the Study of Coal. University of Illinois Geological Survey, Rep. Invest. No. 33.

Bancroft, G. M. 1973. *Mössbauer Spectroscopy: An Introduction for Inorganic Chemists and Geochemists.* New York: Wiley.

Barber, E. G., Jones, G. T., Knight, P. G. K., and Miles, M. H. 1972. *PFA Utilization.* CEGB Publication, London.

Bargstedt, W. 1979. Boiler surface cleaning by impulse generating. *VGB Kraftwerkstechnik,* 59, 648.

Barnhart, D. H., and Williams, P. C. 1956. The sintering test—An index to ash fouling tendency. *Trans. ASME,* 78, 1229.

Barrett, L. R. 1949. Heat transfer in refractory insulating materials, part 1: Texture and insulation power. *Trans. Brit. Ceram. Soc.,* 48, 235.

Barwood, H. L., Curtis, C. W., Guin, J. A., and Tarrer, A. R. 1982. Identification of clay minerals in coal by selective cation saturation. *Fuel,* 61, 463.

Basu, P., and Sarka, A. 1983. Agglomeration of coal ash in fluidized beds. *Fuel,* 62, 924.

Batch, J. H., Street, P. J., and Twamley, C. S. 1970. Temperature measurements of particulate surfaces. *Jour. Phys. E.,* 3, 281.

Baxter, W. A. 1968. Recent electrostatic precipitator experience with ammonia conditioning of power boiler flue gases. *Jour. Air Poll. Contr. Assoc.,* 18, 817.

BCURA. 1963. *The Chemical Composition and the Viscometric Properties of the Slags Formed from Ashes of British Coals.* British Coal Utilization Research Association, Leatherhead, U.K.

Beck, C. W. 1950. DTA curves of carbonate minerals. *Am. Mineralogist,* 35, 985.

Becker, H. B. 1971. The mode of ash fouling on heat transfer to boiler tubes. State Elec. Comm. of Victoria, Australia, Report No. 245.

Becker, H. B., Johnson, T. R., Tyson, G. R., and Wall, T. F. 1979. Ash emissivity measurement using a narrow angle radiometer. Australian Comb. Sci. Conf., Sydney, 8-9 Nov., 1979.

Beer, F. 1961. Wear problems on burning brown coals in Schwandorf. *Brownkohle Warme u. Energie,* 13, 102.

Benjamin, P., and Weaver, C. 1961. The adhesion of evaporated metal films on glass. *Proc. Roy. Soc., Ser. A,* 261, 513.

Bennett, R. P., and Kober, A. E. 1978. Chemical enhancement of electrostatic precipitator efficiency. Symposium Transfer and Utilization of Particulate Control Technology, Denver, Colo.

Bennett, R. P., and Kukin, I. 1977. Process for reducing the fusion point of coal ash. U.S. Patent 4,057,398.

Bennett, S. B., and Bannister, R. L. 1981. Pulverized coal power plants: The next logical step. *Mech. Eng.,* 130, No. 12, p. 18.

Benson, S. A., Karner, F. R., Goblirsch, G. M., and Brekke, D. W. 1982. Bed agglomerates formed by atmospheric fluidized bed combustion of a North Dakota lignite. Am. Chem. Soc. Fuel Div. Proc., Las Vegas.

Berkowitz, J., and Marquart, J. R. 1963. The equilibrium composition of sulfur vapor. *Jour. Chem. Phys.,* 39, 275.

Bertozzi, G., and Soldani, G. 1967. Surface tension of molten salts, alkali sulfate and chloride binary systems. *Jour. Phys. Chem.,* 71, 1536.

Bethell, F. V. 1963. The distribution and origin of minor elements in coal. *Jour. Inst. Fuel,* 36, 478.

Bettelheim, J., and Hann, W. W. 1980. An investigation of water leaching of some British coals. *Jour. Inst. Energy,* 53, 103.

Bibby, D. M. 1977. Composition and variation of pulverized fuel ash obtained from the combustion of New Zealand sub-bituminous coals. *Fuel,* 56, 427.

Bibby, D. M., Macknight, F. J., and Gainsford, G. J. 1978. Fusion temperature of PFA from New Zealand sub-bituminous coals. *Fuel,* 57, 727.

Bickelhaupt, R. E. 1975. Surface resistivity and the chemical composition of fly ash. *Jour. Air Poll. Contr.,* 25, 148.

Bickelhaupt, R. E. 1979. A technique for predicting fly ash resistivity. EPA Report 600/7-79/204.

Bickelhaupt, R. E., Dismukes, E. B., and Spafford, R. B. 1978. Flue gas conditioning for enhanced precipitation of difficult ashes. EPRI Report FP-910.

Bieber, K. H. 1968. Investigations on Thermal Shock of Superheater Tubes Related to Application of Water Sootblowers. *Mitteilungen VGB,* 48, 458.

Bieber, K. H., and Herrmann, W. 1974. Experimental results and operational experience on application of water sootblowers. *Kraftwerkstechnik VGB,* 54, 699.

Bishop, M., and Ward, D. L. 1958. The direct determination of mineral matter in coal. *Fuel,* 37, 191.

Bishop, R. J. 1968. The formation of alkali deposits by a high chlorine coal. *Jour. Inst. Fuel,* 41, 51.

Bishop, R. J., and Holland, C. G. 1964. The properties of ceramic coated tube surfaces and their application to studies of ash deposition. BCURA Inf. Circ. No. 280, National Coal Board, Cheltenham, U.K.

Bitter, J. G. A. 1963. A study of erosion phenomena. *Wear,* 13, Part 1, p. 5, Part 2, 169.

Black, J. A., De Jongh, J. G. V., and Sparnaay, M. J. 1960. Measurements of retarded van der Waals forces. *Trans. Farad. Soc.,* 56, 1597.

Blas, J. G. 1970. Spanish experience with burning low grade coal. *Combustion,* Sept., p. 6.

Blazewicz, A. J., and Gold, M. 1979. High temperature gas side corrosion in coal fired boilers. ASME Paper 79-WA/FH-6.

Bockris, J.O'M., and Lowe, D. C. 1953. An electromagnetic viscometer for molten silicates at temperatures up to 2075 K. *Jour. Sci. Instr.,* 30, 403.

Bockris, J. O'M., Mackenzie, J. D., and Kitcherer, J. A. 1955. Viscous flow in silica and binary liquid silicates. *Trans. Farad. Soc.,* 51, 1734.

Bockris, J. O'M., White, J. L., and Mackenzie, J. D. 1959. *Physico-chemical Measurements at High Temperatures.* Butterworth, London.

Boie, W. 1953. General diagram of fuels. *Technik,* 8, 305.

Boll, R. H., and Patel, H. C. 1960. The role of chemical thermodynamics in analysing gas-side problems in boilers. ASME Paper 60-WA-182.

Bombara, G., Baudo, G., and Tamba, A. 1968. Thermodynamics of corrosion in fused sulphates. *Corr. Science,* 8, 393.

Bonafede, G. 1975. The behaviour of inorganic constituents during the combustion of victorian brown coals. State Electricity Commission, Victoria, Australia, Scien. Dept. Report 315.

Boni, R. E., and Derge, G. 1956. Surface structure of non-oxidizing slags containing sulfur. *Trans. AIME, Jour. Metals,* 8, 59.

Borio, R. W., and Hensel, R. P. 1972. Coal ash composition as related to high temperature fireside corrosion and sulfur oxides emission control. *Jour. Eng. Power,* 94, 142.

Borio, R. W., and Narciso, R. R. 1978. The use of gravity separation techniques for assessing slagging and fouling potential of coal ash. ASME Paper 78-WA/CO-3.

Borio, R. W., Hensel, R. P., Ulmer, R. C., Wilson, E. B., and Leonard, J. W. 1968. Study of means for eliminating corrosiveness of coal to high temperature surfaces of steam generating units, Part 2. *Combustion,* Feb., p. 12.

Boow, J. 1969. Measurements of the viscosity of supercooled slags at 1025 to 1375 K. *Fuel,* 48, 171.

Boow, J. 1972. Sodium and ash reactions in the formation of fireside deposits in pulverized fuel fired boilers. *Fuel,* 51, 170.

Boow, J., and Goard, P. R. C. 1969. Fireside deposits and their effect on heat transfer in a pulverized coal fired boiler, part 3—The influence of the physical characteristics of the deposit on its radiant emittance and effective thermal conductance. *Jour. Inst. Fuel,* 44, 412.

Boow, J., and Turner, W. E. S. 1942. Viscosity and working characteristics of glasses; I. Viscosity of some commercial glasses at temperatures between 775 and 1675 K. *Jour. Soc. Glass Tech.,* 26, 215.

Boyes, J. M. 1969. Development and use of an abrasion test for cast irons and steels. *Iron and Steel* (Foundary Supplement), 42, 57.

Bratchikov, V. N. 1972. A formula for determining the erosion by ash of economiser tubes. *Teploenergetika (Thermal Engineering),* 19, English Transl., p. 58.

Breisch, E. W. 1978. Method and cost analysis of alternative collectors for low sulfur coal fly ash. Symp. Transfer and Utilization of Particulate Control Technology, Denver, Colo., U.S. Environmental Protection Agency.

Brekke, D. W., and Karner, F. R. 1982. Analysis and characterization of atmospheric fluidized bed combustion agglomerates. U.S. Energy Report DOE/FC/10120-T7. Springfield, Va.

Brennan, J. H., and Reveley, R. L. 1977. Flue gas conditioning with SO_3 to improve precipitator performance. *Am. Power Conf.,* 39, 569. Illinois Institute of Technology.

Brewer, L., and Green, F. T. 1957. Differential thermal analysis of the $Si-SiO_2$ system. *Phys. and Chem. Solids,* 2, 286.

Briner, E., Pamm, G., and Paillard, H. 1948. Study of equilibrium dissociation of calcium, potassium and sodium sulphates in absence and presence of additives. *Helv. Chim. Acta,* 31, 2220.

Brister, P. M., and Bressler, M. N. 1963. Long term experience with steel and alloy superheater tubes in power boiler service. Joint Conf. on Creep, Institute of Mechanical Engineers, London.

British Standard. 1970. The analysis and testing of coal and coke, BS Method 1016, Part 5—Gross calorific value of coal and coke; Part 14—Analysis of coal ash and coke ash; Part 15—Fusibility of coal ash and coke ash. Brit. Stand. Inst., London.

Brook, S. 1962. Modern soot blowing practice. *Elec. Review*, 171, 975.

Brown, D. R., Bozorgmanesh, H., Elias, E., Gozani, T., and Luckie, P. 1978. Moisture determination in coal: Survey of techniques. Argonne Nat. Lab. Report ANL-78-62.

Brown, G. M., Carmichael, A., and Hartmann, H. F. 1962. Bonded deposits from Victorian brown coals. *Journ. Inst. Fuel*, 35, 150.

Brown, H. R., and Swaine, D. J. 1964. Inorganic constituents of Australian coals. *Jour. Inst. Fuel*, 37, 422.

Brown, J., and Read, A. W. 1969. Lead compounds in fireside boiler deposits. CEGB Note SSD/NE/N/E.104, Harrogate, U.K.

Brown, J., Ray, N. J., and Ball, M. 1976. The disposal of pulverized fuel ash in water supply catchment area. *Water Res.*, 10, 1115.

Brown, R. H., Durie, R. A., and Schafer, H. N. S. 1959. The inorganic constituents in Australian coals, Part 1, The direct determination of the total mineral matter content. *Fuel*, 38, 295.

Brown, R. H., Durie, R. A., and Schafer, H. N. S. 1960. The inorganic constituents in Australian coals, Part 2—Combined acid-digestion-low temperature oxidation procedure for determination of total mineral content, water of hydration of silicate minerals and composition of carbonate minerals. *Fuel*, 39, 59.

Brown, R. L., Caldwell, R. L., and Fereday, F. 1952. Mineral constituents of coal. *Fuel*, 31, 261.

Brown, R. W. 1970. The relation between calorific value and ash content for British industrial grade coals. *Jour. Inst. Fuel.*, 43, 461.

Brown, T. D. 1968. The deposition of sodium sulphate from combustion gases. *Jour. Inst. Fuel*, 39, 378.

Brown, T. D., and Ritchie, J. 1968. Fuel ash deposition in naval boilers. *Jour. Inst. Fuel*, 41, 322.

Brown, T. D., Lee, G. K., Reeve, J., and Sekhar, N. 1978a. Improved electrostatic precipitator performance by use of fly ash conditioning agents. *Jour. Inst. Fuel*, 51, 195.

Brown, T. D., Lee, G. K., and Sekhar, N. 1978b. Modification of electrostatic precipitation performance by use of fly ash conditioning agents. ASME Paper 78.

Bryers, R. W. 1976. The physical and chemical characteristics of pyrites and their influence on fireside problems in steam generators. *Jour. Eng. Power*, 98, 517.

Bryers, R. W. 1978. Influence of the distribution of mineral matter in coal on fireside ash deposition. ASME Paper 78-WA/CD-3.

Bryers, R. W., and Taylor, T. E. 1976. An examination of the relationship between ash chemistry and ash fusion temperature in various coal size and gravity fractions using polynomial regression analysis. ASME Paper 75-WA/CD-3.

Buckle, E. R. 1978. Nature and origin of particles in condensation fume: A review. *Jour. Microscopy*, 114, 205.

Bucklow, I. A. 1975. Sprayed coatings of metals and ceramics. *Welding Inst. Res. Bull.*, Feb., p. 41.

Buder, M. K., Huettenhain, H., and Clifford, K. L. 1979. The effects of coal cleaning on power generation economics. *Am. Pow. Conf.*, Vol. 14, p. 610, Chicago.

Burger, R. 1961. Incidents of erosion wear in boilers burning ash rich non-bituminous coals in Rhineland. *Braunkohle Warme Energie*, 3, 82.

Burrows, B. W., and Hills, G. J. 1966. The electrochemistry of deposits of inorganic constituents of fuels at high temperatures. *Jour. Inst. Fuel*, 39, 168.

Buxmann, J., and Schwab, G. 1973. Model studies of large non-bituminous coal fired boilers. *VDI-Forschungsheft* 559, VDI-Verlag, Dusseldorf, Germany.

Cadrecha, M. 1980. Preventing acid corrosion in air heaters. *Power Eng.*, Jan., p. 54.

Cain, C., and Nelson, W. 1961. Corrosion of superheaters and reheaters of pulverized coal fired boilers. *Trans. ASME Jour. Eng. Power, Series A*, 83, 468.

Calleja, J. 1952. New techniques in the study of setting and hardening of hydraulic materials. *Jour. Am. Conc. Inst.*, 48, 525.

Casimir, H. B. G., and Polder, D. 1948. The influence of retardation on the London–van der Waals forces. *Phys. Rev.,* 73, 360.

Castle, G. S. 1980. Mechanisms involved in fly ash precipitation in the presence of conditioning agents–A review. *IEEE Tran. Industry Appl.,* 1A-16, 297.

Caswell, S. A. 1981. Distribution of water soluble chlorine in coals using strains and acetate peels. *Fuel,* 60, 1164.

Cawood, W. 1936. The movement of dust or smoke particles in a temperature gradient. *Trans. Farad. Soc.,* 32, 1068.

CEGB (Central Electricity Generating Board). 1959. Report on the furnace performance tests made at the Brunswick Wharf Generating Station, July, London.

CEGB. 1971. *Modern Power Station Practice,* Vol. 2. Oxford: Pergamon.

CEGB. 1977. Analysis of British coals. Central Electricity Generating Board, Operations Dept., London.

Chambers, A. K., Wynnychkyj, J. R., and Rhodes, E. 1981. A furnace wall ash monitoring system for coal fired boilers. *Jour. Eng. Power,* 103, 532.

Chawla, J. S. 1982. Models for predicting softening temperatures of Bienfait lignite ashes. Ontario Hydro, Canada, Res. Div. Report 82-1-K.

Chironis, N. P. 1981. Britain improves its coal preparation. *Coal Age,* 86, 7.

Chloupek, F. J., and McKay, G. A. 1981. A fuel suppliers purview. Seminar, The Use of Coal in Oil Design Utility Boilers. EPRI Report WS-80-141, Palo Alto, Calif.

Chojnowski, B., and Chew, P. E. 1978. Getting the best out of rotary air heaters. *CEGB Research,* No. 7, p. 14.

Clark, J. E. 1966. Magnesium oxide slurry additives to improve cleaning of recovery boilers. *TAPPI,* 49, 50A.

Clements, J. F. 1966. Characteristics of refractory materials. *Trans. Brit. Ceram. Soc.,* 65, 479.

Clifford, K. L., McGowin, C. R., Buder, M. K., and Heuttenhain, H. 1980. Combined coal cleaning and flue gas desulfurization for SO_2 emissions. Am. Power Conf., Chicago, Ill.

Coats, A. W., Dear, D. J. A., and Penfold, D. 1968. Phase studies on the systems Na_2SO_4–SO_3, K_2SO_4–SO_3 and Na_2SO_4–K_2SO_4–SO_3. *Jour. Inst. Fuel,* 41, 129.

Cole, R. M. 1978. Economics of coal cleaning and flue gas desulfurization for compliance with revised NSPS for utility boilers. Paper, EPA Symposium on Coal Cleaning, Hollywood, Florida.

Cook, M. C., and Furman, R. C. 1981. Oil replacement options being considered by FP and L. Seminar, The Use of Coal in Oil Design Utility Boilers. EPRI Report WS-80-141, Palo Alto, Calif.

Cook, R. E. 1975. Sulfur trioxide conditioning. *Jour. Air. Poll. Contr.,* 25, 156.

Cooley, S. A., and Ellman, R. C. 1979. Analysis of coal and ash from lignites and sub-bituminous coals of North Dakota and eastern Montana. Grand Forks Energy Research Center Report GFETC/R179/1, Grand Forks, North Dakota.

Cooling, D. R. 1965. The hardness of coal and its associated minerals. *British Coal Utilization Research Association Bulletin,* 29, 409. NCB, Chelternham, U.K.

Corder, W. C., Batchelder, H. R., and Goldberger, W. M. 1973. The Union Carbide/Battelle coal gasification. Development Unit Design, 5. Pipeline Gas Symp., p. 107, Chicago, Ill.

Corey, R. C. 1974. Measurements and significance of the flow properties of coal ash slag. U.S. Bur. Min. Bull. No. 618.

Corey, R. C., and Cohen, P. 1950. Furnace heat absorption in Paddy's Run pulverized coal fired steam generator using turbulent burners. *Trans. Am. Soc. Mech. Engrs.,* 72, 925.

Corey, R. C., Cross, B. J., and Reid, W. T. 1945. External corrosion of furnace wall tubes, Part 2, Significance of sulfate deposits and sulfur trioxide in corrosion mechanism. *Trans. ASME,* 67, 289.

Cross, N. L., and Picknett, R. G. 1963. Particle adhesion in the presence of liquid film. Conf. Mechanism of Corrosion by Fuel Impurities, p. 383, Marchwood, U.K., Butterworth, London.

Crossley, H. E. 1952. External boiler deposits. *Jour. Inst. Fuel,* 25, 221.

Crossley, H. E. 1963. The Mechett Lecture: A contribution on the development of power stations. *Jour. Inst. Energy,* 36, 228.

Crossley, H. E. 1967. External fouling and corrosion of boiler plant: A commentary. *Jour. Inst. Fuel,* 40, 342.

Crumley, P. H., and McCamley, W. 1958. Behaviour of chloride in coal during combustion. Conf. Science in Use of Coal, Paper 32, D 43, Sheffield University, U.K.

Crystal, J. T. 1970. The removal of sodium from lignite by ion exchange. Master's thesis, University of Grand Forks, North Dakota, N. Dakota.

CSIRO. 1969. Electrostatic precipitation of fly ashes from some New South Wales coals. Coal Research on CSIRO Report No. 37, Melbourne, Australia.

Cumming, J. W. 1980. Communication: The electrical resistance of coal ash at elevated temperatures. *Jour. Inst. Energy,* 53, 153.

Cutler, A. J. B. 1971. The effect of oxygen and SO_3 on corrosion of steels in molten sulphates. *Jour. Appl. Electrochem.,* 1, 19.

Cutler, A. J. B., and Grant, C. J. 1973. Corrosion of iron and nickel base alloys in alkali sulphate melts. *Conf., Deposition and Corrosion in Gas Turbines,* eds. A. B. Hart and A. J. B. Cutler, p. 178. London: Applied Science.

Cutler, A. J. B., and Grant, C. J. 1975. Corrosion of iron and nickel base alloys in alkali sulphate melts. *Conf. Metal-Slag-Gas Reactions,* p. 591. Electrochem. Soc., Princeton, N.J.

Cutler, A. J. B., and Raask, E. 1981. External corrosion in coal fired boilers: Assessment from laboratory data. *Corr. Science,* 21, 789.

Cutler, A. J. B., Halstead, W. D., Laxton, J. W., and Stevens, C. G. 1970. The role of chloride in the corrosion caused by flue gases and their deposits. ASME Paper 70-WA/CD-1.

Cutler, A. J. B., Hart, A. G., Laxton, J. W., and Stevens, C. G. 1974. Current problems in the CEGB caused by fire side corrosion and recent research into solutions. *VGB Krafwerkstechnik,* 54, 611.

Cutler, A. J. B., Flatley, T., and Hay, K. A. 1978. Fire side corrosion. CEGB Research No. 8, p. 13. London.

Cutress, J. O., and Peirce, J. T. 1965. Combustion of pulverized anthracite at Tir John Power Station. *Jour. Inst. Fuel,* 38, 54.

Dainton, A. D. 1979. The combustion of coal in coal. In *Modern Coal Processing: An Introduction,* eds. G. J. Pitt and G. R. Millward, p. 109. London: Academic.

Dalmon, J. 1980. Electrostatic precipitators for large power station boilers. *Jour. Electrostatics,* 8, 309.

Dalmon, J., and Raask, E. 1972. Resistivity of particulate coal minerals. *Jour. Inst. Fuel,* 45, 201.

Dalmon, J., and Tidy, D. 1972. A comparison of chemical additives as aids to the electrostatic precipitation of fly ash. *Atmos. Envir.,* 6, 721.

Dantuma, R. S. 1928. Measurement of coefficient of internal friction (viscosity) of molten salts. *Zeitschr. Anorg. Allgem. Chem.,* 175, 1.

Davison, R. L., Natusch, D. F. S., and Wallace, J. R., and Evans, C. A. 1974. Trace elements in fly ash-dependence of concentration on particle size. *Env. Scien. Techn.,* 8, 1107.

Daybell, G. N. 1967. The relationship between sodium and chlorine in some British coals. *Jour. Inst. Fuel,* 40, 3.

Daybell, G. N., and Gillham, E. W. F. 1959. Removal of chlorides from coal by leaching: I. Laboratory experiments. *Jour. Inst. Fuel,* 32, 589.

Daybell, G. N., and Pringle, W. J. S. 1958. The mode of occurrence of chlorine in coal. *Fuel,* 37, 283.

De Haller, F. S. 1939. *Data Book for Materials Testing,* p. 471. Berlin: Springer-Verlag.

Demeter, J., McCann, C. R., Bellas, G. T., Ekmann, J. M., and Bienstock, D. 1978. Combustion of coal-oil slurry in a 100-HP firetube boiler. *Combustion,* April, p. 31.

Dering, I. S., Dubrovskii, V. A., and Dik, E. P. 1972. Sintering of different fractions of fly ash from brown coal. *Teploenergetika (Thermal Engineering),* 19, No. 12, English Transl., p. 70.

Derjaguin, B. V., Abrikosova, I. I., and Lifshitz, E. M. 1956. Direct measurement of molecular attraction between solids separated by a narrow gap. *Quart. Rev. Chem. Soc.,* 10, 295.

Desrosiers, R. E., Riehl, J. W., Ulrich, G. D., and Chiu, A. S. 1978. Submicron fly ash formation in coal fired boilers. 17th Int. Conf. on Combustion, Leeds, U.K., Combustion Institute, Pittsburgh, Pa.

Deurbrouck, A. W. 1972. Sulfur reduction potential of the coals of the United States. U.S. Bureau of Mines Report of Investigation 7633.

Deutsch, W. 1922. Movement of charge carriers in cylindrical conductors. *Ann. Physic.,* 68, 335.

Dietzel, A. 1934. Explanation of adhering problem in sheet iron enamels. *Emailwaren-Industrie,* 11, 61.

Dik, E. P., Surovitskii, V. D., Soboleva, A. N., and Kuskova, Y. Y. 1977. The formation of deposits with concentrations of iron oxides on boiler heating surfaces. *Teploenergetika (Thermal Engineering),* 24, No. 9, English Transl., p. 30.

Dimmer, J. P. 1980. Influence of coal mineral matter on slagging of utility boilers. EPRI Report CS-1418/RP 736, Palo Alto, Calif.

DIN. 1976. Determination of ash fusion behaviour. German Standard, DIN 51 730.

Dismukes, E. B., 1975. Conditioning fly ash with ammonia. *Jour. Air Poll. Contr.,* 25, 152.

Dixit, S. N. 1978. Fuel additive treatment in vapor phase. *Power Engr.,* 82, 62.

Dixit, S. N., and Cuisia, D. G. 1977. Additives for coal. *Combustion,* 49, No. 6, 32.

Dixon, K. 1958. Spectrochemical methods of analysis as applied to mineral matter associated with coal. *Analyst,* 83, 362.

Dixon, K., Skipsey, E., and Watts, J. T. 1964. The distribution and composition of inorganic matter in British coals: Part 1—Initial study of seams from the East Midlands Division of the National Coal Board. *Jour. Inst. Fuel,* 37, 485.

Dixon, K., Skipsey, E., and Watts, J. T. 1970a. The distribution and composition of inorganic matter in British coals: Part 2—Alumino-silicate minerals in the coal seams of the East Midlands coalfields. *Jour. Inst. Fuel,* 43, 124.

Dixon, K., Skipsey, E., and Watts, J. T. 1970b. The distribution and composition of inorganic matter in British coals: Part 3—The composition of carbonate minerals in the coal seams of the East Midlands coalfields. *Jour. Inst. Fuel,* 43, 229.

Dolezal, R. 1967. *Large Boiler Furnaces.* Amsterdam: Elsevier.

Dolinskaya, L. A., and Gritsenko, G. L. 1980. The structure and properties of superheater tubes with chromized outside surface. *Teploenergetika (Thermal Engineering),* 27, No. 3, 45, English Transl., p. 148.

Dooher, J., Jakatt, S., Batra, S. K., Dunn, R., Wood, A., and Tsai, T. 1978. Evaluation of mixing equipment and flow properties for the NEPSC Coal-Oil Mixture Demonstration Plant. Conf., Coal-Oil Mixture Combustion, p. 207, St. Petersburg Beach, Fla., U.S. Department of Energy.

Doxey, G., Hodgkinson, G. E., and Woodhouse, G. 1967. Some uses of calcium chloride as an additive to boiler coals. *Jour. Inst. Fuel,* 40, 531.

Drenker, S. G. 1981. Seminar, The Use of Coal in Oil-Design Utility Boilers. EPRI Report WS-80-141, Palo Alto, Calif.

Drummond, A. R., Boar, P. L., and Deed, R. G. 1978. Rational analysis of fireside deposits from brown coal fired boilers. *Ash Deposition and Corrosion due to Impurities in Combustion Gases,* ed. R. W. Bryers, p. 243. Washington, D.C.: Hemisphere.

Drury, M. D., Perry, K. P., and Land, T. 1951. Pyrometers for surface temperature measurements. *Jour. Iron and Steel Inst.,* 169, 245.

Duck, N. W., and Himus, G. W. 1951. On arsenic in coal and its mode of occurrence. *Fuel,* 30, 267.

Dunderdale, J., Durie, R. A., Mulcahy, M. F. R., and Schafer, H. N. S. 1963. Studies relating to the behaviour of sodium during the combustion of solid fuels. *Conf., Mechanism of Corrosion by Fuel Impurities,* p. 139, Marchwood, U.K. Butterworth, London.

Dunham, K. C. 1970. Mineralization by deep formation waters: A review. *Trans. Inst. Min. Metal.,* 79, B127.

Dunn, R. M. 1981. Review of Salem Harbor Coal-Oil Mixture Project. Seminar, The Use of Coal in Oil-Design Utility Boilers. EPRI Report WS-80-141, Palo Alto, Calif.

Durie, R. A. 1961. The inorganic constituents in Australian coals, Part 3, Morwell and Yallourn brown coals. *Fuel,* 40, 107.

Durie, R. A., and Swaine, D. J. 1971. Inorganic constituents in coal. Coal Research in CSIRO, Australia Report No. 45, p. 9.

Dutcher, R. R., White, E. W., and Spackman, W. 1964. Elemental ash distribution in coal components—Use of the electron probe. Ironmaking Conf., Iron and Steel Div., Am. Inst. Mining Eng., New York, p. 463.

Duzy, A. F. 1965. Fusibility-viscosity of lignite type ash. ASME Paper 65-WA/FU-7.

Duzy, A. F., and Walker, J. B. 1965. Utilization of solid fuels having lignite type ash. U.S. Bureau of Mines, Inf. Circ. 8304.

Eckert, E. R. G., Hartnett, J. P., and Irvine, R. F. 1956. Measurement of total emissivity of porous materials in use for transpiration cooling. *Jet Prop.*, 26, 280.

Edgcombe, L. T. 1956. The state of combination of chlorine in coal. Extraction of coal with water. *Fuel*, 35, 38.

Edwards, A. M., Evans, G. J., and Howes, L. S. 1962a. Operational trial of superheater steels in a CEGB pulverized fuel fired boiler burning Yorkshire coal. *Jour. Inst. Fuel*, 35, 121.

Edwards, A. M., Jackson, P. J., and Howes, L. S. 1962b. Operational trial of superheater steels in a CEGB pulverized fuel fired boiler burning East Midlands coal. *Jour. Inst. Fuel*, 35, 16.

Edwards, G. R., Evans, T. M., Robertson, S. D., and Summers, C. W. 1980. Assessment of the standard method of test for the grindability of coal by the Hardgrove Machine. *Fuel*, 59, 826.

Effertz, P. H., and Wiume, D. 1979. Mechanism and products of high temperature corrosion of superheater tubes in coal fired boilers. *VGB Kraftwerkstechnik*, 59, 595.

Einstein, A. 1906. Contribution to the theory of Brownian movement. *Ann. Physik*, 19, 371.

Einstein, A. 1924. Contribution to the theory of radiation. *Zeitschr. Phys.*, 27, 1.

Eitel, W. 1954. *The Physical Chemistry of the Silicates*. Chicago, Ill.: University of Chicago Press.

Ellery, A. R., Johnson, T. R., and Newton, J. D. 1973. Investigation into the likelihood of thermal fatigue damage to furnace and superheater tubes caused by on-load water deslagging. ASME Paper 73-WA/CD5.

Elliott, J. F., and Gleiser, M. 1960. *Thermochemistry for Steelmaking*. Reading, Mass.: Pergamon.

Endell, K., and Zauleck, D. 1950. Relationship between chemical composition and viscosity of coal ash slags in cyclone fired boilers. *Bergbau and Energiewirt.*, 3, 70.

Endell, K., Heidtkamp, G., and Hax, L. 1936. Fluidity of calcium silicates and ferrites and basic slags at temperatures up to 1900 K. *Arch. Eisenhuttenwesen*, 10, 85.

Engel, P. A. 1978. *Impact Wear of Materials*, Tribology Series 2. Amsterdam: Elsevier.

Epstein, P. 1929. Contribution to the theory of radiation. *Zeitschr. Phys.*, 54, 537.

Erdos, E. 1973. Sulphide formation in nickel base superalloys. *Deposition and Corrosion in Gas Turbines*, p. 115. London: Applied Science.

Eubanks, A. G., and Moore, D. G. 1955. Effect of oxygen content of furnace atmosphere of adherence of vitreous coatings to iron. *Jour. Am. Ceram. Soc.*, 36, 410.

Evans, H. W. (ed.). 1978. Conf., Surface Treatments for Protection Spring Review Course, Series 3, No. 10, Inst. Metallurgists, London.

Eydam, H. 1978. Service experience and prospects of double reheater systems. *VGB Kraftwerkstechnik*, 58, 884.

Fairhurst, W., and Rohrig, K. 1976. Use of abrasion-resistant high chromium molybdenum cast irons in mineral processing. *Mine and Quarry*, 5, 40.

Fehndrich, W. 1969. Erosion tests with boiler tubes. *VGB Mitteilungen*, 49, 58.

Fenton, G. W., Adams, H. F., and Rumsby, P. L. 1962. Mapping and appraisal of characteristics of British coal seams. *Min. Engr.*, April, p. 454.

Fichte, R. 1953. *Thermodynamic Properties of Metal Chlorides*. Berlin: VEG Verlag Technik.

Field, M. A. 1970. Measurement of the effect of rank on combustion rates of pulverized coal. *Comb. and Flame*, 14, 237.

Field, M. A., Gill, D. W., Morgan, B. B., and Hawksley, P. G. W. 1967. Combustion of pulverized coal. BCURA Publication, Leatherhead, U.K.

Fielder, R. S. 1958. Ammonia injection trial, Whitebirk Power Station: Determination of sulphur trioxide in the flue gas and analysis of air heater deposits. CERL Report No. 774, Leatherhead, U.K.

Fincham, C. J. B., and Richardson, F. D. 1954. Behaviour of sulphur in silicate and aluminate melts. *Proc. Royal Soc.*, A223, 40.

Finkelman, R. B., and Stanton, R. W. 1978. Identification and significance of accessory minerals from a bituminous coal. *Fuel*, 57, 763.

Fisher, G. L., Chang, D. P. Y., and Brummer, M. 1976. Fly ash collected from electrostatic precipitators: Micro-crystalline structures and the mystery of the spheres. *Science*, 192, 553.

Flagan, R. C. 1978. Submicron particles from coal combustion. 17th Symp., Combustion, Leeds, U.K.

Flagan, R. C., and Friedlander, S. K. 1976. *Particle Formation in Pulverized Coal Combustion— A Review, Recent Developments on Aerosol Science.* New York: Wiley.

Flatley, T., King, C. W., Latham, E. P., and Mortimer, D. 1975. Potential materials to combat fireside corrosion in coal fired boilers. Conf., Materials in Power Plant Inst. Metallurgy Spring Course, p. 63.

Flatley, T., Latham, E. P., and Morris, C. W. 1981. Co-extruded tubes to improve resistance to fuel ash corrosion in U.K. utility boilers. *NACE Jour. Materials Performance*, 20, May, 12.

Ford, C. T., and Boyer, J. F. 1978. Effect of coal cleaning on fugitive elements: A progress report. EPA Symposium on Coal Cleaning, Hollywood, Florida.

Foster, G. G., and Toft, L. H. 1962. Wastage trials of superheater steels in CEGB power stations at Westwood and Bromborough. *Jour. Inst. Fuel*, 35, 28.

Francis, W. 1961. *Coal, Its Formation and Deposition.* London: Arnold.

Francis, W., 1965. *Fuels and Fuel Technology, Vol. I.* London: Pergamon.

French, B. M., and Rosenberg, P. E. 1965. Siderite ($FeCO_3$)–Thermal decomposition in equilibrium with graphite. *Science*, 147, 1283.

Frenkel, J. J. 1945. Viscous flow of crystalline bodies under the action of surface tension. *Jour. Phys. (Moscow)*, 9, 385.

Frenske, A. 1961. Abrasion wear of brown coal mills. *Braunkohle Warme Energie*, 13, 95.

Frossling, N. 1938. Evaporation of falling droplets. *Beitr. Geophys.*, 52, 170.

Fruehan, R. J., and Martonik, L. J. 1973. The rate of oxidation of metals and oxides, Part 3, The rate of chlorination of Fe_2O_3 and NiO in Cl_2 and HCl. *Met. Trans.*, 4, 2793.

Fulcher, G. S. 1925. Analysis of recent measurements of viscosity of glasses. *Jour. Am. Ceram. Soc.*, 8, 339.

Fullard, K. 1971. The computation of temperature distribution and thermal stresses using finite element techniques. Conf., Structure Mechanical Reactor Technology, 6, Part M, Berlin.

Furr, A. K., Parkinson, T. F., Hinrichs, R. A., Van Campen, R. D., Bache, C. A., Gutenman, W. H., St. John, L. E., Pakkala, I. S., and Lisk, D. J. 1977. National survey of elements and radioactivity in fly ash absorption of elements by cabbages grown in fly ash/soil mixtures. *Envir. Science Techn.*, 11, 1194.

Gal-or, B. 1980. Plasma spray coating processes: Physico-mathematical characterization. *ASME Jour. Pow.*, 102, 589.

Gat, N., and Tabakoff, W. 1980. Effects of temperature on the behaviour of metals under erosion by particulate matter. *Jour. Test. Eval.*, 8, 177.

Gauger, A. W., Barrett, E. P., and Williams, F. J. 1934. Mineral matter in coal—A preliminary report. *Trans. AIME*, 108, 226.

Genck, C. R. 1964. Soot blower systems using steam or air. *Combustion*, 35, 36.

Gibb, W. H. 1981. The slagging and fouling characteristics of coals, Part 1, Ash viscosity measurements for the determination of slagging propensity. *Power Ind. Research*, 1, 29.

Gibb, W. H. 1982. The role of calcium in the slagging and fouling characteristics of bituminous coal. ASME Joint Power Conf., Denver, Colo.

Gibb, W. H., and Angus, J. G. 1983. The release of potassium from coal during bomb combustion. *Jour. Inst. Energy*, 56, 149.

Gladney, E. S., Small, J. A., Gordon, G. E., and Zoller, W. H. 1976. Composition and size distribution of in-stack particulate material at a coal-fired power plant. *Atmos. Envir.*, 10, 1071.

Gleit, C. E. 1963. Electronic apparatus for ashing biologic specimens. *Am. Jour. Med. Electronics,* 2, 112.

Gleit, C. E., and Holland, W. D. 1962. Use of electrically excited oxygen for the low temperature decomposition of organic substances. *Anal. Chem.,* 34, 1454.

Glover, G. M. 1969. Flow and physical properties of potassium sulphate and coal ash slag. In *Open Cycle MHD Power Generation,* eds. J. B. Heywood and C. J. Womack. Oxford: Pergamon.

Gluckman, M. J., Yerushalmi, J., and Squires, A. M. 1976. Defluidization characteristics of sticky or agglomerating beds. *Fluidization Technology,* vol. 2, ed. D. L. Keairns, pp. 395–422. Washington, D.C.: Hemisphere.

Gluskoter, H. J. 1965. Electronic low-temperature ashing of bituminous coals. *Fuel,* 44, 285.

Gluskoter, H. J. 1967. Clay minerals in Illinois coal. *Jour. Sedim. Petrol.,* 37, 205.

Gluskoter, H. J. 1975. Mineral matter and trace elements in coal. In *Trace Elements in Coal,* ed. S. P. Babu. Am. Chem. Soc., Publication Series 141, Washington, D.C.

Gluskoter, H. J. 1977. Inorganic sulfur in coal. *Energy Sources,* 3, 125.

Gluskoter, H. J., and Lindahl, P. C. 1973. Cadmium: Mode of occurrence in Illinois coals. *Science,* 181, 246.

Gluskoter, H. J., and Rees, O. H. 1964. Chlorine in Illinois coal. Illinois Geological Survey Circ. 372, Urbana, Ill.

Gluskoter, H. J., and Ruch, R. R. 1971. Chlorine and sodium in Illinois coal as determined by neutron activation analyses. *Fuel,* 50, 65.

Gluskoter, H. J., and Simon, J. A. 1968. Sulfur in Illinois coals. Illinois State Geological Survey Circ. 432, Urbana, Ill.

Goard, P. R. C. 1966. Application of hemispherical surface pyrometers to the measurement of the emissivity of platinum—A low emissivity material. *Jour. Sci. Inst.* 43, 256.

Goblirsch, G. M., and Sondreal, E. A. 1979. Low rank coal atmospheric fluidized bed combustion technology, p. 30. Lignite Symp., Grand Forks, N. Dak.

Godel, A., and Cosar, P. 1967. The scale-up of a fluidized bed combustion system to utility boilers. *Am. Inst. Chem. Eng., Symp. Ser.,* 67, 210.

Godridge, A. M., and Morgan, E. S. 1971. Emissivities of materials from coal and oil fired water tube boilers. *Jour. Inst. Fuel,* 44, 207.

Godridge, A. M., and Read, A. W. 1976. Combustion and heat transfer in large boiler furnaces. *Prog. Energy Comb. Sci.,* 2, 83.

Godridge, A. M., Bradzioch, S., and Hawksley, P. G. W. 1962. A particle size classifier for preparing graded sub-sieve fractions. *Jour. Sci. Instrum.,* 39, 611.

Goldberg, H. J., and Kittle, P. A. 1979. The impact of elevated temperatures on the reactivity of alkaline additives used to control cold end boiler corrosion. *Jour. Inst. Energy,* 52, 27.

Goldberger, W. H. 1967. Collection of fly ash in a self-agglomerating fluidized bed coal burner. ASME Paper 67-WA/FU-3.

Golovin, V. N. 1964. Investigation of tube fouling in a TP-90 boiler. *Teploenergetika (Thermal Engineering),* 11, No. 3, English Trans., p. 23.

Golovin, M. N., and Putman, A. A. 1962. Inertial impaction on single elements. *Ind. Eng. Chem. Fundamentals,* 1, 265.

Gooch, J. P., Bickelhaupt, R. E., Marchant, G. M., Mcranie, R. D., Sparks, L. E., and Altman, R. F. 1981. Improvement of hot side precipitator performance with sodium conditioning—An interim report. *Jour. Air Poll. Contr. Assoc.,* 31, 291.

Good, W. D. 1962. The heat of formation of silica. *Jour. Phys. Chem.,* 66, 380.

Goodier, J. N. 1936. Slow viscous flow and elastic deformation. *Phil. Magazine,* 22, 678.

Goodwin, J. E., Sage, W., and Tilly, G. P. 1969. Study of erosion by solid particles. *Proc. Inst. Mech. Engs.,* 184, Part 1, No. 15, 279.

Grant, K., and Weymouth, J. H. 1962. The nature of inorganic deposits formed during the use of Victorian brown coal in large industrial boilers. *Jour. Inst. Fuel,* 35, 154.

Grant, K., and Weymouth, J. H. 1964. The influence of the inorganic constituents in Victorian brown coals on the fouling properties of the coals when used in gas generators and industrial boilers. Symp. Inorganic Constituents of Fuel, Paper No. 10, Inst. Fuel, Melbourne, Australia.

Gray, M. 1981. The effects of a volatile fireside manganese additive on coal fired utility boiler operation. Workshop Proc. Application of Fireside Additives to Utility Boilers. EPRI Report WS-80-127, Palo Alto, Calif.

Gray, R. J., and Moore, G. F. 1974. Burning the sub-bituminous coals of Montana and Wyoming in large utility boilers. ASME Paper 74-WA/FU-1.

Grayson, M., and Eckroth, D. 1978. *Kirk-Othmer Encyclopedia of Chemical Technology, Vol. 4.* New York: Wiley.

Greene, R. P., and Gallagher, J. M. 1980. *Future Coal Prospects: Country and Regional Assessment.* Cambridge, Mass.: Ballinger.

Greer, R. T. 1977. Coal microstructure and pyrite distribution. *Am. Chem. Soc. Symp. Series,* Vol. 64, paper No. 1, p. 3.

Gregg, S. J. 1953. The production of active solids by thermal decomposition, Part 1: Introduction. *Jour. Chem. Soc.,* p. 3940.

Griffin, E. M., and Profita, G. D. 1965. Design and construction of the steam generator and related equipment. *Am. Power Conf.,* 27, 571.

Gronhovd, G. H., Wagner, R. J., and Wittmaier, A. J. 1967. A study of the ash fouling tendencies of a North Dakota lignite as related to its sodium content. *Trans. Soc. Min. Engs.,* Sept., p. 313.

Gronhovd, G. H., Harak, A. E., and Paulson, L. E. 1968. Ash fouling studies on North Dakota lignite. U.S. Bureau Mines Report IC.8376, p. 76.

Gronhovd, G. H., Harak, H. E., and Tufte, P. H. 1970. Ash fouling and air pollution studies using a pilot test furnace. U.S. Bureau of Mines Inf. Circ. 8471, p. 69.

Gronhovd, G. H., Tufte, P. H., and Selle, S. J. 1973. Some studies on stack emissions from lignite fired power plants. Lignite Symp., Grand Forks, N. Dakota.

Gruner, E., and Bousquet, G. 1911. *A General Atlas of Coals.* Paris: Comité des Houillères de France.

Grunert, A. E., Skog, L., and Wilcoxson, L. S. 1947. The horizontal cyclone burner. *Trans. ASME,* 69, 613.

Gumz, W., Kirsch, H., and Mackowsky, M-T. 1958. *Slagging Studies; Investigations on Minerals in Fuel and Their Role in Boiler Operation.* Berlin: Springer-Verlag.

Gurvich, A. M., and Prasolov, R. S. 1960. Some properties of ash deposits on the wall tubes of steam tubes. *Teploenergetika (Thermal Engineering),* 7, English Transl., p. 80.

Hadrill, H. F. J. 1979. The performance of the 2000 MW coal fired Ratcliffe power station of CEGB with particular reference to the behaviour of dry bottom furnaces. *VGB Kraftwerkstechnik,* 59, 273.

Halstead, W. D., and Raask, E. 1969. The behaviour of sulphur and chlorine compounds in pulverized coal fired boilers. *Jour. Inst. Fuel,* 42, 344.

Haque, R., Dutta, M. L., and Chakrabarti, R. K. 1979. Fluidized bed combustion of high sulphur coals. *Jour. Inst. Energy,* 52, 173.

Hardgrove, R. M. 1932. Grindability of coal. *Trans. ASME,* 54, 37.

Harrison, W. N., Richmond, J. C., Pitts, J. W., and Brenner, S. G. 1952. Radioisotope study of cobalt in porcelain enamel. *Jour. Am. Ceram. Soc.,* 35, 113.

Hazard, H. R., Barrett, R. E., and Dimmer, J. P. 1979. Coal mineral matter and furnace slagging. *Am. Power Conf.,* Chicago, Ill. 41, p. 610.

Hedley, A. B. 1967. Factors affecting the formation of sulphur trioxide in flame gases. *Jour. Inst. Fuel,* 40, 142.

Hedley, A. B., Brown, T. D., and Shuttleworth, A. 1965. Available mechanisms for deposition from a combustion gas stream. ASME Paper No. 65-WA/CD-4.

Hein, K. 1976. Characteristics of Rhine brown coals with respect to their combustion and fouling behaviour in thermal power stations. ASME Paper 76-WA/CD-3.

Hein, K. 1979. Chemical and mineralogical aspects of fouling of heat exchange surfaces in non-bituminous coal fired boilers. *VGB Kraftwerkstechnik,* 59, 576.

Hendry, A., and Lees, D. J. 1980. Corrosion of austenitic steels in molten sulphate deposits. *Corrosion Sci.,* 20, 383.

Herrmann, W., and Krause, W. 1965. Flow studies on a three dimensional model of non-bituminous coal fired boiler. *VGB Mitteilungen*, 45, 135.

Heywood, J. B., and Womack, C. J. 1969. *Open Cycle MHD Power Generation.* Oxford: Pergamon.

Hicks, D., and Nagelschmidt, G. 1943. Chronic pulmonary disease in South Wales coal-mines, environmental studies, the chemical and X-ray diffraction analysis of the roof and clod of some South Wales seams and of the mineral matter in the coal. Special Report 244, Sect. E and F, Med. Res. Council, London.

Hills, G. J. 1963. Corrosion of metals by molten salts. *Conf., Mechanism Corrosion,* p. 583, Marchwood, U.K., Butterworth, London.

Hilsenrath, J. 1960. *Tables of Thermodynamic and Transport Properties of Air, Argon, Carbon Monoxide, Carbon Dioxide, Hydrogen, Nitrogen and Steam.* Oxford: Pergamon.

Hiorns, F. J., and Parish, B. M. 1966. Investigations of the wear of coal pulverizers; Part 1, General considerations and description of a laboratory method. *Jour. Inst. Fuel,* 39, 126.

Hirsch, M., and Rasch, R. 1968. Contribution for formation of iron chloride from reaction between HCl in flue gas and iron oxide in fly ash and on heat exchange surfaces. *Aufbereitungs-Technik,* 9, 614.

Hocke, H. 1972. Wear-resistant materials for steelworks plant handling coke and sinter. British Steel Pub., PE/B/5/72, London.

Hodgman, C. D. (ed.). 1962. *Handbook of Chemistry and Physics,* Cleveland, Ohio: The Chemical Rubber.

Hodges, N. J., Ladner, W. R., and Martin, T. G. 1983. Chlorine in coal: A review of its origin and mode of occurrence. *Jour. Inst. Energy,* 56, 158.

Hoffman, L., and Deurbrouck, A. W. 1977. Engineering economic analysis of coal preparation with SO_2 cleanup processes for keeping higher sulfur coals in the energy market. EPA Conf., Coal Cleaning, Triangle Park, N. Carolina.

Hoffmann, A., Klein, U., Michalik, W., and Wandslebe, J. 1980. Erosion wear in pipes and bends on pneumatic and hydraulic transport. *Energietechnik,* 30, 346.

Hoffmann, K. A., and Hoffmann, U. R. 1948. *Inorganic Chemistry.* Braunschweig, W. Germany: Vieweg.

Hoke, R. C., and Ernst, M. 1980. Control of particulate emissions from the pressurized fluidized bed combustion of coal. Conf., Fluidized Combustion: Systems and Applications, Inst. Energy, London.

Holland, L. 1964. *The Properties of Glass Surfaces.* London: Chapman and Hall.

Honda, H., and Sanada, Y. 1957. Microhardness of carbonized coal. *Fuel,* 306, 403.

Honea, F. I., and Persinger, M. M. 1981. Ash related outages and curtailments for utility boilers burning low rank coals. Am. Pow. Conf., Chicago, Ill.

Honea, F. I., Selle, J. J., and Sondreal, E. A. 1979. Factors affecting boiler tube fouling from Western Gulf Coast lignites and sub-bituminous coals. Intern. Corrosion Forum, Atlanta, Ga.

Hosegood, E. A., and Raask, E. 1964. Reactions and properties of alkalis and ash in pulverized fuel combustion. CERL Note No. RD/L/N 49/64, Leatherhead, U.K.

Hoy, H. R., Roberts, A. G., and Wilkins, D. M. 1962. Some investigations into the gasification of solid fuel in a slagging fixed-bed gasifier. Conf., Gasification Processes, pC-11, Hastings, U.K.

Hoy, H. R., Roberts, A. G., and Wilkins, D. M. 1965. Behaviour of mineral matter in slagging gasification processes. *Jour. Inst. Gas Engrs.,* 5, 444.

Huffman, G. P., Huggins, F. E., and Dunmyre, G. R. 1981. Investigation of the high temperature behaviour of coal ash in reducing and oxidizing atmosphere. *Fuel,* 60, 585.

Hulett, L. D., and Weinberger, A. J. 1980. Some etching studies of the microstructure and composition of large alumino-silicate particles in fly ash from coal-burning power plants. *Envir. Science Technol.,* 14, 965.

Hull, A. W., Burger, E. E., and Navias, L. 1941. Glass to metal seals. *Jour. Appl. Phys.,* 12, 698.

Hurst, R. C., Johnson, J. B., Davies, M., and Hancock, P. 1973. Sulphate and chloride attack of nickel based alloys and mild steels. *Conf., Deposition and Corrosion in Gas Turbines*, p. 143. London: Applied Science.

Ingraham, T. R., and Marier, P. 1963. Kinetic studies on thermal decomposition of $CaCO_3$. *Jour. Can. Chem. Engng.*, 41, 170.

International Critical Tables. 1928. *Surface Tension of Molten Sulfates, Vol. 4*, p. 443. New York: McGraw-Hill.

Ion, D. C. 1980. *Availability of World Energy Resources*. London: Graham and Trotman.

Israelachvili, J. N., and Tabor, D. 1973. Van der Waals forces: Theory and experiment. *Prog. Surf. Membr. Sci.*, 7, 1.

Jackson, A. W. 1980. Lignite slagging tests–Phase 3. Ontario Hydro, Canada, Res. Div. Report 80-454-K.

Jackson, P. J. 1963. The physiochemical behaviour of alkali-metal compounds in fireside boiler deposits. *Conf. Mechanism of Corrosion by Fuel Impurities*, p. 484. Marchwood, U.K., Butterworth, London.

Jackson, P. J., and Duffin, H. C. 1963. Laboratory studies of the deposition of alkali-metal salts from flue gas. *Conf., Mechanism of Corrosion by Fuel Impurities*, p. 427. Marchwood, U.K., Butterworth, London.

Jackson, P. J., and Raask, E. 1957. A trial of furnace steam injection at Rye House Power Station. CERL Report No. 752, Leatherhead, U.K.

Jackson, P. J., and Raask, E. 1961. A probe for studying the deposition of solid material from flue gas at high temperatures. *Jour. Inst. Fuel*, 34, 275.

Jackson, P. J., and Ward, J. M. 1956. Operational studies of the relationship between coal constituents and boiler fouling. *Jour. Inst. Fuel*, 29, 154.

James, W. C., and Fisher, A. G. 1967. A new fuel additive for coal fired boilers. *Jour. Inst. Fuel*, 40, 170.

JANAF. 1963. *Thermodynamical Tables*. Dow Chemicals, Midlands, Michigan.

JANAF. 1971. *Thermochemical Tables*. U.S. National Bureau of Standards, Publication NSRDS-NBS37, Washington, D.C.

Jensen, E. 1942. Phyrrhotite–Melting relations and composition. *Science*, 240, 695.

Johnson, P. D. 1952. Absolute optical adsorption from diffuse reflectance. *Jour. Am. Opt. Soc.*, 42, 978.

Johnson, P. D. 1964. Optical absorption and diffuse reflectance of powders. *Jour. Appl. Phys.*, 35, 334.

Johnston, W. D., and Chelko, A. 1966. Oxidation-reduction equilibria in molten $Na_2O \cdot 2SiO_2$ glass in contact with metallic copper and silver. *Jour. Am. Cer. Soc.*, 49, 562.

Johnstone, H. F., and Roberts, M. H. 1949. Deposition of aerosol particles from moving gas streams. *Ind. Eng. Chem.*, 41, 2417.

Jonakin, J., Reese, J. T., and Rice, G. A. 1959. Fireside corrosion of superheater and reheater tubing. ASME Paper 59-F4-5.

Juranek, G., Kirsch, H., Nemetschek, T., and Schliephake, R. W. 1958. Contribution to the knowledge of mineral substances in particular clay minerals found in Ruhr coals. *Rev. Industrie Min.*, July, p. 191.

Kaakinen, J. W., Jorden, R. M., Lawasani, M. H., and West, R. E. 1975. Trace element behaviour in coal-fired power plant. *Envir. Scien. Tech.*, 9, 856.

Kanowski, S., and Coughlin, R. W. 1977. Catalytic conditioning of fly ash without addition of SO_3 from external source. *Envir. Science Techn.*, 11, 67.

Karr, C. 1979. *Analytical Methods for Coal and Coal Products*, Vols. 1–3. New York: Academic.

Kaye, G. W. C., and Laby, T. H. 1978. *Tables of Physical and Chemical Constants*. London: Longman.

Keck, W. J. 1963. Regenerative air preheaters can be water washed effectively. *Power Eng.*, May, p. 47.

Keller, P. 1969. Occurrence, origin and phase changes of β-FeOOH in rust. *Werkstoffe Korrosion*, 20, 102.

Kelley, J. E., Roberts, T. D., and Harris, H. M. 1964. A penetrometer for measuring the absolute viscosity of glass. U.S. Bureau of Mines Rep. No. 90, Washington, D.C.

Kelley, K. K. 1960. Contribution to the data on theoretical metallurgy. U.S. Bureau Mines Bull. No. 584, Washington, D.C.

Kelley, K. K. 1962. Heats and free energy of formation of anydrous silicates. U.S. Bureau of Mines Rep. No. 5901, Washington, D.C.

Kelley, K. K., and King, E. G. 1961. Contributions to the data on theoretical metallurgy. U.S. Bureau of Mines Bull. No. 592, Washington, D.C.

Kemezys, M., and Taylor, G. H. 1964. Occurrence and distribution of minerals in some Australian coals. *Jour. Inst. Fuel,* 37, 389.

Kent, C. R., and Champion, A. O. 1964. The influence of carbonate minerals on the ash fusion characteristics of coals. Inst. Fuel Symp., Inorganic Constituents in Fuel, Melbourne, Australia.

Kerekes, Z. E., Bryers, R. W., and Sauer, A. R. 1978. The influence of heavy metals Pb and Zn on corrosion and deposits in refuse fired steam generators. In *Ash Deposits and Corrosion due to Impurities in Combustion Gases,* ed. R. W. Bryers, p. 455. Washington, D.C.: Hemisphere.

Khruschov, M. M. 1974. Principles of abrasive wear. *Wear,* 28, 69.

Kilgroe, J. D., Raines, G. E., and Smithson, G. R. 1977. Coal cleaning applications for SO_2 emission control. EPA Conf., Coal Cleaning, Triangle Park, N. Carolina.

King, B. W., Tripp, H. P., and Duckworth, W. H. 1959. Nature of adherence of porcelain enamels to metals. *Jour. Am. Cer. Soc.,* 42, 504.

King, C. W., and Robinson, M. T. 1973. Development and use of composite tube for coal fired boilers. 5th Europ. Congress on Corrosion, Paris.

King, J. G., Maries, M. B., and Crossley, H. E. 1936. Formulae for the calculation of coal analysis to a basis of coal substance free of mineral matter. *Jour. Soc. Chem. Ind.,* 57, 277.

Kirkwood, J. B., and Chatfield, D. R. 1981. The Kwinana coal conversion in Western Australia. Seminar, The Use of Coal in Oil-Design Utility Boilers. EPRI Report WS-80-141. Palo Alto, Calif.

Kirsch, H. 1963. Corrosion in combustion chambers caused by slag attack and flue gases of varying composition. Conf., Mechanism of Corrosion by Fuel Impurities, p. 508, Marchwood, U.K. Butterworth, London.

Kirsch, H. 1964. Influence of coal clay minerals on operation of high temperature systems. *Bergbau-Archiv,* 25, 1.

Kirsch, H. 1965. Slagging and high temperature behaviour of coal ashes. *Tech. Uberwachung,* 6, 203.

Kirsch, H., Pollmann, S., and Ottemann, J. 1967. Reactions of coal arsenic compounds with steel. *Naturwiss.,* 54, 319.

Kirsch, H., Pollmann, S., and Ottemann, J. 1968. The behaviour of arsenic containing fuels in coal fired boilers. *Mitteilungen Grosskesselbesitzer,* 48, 10.

Kiss, L. T., and Lloyd, B. 1975. Thermal decomposition of copper oxychloride. *Jour. Inst. Fuel,* 48, 27.

Kiss, L. T., Lloyd, B., and Raask, E. 1972. The use of copper oxychloride to alleviate boiler slagging. *Jour. Inst. Fuel,* 45, 213.

Kistler, S. S. 1931. Coherent expanded aerogels and jellies. *Nature,* 127, 741.

Kitchener, J. A., and Prosser, A. P. 1957. Direct measurement of the long range van der Waals forces. *Proc. Roy. Soc.,* A 242, 403.

Klein, D. H., Andren, A. W., Carter, J. A., Emery, J. F., Feldman, C., Fulkerson, W., Lyon, W. S., Ogle, J. C., Talm, Y., van Hook, R. I., and Bolton, N. 1975. Pathways of thirty-seven trace elements through coal-fired power plant. *Envir. Scien. Tech.,* 9, 973.

Klomp, J. T. 1972. Bonding of metals to ceramics and glasses. *Am. Ceram. Soc. Bull.,* 51, 683.

Klomp, J. T. 1979. Interfacial reactions between metals and oxides during sealing. *Am. Ceram. Soc. Bull.,* 10, 887.

Koonce, D. F. 1968. Soot blowing for coal fired boilers. *Combustion*, 40, 24.

Koopman, J. G., Marselli, E. M., Jonakin, J., and Ulmer, R. C. 1959. Development and use of a probe for studying corrosion in superheaters and reheaters. *Amer. Power Conf.*, 21, 236.

Koppe, R. H., and Olson, E. A. J. 1979. Nuclear and large fossil unit operating experience. EPRI Report NP-1191, Palo Alto, Calif.

Kordes, E., Zofelt, B., and Proger, H. J. 1951. Reactions in the melting of glass with Na_2SO_4: Immicibility gap in liquid state between Na–Ca-silicates and Na_2SO_4. *Zeitschr. Anorg. Allgem. Chem.*, 264, 255.

Kotwal, R., and Tabakoff, W. 1980. A new approach for erosion prediction due to fly ash. ASME Paper 80-GT-96.

Kozakevitch, P. 1959. Viscosity of lime–alumina–silicate melts between 1875 and 2375 K. Symp., Phys. Chem. Proc. Metallurgy, p. 97, Metallurgical Society, AIME.

Kregg, D. H. 1981. Design features of the Kwihana oil to coal-conversion. Seminar, The Use of Coal in Oil-Design Utility Boilers. EPRI Report WS-80-1411, Palo Alto, Calif.

Krupp, H. 1967. Particle adhesion–Theory and experiment. *Advan. Coll. Interface Sci.*, 1, 111.

Kubaschewski, O., Evans, E. L., and Allcock, C. B. 1967. *Metallurgical Thermochemistry*. Oxford: Pergamon.

Kuczynski, G. C. 1949a. Self diffusion in sintering of metallic particles. *Jour. Metals*, Feb., p. 169.

Kuczynski, G. C. 1949b. Study of the sintering of glass. *Jour. Appl. Phys.*, 20, 1160.

Kuczynski, G. C. 1977. Science of sintering. *Jour. Scien. Sintering*, 9, 243.

Kuczynski, G. C., and Zaplatynski, I. 1955. Sintering of glass. *Jour. Ceram. Soc.*, 39, 349.

Kuhn, J. K., Kohlenberger, L. B., and Shimp, N. F. 1973. Comparison of oxidation and reduction methods in the determination of forms of sulfur in coal. Environm. Geol. Notes No. 66, Illinois Geological Survey, Urbana, Ill.

Kulp, J. L., Kent, P., and Kerr, P. F. 1951. DTA of Ca-Mg-Fe minerals. *Am. Mineralogist*, 36, 648.

Kuper, G., and Werker, E. 1981. Electro-acoustic probe for detecting tube failures in fossil fired boilers. *VGB Kraftwerkstechnik*, 61, 719.

Lai, F. S., Friedlander, S. K., Pich, J., and Hidy, G. M. 1972. The self-preserving particle size distribution for Brownian coagulation in the free-molecule regime. *Jour. Coll. Interface Sci.*, 39, 395.

Laire, C. 1980. Conversion of Ruien power station in coal. *Electricite*, No. 171, 7.

Lane, G. J., Hilliard, P. T., Blackaller, G. S., Tydd, C. B., and Cairns, A. J. 1981. Studies of in-bed corrosion in a pressurized fluidized bed combustor. EPRI Report CS-1935, Palo Alto, Calif.

Lauf, R. J. 1981. Cenospheres in fly ash and conditions favouring their formation. *Fuel*, 60, 1177.

Lebiedzik, J., Burke, K. G., and Troutman, S. 1973. New methods for quantitative characterization of multiphase particulate materials including thickness measurement. Conf., Scanning Electron Microscopy, Chicago, Ill.

Lederman, P. B., Bibbo, P. B., and Bush, J. 1978. Chemical conditioning of fly ash for hot side precipitation. Symp., Transfer Utilization Particulate Control Technology, Denver, Colo.

Lee, C. R., and Bryden-Smith, D. W. 1968. Experience with a simple moisture gauge for embedment in concrete. Conf., Prestressed Concrete Pressure Vessels, Group 1, Paper 54, p. 635, London.

Lees, D. J. 1978. Selection of corrosion resistant coatings for boiler tube application. Conf., Surface Treatment for Protection, Spring Review Course, Series 3, No. 10, 174, Inst. Metallurgist, London.

Leesley, M. E., and Hedley, A. B. 1972. The effect of particle size distribution on the combustion rate of a pulverized anthracite dust cloud. *Jour. Inst. Fuel*, 45, 224.

Lehmann, H., and Mueller, K. H. 1960. DTA of dolomite in vacuum and at atmospheric pressure. *Tonin.-Zeitung Keram. Rundschass*, 84, 200.

Levine, S. R. 1978. High temperature surface protection. Conf., Surface Treatments for

Protection, Spring Review Course, Series 3, No. 10, 157, Inst. Metallurgists, London.

Levy, A., and Merryman, E. L. 1966. Interaction in sulfur oxide/iron oxide systems. ASME Paper 66-WA/CD-3.

Lewis, W. K., Squire, L., and Broughton, G. 1942. *Industrial Chemistry of Colloidal and Amorphous Materials.* New York: Macmillan.

Lifshitz, E. M. 1956. The theory of molecular attractive forces between solids. *Soviet Physics JETP,* 2, 73.

Lightman, P., and Street, P. J. 1968. Microscopic examination of heat treated pulverized coal particles. *Fuel,* 47, 7.

Lillie, H. R. 1929. Measurement of absolute viscosity by use of concentric cylinders. *Jour. Am. Ceram. Soc.,* 12, 505.

Lillie, H. R. 1931. Viscosity of glass between strain point and melting temperature. *Jour. Am. Ceram. Soc.,* 14, 503.

Linton, R. W., Loh, A., and Natusch, D. F. S. 1976. Surface predominance of trace elements in airborne particles. *Science,* 191, 852.

Lipson, C., and Colwell, L. V. 1961. *Handbook of Mechanical Wear.* Ann Arbor, Mich.: University of Michigan Press.

Lissner, A. 1956. *Mineral Matter in Brown Coals,* Sitzungsber. Deutsch. Acad. Wiss., Klasse, Chemie, Geologie u. Biologie, No. 1. Berlin: Akademie-Verlag.

Lissner, A. 1963. Constitution and behaviour of non-bituminous coals. *Forsch. u. Fortschr.,* 37, 357.

Littlejohn, R. F. 1966. Mineral matter and ash distribution in as-fired samples of pulverized fuels. *Jour. Inst. Fuel,* 39, 59.

Locklin, D. W., Krause, H. H., Reid, W. T., and Anson, D. 1980. Electric utility use of fireside additives. EPRI Report CS-1318. Palo Alto, Calif.

Locklin, D. W., Krause, H. H., Reid, W. T., and Anson, D. 1980b. Fireside additive trials in utility boilers—Overview of the EPRI Survey. *Combustion,* Feb., p. 26.

Loeffler, F. J., and Vaka, G. A. 1979. Bed blending of fossil fuels. ASME Paper 79-WA/MH-12.

London, F. 1930. Theory and systematics of molecular forces. *Zeitschr. Physik,* 63, 245.

Lorenz, K., and Kranz, E. 1966. Corrosion by alkali-metal sulfates in steel recuperators and boiler plant. *Werkstoffe und Korrosion,* 17, 97.

Losel, G., and Schmucker, H. 1977. Temperature behaviour of ash particles at combustion chamber exit. *VGB Kraftwerkstecnik,* 57, 839.

Lowe, H. J., Dalmon, J., and Hignett, E. T. 1965. The precipitation of difficult dusts. IEE Colloq. Electrostatic Precipitators, London.

Lowrey, K., and Taylor, J. 1976. Surface coatings as chimney liners. *2d Symp., Chimney Design, Vol. 1,* p. 140. University of Edinburgh, Scotland, U.K.

Lowry, H. H. (ed.). 1945. *Chemistry of Coal Utilization, Vols. I and II.* New York: Wiley.

Lucas, D. H. 1963. The work of CERL in combustion and related fields. *Jour. Inst. Fuel,* 36, 203.

MacDonald, E. J., and Murray, M. V. 1952. The effect of coal quality on the efficiency of a shell boiler equipped with a travelling grate stoker. *Jour. Inst. Fuel,* 28, 479.

MacKenzie, J. D. 1957. The discrete ion theory and viscous flow in liquid silicates. *Trans. Farad. Soc.,* 53, 1488.

MacKenzie, J. K., and Shuttleworth, R. 1949. A phenomenological theory of sintering. *Proc. Phys. Soc.,* B62, 833.

MacKenzie, K. J. D., and Brown, I. W. M. 1975. Devitrification of alumino-silicate glasses containing transition ions. *Phys. Chem. Glasses,* 16, 17.

MacKenzie, R. C. 1957. *The Differential Thermal Analysis of Clays.* Aberdeen, U.K.: Mineral. Soc. London, Central Press.

Malden, P. J., and Meads, R. E. 1967. Substitution by iron in kaolinite. *Nature,* 215, 844.

Maloney, K. L., and Benson, R. C. 1981. Failure cause analysis—Air preheaters. EPRI Report CS-1927, Palo Alto, Calif.

Manfred, R. 1981. Use of coal in oil design boilers. Seminar, The Use of Coal in Oil-Design Utility Boilers. EPRI Report WS-80-141, Palo Alto, Calif.

Manolescu, A. V., and Mayer, P. 1978. Review of coatings for protection against fireside corrosion and erosion. Ontario Hydro Res., Div. Report 78-368-K, Canada.

Mansfield, F. 1973. Simultaneous determination of instantaneous corrosion rates and tafel slopes from polarization resistance measurements. *Jour. Electrochem. Soc.*, 120, 515.

Markley, G. F. 1968. *Progress in Lignite Firing.* Alliance, Ohio: Babcock and Wilcox Co.

Marshall, C. E., and Tompkins, D. K. 1964. Mineral matter in Permian coal seams. Symp., Inorganic Constitutents of Fuel, Paper 4, Melbourne, Australia.

Matting, A., and Witt, W. 1967. Loading strength of glass to metal contacts. *Tonindustrie Zeitung,* 91, 493.

May, M. J., Zetlmeisl, M. J., and Annand, R. R. 1975. High temperature corrosion in gas turbines and steam boilers by fuel impurities. Part 6. A statistical study of the effects of sodium, temperature and additive on corrosion rate and slag friability. ASME Paper 75-WA/CD-5.

Mayer, P., and Manolescu, A. V. 1980. Influence of hydrogen chloride on corrosion of boiler steels in synthetic flue gas. *Nat. Ass. Corrosion Engns. Publ.,* 36, 369.

Mayers, M. A. 1945. Combustion in fuel beds. In *Chemistry of Coal Utilization,* vol. 2, ed. H. H. Lowry, p. 1482. New York: Wiley.

Mazza, M. H., Green, D. A., Parris, M. W., and Newton, G. J. 1978. Mineral characterization of fluidized bed combustion aerosol ash–Montana Rosebud sub-bituminous coal. U.S. Dept. of Energy Morgantown Res. Center Report MERC/TPR-78/1, Morgantown, W.Va.

Mazza, M. H., and Wilson, J. S. 1977. X-ray diffraction examination of coal combustion products related to boiler tube fouling and slagging. *Adv. X-Ray Anal.,* 20, 85.

McCain, J. C., Gooch, J. P., and Smith, W. B. 1975. Results of field measurements of industrial particulate sources and electrostatic precipitator performance. *Jour. Air Poll. Cont. Assoc.,* 25, 117.

McElroy, M. W., Carr, R. C., Ensor, D. S., and Markowski, G. R. 1982. Size distribution of fine particles from coal combustion. *Science,* 215, 13.

McGill, W. A., and Weinbaum, M. J. 1978. Alonized heat exchanger surface tubes give good high temperature service. *Materials Performance,* Feb., p. 16.

McLaughlin, J. 1980. Analysis of coals delivered to Tarong (Australia) power station. Private communication.

McMillan, P. W. 1979. *Glass Ceramics.* London: Academic.

Mei, J. S., Gall, R. L., and Wilson, J. S. 1977. Fluidized bed combustion test of low quality fuels–Anthracite refuse. Morgantown Energy Research Center Report MERC/R1-78/1, Morgantown, W.Va.

Mei, J. S., Grimm, U., and Halow, J. S. 1978. Fluidized bed combustion test of low quality fuels–Texas lignite and lignite refuse. U.S. Dept. of Energy Morgantown Research Center Report MERC/R1-78/3, Morgantown, W.Va.

Menzo, J. P., and Rahoi, D. W. 1979. Overcoming coal ash corrosion in utility boilers. ASME Paper 79-WA/CO-3.

Meringolo, V. 1981. Stop tube corrosion and wear in coal fired power boilers. *Power,* 125, 107.

Mesko, J. E. 1980. Technical and economic evaluation of fluidized bed combustion for industrial steam generation in the U.S.A. *Conf., Fluidized Combustion: Systems and Applications,* p. 1B-7.1, Inst. Energy, London.

Michel, J. R., and Wilcoxson, L. S. 1955. Ash deposits on boiler surfaces from burning central Illinois coal. ASME Paper No. 55-A-95.

Miller, R. N., and Given, P. H. 1978. Variations in Inorganic Constituents of Some Low Rank Coals. In *Ash Deposits and Corrosion due to Impurities in Combustion Gases.* Ed. R. W. Bryers, p. 39. Washington, D.C.: Hemisphere.

Millot, J. O'N. 1958. The mineral matter in coal. Part 1: The water of constitution of the silicate constituents. *Fuel,* 37, 71.

Minchener, A. J., Cooke, M. J., and Wardell, R. V. 1980. Corrosion studies in coal fired fluidized bed combustors by the National Coal Board. Conf., Fluidized Combustion: Systems and Applications, London.

Mitor, V. V., and Konopel'ko, I. N. 1966. A study of the emissivity of solid bodies. *Teploenergetika (Thermal Engineering)*, 13, No. 7, English Transl., p. 92.

Mitor, V. V., and Konopel'ko, I. N. 1969. Experimental investigation of radiation and reflection characteristics of some solid bodies. *Teploenergetika (Thermal Engineering)*, 16, No. 5, English Transl., p. 47.

Mitor, V. V., and Konopel'ko, I. N. 1970. A study of emissivity factor of ash deposits and certain refractory materials. *Teploenergetika (Thermal Engineering)*, 17, No. 10, English Transl., p. 61.

Monkhouse, A. C. 1950. The minor constituents in coal. *Coke and Gas*, 12, 363.

Moore, G. F., and Ehrler, R. F. 1973. Western coals–Laboratory characterization and field evaluations of cleaning requirements. ASME Paper 73-WA/FU-1.

Morey, G. W. 1954. *Properties of Glass*. Monograph Series, No. 124, Reinhold, London.

Morgan, E. S. 1974. Errors associated with radiant heat flux meters when used in boiler furnaces. *Jour. Inst. Fuel*, 47, 113.

Morgan, J., and Mahr, D. 1981. Evaluation of coal blending systems. *Combustion*, April, p. 28.

Morris, E. B., and Schumann, J. L. 1974. Condition fly ash with synthetic SO_3. *Power*, 118, 59.

Mortimer, D., and Raask, E. 1968. External corrosion of furnace wall tubes at High Marnham Power Station. CERL Report RD/L/R 1508, Leatherhead, U.K.

Moss, G. 1978. Principles of fluidized bed processes and application to industry. *Energiespectrum*, 2, 126.

Mott, R. A., and Spooner, C. E. 1940. The calorific value of carbon in coal: The Dulong relation. *Fuel*, 33, Part 1, p. 226; Part 2, p. 242.

Moza, A. K., and Austin, L. G. 1979. A new test for characterizing the slag deposition properties of coal ash: The sticking temperature. *Jour. Inst. Energy*, 52, 15.

Moza, A. K., Austin, L. G., and Johnson, G. G. 1979. Inorganic element analysis of coal particles using computer evaluation of scanning electron microscopy images. Conf. Scanning Electron Microscopy, Chicago, Ill.

Moza, A. K., Shoji, K., and Austin, L. G. 1980. The sticking temperature and adhesion force of slag droplets from four coals on oxidized mild steel. *Jour. Inst. Fuel*, 53, 17.

Mozes, M. S. 1982. Ash fusion characteristics of Bienfait lignite. Ontario Hydro, Canada, Res. Div. Report 82-141-K.

Mulcahy, M. F. R., Boow, J., and Goard, P. R. C. 1966. Fireside deposits and their effect on heat transfer in pulverized fuel fired boiler, Part 1–The radiant emittance and effective thermal conductance of the deposits, Part 2–The effect of the deposit on heat transfer from the combustion chamber considered as a continuous well-stirred reactor. *Jour. Inst. Fuel*, 39, Part 1, p. 385, Part 2, p. 394.

Murchison, D. G., and Westoll, T. S. (eds.). 1968. *Coal and Coal-Bearing Strata*. Edinburgh, U.K.: Oliver and Boyd.

Murphy, P., Piper, J. D., and Schmansky, C. R. 1957. Fireside deposits on steam generators minimized through humidification of combustion air. *Trans. ASME*, 73, 821.

Muyl, G. 1971. Injection of copper oxychloride in 250 MW steam generator. *Rev. Géner. Thermique*, 10, 143.

Napolitano, A., and Hawkins, G. E. 1964. Viscosity of a standard soda–lime–silica glass. *Jour. Res. U.S. Nat. Bur. Stan., Sect. A*, 68, 439.

Natusch, D. F. S. 1979. Toxic potential of sulfate in coal fly ash. Conf. Atmospheric Sulphates, London.

Natusch, D. F. S., Wallace, J. R., and Evans, C. A. 1974. Toxic trace elements: Preferential concentration in respirable particles. *Science*, 183, 202.

NCB. 1979. On-line ash monitor for coal. National Coal Board, London.

Neal, S. B. H. C., and Northover, E. W. 1980. The measurement of radiant heat flux in large boiler furnaces. Part 1. Problems of ash deposition relating to heat flux. *Int. Jour. Heat Mass Transfer*, 23, 1015.

Neal, S. B. H. C., Northover, E. W., and Preece, R. J. 1980. The measurements of radiant heat flux in large boiler furnaces. Part 2. Development of flux measuring instruments. *Int. Jour. Heat Mass Transfer*, 23, 1023.

Neavel, R. C. 1966. Sulfur in coal; Its distribution in the seam and in mine products. Ph.D. Thesis, Pennsylvania State University, State College, Pa.

Neavel, R. C., Nahas, N. C., and Koh, K. K. 1977. Removal of Sodium from Illinois Coal by Water Extraction, Trans. Soc. Min. Eng., AIME, 262, 263.

Nelson, J. B. 1953. Assessment of the mineral species associated with coal. *British Coal Utilisation Research Association, Mon. Bull.*, 17, 41, Leatherhead, U.K.

Nelson, J. E., and Koester, R. D. 1980. Experience with in service water cleaning of boiler fireside deposits. ASME Paper 80-JPGC/Fu-6.

Nelson, W., and Lisle, E. S. 1964. A laboratory evaluation of catalyst poisons for reducing high temperature gas side corrosion and ash bonding in coal fired boilers. *Jour. Inst. Fuel*, 37, 378.

Nelson, W., and Lisle, E. S. 1965. High temperature external corrosion on coal fired boilers: Siliceous additives. *Jour. Inst. Fuel*, 38, 179.

Nettleton, M. A., and Raask, E. 1967. The rate of evaporation of potassium sulphate. *Jour. Appl. Chem.*, 17, 18.

Nicholas, M. G. 1968. The strength of metal/alumina interfaces. *Jour. Mat. Science*, 3, 571.

Nicholas, M. G., Fogran, R., and Poole, D. M. 1968. Adhesion of metal/alumina interfaces. *Jour. Mat. Sci.*, 3, 9.

Nicholls, P., and Reid, W. T. 1940. Viscosity of coal ash slags. *Trans. ASME*, 62, 141.

Norman, T. E. 1959. Martensitic white irons for abrasion resistant coatings. *Trans. Amer. Foundry Soc.*, 67, 242.

Norme Française. 1945. Determination of ash fusibility curves. NF M03-012, Paris.

Northover, E. W. 1977a. Devices for the determination of heat flux and tube metal temperature in highly rated boilers. CEGB Technical Disclosure Bulletin 283, London.

Northover, E. W. 1977b. The CERL Fluxprobe—A portable direct reading radiant heat flux measuring device. CEGB Technical Disclosure Bulletin 286, London.

Northover, E. W. 1978. The CERL Dometer—A radiant heat flux and tube metal temperature measurements system for highly rated boilers. CEGB Technical Disclosure Bulletin 294, London.

Northover, E. W., and Hitchcock, J. A. 1967. A heat flux meter for use in boiler furnaces. *Jour. Sci. Instrum.*, 44, 371.

Norton, F. H. 1935. Measuring viscosity of glass. *Glass Industry*, 16, 143.

Nutkis, M. S., Hoke, R. C., Gregory, M. C., and Bertrand, R. R. 1977. Evaluation of a granular bed filter for particulate control in fluidized bed combustor. 5th Conf., Fluidized Bed Combustion, Washington, D.C.

Oatley, C. W., Nixon, W. C., and Pease, R. F. W. 1965. Scanning electron microscopy. *Adv. in Electronic and Electron Phy.*, 21, 181.

O'Connor, E. F. 1970. Use of copper salts for inhibition of external fouling of peat fired boilers. *Jour. Inst. Fuel*, 43, 449.

Ode, W. H. 1963. Coal analysis and mineral matter. In *Chemistry of Coal Utilization*, Suppl. Vol., ed. H. H. Lowry, p. 202. New York: Wiley.

Oel, H. J. 1960. Sintering and the viscosity of glasses. Paper 10, 7. Ceramic Congress, London.

Oel, H. J., and Gottschalk, A. 1966. The adhesion temperature between glass and metals. *Glastechnische Berichte*, 39, 319.

Oglesby, S., and Nichols, G. B. 1977. Electrostatic precipitation. *Air Pollution*, 4, 220.

O'Gorman, J. V., and Walker, P. L. 1972. Mineral matter and trace elements in U.S. coals. Dept. Interior Office of Coal Res. and Dev. Report No. 61, Interim Rep. No. 2, Washington, D.C.

O'Gorman, J. V., and Walker, P. L. 1973. Thermal behaviour of mineral fractions separated from selected American coals. *Fuel*, 52, 71.

Ondov, J. M., Ragaini, R. C., and Bierman, A. H. 1979. Emissions and particle size distributions of minor and trace elements at two western coal fired power plants equipped with cold side electrostatic precipitators. *Envir. Sci. Tech.*, 13, 947.

O'Neill, H. 1937. The indentation hardness of coal. *Jour. Inst. Fuel*, 10, 351.

Ormerod, W. G. 1981. Automatic trimming of combustion air to a pulverized coal fired boiler using the flue gas carbon monoxide signal. *Jour. Inst. Energy*, 54, 174.

Ormerod, W. G., and Read, A. W. 1979. An improved method of combustion control of coal fired boilers using flue gas carbon monoxide analysis. *Jour. Inst. Energy*, 52, 23.

Ostlie, D., and Stemper, F. 1979. Cleaning regenerative air heaters. *Energy Management*, 123, 101.

Parks, B. C. 1952. Mineral matter in coal, 2. *Conf., Origin and Constitution of Coal*, p. 272, Nova Scotia, Canada: Nova Scotia Department of Mines.

Parr, S. W. 1928. The coal classification. University of Illinois Eng. Exp. Stat. Bull. 180, Urbana, Ill.

Paterson, A. C., and Watson, N. 1979. The National Coal Board's pilot plant for solids pumping at Horden Colliery, 6. *Conf., Hydraulic Transport of Solids in Pipes*, Canterbury, U.K.: British Hydraulic Research Association, Cranfield, U.K.

Patteisky, K., and Teichmuller, M. 1958. Examination of possible application of different methods to study coal rank and coalification processes. Coll. Int. Pet. Char. Rev. de L'Ind. Min., Num. Spec. 121.

Patterson, R. P., Riersgard, R., Parker, R., and Sparks, L. 1978. Flue gas conditioning effects on electrostatic precipitators. Symp., Transfer and Utilization Particulate Control Technology, Denver, Colo.

Pattison, J. R. 1955. The Total Emissivity of Some Refractory Materials above 1175 K, *Trans. Brit. Ceram. Soc.*, 54, 698.

Paulson, L. E., and Fowkes, W. W. 1968. Changes in ash composition of North Dakota lignite treated by ion exchange. U.S. Bureau of Mines Report RI 7176.

Paulson, L. E., and Futch, J. R. 1979. Removal of sodium from lignite by ion exchanging with calcium chloride solutions. *Am. Chem. Soc. Div. Fuel Chem.*, 25, 224.

Pearce, M. L., and Beisler, J. F. 1965. Miscibility gap in the system sodium oxide–silica–sodium sulfate at 1473 K. *Jour. Am. Cer. Soc.*, 48, 40.

Peel, R. B. 1978. Coal utilization and research in Brazil. *Chem. Engineer.*, Oct., p. 736.

Peirce, T. J. 1966. The relationship between the unburnt carbon loss and the combustion conditions in an anthracite fired furnace. *Jour. Inst. Fuel*, 39, 478.

Penney, G. W., and Klinger, E. H. 1962. Contact potentials and the adhesion of dust. *Trans. A.I.E.E.*, 81, 200.

Petersen, H. H. 1978. Conditioning of dust with water-soluble alkali compounds. Symp., Transfer and Utilization Particulate Control Technology, Denver, Colo.

Petrascheck, W. E. 1950. The Angelokastron and Katuna Lignite Basins. Geol. Anagnoriseis, Report 6, Athens, Greece.

Pfefferkorn, G., and Vahl, J. 1963. Investigation of corrosion layers on steel using electron microscopy and electron diffraction. Conf., Mechanism of Corrosion by Fuel Impurities, p. 366, Marchwood, U.K., Butterworth, London.

Philipp, J. 1981. Oil to coal conversion options. Seminar, The Use of Coal in Oil Design Utility Boilers. EPRI Report WS-80-141, Palo Alto, Calif.

Phillips, P. J., and Cole, R. M. 1979. Economic penalties attributable to ash content of steam coals. AIME/SME Meeting, New Orleans, La. (Published in *Mining Engineering*, March 1980, p. 297.)

Pillai, K. K., and Wood, P. 1980. Emissions from a pressurized fluidized bed coal combustor. *Jour. Inst. Energy*, 53, 159.

Pitt, G. J., and Millward, G. R. (eds.). 1979. *Coal and Modern Coal Processing: An Introduction*. London: Academic.

Plumley, A. L., and Roczniak, W. R. 1975. Recovery unit water wall protection—A status report. *Tappi*, 58, 118.

Pollmann, S. 1978. Low temperature corrosion in air heaters. *VGB Kraftwerkstechnik*, 58, 921.

Poole, J. P. 1949. Improved apparatus for measuring viscosity of glasses in annealing range of temperatures. *Jour. Am. Ceram. Soc.*, 32, 215.

Potter, E. C., and Paulson, C. A. J. 1974. Improvement of electrostatic precipitator performance by carrier gas additives and its graphical assessment using an extended Deutsch equation. *Chem. Ind.*, 6, 534.

Powell, H. E. 1965. Thermal decomposition of siderite and consequential reactions. U.S. Bureau of Mines Publ. R.I 6643.

Powell, R. W., Ho, C. Y., and Liley, P. E. 1966. Thermal conductivity of selected materials, Bur. Stand. Publ. NSRDS-NBS8, Washington, D.C.

Prasolov, R. S., and Vainshenker, I. A. 1960. The thermal conductivities and fractional composition of ash deposits on tubes and laboratory ashes from some fuels. *Teplo-energetica (Thermal Engineering)*, 7, No. 3, English Transl., p. 80.

Pringle, W. J. S., and Bradburn, E. 1958. The mineral matter in coal—The composition of the carbonate minerals. *Fuel*, 37, 166.

Procter, N. A. A., and Taylor, G. H. 1966. Microscopic study of boiler deposits formed from Australian brown coals. *Jour. Inst. Fuel*, 39, 284.

Raask, E. 1961a. Formation of superheater and air heater deposits in a cyclone boiler fired with East Midlands coals. CERL Report, Leatherhead, U.K.

Raask, E. 1961b. Deposition of sodium and vanadium compounds in superheater of an oil-fired boiler. CERL Report RD/L/R 1050, Leatherhead, U.K.

Raask, E. 1962. Formation of superheater deposits in coal fired boilers burning East Midlands coals. CERL Report RD/L/R 1166, Leatherhead, U.K.

Raask, E. 1963a. Reactions of coal impurities during combustion and deposition of ash constituents on cooled surfaces. *Conf., Mechanism of Corrosion by Fuel Impurities*, p. 145, Marchwood, U.K., Butterworth, London.

Raask, E. 1963b. Composition of the materials deposited in the superheater of a cyclone-fired boiler burning Cowpen (Northumberland) coal. CERL Note ED/L/N 6/63, Leatherhead, U.K.

Raask, E. 1963c. Investigations of fireside superheater corrosion at Skelton Grange Power Station. CERL Note RD/L/N 93/63.

Raask, E. 1965. Absorption of potassium by slag in a coal fired MHD system. CERL Note RD/L/N 45/65, Leatherhead, U.K.

Raask, E. 1966a. Slag/coal interface phenomena. *Trans. ASME for Power*, Jan., p. 40.

Raask, E. 1966b. Fuel ash impurities in condensed MHD seed. CERL Note RD/L/N 26/66, Leatherhead, U.K.

Raask, E. 1968a. Cenospheres in pulverized fuel ash. *Jour. Inst. Fuel*, 43, 339.

Raask, E. 1968b. The behaviour of potassium silicates in pulverized coal firing. *VGB Mitteilungen*, 48, 348.

Raask, E. 1969a. Tube erosion by ash impaction. *Wear*, 13, 301.

Raask, E. 1969b. Fusion of silicate particles in coal flames. *Fuel*, 48, 366.

Raask, E. 1973. Boiler fouling—The mechanism of slagging and preventive measures. *VGB Kraftwerkstechnik*, 53, 248.

Raask, E. 1974. The thermal and chemical stability of limestone aggregates. *Materiaux et Constructions*, 7, 387.

Raask, E. 1976. Monitoring acid penetration in chimney structures, 2. *Symp., Chimney Design*, Vol. 1, p. 150. Scotland, U.K.: University of Edinburg.

Raask, E. 1978. Examination of ash pellets from Westinghouse fluidized bed gasifier. CERL Memorandum LM/CHEM/207, Leatherhead, U.K.

Raask, E. 1979a. Impact erosion wear caused by pulverised coal and ash, 5. *Conf., Erosion by Solid and Liquid Impaction*, p. 41-1. Cambridge, U.K.: Cavendish Laboratory, Univ. of Cambridge.

Raask, E. 1979b. Sintering characteristics of coal ashes by simultaneous dilatometry–electrical conductance measurements. *Jour. Therm. Anal.*, 16, 91.

Raask, E. 1979c. Erosion caused by ash impaction in coal fired boilers. *VGB Kraftwerkstechnik*, 59, 496.

Raask, E. 1980. Quartz, sulphates and trace elements in P.F. ash. *Jour. Inst. Energy*, 53, 70.

Raask, E. 1981a. Surface properties of pulverized coal ash and chimney solids. *Power Ind. Research*, 1, 233.

Raask, E. 1981b. Flame reactions and deposition of ash particulates in coal fired boilers. *Conf., Gas Borne Particles*, p. 179, University of Oxford: Institution of Mechanical Engineers, London, U.K.

Raask, E. 1981c. Flame imprinted characteristics of ash relevant to boiler slagging, corrosion and erosion. ASME Paper 81-JPGC-FH-1; also *Jour. Eng. for Power*, 104, 858.

Raask, E. 1982a. Sulphate capture in ash and boiler deposits in relation to SO_2 emission. *Prog. Ener. Comb. Sci.,* 8, 261.

Raask, E. 1982b. Coal ash sintering model and the rate measurements. Am. Chem. Soc. Symp., Combustion Chemistry, Las Vegas, Nev.

Raask, E. 1984a. Creation, capture and coalescence of mineral species in coal flames. *Jour. Inst. Energy,* 57, 231.

Raask, E. 1984b. The use of additives in oil fired boiler. *J. Eng. Sci.* King Saud Univ., Saudi Arabia 10(1–2), 95.

Raask, E., and Bhaskar, M. C. 1975. Pozzolanic activity of pulverized fuel ash. *Concrete Research,* 5, 363.

Raask, E., and Goetz, L. 1981. Characterization of captured ash, chimney stack solids and trace elements. *Jour. Inst. Energy,* 54, 163.

Raask, E., and Jessop, R. 1966. Miscibility gap in the potassium sulphate-potassium silicate system at 1575 K. *Phys. and Chem. Glasses,* 7, 200.

Raask, E., and Street, P. J. 1978. Appearance and pozzolanic activity of pulverized fuel ash. Conf., Ash Technology and Marketing, London.

Raask, E., and Wilkins, D. M. 1965. Volatilization of silica in gasification and combustion processes. *Jour. Inst. Fuel,* 38, 255.

Raask, E., Read, A. W., and Walton, E. 1970. Assessment of fireside corrosion and slagging at Eggborough Power Station. CERL Note RD/L/N 92/69, Leatherhead, U.K.

Radmacher, W. 1949. Determination of fusion characteristics of solid fuel ashes. *Brennst-Chem.,* 38, 377.

Radmacher, W., and Mohrhauer, P. 1955. Direct determination of mineral content of coal. *Brennst-Chem.,* 36, 236.

Radway, J. E., and Boyce, T. R. 1978. Reduction of coal ash deposit with magnesia treatment. In *Ash Deposits and Corrosion Due to Impurities in Combustion Gases,* ed. R. W. Bryers, Washington, D.C.: Hemisphere.

Radway, J. E., and Rohrbach, R. R. 1976. SO_3 control and steam generator emissions. ASME Paper 76/-WA/APC-9.

Rahmel, A. 1963. Influence of calcium and magnesium sulphates on the high temperature oxidation of austenitic chrome-nickel steels in the presence of alkali sulphates and sulphur trioxide. *Conf., Mechanism of Corrosion by Fuel Impurities,* p. 556, Marchwood, U.K., Butterworth, London.

Rahmel, A. 1968. Thermodynamic aspects of corrosion in molten sulphates: Construction of $E/\log pSO_3$ diagrams. *Electrochimica Acta,* 13, 495.

Rahmel, A., and Jaeger, W. 1960. Investigations of potassium sulphate–iron sulphate system. *Ziet. Anorg. Allg. Chem.,* 303, 90.

Rahoi, D. W., and DeLong, J. F. 1980. Evaluation of Incoclad 671/800H tubing after nine years of reheater service in a coal fired utility boiler. ASME Paper 80-WA/FU-6.

Ramanan, T., and Chaklader, A. C. D. 1975. Electrical resistivity of hot-pressed compacts. *Jour. Am. Ceram. Soc.,* 58, 476.

Ramsden, A. R. 1969. A microscopic investigation into the formation of fly ash during combustion of a pulverized bituminous coal. *Fuel,* 48, 121.

Rao, C. P., and Gluskoter, H. J. 1973. Occurrence and distribution of minerals in Illinois coals. Illinois State Geological Survey, Circular No. 476, Urbana, Ill.

Rayner, J. E., and Marskell, W. G. 1963. The distribution of mineral matter in pulverized fuel and solid products of combustion. *Conf., Mechanism of Corrosion by Fuel Impurities,* p. 113, Marchwood, U.K., Butterworth, London.

Recknagel, J. 1967. Distribution of sand in flue gas of sub-bituminous coal fired boilers. *VGB Mitteilungen,* 47, 127.

Reese, J. T., and Greco, J. 1968. Experience with electrostatic fly ash collection equipment serving steam electric generating plant. *Jour. Air Poll. Control Assoc.,* 18, 52.

Reese, J. T., Jonakin, J., and Koopman, J. G. 1961. How coal properties relate to corrosion of high temperature boiler surfaces. *Am. Power Conf.,* 23, 391.

Rehmat, A., Vora, M. K., and Sandstrom, W. A. 1978. Low BTU gas from the IGT ash agglomeration gasification process, 13. Inter Society Energy Conversion Engineering Conf., San Diego, Calif.

Rehn, I. M. 1980. Corrosion problems in coal-fired boiler superheater and reheater tubes. EPRI Report CS-1653, Palo Alto, Calif.

Reid, W. T. 1971. *External Corrosion and Deposits*. New York: Elsevier.

Reid, W. T., and Cohen, P. 1944. The flow characteristics of coal ash slags in the solidification range. *Jour. Eng. Power, Trans. ASME, Series A*, 66, 83.

Reid, W. T., Cohen, P., and Corey, R. C. 1948. An investigation of the variation in heat absorption in a pulverized coal fired water cooled steam boiler furnace. *Trans. Am. Soc. Mech. Engns.*, 70, 569.

Rekus, A. F., and Haberkorn, A. R. 1966. Identification of minerals in single particles of coal by the X-ray powder method. *Jour. Inst. Fuel*, 39, 474.

Rhodes, J. R. 1978. Simple neutron gamma spectrometer for on-stream analysis of fossil fuels. Argonne Nat. Lab. Rep. ANL-78-62, p. 496; Symp., Instrument Control for Fossil Demonstration Plant, Newport Beach, Calif.

Rice, R. L., Shang, J. Y., and Ayers, W. J. 1980. Fluidized bed combustion of North Dakota lignites. Fluidized Bed Combustion Conf., p. 963. Atlanta, Ga.

Richards, L. 1978. Conversion of coal—Fact or fiction. *Combustion*, April, p. 7.

Richardson, F. D., and Billington, J. C. 1956. Copper and silver in silicate slags. *Bull. Inst. Min. Metal.*, No. 953.

Richardson, F. D., Jeffes, J. H. E., and Withers, G. 1960. The thermodynamics of substances of interest in iron and steel making, Part 2, Compounds between oxides. *Jour. Iron and Steel Inst.*, 166, 213.

Rieke, R., and Steinbock, H. 1933. Introduction of copper oxide into low melting ceramic enamel. *Berichte Deutsch. Keram. Ges.*, 14, 547.

Rigby, G. R. 1953. *The Thin-Section Mineralogy of Ceramic Materials*. Brit. Ceram. Res. Assoc., Stoke-on-Trent, U.K.

Roberts, A. G., Raven, P., Phillips, R. N., Barker, S. N., Pillai, K. K., and Wood, P. 1980. Pressurized fluidized bed combustor—1000 Hour programme. Conf., Fluidized Combustion Systems and Applications, Paper 3.1, Inst. Energy, London.

Roberts, J. P. 1950. The mechanism of sintering. *Metallurgia*, 42, 123.

Robijn, P., and Angenot, P. 1963. Emissive properties of refractory materials. *Verres et Refractaires*, 17, 3.

Robinson, H. A., and Peterson, C. A. 1944. Viscosity of recent container glass. *Jour. Am. Ceram. Soc.*, 27, 129.

Rogers, S. E., and Lemmon, A. W. (eds.). 1979. Symp., Coal Cleaning to Achieve Energy and Environmental Goals, Hollywood, Fla., U.S. Environmental Protection Agency.

Rolker, J. 1973. Monitoring acid dewpoint of SO_3 containing flue gas by TMLA method. *Chemie-Ing. Techn.*, 45, 698.

Rosenblatt, P., and LaMer, V. K. 1946. Motion of a particle in a temperature gradient; Thermal repulsion as a radiometer phenomenon. *Phys. Rev.*, 70, 385.

Rost, F., and Ney, P. 1956. Constituent composition of boiler slag. *Brennstoff-Chemie*, 37, 201.

Rowell, R. L., Vasconcellos, S. R., Ford, J. R., Lindsey, E. E., Glennon, C. N., Tsai, S. Y., and Batra, S. K. 1978. Investigation and measurement of stability of coal-oil mixtures. Conf., Coal-Oil Mix. Combustion, p. 288, St. Petersburg Beach, Fla., U.S. Department of Energy.

Ruby, J. D., and Huettenhain, H. 1981. Western sub-bituminous coals and lignites. EPRI Report CS-1768, Palo Alto, Calif.

Ruch, R. R., Gluskoter, H. S., and Shimp, N. F. 1973. Occurrence and distribution of potentially volatile trace elements in coal. Illinois State Geological Survey—Envir. Geol. Note, 61, p. 43.

Rudenko, I. M., and Spodyryak, N. T. 1977. An air heater made of corrugated tubes. *Teploenergetika (Thermal Engineering)*, 24, No. 6, English Transl., p. 12.

Russel, J. E. 1979. Surface mining technology. Practices and plans for Gulf Coast lignite, 10. Biennial Lignite Symposium, May.

Russel, S. J., and Rimmer, S. M. 1979. Analysis of mineral matter in coal, coal gasification ash, and coal liquefaction residues by scanning electron microscopy and X-ray diffraction. In *Analytical Methods for Coal and Coal Products*, Vol. 3, ed. C. Karr. New York: Academic.

Sage, W. 1969. The erosive properties of pulverized fuel ash. National Gas Turbine Estab. Report, Pystock, U.K.

Sage, W. L., and McIlroy, J. B. 1960. Relationship of coal-ash viscosity to chemical composition. *ASME Jour. Eng. Power*, April, 145.

Saltsman, R. D. 1968. The removal of pyrite from coal. ASME Paper 68-WA/FU-2.

Sandstrom, W. A., Rehmat, A. G., and Bair, W. G. 1976. The gasification of coal chars in a fluidized bed ash agglomeration gasifier, 69. AIChE Meeting, Chicago, Ill.

Sarjeant, M. 1973. Sootblowers–Jet aerodynamics and nozzle design. CEGB Report No. R/M/R 188, Marchwood, U.K.

Sarjeant, M. 1974. Sootblowers–Deposit removal, erosion and economics. CEGB Report No. R/M/N 748, Marchwood, U.K.

Sarofim, A. F., Howard, J. B., and Padia, A. S. 1977. The physical transformation of the mineral matter in pulverized coal under simulated combustion conditions. *Comb. Scien. Techn.*, 16, 187.

Saunders, K. G. 1980. Microstructural studies of chlorine in some British coals. *Jour. Inst. Energy*, 53, 109.

Sawyer, J. C. 1961. Catastrophic oxidation of stainless steel in the presence of lead oxide. *Trans. Metal. Soc. AIME*, 221, p. 63.

Schliesser, S. P. 1978. Sodium conditioning test with mobile electrical precipitator. Symp., Transfer and Utilization Particulate Control Technology, Denver, Colo.

Schmitt, R. W. 1981. The 1980 Robens Coal Science Lecture: Coal-based electricity in the United States. *Jour. Inst. Energy*, 54, 63.

Schneider, A. 1962. Possible mechanism of corrosion in oxidizing atmosphere of cyclone fired boilers related to boiler wall temperatures. *VGB Mitteilungen*, 40, 111.

Schoen, J. 1956. Effect of temperature on the liberation of Cl and alkalis from coal. *BCURA Abstracts*, July, p. 1, Leatherhead, U.K.

Schrader, K. 1970. Improvement of the efficiency of electrostatic precipitation by injecting SO_3 into the flue gas. *Combustion*, 42, 22.

Schuyer, J., and van Krevelen, D. W. 1954. Chemical structure and properties of coal, Part 4–Calorific value. *Fuel*, 33, 384.

Schwab, G. M., and Philinis, J. 1947. Reduction of iron pyrites: Thermal decomposition, reduction by hydrogen and aerial oxidation. *Jour. Am. Chem. Soc.*, 69, 2588.

Schwob, Y. 1950. DTA of carbonates. Publication No. 22, Centre d'Études des Recherches de l'Industrie des Liants Hydrauliques, Paris.

Sedor, P., Diehl, E. K., and Barnhart, D. H. 1960. External corrosion of superheaters in boilers firing high alkali coals. *Trans. ASME, Series A, Eng. Power*, 82, 181.

Sekhar, N. 1978. Lignite combustion tests. Ontario Hydro, Canada, Res. Div. Report 78-414-H.

Sekhar, N. 1980. Lignite combustion tests–Phase 2, Ontario Hydro, Canada, Res. Div. Report 80-429-K.

Sell, W. 1931. Dust deposition on single targets and air filters, Forschungsheft 347. *Ver. Detusch. Ing.*, Issue B, 2, August, Berlin.

Selle, S. J., Tufte, P. H., and Gronhovd, G. H. 1972. A study of the electrical resistivity of fly ashes from low sulfur western coals using various methods, 65. Meet., Air Poll. Control Assoc., p. 18.

Sen, P., and Roy, A. N. 1962. The nature and distribution of mineral matter in some Indian coals. *Jour. Mines Metals Fuels*, 10, July, 7.

Seyler, C. A. 1900. The chemical classification of coals. *Proc., South Wales Inst. Eng.*, 22, 112.

Shaw, J. T. 1961. The BCURA slag viscometer. BCURA Members Circular, No. 249, Nat. Coal Board, Cheltenham, U.K.

Shearer, J. A., Johnson, I., and Turner, C. B. 1979. Effects of sodium chloride on limestone calcination and sulfation in fluidized bed combustion. *Envir. Sci. Techn.*, 13, 1113.

Sheldon, G. L., and Finnie, I. 1966. On ductile behaviour of nominally brittle materials during erosive cutting. *Trans. ASME, Ser. B.,* 88, 40.

Shenker, J. D. 1980. A little water cuts boiler-cleaning costs. *Power,* 124, 61.

Shenker, J. D., White, A. R., and Ziels, B. D. 1981. Water tempered cleaning medium for sootblowers. *Jour. Eng. Power,* 103, 561.

Shimp, N. F., Helfinstine, R. J., and Kuhn, J. K. 1975. Determination of forms of sulfur in coal. *Div. Fuel Chem. Am. Chem. Soc.,* 20, 99.

Shirley, H. T. 1956. Effects of sulphate/chloride mixtures in fuel ash corrosion of steels and high nickel alloys. *Jour. Iron Steel Inst.,* 182, 144.

Shirley, H. T. 1963. Sulphate-chloride attack on high alloy steels and nickel-based alloys. *Conf., Mechanism of Corrosion by Fuel Impurities,* p. 617, Marchwood, U.K., Butterworth, London.

Shreir, L. L. 1976. *Corrosion,* Vol. 2. London: Newnes-Butterworth.

Siegill, J. H. 1976. Defluidization phenomena in fluidized beds of sticky particles at high temperatures. Ph.D. dissertation, New York Univ., New York, N.Y.

Simons, E. L., Browning, G. V., and Liebhafsky, H. A. 1955. Sodium sulphate in gas turbines. *Corrosion,* 2, 17.

Simons, H. S., and Jeffery, J. W. 1960. An X-ray study of pulverized fuel ash. *Jour. Appl. Chem.,* 10, 328.

Skipsey, E. 1974. Distribution of alkali chlorides in British coal seams. *Fuel,* 53, 258.

Skipsey, E. 1975. Relations between chlorine in coal and the salinity of strata waters. *Fuel,* 54, 121.

Slepow, L. D., and Mendlessohn, A. S. 1981. Florida Light and Power Co., Sanford coal-oil mixture demonstration project. Seminar, Use of Coal in Oil-Design Utility Boilers. EPRI Report WS-80-141, Palo Alto, Calif.

Smeggil, J. G., Bornstein, N. S., and DeCrescente, M. A. 1977. The effect of NaCl(g) on the Na_2SO_4-induced hot corrosion of NiAL. *Conf., Ash Deposits and Corrosion due to Impurities in Combustion Gases,* ed. R. W. Bryers, p. 271. Washington, D.C.: Hemisphere.

Smith, E. J. D. 1956. The sintering of fly ash. *Jour. Inst. Fuel,* 29, 1.

Smith, M. C. G. 1952. A theoretical note upon the mechanism of turbine blade fouling. National Gas Turbine Est., Memorandum M 145, Pyestock, U.K.

Smith, R. D. 1980. The trace element chemistry of coal during combustion and the emissions from coal-fired plants. *Prog. Ener. Comb. Sci.,* 6, 53.

Smith, R. D., Campbell, J. A., and Nielson, K. K. 1979. Characterization and formation of submicron particles in coal-fired plant. *Atmos. Environm.,* 13, 607.

Sondreal, E. A., and Ellman, R. C. 1975. Fusibility of ash from lignite and its correlation with ash composition. ERDA Report GFERC/RI-75/1.

Sondreal, E. A., Kube, W. R., and Elder, J. L. 1968. Analysis of the Northern Great Province lignites and their ash: A study of variability. U.S. Bureau of Mines, Report No. 7158, Washington, D.C.

Sondreal, E. A., Selle, S. J., Tufte, P. H., Menze, V. H., and Laning, V. R. 1977. Correlation of fireside boiler fouling with North Dakota lignite ash characteristics and powerplant operating conditions. *Am. Pow. Conf.,* Chicago, Ill.

Sondreal, E. A., Gronhovd, G. H., Tufte, P. H., and Beckering, W. 1978. Ash fouling studies of low-rank western U.S. coals. *Conf., Ash Deposition and Corrosion due to Impurities in Combustion Gases,* ed. R. W. Bryers, p. 85. Washington, D.C.: Hemisphere.

Sparnaay, M. J., and Jochems, P. W. J. 1960. Measurement of attraction between solid bodies, 3. *Cong., Surface Activity,* Vol. 2, p. 375, Cologne, W. Germany.

Squires, A. M. 1970. Clean power from coal. *Science,* 169, 821.

Squires, A. M. 1972. Clean power from dirty fuels, *Scient. Am.,* 227, Oct., p. 26.

Stach, E., Taylor, G. H., Mackowsky, M.-T., Chandra, D., Teichmuller, M., and Teichmuller, R. 1975. *Coal Petrology.* Berlin: Borntraeger.

Stairmand, C. S. 1950. Dust collection by impingement and diffusion. *Trans. Inst. Chem. Eng.,* 28, 130.

Staley, H. F. 1934. Electrolytic reactions in vitreous enamels and their reaction to adherence of enamels to steel. *Jour. Am. Ceram. Soc.,* 17, 163.

Stallmann, J. J., and Neavel, R. C. 1980. Technique to measure the temperature of agglomeration of coal ash. *Fuel,* 59, 584.

Stauffer, C. H. 1981. Economic replacement of oil and gas with coal in the electric utility sector. Seminar, The Use of Coal in Oil Design Utility Boilers, Palo Alto, Calif.

Stauffer, W. A. 1958. Abrasion of hydraulic plant by sandy water. *Schweitzer Archiv für Angew. Wiss. Technik,* 24, 3.

Steel, J. S., and Brandes, E. A. 1963. Growth and adhesion of oxides in furnace deposits and their influence on subsequent deposition of ash particles as combustion products. *Conf., Mechanism of Corrosion by Fuel Impurities,* p. 374, Marchwood, U.K., Butterworth, London.

Stein, H., Gelsdorf, G., and Schalb, M. 1979. Ceramic lining for studded tubes in refuse fired boilers. *VGB Kraftwerkstechnik,* 59, 332.

Stewart, F., and Hall, A. W. 1978. On-line monitoring of major ash elements in coal conversion processes, 13. Conf., Inter Society Energy Conversion Engineering, San Diego, Calif.

Stewart, O. W., and Shou, J. K. 1975. Clean coal desulfurization and blending. Progr. Report PD-9, University of Kentucky, Lexington.

Stinespring, C. D., and Stewart, G. W. 1981. Surface enrichment of alumina-silicate minerals and coal combustion ash particles. *Atm. Envir.,* 15, 307.

Street, P. J., Weight, R. P., and Lightman, P. 1969. Further investigations of structural changes occurring in pulverized coal particles during rapid heating. *Fuel,* 48, 343.

Stringer, J. 1981. Control of fan erosion in coal fired power plants: Phase 1. EPRI Report CS 1649-4, Palo Alto, Calif.

Sur, G. N. 1979. Design of coal fired boilers. *Riv. Combust.,* 33, 229.

Suydam, C. D., and Duzy, A. F. 1977. Economic evaluation of washed coal for the Four Corners Generating Station. ASME Paper, Winter Annual Meeting, Atlanta, Ga.

Swaine, D. J., and Taylor, G. F. 1970. Arsenic in phosphatic boiler deposits. *Jour. Inst. Fuel,* 43, 261.

Swanson, V. E., Medlin, J. H., Hatch, J. R., Coleman, S. L., Wood, G. H., Woodruff, S. D., and Hildebrand, R. T. 1976. Collection, chemical analysis and evaluation of coal samples. U.S. Dept. of the Interior Geological Survey Report 76-468.

Szulakowski, W. 1968. The influence of finely ground silica on the ash fusibility of different coals. *Jour. Inst. Fuel,* 41, 25.

Tabakoff, W., Kotwal, R., and Hamed, A. 1979. Erosion study of different materials affected by coal ash particles. *Wear,* 52, 161.

Tabor, D. 1971. Surface forces. *Chem. Ind.,* No. 35, p. 969.

Tabor, D., and Winterton, R. S. H. 1968. Surface forces: Direct measurement of normal and retarded van der Waals forces. *Nature,* 219, 1120.

Tassicker, O. J. 1975. Some aspects of electrostatic precipitator research in Australia. *Jour. Air Poll. Contr.,* 25, 122.

Taylor, G. I. 1940. Notes on possible equipment and techniques for experiments on icing of aircraft. Res. Memorandum 2024, National Gas Turbine Est., Pyestock, U.K.

Taylor, M. A., and Arulanandan, K. 1974. Relationship between electrical and physical properties of cement pastes. *Cement Conc. Res.,* 4, 881.

Thiessen, R. 1919. Occurrence and origin of finely disseminated sulfur compounds in coal. *Trans. ASME,* 63, 913.

Tilly, G. P. 1969. Erosion cuased by airborne particles. *Wear,* 14, 63.

Trenkler, H. 1967. Corrosion resistance of different alloys in cyclone fired boilers. *VGB Mitteilungen,* 47, 419.

Truscott, G. F. 1970. A literature survey on abrasive wear in hydraulic machinery. Brit. Hydromechanics Res. Assoc. Report No. 1N 1079, Wallingford, U.K.

Tufte, P. H., and Beckering, W. 1975. A proposed mechanism for ash fouling burning Northern Great Plains lignite. *ASME Jour. Eng. for Power,* July, p. 407.

Tufte, P. H., Gronhovd, G. H., Sondreal, E. H., and Selle, S. J. 1976. Ash fouling potentials of Western sub-bituminous coals as determined in a pilot plant test furnace. Am. Power Conf., Chicago, Ill.

Turkdogan, E. T., and Bills, P. M. 1960. A critical review of viscosity of $CaO-MgO-Al_2O_3-SiO_2$ melts. *Am. Ceram. Soc. Bull.*, 39, 682.

Tyul'pin, K. K., Solonouts, M. I., Uporova, V. A., Beketov, B. I., Yuganova, S. A., Tikhonova, V. V., and Duel, N. A. 1980. The resistance to corrosion of steam superheater tubes with an anti-corrosion coating. *Teploenergetika (Thermal Engineering)*, 27, No. 3, English Transl., p. 146.

Ulrich, G. D., Riehl, J. W., French, B. R., and Desrosiers, R. E. 1977. Mechanism of submicron fly ash formation in a cyclone coal fired burner. *Conf., Ash Deposits and Corrosion due to Impurities in Combustion Gases*, ed. R. W. Bryers. Washington, D.C.: Hemisphere.

Umpleby, R. A. 1976. A new lining for Drax Power Station chimney. *2d Symp., Chimney Design*, Vol. 1, p. 130. Scotland, U.K.: University of Edinburg.

van Krevelen, D. W. 1953. Physical characteristics and chemical structure of bituminous coal. *Brennst.-Chemie*, 34, 167.

Vasil'ev, A. A., and Dombrovskii, V. I. 1978. Investigation of erosion of air heaters by ash of Ekibastuz coal. *Teploenergetika (Thermal Engineering)*, 25, No. 1, English Transl., p. 32.

Vasil'ev, V. V., and Guzenko, S. I. 1977. Investigating heat transfer with water lancing of boiler heating surfaces. *Teploenergetika (Thermal Engineering)*, 24, No. 9, English Transl., p. 59.

Vaux, W. G., Andersson, C. A., Ranadive, A. Y., and Vojnovich, T. 1981. Design of refractories for resistance to high temperature erosion and corrosion. EPRI Report AP-1955, Palo Alto, Calif.

Villa, H. 1950. Thermodynamic data of the metallic chlorides. *Jour. Soc. Chem. Ind.*, No. 1, p. 9.

Vivenzio, T. A. 1980. Impact of cleaned coal on power plant performance and reliability. EPRI Report CS-1400, Palo Alto, Calif.

Volborth, A. 1979. Problems of oxygen stoichiometry in analysis of coal and related materials. In *Analytical Methods for Coal and Coal Products*. New York: Academic.

Vora, M. K., Mason, D. M., Rehmat, A., and Sandstrom, W. A. 1980. Ash agglomerates from coal. *Energy Communications*, 6, 195.

Vulis, L. A., and Terekhina, N. N. 1955. Propagation of a turbulent gas jet in a medium with a different density. *Jour. Tech. Phys.*, 26, 1249.

Walker, A. B. 1971. Effects of desulfurization dry additives on the design of coal-fired boiler particulate emission control systems. *Canadian Min. and Met. Bulletin*, p. 85.

Wall, T. F., Lowe, A., Wibberley, L. J., and Stewart, I. McC. 1979a. Mineral matter in coal and the thermal performance of large boilers. *Prog. Energy Comb. Scien.*, 5, 1.

Wall, T. F., Mai-Viet, T., Becker, H. B., and Gupta, R. P. 1979b. Fireside deposits and their effect on heat transfer in P. F. boiler: The emissivity and thermal conductivity of deposits and their components. Victoria State Electricity Commission Report GO/79/11, Melbourne, Australia.

Wandless, A. M. 1957. British coal seams: A review of their properties with suggestions for research. *Jour. Inst. Fuel*, 30, 541.

Wandless, A. M. 1959. The occurrence of sulphur in British coals. *Jour. Inst. Fuel*, 32, 258.

Warne, S. S. J. 1965. Identification and evaluation of minerals in coal by differential thermal analysis. *Jour. Inst. Fuel*, 38, 207.

Warne, S. S. J. 1970. The detection and identification of the silica minerals quartz, chalcedony, agate and opal by differential thermal analysis. *Jour. Inst. Fuel*, 43, 240.

Warne, S. S. J. 1975. An improved differential thermal analysis method for identification and evaluation of calcite, dolomite and ankerite in coal. *Jour. Inst. Fuel*, 48, 142.

Warne, S. S. J. 1979. Differential thermal analysis of coal minerals. In *Analytical Methods for Coal and Coal Products*, Vol. 3, ed. C. Karr. New York: Academic.

Warren, W. R. 1955. Static pressure variation in compressible free jets. *Jour. Aeronaut. Sci.*, 22, 205.

Watson, K. S. 1976. Australia experience with flue gas conditioning. EPA Publ. 600/7-76-010.

Watson, K. S., and Blecher, K. J. 1966. Further investigation of electrostatic precipitators for large pulverized fuel fired boilers. *Jour. Air Water Poll.*, 10, 518.

Watt, J. D. 1968. The physical and chemical behaviour of the mineral matter in coal under conditions met in combustion plant. Part 1, The occurrence, origin, identity, distribution, and estimation of the mineral species in British coals, Part 2, Thermal decomposition of the mineral species: Volatilization of fouling components; Sintering and melting of coal ashes; Ash fusibility. BCURA Literature Survey, National Coal Board, Cheltenham, U.K.

Watt, J. D., and Fereday, F. 1969. The flow properties of slags formed from the ashes of British coals, Part 1, Viscosity of homogeneous liquid slags in relation to slag composition. *Jour. Inst. Fuel*, 42, 99.

Watt, J. D., and Thorne, D. J. 1965. Composition and pozzolanic properties of pulverized fuel ashes, Part 1, Composition of fly ashes from some British power stations and properties of their component particles, Part 2, Pozzolanic properties of fly ashes as determined by crushing strength tests on lime mortars. *Jour. Appl. Chem.*, 15, Part 1, p. 585, Part 2, p. 595.

Weast, R. C. (ed.). 1977. *Handbook of Chemistry and Physics.* Cleveland: CRC Press.

Webb, T. L., and Heystek, H. 1957. The carbonate minerals–D.T.A. investigation of clays. Mineralogical Society, London.

Webner, H. L. 1969. Cleaning of furnace walls by sandblasting. Monongahela Power Co., Allegheny Power System Communication, Fairmont, W. Va.

Werthaum, M., and Sogndal, C. 1982. Sootblowing by audible sound. ASME 103. Winter Annual Meet., Nov., Phoenix, Ariz.

Weyl, W. A. 1953. Wetting of solids as influenced by polarizability of surface ions. In *Structure and Properties of Solid Surfaces,* eds. R. Gomer and C. S. Smith, p. 147. Chicago, Ill.: University of Chicago Press.

White, H. J. 1963. *Industrial Electrostatic Precipitation.* Oxford: Pergamon.

White, J. F. 1939. Silica aerogel–Effect of variables on its thermal conductivity. *Industrial Eng. Chem.,* 31, 827.

Wibberley, L. J., and Wall, T. F. 1982. Alkali-ash reactions and deposit formation in pulverized-coal-fired boilers, Part 1, The thermodynamic aspects involving silica, sodium, sulphur, and chlorine, Part 2, Experimental aspects of sodium silicate formation, and formation of deposits. *Fuel,* 61, Part 1, p. 87; Part 2, p. 93.

Wilhelm, B. W., Simon, J. J., and Nelson, S. E. 1975. The effect of water jet lancing on furnace wall tubes. Am. Power Conf., Chicago, Ill.

Williams, E. G., and Keith, L. M. 1963. Relationship between sulphur in coals and the occurrence of marine roof beds. *Econ. Geol.,* 58, 720.

Winegartner, E. C. (ed.). 1974. *Coal Fouling and Slagging Parameters.* ASME Research Committee on Corrosion and Deposits from Combustion Gases, ASME Publ.

Winegartner, E. C., and Rhodes, B. T. 1975. An empirical study of the relation of chemical properties of ash fusion temperatures. *Trans. ASME Jour. Eng. Power, Ser. A.,* 97, 395.

Wolfe, J. D. 1962. A new method for the study of the fusion behaviour of Brown coal ash. *Jour. Inst. Fuel,* 35, 448.

Wright, J. B. 1972. *Second Level Course Geochemistry, Unit 1 Geochemical Data.* Bletchley, England: The University Press.

Wynnyckyj, J. R., and Rhodes, E. 1981. Mechanisms of furnace fouling. Seminar, Advancement in Heat Exchangers, Paper 17, Dubrovnik, Yugoslavia.

Yerushalmi, J., Kolodney, M., Graff, R. A., Squares, A. M., and Harvey, R. D. 1975. Agglomeration of ash in fluidized beds gasifying coal: The Godel phenomenon. *Science,* 187, 646.

Young, R. C., Harwig, F. J., and Norton, C. L. 1964. Effect of various atmospheres on thermal conductance of refractories. *Jour. Am. Ceram. Soc.,* 47, 205.

Zetlmeisl, M. J., May, W. R., and Annand, R. R. 1975. High temperature corrosion in gas turbines and steam boilers by fuel impurities, Part 5, Lead containing slags. *Trans. ASME Jour. Eng. Power,* July, p. 441.

Zimm, B. H., and Mayer, J. E. 1944. Vapour pressures, heats of vaporization and entropies of some alkali halides. *Jour. Chem. Phys.,* 12, 362.

Zinc, J. B., Washington, W. D., and Peterson, M. J. 1967. Spectrochemical analysis of coal ash. U.S. Bureau of Mines Rep. of Investigations 6985, Washington, D.C.

INDEX

Abrasion wear:
  ash transport lines, 385–386
  coal plant, 56–57
  remedial measures, 56–58
Abrasive index:
  and ash particle size, 366–368
  and ash quartz content, 366–369
  British coals, 48–49
  coal/oil dispersion ash, 386–387
  fluidized combustion ash, 392–393
  pulverized coal ash, 369
  pyrite, 48
  quartz, 48, 366–369
  U.S. coals, 48–49
Absorptance:
  ash, 219
  measurements, 219–220
Acid deposition:
  in air heaters, 400–402
  and coal sulfur content, 404
  maximum rate, 400, 404
  and oxygen in flue gas, 400–402
  in pulverized-coal-fired boilers, 400–402
  and SO_3 in flue gas, 402–404
  and temperature, 400–402
Acid dewpoint, 398–399, 404
Acid penetration:
  in concrete, 416–418
  monitoring, 416–418
  protective lining, 416

Acidic oxides:
  in basic/acid oxide ratio, 162–165
  in flue gas, 294, 402–404
  pyrochemical, 162
Additive application:
  aims, 283–284
  in chain grate boilers, 302–303
  costs, 296
  in cyclone boilers, 299, 307–308
  dosage, 299, 301–303
  in fluidized bed combustion, 433
  intermittent, 310–311
  in pulverized coal boilers, 289–300, 301–310
Additives:
  ammonium bisulfate, 296
  ammonium sulfate, 296, 299
  for ash precipitation, 293–299
  boron, 308
  calcium, 288–289, 298, 508–510
  for corrosion prevention, 357–359, 415
  fluxing agents, 307–308
  iron, 287–288, 308–310
  for slag devitrification, 300–317
  for slag flow, 308, 310
  for slagging prevention, 300–311
  sodium, 288, 397–399
  SO_3, 290–299
  sulfamic acid, 290–291, 296, 299
  sulfuric acid, 290–291, 296

Additives (*Cont.*):
  and surface resistivity, 284–299
  triethylamine, 297, 299
Adhesion:
  air heater deposits, 406–407
  chemical compatibility, 175–177
  high temperature deposits, 173–174
  temperature environment, 174, 181–186
  thermal compatibility, 175–177
Adhesive bond:
  air heater deposits, 404–407, 412
  and ash composition, 177, 182
  chemical, 178, 180
  condensable salt effect, 173
  deposit on austenitic steels, 184,
     248–249
  deposit on ferritic steels, 184
  iron oxide index, 176–177
  measurements, 180–186
  mechanical, 178–180
  silicates, 184
  strength, 181–186
  sulfates, 172
  and sulfuric acid reactions, 404–408
Adhesive forces:
  electrostatic, 171, 193–194
  surface tension, 171–172, 198–199
  van der Waals, 169–171
  work equation, 172
Aerosol particles:
  number prediction, 108–110
  silicate fume, 106–108
  sulfate fume, 108–113
Air heater:
  acid/ash reactions, 405–407
  acid bonded deposits, 406
  air inlet temperature, 399, 409, 412–414
  ash plugging, 404–405, 412–414
  baked deposits, 406–407
  corrosion, 404, 409, 412–416
  corrosion preventive measures, 411–416
  corrosion resistant coatings, 415–416
  deposit analysis, 415–416
  erosion wear, 416
  fouling, 408–411
  fouling preventive measures, 411–416
  sootblowing, 412–415
  SO_3 suppressive additives, 415
Air heaters:
  recuperative, 397–399, 409–410, 416
  regenerative, 397–399, 409–416
  temperature regime, 397–399
Albite particle resistivity, 285–286

Alkali-metal sulfates:
  capture by ash, 122
  in corrosive deposits, 320–323, 336–337
  decomposition by silicates, 211
  deposition, 201, 205
  enrichment on ash surface, 111–113
  equilibrium with silicates, 97–99
  formation, 92–94, 112–113
  fume, 108–116
  molar ratio, 204
  phase separation, 97–99
Alkalis:
  in ash, 12–13, 98, 265, 269–272
  distribution in coal, 21, 265
  distribution in molten phases, 97–100
  enrichment in deposits, 209–213
  and sintering, 149–152, 268–272
  and viscosity, 127–128, 269
Alumina:
  abrasion wear, 47
  in ash, 11–13, 262–263
  erosion wear, 55
  hardness, 45
  for quartz calculation, 17–18
Aluminosilicates, 15–17
  analysis, 10, 15
  composition, 14, 65
  dehydration, 61–62, 285
  density, 14
  hardness, 45
  particle spheroidization, 61–67
  species in coal, 14–16, 262–264
Ammonium additives, 290–291, 296–297,
  299
Analysis:
  ash, 10–11
  ash monitor, 11
  coal, 10, 19
  coal strata, 16, 31
  colorimetric, 10
  density separation, 30–35
  deposit, 100, 321
  infrared absorption, 17
  microscopic, 10, 28, 64, 75–76,
    114–115, 193, 267, 431
  mineralogical, 9–11, 15, 19–21
  neutron activation, 11
  on-line, 11
  single particle, 10, 34
  soluble constituent, 100, 212
  thermal, 10, 61–62
  X-ray diffraction, 10, 17
  X-ray fluorescence, 10

Ankerite, 19, 94–95, 104–150, 262
Antimony additive, 358
Arsenic:
  in corrosive deposits, 335–337
  volatilization, 335–336
Ash abrasiveness (see Abrasive
    characteristics)
Ash adhesion (see Adhesion)
Ash agglomerates:
  formation in fluidized gasifier, 434–437
  properties, 435–436
Ash agglomeration:
  in burning coal particles, 113–155
  on coke particle surface, 114
  and sintering measurements, 155
Ash alkalinity:
  basic/acidic oxide ratio, 162–165
  and sulfur capture, 251
  water dispersion, 294–295
Ash analysis data:
  anthracite, 263
  bituminous coals, 12, 65, 77, 182, 207,
    263, 304
  British coals, 12, 65, 77, 207, 304
  lignite, 13, 182, 262
  sub-bituminous coals, 13, 182, 262
  U.S. coals, 12–13, 182, 262–263
Ash burden:
  in coal gas turbine, 392
  cyclone boilers, 116, 375
  pulverized coal boilers, 166, 208, 373
Ash cenospheres:
  formation, 71–77
  gas evolution, 72–73
  properties, 74
Ash composition (see Ash analysis data)
Ash discharge from boiler:
  bed ash, 39
  fly ash, 39
  and laboratory prepared ash, 36–39
  sintered clinker, 38–39
  slag, 39
Ash dispersion in water:
  acidic dip, 294–295
  alkalinity, 294–295
  pH, 294–295
Ash erosion (see Boiler erosion)
Ash fouling (see Boiler fouling)
Ash fusion:
  and basic/acidic oxide ratio, 162
  and slagging index, 159
  stages, 158
  tests, 154, 156–160

Ash particle size:
  cyclone boiler, 117–118
  fluidized bed, 430
  pulverized coal, 117–118
Ash plerospheres:
  formation, 74–77
  properties, 74
Ash preparation:
  high temperature, 9
  low temperature, 9
Ash related problems, 421–437
  and coal cleaning, 423–426
  in converted boilers, 426–430
  summary, 422
Ash resistivity (see Resistivity)
Ash sintering:
  bituminous coal, 147–148, 152–155
  and calcium content, 270–272
  lignite, 155, 266–272
  and sodium content, 150–152, 266–272
  sub-bituminous coal, 149–150, 266–272
  U.S. coal, 147–149, 150, 152–153,
    266–272
  (See also Sintering)
Ash type:
  bituminous coal, 163
  lignite, 163
Austenitic steels:
  alloy composition, 332, 349, 351
  corrosion resistance, 348–355
  erosion measurement, 55
  nonadhering surface, 183–184, 248–249
  temperature limitation for corrosion,
    348–349
  temperature limitation for oxidation,
    348–349

Basic oxides:
  pyrochemical, 162–166
  ratio to acidic, 162–166, 272
Bentonite dissolution in slag, 124–126
Bituminous coals (see Coal; Coals)
Boiler erosion:
  carbon steel wear, 54–55, 371–374
  and coal cleaning, 382
  and coal quartz content, 379
  and metal temperature, 372–373
  and load reduction, 382, 384
  remedial measures, 383–386
  and sootblowing, 383–384
  temperature regime, 369–377
  tube wear, 369–381

Boiler erosion (*Cont.*):
    velocity limitation, 380–381
    velocity parameter, 371–372
Boiler deposits (*see* Deposits)
Boiler fouling:
    bituminous coals, 162–166
    British coals, 163
    definition, 161
    lignites, 162–166, 266–272
    sub-bituminous coals, 162–166, 266–272
    U.S. coals, 162–166, 266–272
Boiler slagging:
    and ash content, 249–250
    bituminous coals, 164–166, 250, 275–280
    British coals, 161
    and coal blending, 251
    and coal cleaning, 249–251
    on-load monitoring, 252
    overall rate, 241, 249–250, 300, 307
    remedial measures, 239–260
    U.S. coals, 164–166, 275–280
Boilers:
    converted, 386–390, 428
    cyclone, 6, 239–240, 351, 375–377,
        400–401, 428
    design developments, 5–7, 28–29,
        241–242
    fluidized bed, 7, 430–431
    plant failures, 7, 56–59, 332–333,
        351–352, 377–380, 408–411
    pulverized coal, 5–7, 27–29, 239–260,
        302–303, 351, 377–380, 400–402,
        428
    slag-tap, 6
    stoker, 5
Bonding (*see* Adhesion)
Burner:
    combination, 253
    quarl deposits and materials, 246–248
    zone heat flux and slagging, 244–245,
        253

Calcium:
    additives, 297–298, 308–310, 432–433
    in bituminous coals, 12–13, 37
    in lignites, 13, 265
    mode of occurrence, 264–266
    and sintering, 162, 166, 270–272
    in sub-bituminous coals, 13, 38–39, 265
    in U.S. coals, 12–13, 264–266
Calcium carbonates:
    dissociation, 95–96
    occurrence in coal, 16, 19–20, 262–263

    sulfation, 96
    thermal analysis, 94–96
Calcium chloride:
    additive, 309
    occurrence in coal, 21
    sulfation, 94
    volatilization, 94
Calcium oxide:
    formation, 95–96
    structure, 96
Calcium silicates:
    crystalline species, 213, 270
    and sintering, 270
Calcium sulfate:
    in ash, 38–39
    in coal, 16, 262–263
    in deposit, 322–323, 406
    fume, 104–106, 112
Calcite hardness, 45
Calorific value:
    and ash content, 25–26
    British coals, 25
    and coal rank, 24
    and coal water content, 25–26
    net, 27
Captive surfaces:
    depth of penetration, 198
    fused slag, 198–199
    molten sulfate, 198–199
    threshold velocity, 198–199
Capture (*see* Particle capture)
Carbon content:
    and coal rank, 24
    fly ash, 37–38
    slag agglomerates, 436–437
Carbon loss:
    in anthracite combustion, 28–29
    in cyclone boilers, 241
    in pulverized-coal-fired boilers, 28
Carbon steel (mild steel):
    corrosion, 332–333, 351
        temperature limit, 349
    erosion wear, 55
    hardness, 55
    oxidation temperature limit, 349
Carbonates:
    inclusion in coal, 19–20
    mineral species in coal, 19–20
    sulfation, 96
    thermal dissociation, 94–96, 104–106,
        275
    (*See also* Calcium carbonates; Ankerite;
        Siderite)
Cenospheres (*see* Ash cenospheres)

Chimney emission solids (*see* Emission
    solids)
Chloride:
  in coal water, 20–21
  deposition, 202–203
  enhanced corrosion, 330–334
  sulfation, 92–94
  volatilization, 89–92
Chlorine in coal:
  blended, 348
  and boiler fouling, 163
  British, 20–21
  and corrosion, 330–334
  mode of occurrence, 20–22
  origin, 20–21
  and sulfate deposition, 202–203
  U.S. coals, 20
  washed, 346–348
Chlorite:
  analysis, 65
  in coal, 14, 262–263
  particle spheroidization, 65–66
Clay minerals (*see* Aluminosilicates)
CO in flue gas:
  and boiler slagging, 253–254
  and corrosion, 356
  monitoring, 253–254
Coal:
  abrasive characteristics, 43–49
  abrasive index, 48–49
  adventitious mineral matter, 29
  ash (*see* Fluidized bed ash; Fly ash)
  ash content, 25, 31
    and mineral matter, 35–36
  calorific value, 24–28
  carbon content, 24–25
  chlorine content, 25
  combustibility, 27–28
  grindability, 41–43
  hardness, 41–43
  hydrogen content, 24–25
  inherent mineral matter, 29
  inherent moisture, 24
  moisture content, 24–26
  nitrogen content, 24
  oxygen content, 24–25
  quality, 23–39
  rank, 24
  sulfur content, 25, 31
  volatile matter, 24–25
  wear index, 48–49
Coal blending, 26, 250–251, 346–348
Coal cleaning:
  and ash content, 25

  and ash related problems, 423–426
  and boiler slagging, 249–250
  and boiler erosion, 382, 425–426
  and chlorine content, 346–348
  for converted boilers, 386–387
  and corrosion, 346–348, 425
  and mill wear, 425
  and precipitator performance, 425
  and sodium content, 250, 272
Coal mills:
  performance, 42–43
  wear, 43, 45
Coal minerals:
  abrasiveness, 43–49
  distribution, 11–22, 30–35, 261–266
  flame changes, 61–83, 85–96, 103–115
  (*See also* Mineral matter in coal)
Coal gas turbines:
  erosion wear, 390–395
  residual ash, 390–395
Coalescence (*see* Particle coalescence)
Coals:
  bituminous, 12, 24, 263
  British, 12, 25, 49
  high ash, 49, 251
  high calcium, 13, 265
  high chlorine, 20–21
  high iron, 272–277
  high pyrite, 272–277
  high quartz, 18, 49, 51, 262–264,
    369, 379
  high rank, 23–24
  high silica, 12–13
  high sodium, 13, 265
  high sulfur, 19, 251, 272–277
  low ash, 49, 386–389
  low rank, 13, 24, 251, 264–265
  low silica, 12–13
  low sulfur, 251
  sub-bituminous, 13, 262–265
  U.S., 12–13, 49, 261–266
Coextruded duplex tubes:
  composition, 351
  corrosion resistance, 350–352
  furnace wall, 351–352
  superheater, 350–352
Coke residue particles:
  in fly ash, 28, 114
  inherent ash, 114
  porosity, 28
  specific surface, 28
Combustion:
  carbon loss, 28
  and coal fineness, 28

Combustion (*Cont.*):
  efficiency, 240–241, 243–245
  low volatile coals, 28–29
Combustion chamber:
  burner layout, 244
  heat flux, 235–236, 243, 245–246,
    252–253
  slagging (*see* Boiler slagging)
Combustion control:
  CO monitoring, 253–254
  coal/air mixing, 251–252
  for corrosion prevention, 356–357
  excess air, 252
  for slagging prevention, 251–254
Complex sulfates, 317–323, 328–329,
    337, 341
  alumina dissolution, 317–320
  analysis, 320–324
  corrosion propensity, 321, 323
  formation, 316–320
  iron dissolution, 317–320
  potassium enrichment, 319–323
  thermal stability, 319, 323
Compressed air sootblowers, 254–255
Conductance measurements:
  ash sintering, 143–152
  chimney acid, 416–419
  deposit adhesion, 184–186
Conductivity (*see* Thermal conductivity)
Converted boilers:
  ash abrasive characteristics, 386–387,
    389–390
  ash related problems, 426–430
  capacity limitation, 386–388
  coal choice, 386–387
  coal/oil dispersion, 388–390, 428–429
  coal/water slurry, 428–429
  combustion chamber size, 386–387
  erosion wear, 386–390
  flue gas velocity limitation, 387–389
Copper oxychloride additive, 301–307
  dosage rates, 301–303, 307
  gas bubble formation, 306
  nucleation, 304–307
  plant experience, 301–303
  slag crystallization enhancement, 304–306
  volatilization, 306
Copper species in coal, 19, 307
Copper sulfate additive, 301
Corrosion:
  air heater, 406–408
  and arsenic enrichment, 335
  austenitic steels, 327–328, 331, 345,
    349–351

  bell-shape rate curve, 329–337
  carbon (mild) steel, 332, 345–346, 349,
    351
  coextruded tubes, 350–352
  chloride-enhanced, 330–334, 407–408
  electrochemical, 325–329
  ferritic steels, 345, 349–350
  and fuel impurities, 313–316
  furnace wall, 332, 345–346, 349, 351, 356
  in HCl, 330–332
  and lead enrichment, 337–342
  mechanisms, 324–329, 330–334, 337, 341
  nickel alloys, 334, 350–351
  reactions, 313–316, 324–329, 330, 332
  in reducing atmosphere, 332–334, 356
  reheater, 350–351
  and slagging, 356–357
  and steel chromium content, 349–352
  superheater, 334–345, 349–353, 357
  temperature limitation, 314, 349
  temperature regime, 313–316
  and trace metals in coal, 334–341
Corrosion prevention:
  and additives, 357–359
  air heater, 411–416
  boiler design, 356–357
  coal blending, 348
  coal washing, 346–348, 425
  coatings, 353–355
  coextruded tubes, 350–352
  combustion control, 356–357
  overview, 355
  shields, 352–353
Corrosive deposit propensity:
  and acid-soluble alkalis, 342
  and chlorine content, 343–346, 347–348
  and flame released potassium, 342–343
  and flame volatile alkalis, 343
  and furnace exit temperature, 356–357
  and potassium/sodium ratio, 321
  and SO_3 excess, 323
  and water-soluble potassium, 342
Corrosive deposits:
  aluminum sulfate content, 318–319, 322–323
  arsenic rich, 335–337
  chemical analysis, 320–321
  complex sulfates, 317–320
  conductance measurements, 323–324
  crystalline species, 321–322
  formation, 316–320
  iron sulfates, 317–323, 328
  lead rich, 341
  potassium sulfates, 317–319, 322–323,
    341

Corrosive deposits (*Cont*.):
    pyrosulfates, 317–318
    in reducing atmosphere, 340
    sodium sulfates, 317–320, 322–323, 341
    sulfide, 340
    thermal analysis, 318–319, 323–324
    thermal stability, 318–320, 323, 328–329
Crack formation:
    in oxide layer, 333
    in slag, 304–305, 307
Cyclone boiler:
    additives, 307–308, 310
    ash characteristics, 115–118
    corrosion, 336
    slag flow, 132

Density:
    cenospheres, 74
    coal minerals, 14, 16
    deposit, 227–229
    and particle aerodynamic diameter, and
        thermal conductivity, 224–225
Deposit:
    adhesion, 173–174
    bond strength (*see* Adhesive bond)
    crystalline species, 270
    formation, 205–214, 300, 316–320,
        404–408
    friability index, 309
    minor constituents, 214–215
    potassium enrichment, 202–204, 209–211,
        319–320
    sodium enrichment, 202–204, 209–211
    sulfur enrichment, 202–204
    (*See also* Deposits)
Deposit removal:
    load reduction, 252
    novel suggestions, 259–260
    off-load cleaning, 259
    sootblowing, 254–257
    water jet sootblowers, 257–258
Deposition:
    acid in air heater, 399–402, 404, 407
    alkali-metal compounds, 201–205
    chloride, 203
    efficiency, 195–197
    electrophoresic, 194
    inertial impaction, 193–197
    iron rich species, 276
    mass transfer equation, 189
    mechanisms, 189–199
    particle diffusion, 189–191
    probe, 200

    rate measurements, 200–208
    silicates, 205–213
    sulfates, 201–205
    thermophoresic, 191–193
    vapor diffusion, 189–191
Deposits:
    bulk density, 227–229
    air heater, 404–408
    combustion chamber, 161
    composition, 212–214
    corrosive (*see* Corrosive deposits)
    layer structured, 211–213
    monolithic, 205–213
    probe, 200–214
    superheater, 133, 174, 205–214, 256,
        321–323, 336–337
    water soluble, 212–213, 321–323, 406
Dolomite:
    additive, 298, 422–423
    dissociation, 95
    occurrence in coal, 16, 19–20, 94,
        263
Dust burden (*see* Ash burden)

Electrical precipitation (*see* Precipitation)
Electrophoresic deposition, 194, 260
Emission solids:
    composition, 112
    particle surface characteristics, 112
    sulfates, 112
Emissivity:
    ash, 218
    deposit, 218
    monochromatic, 217–218
Emittance:
    ash, 222–224
    cyclic variations, 223–224
    deposits, 222–224
    glasses, 222
    initial deposit, 223
    and iron content, 223
    measurements, 217–220
    and temperature, 222–223
    and thermal conductance, 226
Erosion:
    and ash characteristics, 365–369
    cutting, 51–52
    deformation, 52
    and impaction angle and velocity, 54
    and particle hardness, 365–366
    and particle shape, 52–53
    and particle size, 52, 366–368
    rate equations, 51, 54

Erosion wear:
    austenitic steel, 55
    carbon steel, 55
    in coal gas turbines, 394–395
    composite materials, 55
    coatings, 385
    in converted boilers, 386–390, 427–428
    in cyclone boilers, 375–377, 428
    economizer tubes, 370, 376–377, 379–384
    fans, 57–58
    in fluidized combustion, 391
    hardened steels, 55
    in pulverized-coal-fired boilers,
            369–371, 377–380, 422, 425
    pulverized coal pipes, 57–59
    refractory materials, 55
    reheater tubes, 371, 376–380
    remedial measures, 57–59, 382–385
    superheater tubes, 370–371, 376–384
    and tube life, 377–380, 388
    and velocity parameter, 51–54
            371–377, 379–381, 388–390, 394

Ferritic steels:
    composition, 349–350
    corrosion temperature limitation, 349
    deposit adhesion, 184
    oxidation temperature limitation, 349
Flame borne particles:
    capture, 111–113
    coalescence, 113–155
    creation, 103–110
    fragmentation, 103–105
    heating rate, 103–106
    spheroidization, 61–67
Flue gas:
    acidity, 294, 402–404
    acidity and coal calcium content, 294,
            402–404
    acidity and coal sulfur content, 296,
            403–404
    alkali-metal vapor concentration, 92
    CO concentration, 253–254, 356
    HCl concentration, 91–92
    lead vapor concentration, 339–340
    SO_2 concentration, 92
    SO_3 concentration, 92, 294, 320, 329, 358
Fluidized bed ash:
    abrasive characteristics, 392–395
    general characteristics, 430–431
    particle size, 391–393
    sintering propensity, 432–434
    slag agglomerate formation, 434–437

Fluidized bed combustion:
    ash retention, 39, 430
    cogeneration gas turbine wear, 392
    erosion wear tests, 392–395
    mineral matter changes, 431
    and pulverized coal firing, 430–431
    temperature regime, 430–431
Fly ash (Pulverized-coal ash):
    abrasive characteristics, 365–369
    abrasive index, 366–368
    analysis (see Ash analysis data)
    bituminous coal, 116
    carbon content, 300
    crystalline species, 193
    erosion propensity: and maximum flue
            gas velocity, 380–381
        and quartz content, 369
        and silica content, 363, 369
    fusion tests and magnetite fraction, 276
    lignite, 266–267
    particle hardness, 365
    particle size, 115–118
    particle surface characteristics, 111–112
    size fractions, 368
    sub-bituminous coal, 266–268
    sulfate content, 300
Fouling index:
    bituminous coals, 163–165
    high chlorine coals, 163
    high sodium coals, 163
    lignites, 163–165
    sub-bituminous coals, 162–165
Free energy changes:
    ash constituent oxide formation, 179
    steel alloy oxide formation, 179
    sulfation reactions, 93

Galena (see Lead sulfide)
Gas evolution:
    in cenosphere formation, 72–73
    in slag, 69–70
Gas turbine:
    ash burden, 392
    erosion wear, 390–395
Glass ceramics:
    erosion wear, 55
    nucleation, 301
Glass particles:
    erosion wear, 52–53
    spheroidization, 65
Glass adhesion, 181, 186
Glass sintering, 144, 146
Glass viscosity, 121

Gypsum:
  in coals, 16, 262–263
  in lignite, 262

Hardgrove index:
  measurements, 41
  and mill output, 42
Hardness:
  alumina, 45
  ash particles, 365
  austenitic steel, 55
  carbon steel, 55
  carbonates, 45
  coal minerals, 45
  coal substance, 41–42, 45
  hardened steels, 55
  microhardness measurements, 41–42
  Nihard steel, 55
  pyrite, 45
  quartz, 45
  silicates, 45
HCl:
  and air heater corrosion, 407–408
  flue gas concentration, 91–92
  and high temperature corrosion,
      330–334
Heat flux:
  and boiler slagging, 245
  in burner zone, 243–245
  in coal fired boilers, 234–236
  measurements, 232–236
  monitoring for slagging, 252
  in oil fired boilers, 235–254
  probes, 232–236
Heat transfer:
  and ash deposits, 230–232
  convective, 225, 230
  and emittance, 231–232
  Knudsen effect, 231
  radiative, 225, 230
Heating microscope, 71, 98
Hematite, 14
High calcium coals:
  and boiler fouling, 166, 270–272, 422
  and flue gas acidity, 404
  and sulfur (sulfate) capture, 37, 251,
      422
High chlorine coals:
  and boiler fouling, 163, 422
  and corrosion, 343–346, 422
High silica (quartz) coals:
  boiler erosion, 379, 422
  mill wear, 45–46, 51, 422

High sodium coals:
  and boiler fouling, 163–165, 422
  and corrosion, 343, 422
High sulfur coals:
  and boiler slagging, 165, 277–280
  and corrosion, 346, 404
  and precipitator performance, 296
  and flue gas acidity, 404

Illite:
  analysis, 65
  in deposit formation, 166
  hardness, 45
  occurrence in coal, 14–15, 262–263
  particle spheroidization, 66
Impaction (see Particle impaction)
Interfaces:
  deposit and boiler tube, 259
  slag and coal, 67–71
  sulfate and silicate, 99
Ion exchange:
  in coal minerals and substance,
      265–266
  in coal washing, 272
Iron:
  enrichment in adhesive deposits, 176–177,
      184–185
  enrichment in slag, 276–278
  occurrence mode in coal, 272–275
Iron carbonates (see Ankerite; Siderite)
Iron chloride:
  in air heater deposits, 407–408
  in furnace wall deposits, 332
  vapor pressure, 332
Iron oxides:
  adhesive index, 177
  in ash, 177
  in coal, 14
  formation, 87–88
  magnetite in ash, 276
  pyritic origin, 278–279
  solubility factors, 176–177
  solubility limit, 177
Iron sulfate:
  in air heater deposits, 405–407
  in coal, 19
  in high temperature deposits, 318–323,
      328
  thermal stability, 319–320

Jarosite, 263

Kaolin:
  dehydration, 61–62, 285
  hardness, 45
  occurrence in coal, 41, 262–263
Knudsen heat transfer effect, 227

Lead:
  capture by ash, 339–340
  chloride, 338–340
  corrosive deposits, 340–341
  enrichment, 337
  flue gas concentration, 339–340
  occurrence in coal, 16, 334, 339
  oxide, 338–341
  silicate formation, 339
Lead refinery deposit:
  composition, 338
  fusion characteristics, 338
Lead sulfate:
  in corrosive deposits, 341
  in metal refinery deposits, 338
  in refuse-fired boiler deposits, 340–341
Lead sulfide:
  in coal, 16, 337
  in deposits, 335, 341
  fume formation, 338
  melting characteristics, 337
Lignite:
  ash content, 265
  calcium content, 262, 265
  ion exchange, 265–266
  mineral matter, 262
  mineral retention, 264–266
  sodium content, 163, 264–265
  sulfur content, 273–274
Lignite ash:
  composition, 262–264
  definition, 163
  fouling index, 162–163
  sintering characteristics, 266–272
Limestone additive, 422–423
Liquid phase:
  adhesion equation, 172
  cohesion equation, 172
  sulfate/silicate miscibility, 98–100
  wetting in adhesion, 172–173
Low rank coals:
  calorific value, 24
  combustibility, 27
  volatile matter content, 24

Magnesium:
  additives, 298–299, 309–311, 358–359

in carbonates, 16, 19–20
in silicates, 14
Magnesium sulfate:
  in coal, 16
  in deposit, 358
Magnetite:
  in ash, 276
  in coal, 14
  density, 14
  from pyrite, 278–279
Material wear:
  abrasion measurements, 56–57
  erosion measurements, 54–55
  remedial applications, 56–59
Melting temperature:
  oxides, 14
  silicates, 14, 139
  sulfates, 16, 318
  sulfides, 16, 86–87, 337
Metal coatings (*see* Corrosion prevention)
MHD:
  potassium rich slags, 128
  silica volatilization, 82
Mild steel (*see* Carbon steel)
Mineral matter:
  analysis, 9–11
  earth's crust, 11–12
  geological environment, 11–12
  igneous rock, 12
Mineral matter in coal:
  adventitious, 29
  and ash content, 35–36
  distribution, 15–17, 30–35
  epigenetic, 29
  influence on boiler design, 5–8,
      239–245, 248–249, 356–357,
      379–381, 386–390
  inherent, 29
  high rank, 263
  low rank, 261–265
  separation, 30
  species, 14, 16
  syngenetic, 29
Mineral particles:
  abrasiveness, 45–48
  fragmentation, 103–105
  in pulverized coal, 30–34
  spheroidization, 61–67
Minor constituents (*see* Trace elements)
Montmorillonites in coal, 14, 263
Muscovite:
  analysis, 65
  dissolution in slag, 124–126
  hardness, 45
  occurrence in coal, 14

Muscovite (*Cont.*):
    potassium release, 79
    resistivity, 287
    spheroidization, 66

Nihard materials:
    erosion wear, 55
    hardness, 55
    use in coal plant, 56–57
Nucleation:
    with additives, 300–307
    on ash particle surface, 300, 304–305
    in slag, 304

Oil-fired boilers:
    combustion chamber size, 231, 426
    conversion to coal, 386–390, 426–430
    deposit emittance, 219
    heat flux, 235–236
Orthoclase:
    hardness, 45
    occurrence in coal, 14
    potassium release, 79
Oxidation:
    and corrosion, 313–316, 330–331
    and erosion, 373, 378
    of metals, 313–316, 330–331
    of pyrite, 85–89
Oxide layers:
    dissolution in sulfate, 328
    porous, 333
    protective, 333
Oxide minerals in coal, 14

Particle capture:
    on boiler walls, 116
    in deposition, 189–197
    in electrical precipitators, 283, 284–299
    fume by ash, 111–113, 268
    by fused slag, 198–199
    by molten sulfate, 198–199
Particle coalescence:
    in burning coal particles, 113–115
    inherent ash, 113–115
    plerosphere formation, 74, 77
    summary, 116
Particle diffusivity:
    Einstein equation, 190
    Stokes-Einstein equation, 191
Particle fragmentation:
    by gas evolution, 104
    by thermal shock, 103–106

Particle impaction:
    angle, 53–55
    collection efficiency, 194–196
    inertial, 194–197
    velocity, 53–55
Particle shape:
    boiler grit, 373
    carbonate residues, 105
    cenospheres, 75
    changes and ash erosion wear, 361–365
    changes in flame, 61–67
    coal minerals, 34, 64
    coke residue, 28, 114
    fly ash, 64, 111, 193, 267–268
    fume, 105, 108, 111, 268
    kaolin, 46
    mica, 46
    plerospheres, 76
    pyrite residue, 46, 87, 105
    quartz, 46
    sinter bridged ash, 115
    surface fused glass, 64
Particle size:
    aerodynamic, 30
    and erosion wear, 366–369
    fluidized combustion ash, 393, 434
    fluidized combustion coal, 430
    fly ash, 117–118, 368–369
    minerals in pulverized coal, 32
    pulverized coal, 28, 430
Phagioclase, 14, 270
Phase diagrams:
    iron oxide/sulfide, 87
    potassium-iron/sodium-iron sulfates, 318
    potassium silicate/sulfate, 99
Phosphates:
    in boiler deposits, 214
    in coal, 214
Phosphorus:
    enrichment in boiler deposits, 214
    occurrence in coal, 214
    volatilization, 214
Plerospheres, 74–77
Pore size:
    cenospheres, 71–73
    insulating bricks, 225
    in slag, 305–306
    and thermal conductivity, 225
Porosity:
    coal, 21
    coke particles, 27–28
    deposits, 208, 227
    and heat transfer, 225
    slag, 208–209, 227–228
    and thermal conductivity, 224–225

Potassium:
    in ash, 12–13, 65, 182, 207, 262–263
    corrosion propensity, 321, 342–343
    in deposit formation, 164–166, 201–205
    soluble, 78–79, 342–343
Potassium aluminosilicates:
    analysis, 65
    in coal, 14–15, 262–263
    in deposit formation, 166, 319–320
    particle spheroidization, 65–66
Potassium release, 77–83, 113, 343
Potassium sulfate:
    in corrosive deposits, 316–323, 341,
        358–359
    deposition, 201–205
    enrichment in deposit, 209
    miscibility with silicates, 98–99
    surface tension, 172, 199
    viscosity, 199
Precipitation:
    and ash acid resistivity, 289
    and ash bulk resistivity, 288–289
    and ash surface resistivity, 283–284,
        289–299
    Deutsch equation, 289
    efficiency, 289–291
    flashover voltage, 290
    migration velocity, 289–290
    and sulfur in coal, 291, 296
Precipitation additives (*see* Additives)
Probes:
    acid deposition, 400
    corrosion, 344
    high temperature deposition, 200
Pulverized coal:
    abrasive characteristics, 43–49
    ash content, 35
    ash rich fractions, 35
    erosion wear, 58–59
    fineness and combustibility, 28–29
    firing, 5–7, 27–29, 61, 239–240,
        428, 431
    mineral matter distribution, 30–35
    size distribution, 30
Pulverized coal ash (*see* Fly ash)
Pyrite:
    in British coals, 19, 47
    and coal rank, 274
    and coal sulfur, 273–274
    dissociation on heating, 85–89
    distribution in coal, 32
    gas evolution, 85–86
    hardness, 45
    melting, 86–87

    oxidation, 85–89
    particle size, 32, 165
    slagging propensity, 165
    in U.S. coals, 19, 262–263, 273–275
Pyritic iron:
    and coal age, 274
    slagging propensity, 277–280
    in U.S. coals, 273–274
Pyritic sulfur:
    in British coals, 19
    in U.S. coals, 273–274

Quartz:
    abrasiveness, 48
    in ash, 193, 300, 363, 368, 379, 393
    analysis, 65
    in bituminous coals, 14–15, 17–18, 45, 263
    in British coals, 17, 45
    and coal plant wear, 45–46, 49, 51
    and erosion wear, 52–55, 365–369
    estimation formula, 17
    hardness, 45, 365
    native, 65
    particle size in pulverized coal, 32
    in sub-bituminous coals, 18, 262, 379
    in U.S. coals, 18, 262–263
    vitrification, 66, 362–363

Radiation flux (*see* Heat flux)
Radiation force:
    equation, 191
    on particles, 191–192
Radius of curvature:
    at particle contact, 52, 62–63
    pipe bend, 58
Rankine cycle:
    fluid pressure, 242
    fluid temperature, 242
Reducing atmosphere:
    and boiler slagging, 251–254
    and corrosion, 331–334, 356
Reflectance:
    coke/magnetite mix, 220–221
    measurements, 219–220
    and particle size, 220–221
    and refractive index, 220–221
Reflectivity, 217
Refractory materials:
    abrasion wear, 57
    erosion wear, 55
Refuse-fired-boiler deposits, 341
Reheater erosion, 371

Resistivity:
    and additives, 286–288, 290–291
    bulk, 288–289
    coal minerals, 284–288
    fly ash, 283–284, 288, 290–291
    and sodium in ash, 288
    and SO$_3$ adsorption, 294–295
    surface, 283–284, 289–299
Rutile, 262–263

Salinity:
    coal seam water, 21
    ground water, 21
    sea water, 21
Scanning electron microscope examination:
    ash particles, 63, 75–76, 114–115,
        193, 267, 333, 364
    coal mineral particles, 10, 64
    coke particles, 28, 114
Siderite:
    dissociation on heating, 94–95
    fume formation, 104–105
    hardness, 45
    occurrence in coal, 16, 19–20, 94, 263
Silica in coal minerals:
    combined, 17–18
    free, 17–18
    total, 17–18
Silica fume:
    formation, 106–108
    particle size, 107–108
Silica ratio of slag, 130, 161
Silica volatilization, 81–82
Silicate minerals (*see* Aluminosilicates)
Silicate particles:
    alkali enrichment, 78–80, 97–98,
        266–269
    spheroidization in flame, 61–67
    sulfate cpature, 111–113
    vitrification, 62, 275, 361–362
Silicate structure, 127–128
    and viscosity, 127–128
Siliceous additives, 359
Silicon carbide:
    in burner quarls, 246–248
    deposit adhesion, 248
    erosion wear, 55
    thermal conductivity, 248
Silicon monoxide:
    oxidation to fume, 106–108
    vapor pressure, 80–83
Sinter bonding:
    degree, 138

    and diffusion, 138
    Frenkel's equation, 137
    model, 137–140
    and particle size, 138
    and surface tension, 140
    and viscous flow, 138–140
    by volatilization route, 138
    (*See also* Ash sintering)
Sintered ash:
    and fusion tests, 154
    strength measurements, 152–154
Sintered deposits (*see* Deposit; Deposits)
Sintering measurements:
    activation energy, 143
    conductance, 143–152
    furnace construction, 144–145
    isothermal, 147–148
    shrinkage, 141–152
    and sodium content, 149–150
    summary of methods, 156
Slag:
    carbon transfer, 72–73
    interface phenomena, 67–71
    iron enrichment, 276–277
    particle capture, 198–199
    in pulverized-coal-boiler, 209–210
    surface tension, 68–69
    wetting, 67–68
Slag agglomerates:
    carbon content, 436–437
    formation in fluidized gasifier, 434–437
    particle size, 435
    porosity, 436
Slag crystallization:
    additive induced, 304–307
    and gas bubbles, 306
    in sintering measurements, 306
    species, 124–125, 213
    in viscosity measurements, 126, 129,
        305–306
Slag flow:
    critical temperature, 129
    Newtonian, 124
    non-Newtonian, 124
    and viscosity, 132
Slag globules:
    and cenosphere formation, 71–72
    coal particle ejection, 69–70
    on coke surface, 69, 114
    and gas evolution, 71–73
Slag preventive measures:
    boiler design, 239–246
    coal blending, 250–251
    coal cleaning, 249–250

Slag preventive measures (*Cont.*):
  combustion control, 251–254
  and deposit adhesion, 249
  novel suggestions, 259–260
  summary, 254
Slag viscosity (*see* Viscosity)
Slagging index (propensity):
  ash composition, 164–165
  ash fusion tests, 159
  basic/acidic oxide ratio, 162
  pyritic iron content, 280
  silica ratio, 161
Soda glass:
  sintering measurements, 140, 146–147
  viscosity, 121, 146–147
Sodium:
  and ash resistivity, 288–289
  capture by ash, 78–80
  enrichment in boiler deposit, 266–269
  flue gas concentrations, 92
  in lignites, 264–265
  occurrence mode in coal, 12–13, 264–266
  volatilization, 266
  water-soluble, 164
Sodium chloride:
  in coal water, 20–21
  and corrosion, 330, 334
  sulfation, 92–94
  vapor pressure, 93
  volatilization, 89–91
Sodium silicates:
  in coal, 14
  in deposit, 213
  miscibility with sulfate, 97–98
Sodium sulfate:
  on ash surface, 112
  in coal, 16, 261–262
  in corrosive deposit, 317–318,
    322–323
  deposition, 201–204
  formation, 92–94
  miscibility with silicates, 98
  surface tension, 172, 199
  vapor pressure, 93
  viscosity, 199
Sootblowers:
  acoustic, 259
  air heater, 412, 414–415
  ash entrainment, 255
  compressed air, 254–257
  furnace wall, 256
  impact pressure, 255
  steam, 254–257
  superheater, 256

  water jet, 258–278
  water tempering, 257
SO_2:
  concentration in flue gas, 92
  oxidation, 294–295, 402–403
SO_3:
  adsorption on ash, 294–295
  and ash resistivity, 289
  concentration in flue gas, 294, 320, 404
  formation, 294, 320
  precipitation additive, 290–296, 299
  role in corrosion, 325–329
Specific gravity (*see* Density)
Specific surface:
  ash, 295
  coke residue, 27
Sub-bituminous coals:
  abrasive, 51
  ash content, 51, 265
  blended, 26
  calcium content, 262, 265
  fouling index, 162–165
  sodium content, 262–265
  sulfur content, 273–274
Sulfamic acid additive, 290–291, 296, 299
Sulfate fume:
  capture by ash, 111–113
  concentration in flue gas, 109–110
  deposition, 204
  formation, 108–110
  particle size, 110, 204
Sulfate phases:
  alumina dissolution, 318
  iron dissolution, 318
  low temperature melting, 318
  miscibility with silicates, 98–100
  thermal stability, 97, 318–319, 323–324
Sulfates (*see* Calcium, Iron, Lead, Magnesium,
    Potassium, and Sodium sulfates)
Sulfation:
  carbonates, 96
  chlorides, 92–94
  in fluidized combustion, 432–433
  thermodynamic data, 93
Sulfides:
  in coal, 16, 18–19, 262–263, 307,
    335, 337
  in deposits, 272, 335
  in oxide scale, 332–333, 346
  in slag, 306
Sulfur:
  fixation in coal, 18–19
  occurrence mode, 19, 273–274
  organic, 19

Sulfur (*Cont.*):
  pyritic, 19
  sulfatic, 19
Sulfur content:
  anthracite, 273-274
  bituminous coals, 273-274
  lignites, 273-274
  sub-bituminous coals, 273-274
Sulfur retention in ash:
  with calcium rich coals, 37, 251
  in fluidized combustion, 432-433
Sulfuric acid:
  additive, 290-291, 296
  and air heater corrosion, 406-410
  ash reactions, 405-407
  dewpoint temperature, 399-402, 404, 407
Superheater:
  corrosion, 344, 350, 357
  deposits (*see* Deposits)
  erosion, 370-371, 376-384
  fouling, 161
  sootblowers, 256
  steel alloys, 349-350
  temperature limitations, 349
  temperature regime, 245-246
Surface tension:
  and adhesion, 171-172
  alkali-metal sulfates, 172, 199
  and captive surfaces, 198-199
  measurements, 68
  role in sintering, 137-140
  stress on particle surface, 62-63

Temperature:
  and acid deposition, 399-402, 404, 407
  ash particle, 62
  and ash sintering, 140-155
  and boiler slagging, 241
  and corrosion, 324-325, 329, 345, 349, 407
  and deposit adhesion, 181-185
  and slag viscosity, 124-133
  and sulfur (sulfate) retention, 432-433
  and thermal conductance, 224-228
Temperature cycling:
  and deposit emittance, 223-224
  and pitting corrosion, 329
  in water jet sootblowing, 257-258
Temperature gradient:
  in deposit, 133
  in oxide layer, 240-242
  in slag, 133

Temperature regime for deposit adhesion:
  in combustion chamber, 174
  in convective superheater, 174
  in radiant superheater, 174
Thermal barrier:
  ash, 230-232
  deposits, 230-232
  Knudsen effect, 227
Thermal conductivity:
  air, 226
  ash, 224-229
  and bulk density, 224-225, 227
  deposits, 224-229
  and emittance, 226, 228-229
  oxides, 230
  and particle size, 226
  and pore size, 225
  porous materials, 225
  silicon carbide, 248
  and sintering, 225
  and temperature, 227-229
  tube metal, 230
Thermal cycling:
  for boiler deslagging, 252
  in water jet sootblowing, 257-258
Thermal expansion:
  deposit constituents, 176, 230
  steel alloy oxides, 176, 230
  steels, 176, 230
Thermodynamic reaction data:
  sulfate formation, 93
  sulfide dissociation, 85-86
Thermophoresic deposition, 191-193
Trace elements:
  in coal, 19, 307, 334-335
  in deposits, 335-337, 341
  enrichment in deposits, 214-215
  volatilization, 214
Tungsten carbide:
  wear measurements, 55, 57
  use in coal plant, 58

U.S. coals:
  anthracites, 18, 263
  ash adhesion, 181-182
  ash fouling propensity, 160-166, 266-272
  ash sintering, 149, 152-153
  ash slagging propensity, 160-166, 275, 280
  bituminous, 12, 263
  high calcium, 265
  high iron, 271-275, 279
  high quartz, 18, 48-49, 262-264

U.S. coals (*Cont.*):
  high sodium, 265
  high sulfur, 273–275, 279
  lignites, 13, 262
  specific problems, 261–280
  sub-bituminous, 18, 262

Vapor pressure:
  iron chloride, 332
  iron sulfide dissociation, 85–86
  silicon monoxide, 81–83
  sodium chloride, 93
  sodium sulfate, 93
Vickers hardness (*see* Hardness)
Viscosity:
  and additives, 124–126, 305
  and alumina content, 127–128
  and ash composition, 127–131
  and ash fusion tests, 158
  and calcium content, 130–131
  critical temperature, 124, 130–131
  and crystallization, 124
  empirical formulas, 130–131
  and iron content, 128–129
  measurements, 121–126
  Newtonian melts, 124
  non-Newtonian melts, 124–125
  in oxidizing atmosphere, 128–129
  and potassium content, 127–128
  in reducing atmosphere, 128–129
  and silica ratio, 130
  sintering parameter, 131
  soda glass, 121, 147
  sulfates, 199
Viscosity gradient:
  in furnace deposit, 132–134
  in superheater deposit, 132–134
Viscosity ranges:
  for ash particle spheroidization, 63, 132

  for cenosphere formation, 71, 132
  for sintering, 132
  for slag flow, 132, 159
  for slag formation, 131
Viscous flow, 131
Volatile matter:
  in coal, 24
  and coal combustibility, 27–29
Volatilization:
  alumina, 107
  arsenic, 335–336
  chlorides, 89–94
  lead, 338–340
  potassium, 77–80
  silica, 80–83
  sodium, 89–94

van der Waals forces:
  ratio to gravitational force, 171–172
  retarded, 170–171
  short-distance, 170–171

Water jet sootblowers:
  efficacy, 257–258
  temperature cycling, 257–258
Wetting:
  and adhesion, 171–173
  molten sulfate, 68
  and particle capture, 198–199
  slag, 67–68
Work of adhesion:
  equation, 172
  silicates, 172–173
  sulfates, 172–173
Work of cohesion:
  equation, 172
  silicates, 172–173
  sulfates, 172